Proceedings of the Cambridge Unsteady Flow Symposium 2024

James C. Tyacke · Nagabhushana Rao Vadlamani
Editors

Proceedings of the Cambridge Unsteady Flow Symposium 2024

Springer

Editors
James C. Tyacke
Department of Mechanical and Aerospace Engineering
Brunel University London
Uxbridge, UK

Nagabhushana Rao Vadlamani
Department of Aerospace Engineering
Indian Institute of Technology Madras
Chennai, Tamil Nadu, India

ISBN 978-3-031-69034-1 ISBN 978-3-031-69035-8 (eBook)
https://doi.org/10.1007/978-3-031-69035-8

This Springer imprint is published by the registered company Springer Nature Switzerland AG
The registered company address is: Gewerbestrasse 11, 6330 Cham, Switzerland

Organising Committee

Paul Tucker (Cambridge University, UK)—Chairman
Will Trojak (IBM Research, UK)—Secretary
James C. Tyacke (Brunel University London, UK)—Editor
Nagabhushana Rao Vadlamani (Indian Institute of Technology Madras, India)—Editor
Rob Watson (Loughborough University, UK)—Treasurer
Felix Möller (DLR, Germany)—Scheduling
Zhong-Nan Wang (University of Birmingham, UK)—Attendance
Jiahuan Cui (Zhejiang University, China)—Website

Preface

This book covers the state-of-the-art research on the numerical modelling of unsteady turbulent flows presented at the first "*Cambridge Unsteady Flow Symposium*". The symposium held at the University of Cambridge, Murray Edwards College, on 4–5 March 2024, brought together internationally leading experts in Computational Fluid Dynamics (CFD). Around 60 participants from both academia and industry attended the 2-day Symposium. A wide range of topics related to CFD were covered, including but not limited to, large-eddy simulations, unsteady flows in aerospace, high order methods, and mesh generation. Keynotes delivered by Prof. P. Vincent, Prof. P. Roe, Prof. L. Davidson, and Prof. L. Enghardt promoted intriguing discussions on the current status and future possibilities of exploiting unsteady CFD for real-world problems.

The book comprises papers submitted by Keynote speakers and four chapters comprising 27 papers in total, each being peer-reviewed by two experts. The first part on "Methods in Turbulence" comprises five papers that explore different post-processing techniques, modal analysis, and data-driven modelling of turbulent flows. The second part, comprising ten papers, explores unsteady flows in canonical wall-bounded flows, static and flapping airfoils, and Rayleigh Bénard Marangoni convection induced by evaporative cooling. This part also includes the test cases of industrial relevance such as ribbed ducts, wind turbine blades, pneumatic shock absorbers, noise control with porous coatings, and jet noise control from Biconical nozzles. Seven papers in the third part focus on the unsteady flows encountered in various "Turbomachinery" components such as compressor disc cavities, high-speed low pressure turbines, cutback trailing edges, and tip leakage flows. Installed jet noise with wing-flap-fuselage configurations, and flow control over aircraft intakes are also discussed in this part. The final part titled "Beyond Navier-Stokes" consists of three papers. These parts address novel simulation strategies to capture complex flows using the Boltzmann equation and magnetohydrodynamic turbulence in liquid metal flows and explore efficient methods to decompose high-order partial differential equations. These parts collectively explore different facets of unsteady flows encountered in both canonical flows and in industry, contributing to the advancement of the field and its practical applications.

Uxbridge, UK James C. Tyacke
Chennai, India Nagabhushana Rao Vadlamani

Acknowledgements We give special thanks to the CUFS 2024 keynote speakers Prof. P. Vincent, Prof. P. Roe, Prof. L. Davidson, and Prof. L. Enghardt not only for their interesting talks but also the valuable discussions that followed. Thanks also to F. Witherden and J. Su for their interesting presentations.

A final thank you to all the hard working committee members, reviewers, and speakers making CUFS 2024 a success!

Keynote Presentations/Papers

Keynote 1

Professor Peter Vincent

Department of Aeronautics, Imperial College London, London, UK
Title: Taking Scale-Resolving Simulations from Academia to Industry

Biography: Peter is a professor in the Department of Aeronautics at Imperial College. He works at the interface between mathematics, computing, fluid dynamics, and aeronautical engineering. He holds a 1st class undergraduate degree from the department of Physics at Imperial College (graduating top of the year), a Ph.D. from the department of Aeronautics at Imperial College in the field of Computational Fluid Dynamics, and previously served as a postdoctoral scholar in the department of Aeronautics and Astronautics at Stanford University. He is a 2016 Gordon Bell Prize finalist, a 2016 Philip Leverhulme Prize winner, and has held two INCITE Awards from the Department of Energy. He plays an active role in his community, co-leading the open source PyFR project and acting as a Principal Editor for Computer Physics Communications. He is also a fellow of the Royal Aeronautical Society.

Keynote 2

Emeritus Professor Philip Roe, FAIAA

Department of Aerospace Engineering, University of Michigan, Ann Arbor, MI, USA
Title: Musings of a Computational Philosopher

Biography: Philip Roe was born in Derby, England, in 1938, and educated at Derby School and Cambridge University. He went to work at the Royal Aircraft Establishment, Bedford, where he was part of the team that laid the foundations of waverider aerodynamics. After the oil crisis interrupted that line of research, he transferred in 1974 to the new field of Computational Fluid Dynamics, where he was commissioned to “find something new”. The outcome was his 1981 paper on the “Roe Riemann solver” which is still cited almost twice per day. However, his paper here describes his attempts to go beyond it. He moved to Cranfield University in 1984 and to the University of Michigan in 1990. He has graduated 35 Ph.D. students and held visiting appointments in France, Italy, Germany, and India. He retired in 2023, but cannot stop thinking about CFD.

Keynote 3

Professor Lars Davidson

Division of Fluid Dynamics, Department of Mechanics and Maritime Sciences, Chalmers University of Technology, Sweden
Title: Using Machine Learning for Improving Wall-Functions and Turbulence Models

Biography: Dr. Lars Davidson is a professor at Chalmers University of Technology, Department of Mechanics and Maritime Sciences, and he has been head of the division of Fluid Dynamics since 2005. The general subject of his research is computation of fluid flow and heat convective transfer using finite volume computations. He is working with in-house finite volume codes (pyCALC-LES, CALC-LES), which have been developed by himself. Turbulence modelling for fluid flow and heat transfer is the focus of his research. A large part of the research is concentrated on Large Eddy Simulations (LES). Lately, he has been focused on hybrid LES/RANS, an approach to make it possible to use LES for wall-bounded flows at high Reynolds numbers. In 2022, he started to work on Machine Learning to improve wall functions and turbulence models.

Keynote 4

Professor Lars Enghardt

Director of the Institute of Electrified Aero Engines at German Aerospace Center (DLR), Professor for Electrified Aero Engines at BTU Cottbus
Title: Recent Developments in Turbomachinery Aeroacoustic Research at DLR

Biography: Professor Enghardt has been working in the area of aeroacoustic research for more than 26 years. He joined the DLR—German Aerospace Center in the year 1997 as a postdoc, started to head the group of turbomachinery acoustics in the year 2004, and became acting head of the department of engine acoustics in 2006. In 2009, he was appointed as a professor for the Chair of Turbomachinery and Thermoacoustics at the Technical University of Berlin, simultaneously with the leadership of the DLR Department of Engine Acoustics. In 2023, he was appointed as a professor for the Chair of Electrified Aero Engines at the Brandenburg Technical University in Cottbus, connected with the Directorship of the new DLR Institute of Electrified Aero Engines at the same site. He has quite a record in European research, amongst others being a member of the X-Noise Community from the very beginning. He represented the DLR in several European Research projects, in particular coordinating the projects PROBAND, FLOCON, and TurboNoiseBB, each with 15 European partners, including the major players of the aero engine industry, Rolls-Royce, MTU, SAFRAN Aero Engines, and Airbus. Under his coordination, the EU project SENECA on supersonic transport noise started in January 2021. As a University professor in Berlin, he gave two lectures on engine acoustics. In Cottbus, he gives one lecture on aircraft noise. In the last three years, he was the speaker of the DLR committee of aircraft acoustics.

Contents

Exploring Unsteady Flows

Turbomachinery

About the Editors

Dr. James C. Tyacke is a senior lecturer in Aerospace Engineering at Brunel University London and co-director of the Aerospace Research Centre (Numerical Methods). He received his Ph.D. from Swansea University (2009) on unsteady modelling of heat transfer in electronics before joining the University of Cambridge Whittle Laboratory. His key interests include tackling complex geometry flows using multi-fidelity modelling, heat transfer, aeroacoustics, and high performance computing. He has developed frameworks for aeroengine turbine flow modelling and installed jet aeroacoustics, recognised in Journal highlights and receiving the best use of ARCHER award 2017.

Dr. Nagabhushana Rao Vadlamani is currently working as an associate professor in the propulsion group of the Department of Aerospace Engineering at the Indian Institute of Technology, Madras. He pursued his Ph.D. from the Whittle Laboratory, University of Cambridge (2014) and Masters degree in thermal engineering from IIT Madras (2009). His research interests include eddy-resolving simulations in turbomachines, transition to turbulence, high order methods and high performance computing, and data-driven modelling. He was the recipient of the Bowring research fellowship from St. Catharine's college, Cambridge (2015–18) and ASME best paper award from International Gas Turbine Institute, IGTI in 2014.

Musings of a Computational Philosopher

Philip Roe

Abstract "Philosophy's role is...to ask the right questions" (Slavoj Žižek), because there is little value in answering a misguided question. In this lecture I will try to ask the right questions concerning the correct computation of hyperbolic conservation laws. For example, I will argue that accuracy by itself is merely a low-frequency property of little value unless accompanied by bandwidth, robustness and stability. These properties are obtained by ensuring the correct flow of the correct information. The tools to achieve this are careful analysis of the governing equations and the use of appropriate stencils; several examples will be given. The approach throughout will be multidimensional. One dimensional analogies will be made only when appropriate. The recommendations that emerge are frequently not in accord with common practice.

1 Introduction

According to the political philosopher Slavoj Žižek, "Philosophy's role today is not to provide solutions, but to help us ask the right questions and avoid formulating problems in the wrong way" [1]. He adds that questions are actually more important than answers, because there can only be limited value in giving good answers to misguided questions. According to the educational philosopher Matthew Lipman, a good answer to a good question is "like a candle in the dark" illuminating a small part of the room, and hopefully guiding our way to the next candle [2]. And Albert Einstein is said to have remarked[1] that "If I had one hour to solve a problem, and my life depended on it, I would spend the first fifty-five minutes determining the right

[1] This appears in many anthologies of quotations, but I cannot guarantee its authenticity. However, as Studs Terkel once genuinely said (*The Guardian, March 1, 2002*), "I like quoting Einstein, nobody dares contradict you".

CUFS-2024-PAPER ID, Cambridge, United Kingdom, 4-5 March 2024.

P. Roe (✉)
Emeritus Professor, Aerospace Engineering, University of Michigan, Ann Arbor, USA
e-mail: philroe@umich.edu

J. C. Tyacke and N. R. Vadlamani (eds.), *Proceedings of the Cambridge Unsteady Flow Symposium 2024*, https://doi.org/10.1007/978-3-031-69035-8_1

questions to ask". In Computational Fluid Dynamics, there are many questions to be asked, concerning for example the nature of the grid, the discrete variables to be used, their placement on the mesh, the manner of discretization, the means of advancing in time, perhaps even the selection of hardware. For all of these, there are numerous options and I want to take a philosopher's stance in choosing between them.

Philosophers assign levels of priority to different types of question. For example, much has been written in response to the question, "How may I lead a good life?" But the more fundamental question is "What is the nature of a good life?" Is it one leading to wealth and power, spiritual fulfillment, or service to others? It is very important to ask questions in the correct order, and I shall try to do so here.

In designing or selecting a CFD algorithm, perhaps the most fundamental question that can be asked is "What information should be used to find the solution?" Most simply, the stencil choice. We learn the Courant-Friedrichs-Levy (CFL) rule very early in our studies. *The discrete domain of dependence must include the exact domain of dependence.* Although this is a mathematical theorem that may easily be proved, I prefer to regard it merely as an expression of common sense. How indeed could we obtain a correct answer without accounting for all relevant information?

I want to point out that there is another expression of common sense that is seldom stated as vigorously as I think it should be. *There is no value to information from outside the exact domain of dependence.* In fact, such information could be assigned completely at random. Taking both statements of common sense together, we arrive at this conclusion:

A CFD algorithm should use all of the information on which the solution depends, and should use only that information.

This is, admittedly, a counsel of perfection, feasible in only a few special cases. Linear hyperbolic problems in (x, y) or (x, t) can be solved exactly on any size of domain by the method of characteristics, and the order of accuracy is infinite. For whatever resolution is imposed by the data, the accuracy does not depend on the mesh spacing. But for more complex cases, unless special data allows exact integration, a numerical solution must be found by interpolation in a mesh that extends outside the exact domain of dependence. In fact, we can restate the CFL condition thus;

In most practical cases, a CFD algorithm must use information on which the solution does not depend

I will devote this lecture to the consequences of this glum paradox. I will first show that the paradox allows us to understand why certain approaches to high-order accuracy are bound to prove disappointing. Then I will discuss recently-developed methods that put great stress on correct stencil choice. I will not however claim that any of these represents the best possible version. Many of the ideas that arise are different from, and sometimes counter to, the current orthodoxy in CFD. Even newer ideas may be just around the corner.

2 Lessons from Linear Advection

Traditionally, the study of hyperbolic partial differential equations has profited greatly from the seemingly simple example of linear advection in one dimension,

$$\partial_t u + a\partial_x u = 0. \tag{1}$$

Most of what is valuable can be derived from the simplest situation where the grid is a uniform sectioning of the x-axis by nodes at $x_j = j\Delta x$, and time is advanced in uniform increments $t^n = n\Delta t$.

2.1 *Finite-Difference Schemes*

In the earliest years, almost all discrete methods were finite-difference schemes and the quantity u_j^n located at node j and time-level n was taken to be an approximation of the exact value $u(j\Delta x, n\Delta t)$. The differential operators were replaced by finite-difference operators, and the outcome was a set of schemes having the form

$$u_j^{n+1} = \sum_{k=k_{\min}}^{k=k_{\max}} c_k u_{j+k}^n. \tag{2}$$

Iserles and Strang [3] established the optimum members of this class by finding the stencil ($k_{\min/\max}$), and the weights c_k, that minimize the truncation error. The conclusions are easy to state, but depend on the sign of a, that is, on the direction in which information is flowing. As usual, we will take $a > 0$ and the opposite case can be handled by symmetry (Fig. 1).

There are then three possibilities. If we require the solution u_j^{n+1}, then the data must be taken from the set of consecutive values $\{u_J^n\}$ with either (a) $J \in [j-1-m, j+m]$, (b) $J \in [j-1-m, j+1+m]$, or (c) $J \in [j-2-m, j+m]$. In all cases, the optimum scheme coincides with Lagrange interpolation in the optimum

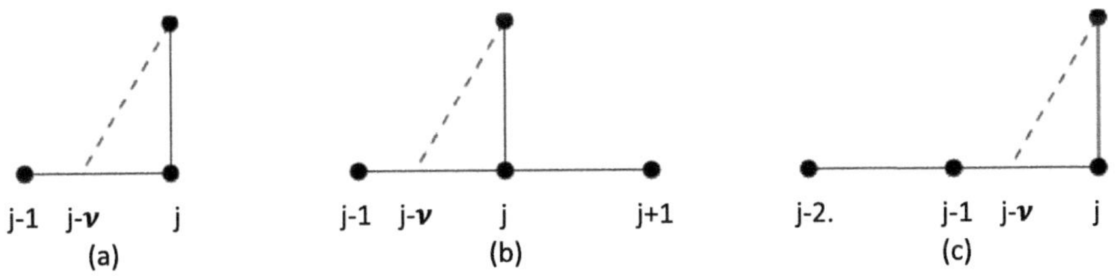

Fig. 1 The three possible types of stencil for an optimal one-dimensional advection scheme. The cases with $m = 0$ are shown. The stencils may have m nodes added at both ends. The dashed line represents the exact domain of dependence $dx/dt = a$

stencil, and is stable precisely for CFL numbers satisfying $0 \leq \nu \leq 1$. Case (a) is the only one where the stencil is centered around the mesh interval containing the exact domain of dependence, and also the only one where the true limits of stability coincide with the rule of thumb known as the CFL rule. The optimal schemes have order of accuracy $p = 2m + 1$ and this is the accuracy of both the dispersion and dissipation errors. In what follows we will find that such a situation is a recurring feature of schemes having good all-round behavior.

In the other two cases, the stencil is not centered around the true domain of dependence and the stability limits do not coincide with the CFL rule. The schemes have even order of accuracy $p = 2m + 2$, but this applies only to the phase error. The amplification error is actually smaller, with $p = 2m + 4$, but this is not in fact beneficial. It means that waves moving with the wrong speed are not strongly damped, and need to be removed by "artificial dissipation". Asymptotic analysis can be used to find an "equivalent equation" for any advection scheme, and thereby demonstrate its response to initial data in the form of a step function. In general [4, 5], even-order schemes produce spurious oscillations only on one side of a captured discontinuity (to the right of it in case(b), to the left in case(c)) with magnitudes exceeding 20% and decaying algebraically. Odd-order schemes produce oscillations on both sides of about 5% magnitude, decaying exponentially.

From here on, I will discuss only odd-order schemes. At least for linear problems, the results are then often useful without limiting. Limiting will be discussed only in one context, where it proved essential. At this time, my philosophy is to develop schemes that make the fewest demands on the limiter. The limiter itself is beyond my present scope.

Linear recurrence relationships such as (2) have solutions

$$u_j^n = \Re\, g^n \exp ij\theta. \tag{3}$$

that may be compared with the exact solution of (1), which is

$$u(j\Delta x, n\Delta t) = \Re \exp i(j - \nu n)\theta. \tag{4}$$

From this we deduce that the exact value of the amplification per time step g is $g(\nu, \theta) = \exp -i\nu\theta$ where θ is the phase change of a sinusoidal signal within one computational cell. The quantity $g(\nu, \theta)^{1/\nu}$ is the amplification after the wave has passed over one cell, and this *normalized* amplification is what I choose to plot. This avoids giving the appearance that small time steps are desirable.

However, the amplification per time step is easy to analyze. For the scheme (2) it is

$$g(\nu, \theta) = \sum_{k=k_{\min}}^{k=k_{\max}} c_k \exp ij\theta. \tag{5}$$

This may be expanded in powers of θ and compared with its exact value. This gives as the condition for exactness

$$\sum_{k=k_{\min}}^{k=k_{\max}} c_k\left(1 + ik\theta + \frac{k^2\theta^2}{2} + \cdots \frac{(ik\theta)^m}{m!}\cdots\right) = 1 - i\nu\theta - \nu^2\theta^2 + \cdots \frac{(-i\nu\theta)^m}{m!}\cdots \tag{6}$$

and therefore the conditions for accuracy of order p are that the equations

$$\sum_{k=k_{\min}}^{k=k_{\max}} k^q c_k = (-\nu)^q \tag{7}$$

are true for $0 \leq q \leq p$. It is easy to deduce from this that the first p moments of the numerical solution around the location of the exact solution at $x = (j - \nu)\Delta x$ all vanish.

2.2 Polar Dispersion Plots

A visual assessment for the quality of an advection scheme is provided by the polar dissipation plot. We represent the normalized amplification as $g = r \exp i\phi$, where the correct value of r is 1.0, and the correct value of ϕ is θ. Any valid scheme has a polar plot beginning at $\theta = 0, r = 1$. A good scheme has the following characteristics;

1. **Accuracy** Taking θ as a parameter, the plot should then track anti-clockwise around the unit circle, and its deviation from the unit circle is a measure of the truncation error. We expect to find $r = 1 - K\theta^p$, where K is positive and small, and p is "large". Schemes with this behavior are described as "accurate" and those with $p \geq 3$ as "high-order". It is worth bearing in mind that these are low-frequency properties that carry no guarantee of good behavior over the extended frequency range. Also, very high accuracy at very low frequency is "overkill" in most scientific and engineering applications, where the practical limits on accuracy stem chiefly from uncertainty concerning geometric or material properties.
2. **Bandwidth** Inevitably, the tracking will not be perfect. The range of $0 \leq \theta \leq \theta_f$, for which it remains within some tolerance may be taken to define the *resolved scales*. We will not attempt to define a value for θ_f here. It is intended to signify whatever resolution will resolve the features of the problem that concern us and will vary from case to case. Despite this, its magnitude is extremely important. The number of cells needed to achieve a required resolution is proportional to θ_f^{-1} and the cost of conducting a resolved simulation in d space dimensions is proportional to $\theta_f^{-(d+1)}$. It is worth bearing in mind that flow simulations in the future will increasingly be three-dimensional and unsteady. Methods that increase θ_f will be worth developing, and may be described as having high bandwidth [6]. Methods that emphasize bandwidth at the expense of accuracy have been developed, beginning with Tam and Webb [7] in 1999. Their "dispersion preserving

schemes" (DRP) have been influential in aeroacoustics. However, they employ wide stencils.

3. **Stability** The path must remain inside the unit circle to avoid instability.
4. **Damping** When $\theta > \theta_f$, there will be errors. It is not totally clear to me what the best behavior now is, and this judgement has to depend on what behavior turns out to be possible. It is often implied that the locus should move quickly toward the origin and stay there. We do not care what the waves do if they are strongly damped. It is often assumed that $\theta = \pi$ is a barrier (folding frequency, Nyquist barrier, Shannon limit) beyond which the solution loses meaning. I will discuss this question in Sect. 5.1.1.

These four properties characterize what may be called a "healthy" dispersion relationship, and are displayed visually in Fig. 2. I will adopt these diagrams as useful "lightning sketches" of one-dimensional schemes.

The dispersion plots for the Iserles-Strang schemes of order 1, 3, and 5 are shown in Fig. 3. For each order, a rather subjective arrow has been inserted to indicate

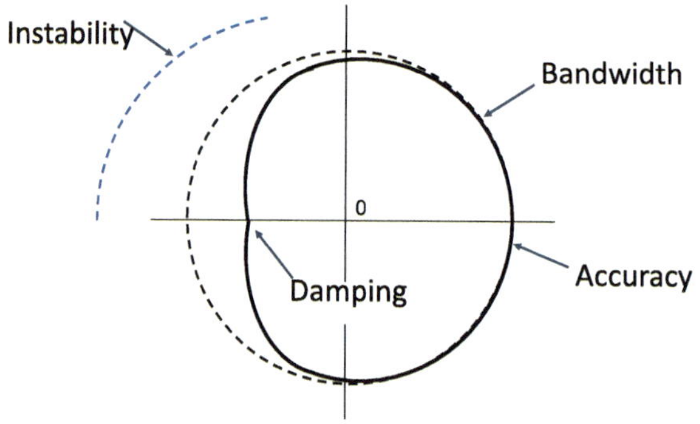

Fig. 2 Characteristics of a healthy dispersion relationship

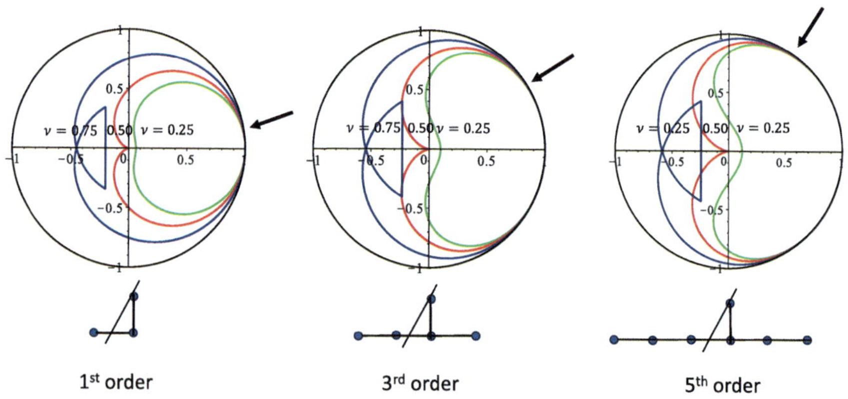

Fig. 3 Polar dispersion plots for the first three Iserles-Strang advection schemes, at $\nu = 0.25, 0.50$ and 0.75

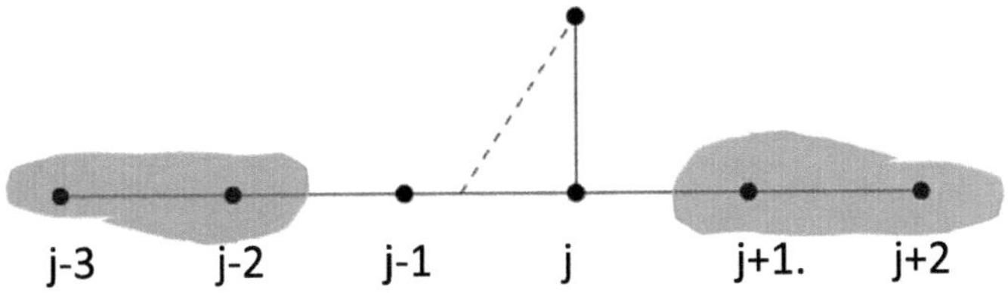

Fig. 4 Stencil for optimum fifth-order finite-difference scheme for linear advection. Notice that most of the data is far from the location of the solution at $x = (j - \nu)\Delta x$

the bandwidth. Although the bandwidth does increase with order, it does so rather slowly, and we will see examples of much faster growth. The explanation for this is that the greater accuracy is being obtained from information well outside the interval $[-\Delta x, 0]$. As explained in the Introduction such information should really have no bearing on the exact solution. Although we appear merely to be interpolating in the data, that is only true for smooth data, and we intend to use these schemes to solve hyperbolic p.d.e.s whose solutions are often not smooth. I find that the process has more the character of *extrapolation*, from outside the true domain of dependence to within it (Fig. 4).

Formally, however, the Iserles and Strang schemes amount to simple Lagrange interpolation in the chosen stencil. I have often taught that interpolation in itself is rather simple. The complexity arises, I said, in linear advection because we successively interpolate in the interpolants, so that errors accumulate. But it is only interpolation if we are using relevant data, which is not the case unless the data is smooth. Advection schemes now seem to me to partake of *extrapolation*, from outside the true domain of dependence toward the one place where data actually matters. In fact I will propose as a general principle that

Problems with the solution of hyperbolic partial differential equations commonly arise from a poor choice of stencil

2.3 Discontinuous Galerkin Schemes

An effort to avoid this is made in the widely-used discontinuous Galerkin method [8]. In its most straightforward version, the same two-point stencil used by the First-order Upwind scheme is employed at all orders within a finite element framework. The degrees of freedom are now associated with the interval $[(j-1)\Delta x, j\Delta x]$ and are the amplitudes of trial functions that provide a polynomial basis. No continuity is enforced between elements, but they are connected through a Riemann solver. The time derivatives of these amplitudes are the unknowns, found by minimizing in L2 the integrated residual over the element. This yields a semi-discrete scheme that is completed with one's favorite time integrator. I employ here fourth-order Runge-Kutta, a fairly universal general-purpose choice. The degrees of freedom can be as

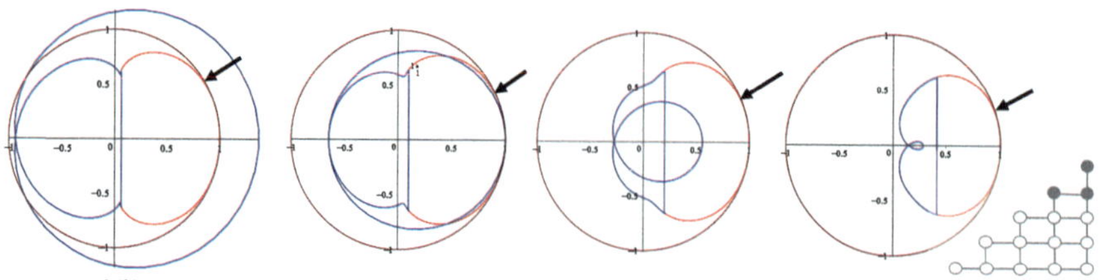

Fig. 5 Dispersion polars for the third-order DG1 method at four values of ν. Time-stepping is carried out with four-stage Runge-Kutta marching. This enlarges the effective stencil as shown at bottom right

numerous as we wish and the order of accuracy correspondingly high. Obviously, raising the accuracy requires processing more information. Here this is done without increasing the stencil by including more information in each cell, in the form of higher-order gradients or trial functions. A large bonus attaching to this is that the procedure needs little modification close to boundaries.

Dispersion plots for the simplest version of the discontinuous Galerkin method (DG1, piecewise linear reconstruction) are provided in Fig. 5 at four different Courant numbers $\nu = 0.50, 0.464, 0.40$ and 0.30. There are some striking contrasts with the finite difference case. To begin with, the dispersion analysis returns an algebraic answer for values of θ greater than the Nyquist limit of $\theta = \pi$. The significance of this will be discussed in Sect. 5.1.1. For now, we will simply accept that it happens. In the finite-difference case the changes of behavior associated with changes in ν were relatively small and very systematic. Here they are wild, especially in the sub-Nyquist frequency range $\theta > \pi$. When $\nu = 0.5$, the scheme is unstable. By lowering ν to 0.464, stability is recovered, but there is a badly inaccurate high-frequency mode that is not damped. Lowering it further to 0.40, we can ensure that all spurious waves are damped by a factor of at least one half for each interval they pass. The resolved waves do behave as hoped, although the bandwidth is slightly less than for finite differences, and the stable CFL number considerably less. In my opinion, the utility of the discontinuous Galerkin method resides in its natural extension to complex geometry.

The explanation for low bandwidth again lies with poor stencil choice. Although the original intention was to contain the stencil, it expands again due to the need for some form of multistep time marching, and this is indicated on the far right of the figure. The additional information that provides formal accuracy is being obtained far away from the exact domain of dependence unless the Courant number is made very small. As a result, practical accuracy is only obtained by taking small time steps. On the other hand, for the finite difference schemes, accuracy actually improves with increasing time steps. I will now show that the best of both worlds is attainable.

2.4 Active Flux Schemes

Dispersion polars for the third-order version of this scheme are displayed in Fig. 6. Stability is achieved for the full range of possible ν. Also, the bandwidth is considerably larger than it is in either of the previous schemes. Recall that this enables large cost savings by enabling coarser grids. Compared with discontinuous Galerkin, the sub-Nyqvist frequencies are well behaved and quite strongly damped except when the Courant number is high. This situation will be discussed in Sect. 5.1.2. In this scheme [9] there are two degrees of freedom per cell, as with discontinuous Galerkin; these are the mean value in the cell $\bar{u}_j$ and the point value at the right-hand edge. Point values are shared between neighboring cells, so the reconstructions are C^1 smooth. A quadratic reconstruction (see Fig. 7) can be made within the jth cell;

$$u_{j-1/2}(\xi) = \bar{u}_j + (\xi - 1/2)(u_{j+\frac{1}{2}} - u_{j-\frac{1}{2}}) \\ + \frac{1}{2}(1 - 6\xi(1-\xi))(u_{j-\frac{1}{2}} - 2\bar{u}_j + u_{j+\frac{1}{2}}), \quad (8)$$

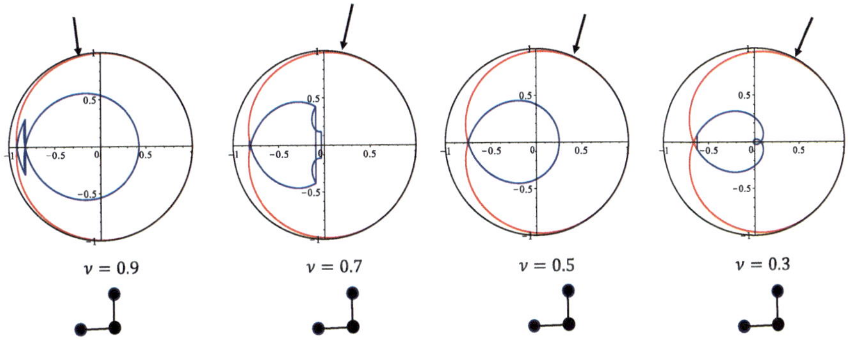

Fig. 6 Dispersion polars for the third-order version of the Active Flux method at three different values of ν

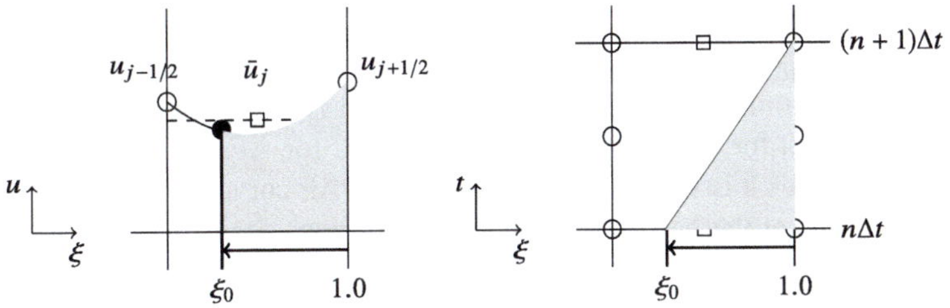

Fig. 7 Integrating the reconstruction to find the new flux and interpolating in the reconstruction to find the new values

where $\xi = \frac{x - x_{j-\frac{1}{2}}}{x_{j+\frac{1}{2}} - x_{j-\frac{1}{2}}} \in [0, 1]$. If $a > 0$, this can be evaluated at $\xi = 1 - \nu$ to find the next value at the right interface, and integrated between $\xi = 1 - \nu$ and $\xi = 1$ to find the flux leaving that interface.

In addition to those properties visible in the dispersion diagrams, these additional properties are noteworthy. As we will show, they can be extended to higher dimensions and higher order.

1. The method is valid for all cells, including the last, so all nodes except the first are accounted for. There is no need for an outflow boundary condition.
2. The first node is covered by the inflow boundary condition.
3. All processes prior to the final update are internal to a single cell. They can be coded as a single loop over cells and communication overheads are eliminated.
4. The method is stable up to the greatest possible time step.
5. The analysis is local to a single cell and therefore applies without modification to irregular grids.
6. It is therefore "quadratically exact", meaning that there is no error on any grid for globally quadratic data.
7. By taking ν to be vanishingly small, we recover a semi-discrete scheme that is equivalent to dG with exact time integration [10]. This limit is simultaneously the least accurate version of AF and the most accurate version of DG1.

This wonderful scheme was first discovered in 1977 by van Leer in [11] and subsquently rediscovered in 2007, both by Popov and Ustygov [12] and by Ii and Xiao [13]. However, none of these authors discovered a good extension to higher dimensions that retains all of the properties listed. This was first achieved by Eymann and Roe [14] in 2013. My work with subsequent students has considerably refined that approach and the outcome will be described in later sections. I note again that this scheme and the DG1 scheme are both third-order accurate and use the same stencil with the same degrees of freedom. The one important difference is that DG1 is semi-discrete and, therefore, compelled to extend its effective stencil to include irrelevant information.

Figure 8 compares the results of this scheme with DG1 on a packet of wave forms. From left to right these are 1. a square wave, 2. a raised cosine, 3. a narrow Gaussian, and 4. a half-ellipse. The plots show the average value in each cell, and also the reconstruction within each cell. The reconstruction is, of course, piecewise linear and discontinuous for DG, but piecewise quadratic for AF. The exact solution is shown, but at times it is scarcely visible behind the AF curve. At the corners of the square wave, and at the base of the others, are spurious oscillations. In no case do these exceed about 6% in magnitude, which is, as already mentioned, a good feature of odd-order schemes. The oscillations due to AF intrude much less into the constant regions between the waves. The AF scheme produces a much "sharper" square wave. There are only two cells in each vertical leg rather than five. This may seem surprising in view of the fact that AF employs a continuous reconstruction. There appears to be a common assumption that the capturing of discontinuities benefits from the use of

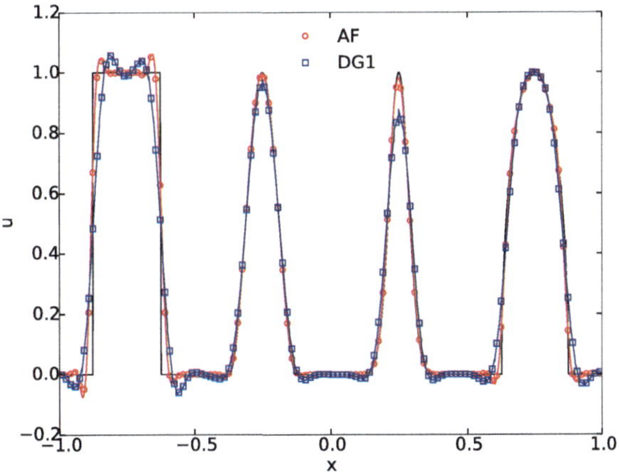

Fig. 8 Linear advection of four waveforms. The curves show the reconstructions assumed by each method

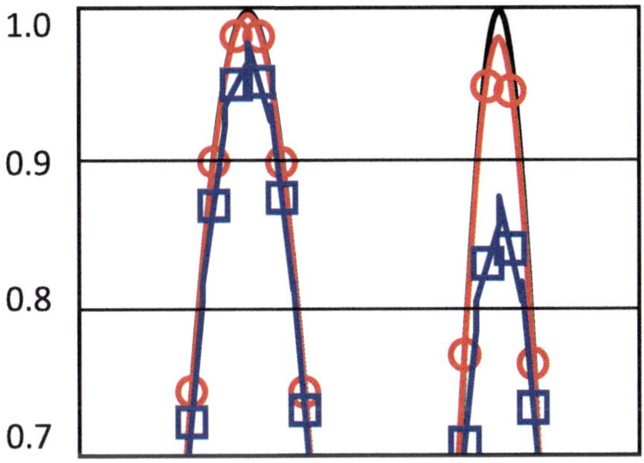

Fig. 9 Enlarged view of the peak regions for cases 2 and 3

discontinuous basis functions, but I know no persuasive argument for this, and it is disproved by this example and later by Fig. 21. The other three waves have smooth extrema of various widths. The extremum of the ellipse is captured to graphical accuracy by both of the methods. Close-ups of cases 2 and 3 are displayed in Fig. 9. For case 2, the peak is captured to about 97% by DG1 and to 99.5% by AF. For case 3, it is captured to about 87% by DG1 and to 98% by AF. In all cases, this is based on the reconstructions within the cell.

2.5 *Lessons Learned*

Here are what I take to be the philosophical points that can be confidently be carried forward from these one-dimensional explorations.

1. Odd-order schemes are superior to even-order schemes for evolutionary problems.
2. Information should be taken from as close as possible to its true point of origin.

3. Semi-discrete schemes must either use irrelevant information or be restricted to small timesteps.
4. Discontinuous solutions can be captured by continuous reconstructions.

Those aspects that can NOT be carried forward are discussed in the next section.

However, to get a rough feel for the potential benefit of departing from current orthodoxy, it can be noted that to obtain comparable results, the mesh spacing for an AF solution need only be about 50% of that for an DG solution. If this is also true in higher dimensions, then for a three-dimensional unsteady flow we will need twice the number of points in each direction, and because of the CFL rule, twice the number of time steps. These will be 2.5 times smaller to ensure stability. Other things being equal, the cost of a DG calculation for such problems will be greater by a factor of $2^4 \times 2.5 = 40$. "Other things" will not, of course, be equal. Preserving the list of valuable properties will require original analysis that may lead to apparently expensive operations. However, we do start out with an enormous potential gain against which to offset these costs.

3 Extensions to Higher Dimension

3.1 Why One Dimension Can Be Misleading

A good model problem should be as simple as possible but no simpler.[2] It should feature, with as few variables as possible, as many of the problematic issues as it can. If any issues are left out, they must be investigated separately.

It is very common to begin the study of multidimensional hyperbolic problems with the multidimensional advection equation

$$\partial_t u + \mathbf{a}\nabla u = 0 \tag{9}$$

with **a** some constant vector. This is satisfactory in one dimension, because all one-dimensional hyperbolic problems, after linearization, reduce to a set of such problems with scalar constant a. However, in higher dimensions, hyperbolic problems

$$\partial_t \mathbf{u} + \sum_k \mathbf{A_k}\nabla\mathbf{u} = 0 \tag{10}$$

only reduce to a set of problems like (9) if they are simultaneously diagonalizable, which almost never happens. More commonly, there are a set of modes, some of which may involve only a single variable, and others that form a small subsystem [16]. Most commonly of all, there is an advective mode, and other modes that reduce

[2] Adaptation of a saying that is again attributed without complete certainty to Einstein. A less "known", but actually authentic and probably more useful version of the same sentiment is "Seek simplicity and distrust it", due to the philosopher Alfred North Whitehead [15].

to a wave equation. Examples include shallow water, Euler, Maxwell and elasticity. Each of these may be written

$$\partial_t \mathbf{u} + \mathbf{v} \cdot \nabla \mathbf{u} + \nabla \cdot \mathcal{F}(\mathbf{u}) \tag{11}$$

by a suitable choice of the conserved variables $\mathbf{u}$ and the "non-advected flux" $\mathcal{F}(\mathbf{u})$, but $\mathbf{v}$ is always the material velocity.

Equation (9) is very far from capturing the richness of real problems. It is simple, but too simple. In particular, it employs the velocity gradient $\nabla \mathbf{v}$ as the driver of changes in both the velocity itself and in other quantities such as pressure. In reality, changes of pressure are driven by the velocity divergence $\nabla \cdot \mathbf{v}$. Only in one dimension are these the same thing, but it has proved irresistable to try building multidimensional methods on one-dimensional foundations. There is a proliferation, too numerous to cite, of "multidimensional Riemann solvers" and "all-Mach Godunov-type schemes" to which I am ashamed to have contributed, but now renounce. My conversion was made possible by realizing that the better model (11) is not hard to implement.

The two special cases are the advective mode $\mathcal{F} = 0$, which is of course universal, and the opposite case $\mathbf{v} = 0$, which I will call the acoustic mode. For linear problems, these two cases commute and may be studied separately. This immediately makes available a set of novel high-order schemes for the long-range propagation of weak waves. These have remarkable properties.

For nonlinear problems, strictly rigorous commutativity breaks down beyond second-order [17]. However numerically convincing third-order results were found for the full Euler equations in [18–20]. A plausible strategy for nonlinear problems is merely to adopt a local linearization within every cell. This does not do great violence to the physics, since the behavior of hyperbolic problems is dominated by local effects.

3.1.1 A Better Model Problem

Although no model problem is likely to be perfect, the one that I have come to like is the div/grad acoustic system, starting with a fruitful collaboration on vorticity-preserving schemes [21] and followed by work with a student [22].

$$\begin{aligned} \partial_t p + c\nabla \cdot \mathbf{v} &= 0, \\ \partial_t \mathbf{v} + c\nabla p &= 0. \end{aligned} \tag{12}$$

We easily deduce that the pressure p satisfies the scalar wave equation

$$\partial_t^2 p - c^2 \nabla^2 p = 0, \tag{13}$$

and if we make a Helmholz decomposition of the vector $\mathbf{v} = \mathbf{v}_1 + \mathbf{v}_2$, with $\nabla \times \mathbf{v}_1 = \nabla \cdot \mathbf{v}_2 = 0$, we find that

$$\begin{aligned} \partial_t^2 \mathbf{v}_1 - c^2 \nabla^2 \mathbf{v}_1 &= 0, \\ \partial_t \mathbf{v}_2 &= 0. \end{aligned} \tag{14}$$

Although it may seem strange that the correct "decomposition" of the first-order system (12) would be the second-order system (13, 14), I see no other way to do it, and it allows access to the large body of classical results regarding the scalar wave equation (13). The solution to the initial-value problem for this was given by Poisson in 1808 [23] and is given in texts such as [24–26]. It is

$$p(\mathbf{x}, t) = p(\mathbf{x}, 0) + ctM_{ct}\partial_t p(\mathbf{x}, 0) + \int_0^{ct} tM_{ct}\partial_{tt} p(\mathbf{x}, 0)dt, \tag{15}$$

where M_R is the spherical mean function that takes the average of some function $\phi(\mathbf{x})$ over a spherical surface of radius R centered on a point $\mathbf{x}$,

$$M_R\phi(\mathbf{x}) = \frac{1}{\frac{4}{3}\pi R^2} \iint \phi dS. \tag{16}$$

This can be immediately applied to the acoustic system (12)

$$p(\mathbf{x}, t) = p(\mathbf{x}, 0) - c^2 tM_{ct}\nabla \cdot \mathbf{v}_1(\mathbf{x}, 0) + c^2 \int_0^{ct} tM_{ct}\nabla^2 p(\mathbf{x}, 0)dt. \tag{17}$$

and, since the solenoidal velocities also obey the scalar wave equation, we have

$$\mathbf{v}_1(\mathbf{x}, t) = \mathbf{v}_1(\mathbf{x}, 0) - c^2 tM_{ct}\nabla p(\mathbf{x}, 0) + c^2 \int_0^{ct} tM_{ct}\nabla^2 \mathbf{v}_1(\mathbf{x}, 0)dt. \tag{18}$$

Because the rotational velocity does not evolve, all three components of the solution can be absorbed into the single equation

$$\mathbf{u}(\mathbf{x}, t) = \mathbf{x}(\mathbf{x}, 0) + ctM_{ct}\partial_t \mathbf{u}(\mathbf{x}, 0) + \int_0^{ct} tM_{ct}\partial_{tt}\mathbf{u}(\mathbf{x}, 0)dt. \tag{19}$$

Since any spherical mean satisfies the Euler-Poisson-Darboux identity [24, 25], there is more than one version of this formula. At first sight, this is off-puttingly complex as the basis for a numerical method, but is actually far less costly than it may appear and has some fascinating properties.

1. The spherical mean function has an infinite series expansion in powers of the Laplacian ∇^2,

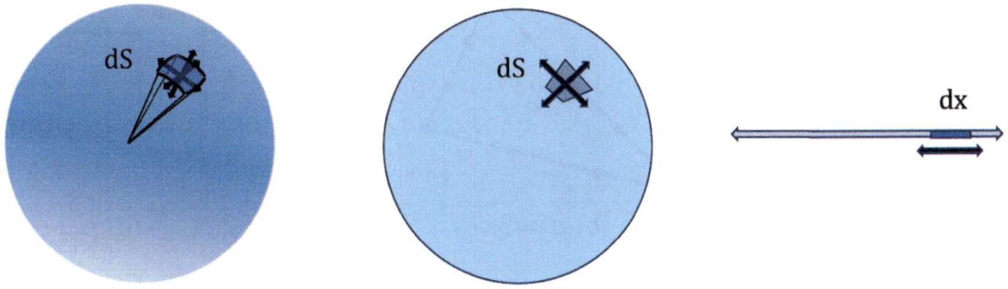

Fig. 10 Illustrating the relationship between solutions to the wave equation in (left) three dimensions, (center) two dimensions, (right) one dimension

$$M_R = 1 + \sum_k \frac{R^k}{(k+1)!} \nabla^{2k} = \frac{\sinh R\nabla}{R\nabla} \tag{20}$$

2. Using this, it is easy to show that (19) gives the full time-series expansion of the solution, with the second term giving all of the even terms, and the first and third terms giving all of the odd ones. If (19) is evaluated with second-order accuracy, the outcome will be a Lax-Wendroff method. Higher-order evaluations (perhaps from higher-order recontructions) will lead to higher-order methods.
3. Since (19) is exact in three dimensions (x, y, z), it will be exact in two dimensions (x, y) if data is taken independent of z. It will also be exact in one dimension x if data is taken independent of (y, z).
4. In two dimensions, the spherical mean becomes an integral over the interior of a circle. We can express the reconstruction in terms of polar coordinates centered on the node in question and use the formula

$$M_R\phi(\mathbf{x}) = \frac{1}{2\pi R} \int_0^{2\pi} \int_0^R \phi(x + s\cos\theta, y + s\sin\theta) \frac{s\,ds}{\sqrt{R^2 - s^2}}. \tag{21}$$

   These integrals have very simple closed-form evaluations.
5. In one dimension we descend further and recover the method of characteristics. This mathematically rigorous "method of descent" contrasts with the empirical "ascents" commonly encountered in computational methods.

A sketch to illustrate the method of descent is provided in Fig. 10. In each case it indicates the physically correct concept. On the left, in three dimensions, we should take account of all three components of the velocity divergence, together with the normal pressure gradient, and integrate these over the sphere of radius $c\Delta t$. In two dimensions, the divergence loses one component. In one dimension, the divergence loses another component. Although the gradients are now along a line, integration by parts shows that only the end point values contribute.

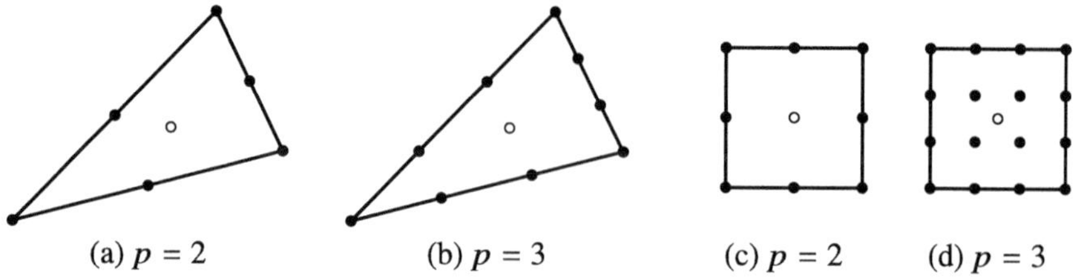

Fig. 11 On triangles, the Lagrange elements of order p place nodes at the point whose areal coordinates are $m/p, n/p, 1 - m/p - n/p, m, n = 0 \cdots p$. These include the centroid if p is even. On quadrilaterals, the Lagrange elements of order p place nodes at the point whose areal coordinates are $m/p, n/p, 1 - m/p - n/p, m + n = 0 \cdots 2p$. These include the centroid if p is even

3.2 Some Musings on Storage in Multiple Dimensions

There is probably no universal "best" storage strategy and I will not try to argue that I have found one. From a background in the aerospace industry, I have a bias toward unstructured grids that can handle intricate geometry.[3] Cartesian grids offer the advantage that the metric terms and nearest-neighbor information are trivial, but must incorporate special devices to handle curved boundaries. I will focus on simplex grids here, but the Active Flux method has been developed on Cartesian grids by other authors. See for example [27, 28] and the many references given there. Many different strategies can be made to work, but cannot be judged by any single criterion. Much more work is needed to discriminate among the possibilities, and what follows is of necessity quite tentative.

Staying for brevity in two dimensions, one approach [29] is to consider the possible degrees of freedom that allow a finite element to support reconstructed data with order of accuracy p. We have seen in one dimension that discontinuous solutions can be captured perfectly well by continuous elements, so I will consider only continuous reconstructions. To achieve pth order accuracy, we must be able to represent complete binomials of degree p (with terms $x^r y^s, r + s \leq p$) and this requires $\frac{1}{2}p(p+1)$ degrees of freedom. This requirement maps quite nicely onto triangular Lagrange elements (Fig. 11 (left)). It does not map so neatly onto quadrilateral elements (Fig. 11 (right)), where the "natural" number of d.o.f. is p^2. Often the reconstruction is then in terms of the extended polynomials with terms $(x^r y^s, r, s \leq p)$. These are always more numerous, especially as the order of accuracy and the number of spatial dimensions increase, but I am not aware of any useful gain in error coefficients.

Although the use of Lagrange elements is very common, I have doubts about using them in this context. Again I have as yet no general analysis to support this, but the grouping of nodes into tightly-bunched units seems very similar to the finite-difference schemes in Sect. 2.1, where the allowable Courant number remains 1.0 if

[3] There are obviously spherically, cylindrically, or conically symmetric objects that are best handled by similarly symmetric grids. Those are not my present concern.

based on the distance between nodes. A great variety of multidimensional methods have been proposed that introduce internal degrees of freedom into the elements, and all of them seem to be characterized by greatly reduced Courant numbers, usually less than $1/2p$. A detailed analysis of the one-dimensional case is given in [30]. This would not be acceptable in a design office because it reduces accordingly the number of design modifications that can be evaluated in a given time. If we were to create some formula by which to measure the productivity of a design office, it would include the achievable Courant number as a factor.

If nodes can be placed on the face of an element, they are available to both elements and therefore offer twice the value. If placed on an edge in three dimensions, they are available to about six elements. Nodes placed at a vertex are available to about six elements in two dimensions and to about 20 in three dimensions. Nodes can carry more than one piece of information. Hermitian methods carry information about the derivatives of the solution. They are commonly used to solve elliptic problems because they impose a high degree of smoothness, but are rather rare in hyperbolic problems because of the popular prejudice against smoothness. Some study of Hermitian methods for hyperbolic problems was started by Takewaki et al. [31] although the initials CIP (for cubic interpolated pseudoparticle) were later repurposed with different meanings. A different line of investigation was begun by Goodrich et al.[32] and has been continued (see [33] and references). Motivation is provided by the fact that, if the additional degrees of freedom needed for high-order can be carried at the fewest number of nodes, then internode communication can be kept to a minimum. This allows the best advantage to be taken of GPU processors [34]. Hermitian methods will be discussed in Sect. 5.2.

3.2.1 An Update Procedure

We are now in a position to give a general multidimensional update procedure that will reproduce the Active Flux method from Sect. 2.4 in the special case of one-dimensional linear advection. This is deliberately left a little vague to allow for future refinement.

1. Set up some discrete projection of the data in terms of point values or cell averages of values or derivatives.
2. Within each cell, use this projection to reconstruct the data.
3. Evolve the data in each cell according to some local linearization of the governing equations. For point values this will be sufficient.
4. For average values, define control volumes made up of point values (the "active fluxes") and integrate to update the averages.

Later sections of this paper will add a bit more flesh to this skeleton, but a simple example is illustrated in Fig. 12.

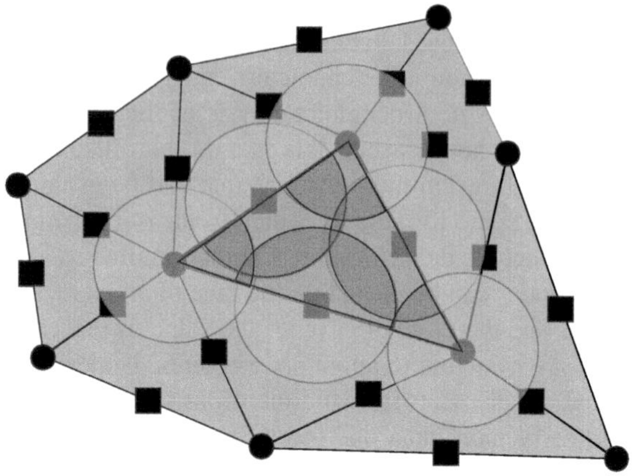

Fig. 12 An example of the update strategy. Data is represented by nodal values at the vertices and edge midpoints of a triangular grid, and a quadratic reconstruction made in each element. The nodal values are updated by integrating (19) over the surrounding disc. Each node receives an update from each neighboring element

3.2.2 Conservation

A very important aspect of a solution strategy is that it must admit a conservation property. Very often that is done by assigning to some nodes the mean values of the conserved quantities in some cell or sub-cell. The simplest example of this is the finite-volume strategy, where the local cell average is the only degree of freedom stored. Alternatively, a non-conservative strategy may be enriched by adding the cell average, as in Fig. 11a, d, or using it to replace a point value, as in Fig. 11b, c. The basis function associated with this is commonly described as a "bubble function" because it vanishes on the perimeter of the element. The storage shown in Fig. 11a was adopted in the first multidimensional extensions of the AF method [?]. It is one example of a "multi-moment" strategy that assigns different aspects of the solution to different classes of node.

However, it is not clear that this strategy is completely trustworthy. Very detailed investigations by Maeng [19] revealed that the results for scalar advection (9) deteriorated if the mesh took the form of square cells divided by consistent diagonals, and the flow direction was very closely aligned with that diagonal. In the case of exact alignment, the method decoupled into two one-dimensional methods, only one of which involved the cell averages, and the overall accuracy became second-order. The situation could be improved by adding a little cross-wind diffusion, but this is not of course really desirable.

The most recent multidimensional examples given in this paper are based on the alternative "residual distribution" (or "fluctation-splitting") approach, pioneered by Roe [35]. In the context of Active Flux methods, it has been refined by He [17]. We suppose that the computational domain is divided into non-overlapping cells of any kind, such as shown in Fig. 13. These cells have areas S_E and are made up of nodes N_j at which we have data $\mathbf{u}_j^n$. From this data we can compute a cell average $\bar{\mathbf{u}}_E$. For each cell we calculate a conservative element residual (also called a fluctuation)

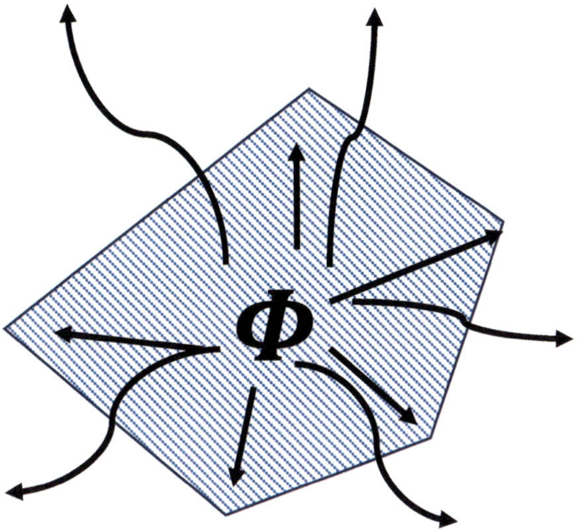

Fig. 13 Schematic of an element residual

$$\mathbf{\Phi}_E = \iint_E \partial_t \mathbf{u}\, ds = \sum_{edges} \bar{\mathbf{F}}\Delta y - \bar{\mathbf{G}}\Delta x. \quad (22)$$

where $\mathbf{F}, \bar{\mathbf{G}}$ are average fluxes defined somehow over each edge. We then compute a nonconservative solution $\mathbf{u}_j^{n+1}$ everywhere by whatever physically-inspired method we please, and define for each cell a discrepancy

$$\delta_E = S_E\big[\bar{\mathbf{u}}_E^{n+1} - \bar{\mathbf{u}}_E^n\big] - \mathbf{\Phi}_E \Delta t. \quad (23)$$

If the conservative and non-conservative solutions have been found from methods of the same formal accuracy, then in smooth regions the discrepancy is a quantity of the same order as the error, and removing it will remove the leading error. If that leading error is of even order, we will be improving the amplitude, and to do this, we can distribute the discrepancy uniformly over the nodes involved. This is the case for all examples here. If the leading error is of odd order, we will be improving the phase, and to do that the distribution should likely be upwinded. These issues are discussed by He [17] but offer much scope for further investigation, especially in linking them to concepts of limiting.

3.3 Nonlinear Advection

Common to many problems is a continuity equation

$$\partial_t \rho + \nabla \cdot (\rho \mathbf{v}) = 0, \quad (24)$$

and a momentum equation

$$\partial_t(\rho \mathbf{v}) + \nabla \cdot (\rho \mathbf{v} \otimes \mathbf{v}) + \cdots = 0, \tag{25}$$

where the ellipsis denotes the physics under discussion. Usually these terms decouple from the remainder of the problem up to second order and may be discussed separately. By themselves they are known as the pressureless Euler equations. They describe a stream of noninteracting particles that retain whatever initial velocity they first had. They are sometimes used (usually together with source terms) by cosmologists to describe the behavior of dust clouds [36] or as a model for dark matter [37]. The density may be eliminated to leave an equation for the velocities alone:

$$\partial_t \mathbf{v} + \nabla \cdot (\mathbf{v} \otimes \mathbf{v}) = 0. \tag{26}$$

This is the most correct generalization of Burgers' equation to higher dimensions. It has a series solution in time that has been used to calculate all of the examples here that have an advective component. For details, the reader is referred to [19].

4 Examples of Nonlinear Acoustics

For the moment, I continue to separate the different effects and focus in this section on nonlinear acoustics as exemplified by the p-system [38]

$$\begin{aligned} \partial_t \rho + \nabla \cdot (\rho \mathbf{v}) &= 0, \\ \partial_t \mathbf{v} + \nabla p(\rho) &= 0, \end{aligned} \tag{27}$$

with an equation of state that we choose to be $p(\rho) = \rho^\gamma/\gamma$. This will be handled by a local linearization within each element; that is, we apply equation (19) with $c^2 = \partial p/\partial \rho$.

4.1 Circular Pressure Wave

The first example of nonlinear acoustics is a spreading wave generated by a Gaussian pressure pulse applied to a stationary flow; $p(x, y, 0) = 1 + 2\exp -50(x^2 + y^2)$ [18]. For this problem $\gamma = 1.4$. The calculation proceeded without difficulty at the maximum geometrically allowable CFL number to $t = 1.25$. By this time the shock has not occurred, so no limiters were applied. Results are shown in Fig. 14, together with those from the DG1 method, using the same grid. Detailed comparisons at various grid sizes are made in [18] but we can summarize here that AF gives much better circular symmetry and better resolved peaks, as evident in the close-ups shown

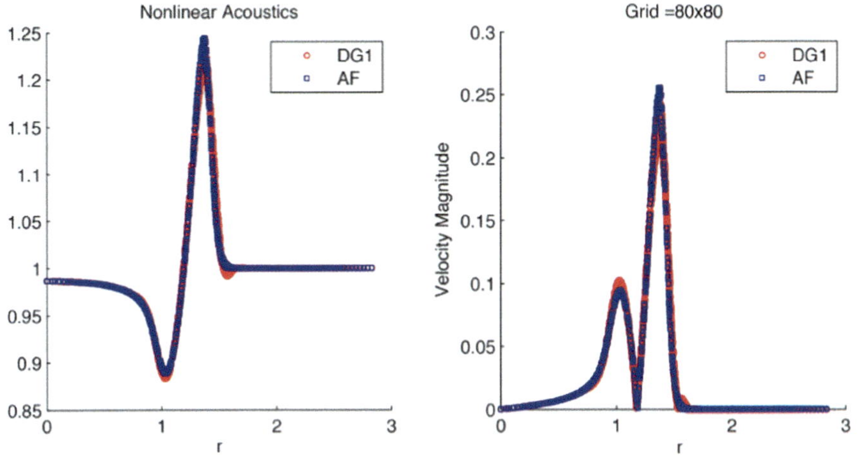

Fig. 14 Results for a nonlinear pressure pulse obtained from the third-order Active Flux method and the third-order DG1 method

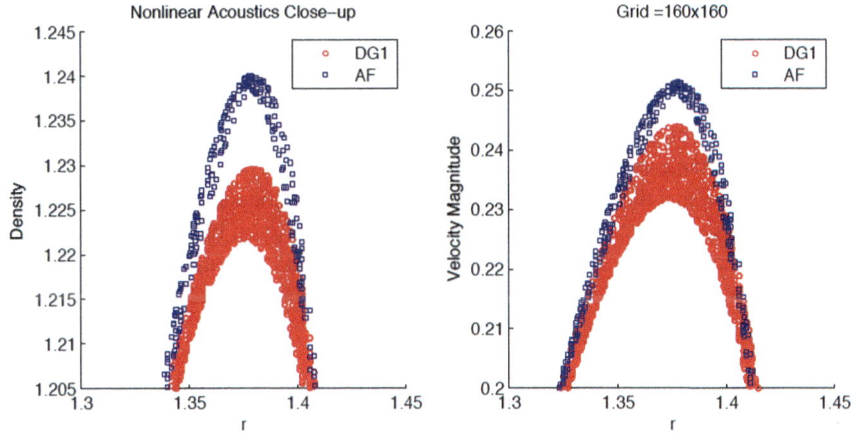

Fig. 15 Close-up of results near to the pressure peak in Fig. 14

in Fig. 15. Accuracy was determined by comparing with a fifth-order DG2 method on a fine grid. Convergence was found to be close to third-order for AF, but about 2.7 for DG. With the DG method, accuracy was found to be sensitive to how the initial conditions were imposed, but no such sensitivity was observed for AF. I conjecture that this may be due to the basis functions for AF being better conditioned.

4.2 Remote Boundary Conditions

The long-range propagation of weak waves is a common issue in topics ranging from the sonic boom to predictions of earthquake damage. Often we wish to say that these waves pass out of the domain without reflection. However, there is no local boundary condition that implies this and the correct global boundary condition is costly [39]. An account of the procedures commonly adopted has been collected in [40]. These divide into Non-reflective Boundary Conditions (NRBCs) that attempt some sort of characteristic analysis at the boundary, and Absorbing Layers (ALs) or Perfectly Matched Layers (PMLs) that damp out the outgoing waves before they reach the boundary. Neither method is perfect and both require that the computation be performed on a large domain. Errors due to spurious reflections are usually tolerated if they amount to no more than 2–3%.

In one dimension there is a simple solution. At the last interface in a finite-volume method any incoming waves are simply set to zero. Even more simply, in a fluctuation-splitting scheme, we set the outer boundary of the domain to be that of the last cell. There are no cells outside the boundary to generate incoming waves. The Active Flux method allows this strategy to be employed in higher dimensions. Nodes on the boundary are updated only by outgoing waves from internal cells. These are handled correctly because their direction is deduced from the gradients of the solution and not from the mesh geometry, as is the case with Godunov-type schemes. The integrals in (19) account for all of the propagation directions that are locally active.

In [18] the same circular acoustic wave was computed on a rectangular domain proportioned so that the wave could leave through the side boundaries but not through the top or bottom (Fig. 16). We do see small reflected waves, but these are within a tolerable 2% amplitude. The ability to use small domains without imposing specific remote boundary conditions will imply considerable savings provided the method can be implemented without excessive expense.

4.3 Preservation of Solution Symmetry

Exploding or imploding shocks may lose spherical symmetry, with large consequences. Maintaining physical symmetry is identified as a vital contribution to the first successful plasma burning in the National Ignition Facility [41]. It is also hard to achieve computationally in the face of so-called "mesh imprinting", when contour lines of the computed solution tend to follow the element boundaries. This is prevalent with Godonuv-type methods that evolve by propagating waves normal to those boundaries. In [22] it was shown that mesh imprinting was greatly reduced by a family of Lax-Wendroff schemes that preserve vorticity [21]. For Active Flux schemes with storage as in Fig. 11c, vorticity preservation was proved by Barsukow [42]. On unstructured grids with storage as in Fig. 11a it was observed computationally by Fan [18]. Although this is an active area of research there is probably much still to be discovered.

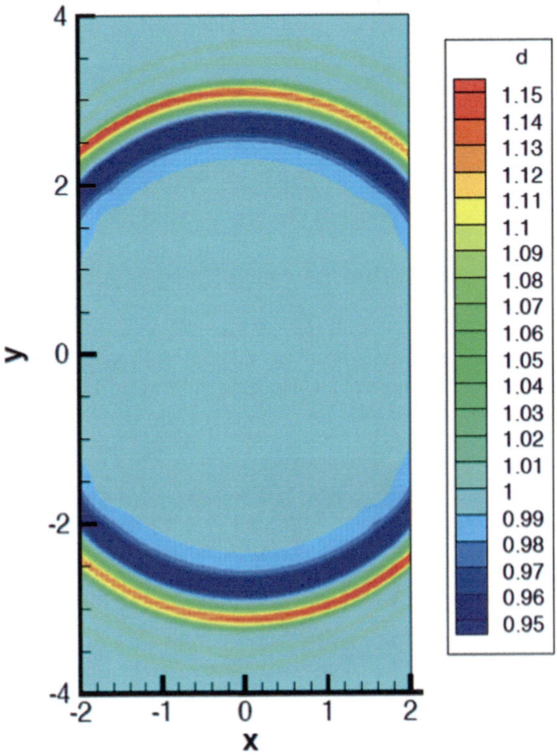

Fig. 16 Pressure distribution after the wave has left from the side boundaries. A very small reflected wave is visible

4.3.1 An Imploding Shockwave

In [20] the p-system was solved with initial data representing a uniform state surrounded by a converging wave:

$$\rho(x, y, 0) = 1, \quad (x^2 + y^2 \leq r_0^2), = 3, \quad (x^2 + y^2 \geq r_0^2)$$
$$v_r(x, y, 0) = 0, \quad (x^2 + y^2 \leq r_0^2), = \frac{-8r_0}{\sqrt{x^2 + y^2}}, \quad (x^2 + y^2 \geq r_0^2) \tag{28}$$

The values of r_0 and γ were taken to be 6 and 2. The calculation was performed on a unstructured triangular grid of average size 0.054 within a large square domain. Since the pressure should theoretically rise without bounds [43], so that no useful bounds are available, a limiter was devised based on the pragmatic axiom of Flux-Corrected Transport [44] that the correction applied to raise a trustworthy low-order scheme to high order may not create new extrema, or exaggerate old extrema. Details can be found in [20]. The calculated density profile is shown in Fig. 17 at $t = 3.00$ and $t = 3.74$. At the earlier time, we see the shock accompanied by large oscillations without limiting. The limiter does partly eliminate these, although the small one at the foot of the shock is almost certainly spurious. Without any limiting, the shock

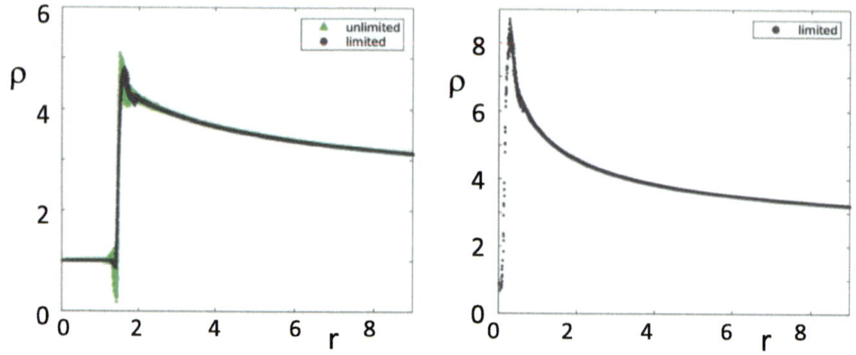

Fig. 17 Calculations of an imploding shockwave: scatter plots of density profiles (left) at $t = 3.0$ and (right) at $t = 3.74$

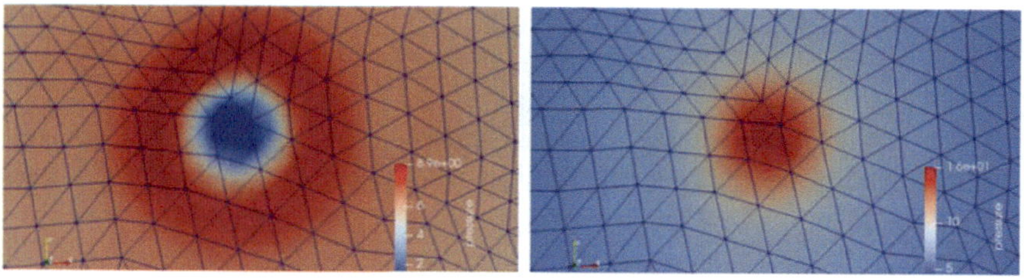

Fig. 18 Calculations of an imploding shockwave: density contours (left) at $t = 3.74$ and (right) at $t = 3.83$. These are consecutive time steps. The wave has changed from an incoming compression to an outgoing rarefaction

is unable to progress further, but with limiting, it continues to converge to $t = 3.74$. At this stage it still displays remarkable cylindrical symmetry, as seen both in the scatter plot in Fig. 17 and the contourplot in Fig. 18 (left).

In the next time step, the velocity in the central region turns radially outward and the wave begins to expand as a rarefaction (Fig. 18 (right)).

It proved quite difficult to obtain comparable results for this problem from any other method. We were able to obtain results from the our discontinuous Galerkin code only by running the fifth-order DG2 version with strong artificial viscosity and either a much weaker shock or extremely small time steps ($\nu = \mathcal{O}(10^{-3})$). I would like to put this problem forward as a challenge to other developers.

4.4 Efficient Implementation of Exact Linear Evolution

The formula (19) is indeed more complicated than a straightforward Riemann solver. The question is how much additional expense is involved? Recent studies by Bai [20] and Samani [29] show that this is much less that might be feared. The formula (19) reduces to integrals of the form

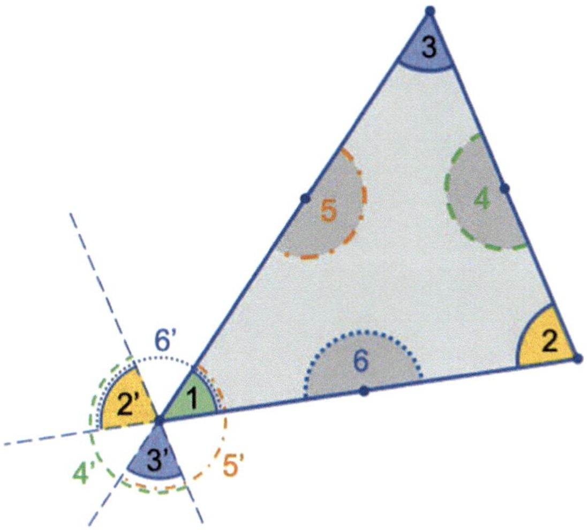

Fig. 19 Domains of integration at the six nodes of a triangular element

$$I_{p_1,p_2} = \int_{\theta_P}^{\theta_Q} \int_0^R \frac{(r\cos\theta)^{p_1}(r\sin\theta)^{p_2}\, dr\, d\theta}{\sqrt{R^2 - r^2}}, \tag{29}$$

with $p_1 + p_2 \in [0, p-2]$ where p is the order of the method. We therefore require $(p-1)(p-2)/2$ such integrals for each node. This is partly offset by the fact that the individual integrals are very simple[4] and that the higher-order versions reuse results from the lower-order versions. For a particular unstructured grid they can be precomputed. The three-dimensional versions are available but not yet published.[5]

For a third-order scheme we only need to integrate linear functions. For this case, Bai [20] found that the integrals at the six nodes of the triangle (Fig. 19) were not independent. She found the relationship

$$\begin{pmatrix} I_1 \\ I_2 \\ I_3 \end{pmatrix} = \begin{bmatrix} 0 & \frac{1}{2} & \frac{1}{2} \\ \frac{1}{2} & 0 & \frac{1}{2} \\ \frac{1}{2} & \frac{1}{2} & 0 \end{bmatrix} \begin{pmatrix} I_4 \\ I_5 \\ I_6 \end{pmatrix} + \begin{bmatrix} \frac{1}{4} - \frac{\Delta\theta_1}{2\pi} & \frac{1}{8} & \frac{1}{8} \\ \frac{1}{8} & \frac{1}{4} - \frac{\Delta\theta_2}{2\pi} & \frac{1}{8} \\ \frac{1}{8} & \frac{1}{8} & \frac{1}{4} - \frac{\Delta\theta_3}{2\pi} \end{bmatrix} \begin{pmatrix} f_1 \\ f_2 \\ f_3 \end{pmatrix} \tag{30}$$

where f_j is the point value of the integrand at node j. It is therefore necessary only to compute the integrals at three of the six nodes. Very likely similar results can be found at higher order.

Although these results can be used to reduce the cost of the method, it is certainly not intuitive how effective this will be. Bai [20] carried out experiments comparing AF and DG at third-order on a translating vortex problem for the full compressible Euler equations. The results are shown in Fig. 20. On the left we see that both methods

[4] They are tabulated up to $p = 5$ in [29] in about half a page.

[5] Iman Samani, private communication.

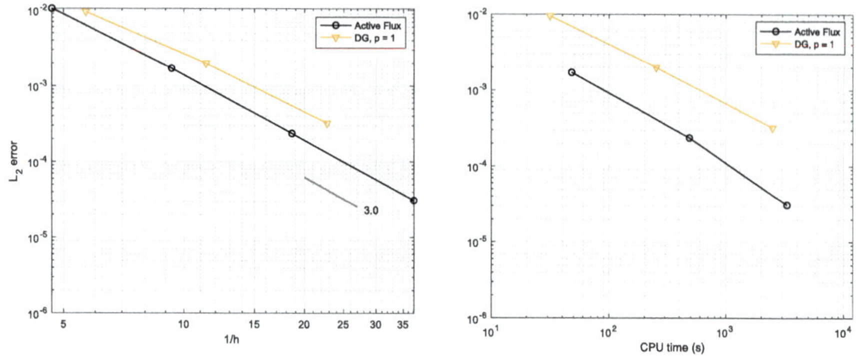

Fig. 20 Comparing the accuracy of Active Flux (AF) with discontinuous Galerkin 1 (DG1) on the basis of mesh size (left) and CPU time (right)

converge at third-order under mesh refinement. The AF results are expected to be both more accurate and more expensive than DG, so on the right the error is plotted against CPU time, and it is found that the AF method is faster by a factor of about seven.

It is of course possible that a method somehow intermediate between these two would do even better. Abgrall [28] has explored some possibilities in one dimension.

5 Beyond Third-Order

Third-order schemes are already a transformative step beyond the usual industrial standard of second-order,[6] but there are applications where it is desirable, and even essential, to go further. One recent and outstanding example of this is the computation of gravitational waves [45].

5.1 Some More Musings

Obviously, increasing accuracy must involve processing more information but doing this by enlarging the stencil implies, as remarked in Sect. 2.5, either using information from outside the true domain of dependence or taking very small time steps. The alternative to extending the mesh is to enrich it by storing more information in each element, as is done for example in Discontinuous Galerkin and Flux Reconstruction methods. The dispersion relationship is then found from an eigenvalue analysis of

[6] The accuracy of many computations is fundamentally restricted by uncertainty in geometrical data or physical constants. Is γ really equal to 1.4?.

the update matrix, which leads to a high-order polynomial for the eigenvalues. This polynomial has complex coefficients and of course there are multiple complex roots.

At third-order we have to solve a quadratic equation, whose roots have two branches corresponding to the choice of sign in the quadratic formula. This always happens with quadratic equations and, in many contexts, it is clear whether one or both of these roots is admissible. However, there is no universal rule applying to all contexts. In the context of analyzing dispersion relationships, it has become customary to refer to the root that gives $g(0) = 1$ as the "principal" root. I am reluctant to describe the other root as "spurious" even though that is appropriate in the case of leapfrog schemes, where the second root describes a wave that propagates in the wrong direction. This wave can only be excited by prescribing inconsistent data for the first two time levels. I will therefore describe the non-principal roots here, which frequently arise from higher-order polynomials, as "secondary". They can be excited by apparently harmless data.

5.1.1 Is the Nyquist-Shannon Barrier Real?

Typically, the secondary roots appear in the dispersion relations at frequencies in excess of π, and are often discarded by appealing to the Nyqvist-Shannon theorem [46–48] (sometimes called the "two points per wavelength" rule) based on the fact that the two function $\sin(\pi + \delta)(t/\Delta t)$ and $\sin(\pi - \delta)(t/\Delta t)$ are not distinguishable if sampled at regular intervals Δt. Somewhat in contradiction to this, and not well-known in the CFD community, is Papoulis' Theorem [49] that the frequency of a noisy sine wave cannot be estimated with less than two pieces of information per wavelength. The situation envisaged is that the signal is being sampled on more than one channel ("multi-channel sampling"). Each channel samples simultaneously a difference piece of information, defined to be "a linear functional of the solution". For the purposes of the theorem it does not matter what these are, as long as they are independent. It should therefore be possible to undertake a CFD calculation with fewer than two points per wavelength (perhaps fewer than one!) if the sampling is sufficiently rich. Although this is speculation, I offer evidence later that the straightforward Nyquist-Shannon barrier can be broken.

5.1.2 Are the Secondary Roots Useful?

For the schemes discussed here, the secondary roots are excited by apparently normal initial conditions and are often responsible for the stability bound on the CFL number. This appears for example in Fig. 5. Also, in the complex plane, the principal and non-principal roots are often (but not always) smoothly connected at a frequency that is close to a "round number" like $3\pi/2$. They can be identified by the choice of signs in the cubic and quartic formulas, but this seems to have no deep significance and is anyhow not available at higher order. There seems to be no general agreement about the meaning of the secondary roots, although the matter is discussed, for example in [50]. They cannot contribute to the formal accuracy of the method because that concept applies only to the resolved frequencies, but it seems quite possible that they may contribute to what I call its "utility".

It is common to suppose that the scales of a turbulent flow can be divided into "resolved" and "unresolved" frequencies when modeled at some grid spacing by some algorithm. This is the basis for Large Eddy Simulation where the unresolved scales are sometimes handled with some turbulence model [51] but sometimes allowed to dissipate numerically [50, 52]. The "cut-off frequency" is often taken to be the mesh spacing, but this can be very optimistic. It is also often assumed that the cut-off is sharp, but this is very hard to achieve. The dispersion diagrams scattered throughout this paper reveal that the complex amplification often pulls away smoothly from the unit circle and remains "semi-resolved" over some range of frequencies. It is far beyond the scope of these musings to determine whether this may be beneficial.

5.2 *Hermitian Methods*

Hermitian integration works with the function derivatives as well as with the function values. It is widely used in the solution of ordinary differential equations and elliptic partial differential equations because it imposes a desirable smoothness on the results. For the opposite reason, it is usually shunned in the hyperbolic case because of a belief that discontinuities should be allowed and even encouraged. A counterargument, that I think prevails, is that they allow the stencil to be made more compact, so that captured discontinuities are even better resolved. To examine this, I will now return to linear advection.

5.2.1 Linear Advection Again

There are many possible ways to include gradient information [53], but there must always be some non-gradient information to allow a conservation property. The version that we employ here (not considered by [53]) was proposed in [6] and employs the same compact stencil as the Active Flux method described in Sect. 2.4, with a cell average value $\bar{u}_j$ but the data at the $(j + 1/2)$ interface is now a vector $\mathbf{u}$ of length $p + 1$ whose k-th component is the solution derivative of order $k - 1$. There are now $2p + 3$ degrees of freedom available in a cell (but only $p + 2$ d.o.f. stored per cell), and so the reconstruction is polynomial of order $2p + 2$. The order of the scheme is $2p + 3$. The regular Active Flux method corresponds to $p = 0$. The flux between cells is again calculated by integrating between $\xi = 1 - \nu$ and $\xi = 1$. All of the derivative values are found by interpolation at $\xi = 1 - \nu$, as in Fig. 7. These methods have been found to remain stable until $\nu = 1$. Results from the fifth-order method $p = 1$ and

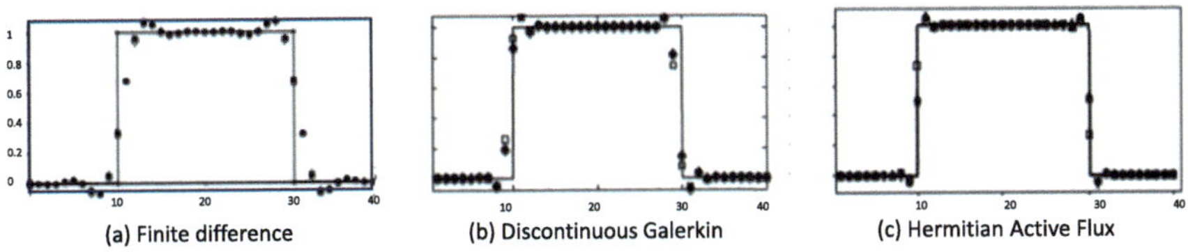

Fig. 21 Linear advection results for a square wave by three different methods. Results are shown at three different CFL numbers (0.25 (open circle), 0.50 (filled circle), and 0.75 (diamond)) but the symbols very often cannot be distinguished because they overplot

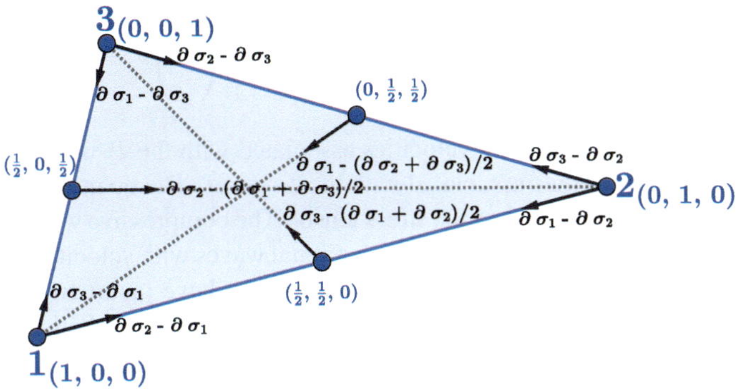

Fig. 22 A novel two-dimensional finite element

the seventh-order method $p = 2$ were presented in [6] and [29]. Here only the results for a square wave from the fifth-order method are shown in Fig. 21.

Proceeding from left to right, we see substantial improvements in (a) the width over which the jumps are captured, (b) the extent to which the solution near the jump is contaminated, and (c) the amplitude of the overshoots. These demonstrate the value of a compact stencil. From (a) to (b) we see the value of reducing the stencil in space, and from (b) to (c) the value of eliminating intermediate time steps. Both the finite-difference scheme and the Hermitian Active Flux scheme are stable up to $\nu = 1$, but the discontinuous Galerkin scheme is restricted to $\nu = 0.271$.

5.3 *Hermitian Active Flux Scheme in Two Dimensions*

Figure 22 shows what I think is a novel two-dimensional finite element, supporting a quartic reconstruction through 15°C of freedom. It uses 6 point values at the vertices and midpoints, and 9 first derivatives. There are two at each vertex, directed along the edges that meet there, and one at each midpoint, directed toward the center. Details are given in [29], including the corresponding 35 d.o.f. element in three dimensions, although no three-dimensional results are yet available. Two dimensional results for linear acoustics were shown in [54], including cases with more than one material. Test on smooth data showed better than fifth-order accuracy. In his thesis [29], Samani applies this element to a model of linear elasticity that separates compressive (P) and torsional (S) waves [55],

$$\partial_t \begin{pmatrix} \sigma \\ u_P \\ v_P \end{pmatrix} = \begin{bmatrix} 0 & a_P^2\partial_x & a_P^2\partial_y \\ \partial_x & 0 & 0 \\ \partial_y & 0 & 0 \end{bmatrix} \begin{pmatrix} \sigma \\ u_p \\ v_p \end{pmatrix}, \tag{31}$$

$$\partial_t \begin{pmatrix} \omega \\ u_S \\ v_S \end{pmatrix} = \begin{bmatrix} 0 & -a_S^2\partial_y & a_S^2\partial_x \\ -\partial_y & 0 & 0 \\ \partial_x & 0 & 0 \end{bmatrix} \begin{pmatrix} \omega \\ u_p \\ v_p \end{pmatrix}, \tag{32}$$

where u_P, v_P are the solenoidal velocities associated with the P-waves, and u_S, v_S are the rotational velocities associated with the S-waves. The quantities $\sigma = \sigma_{xx} + \sigma_{yy}$ and $\omega = \sigma_{xy}$ are invariants of the stress tensor. The compressive waves propagate with velocity $a_P = \sqrt{(\lambda + 2\mu)/\rho}$ and the torsional waves with velocity $a_S = \sqrt{\lambda/\rho}$, where λ, μ are the Lamé coefficients. Essentially we have (apart from some sign changes) two copies of the acoustic system, and apply the method to both. The domain is a 1×2 rectangle divided into two square subdomains, each occupied by a different elastic material. In each subdomain there is a quite coarse, fully unstructured grid of about 1800 triangular elements. On the left, we have $\lambda/\rho = 1$, $\mu/\rho = 1$, so $a_P = \sqrt{3}$, $a_S = 1$. On the right, $\lambda/\rho = 2$, $\mu/\rho = 1$ so $a_P = \sqrt{5}$, $a_S = 1$. At $t = 0$ a small symmetrical perturbation, about four cells in diameter, that generates both types of wave is introduced at the center of each subdomain.

5.4 *Results*

Results before the waves have left their "starting boxes" are shown in Fig. 23. All the waves are clean and well separated. Results in [29], but not shown here, verify that, when the initial disturbance generates only one type of wave, the other type never appears. In Fig. 24 the P-waves from both boxes have either left the domain altogether or passed into the other box. Waves that have left have not created any visible reflection. Waves passing into the other box have adjusted their speed. The S-waves have started to cross the interface. This creates vorticity, visible only as a surface wave within the interface layer one cell deep.

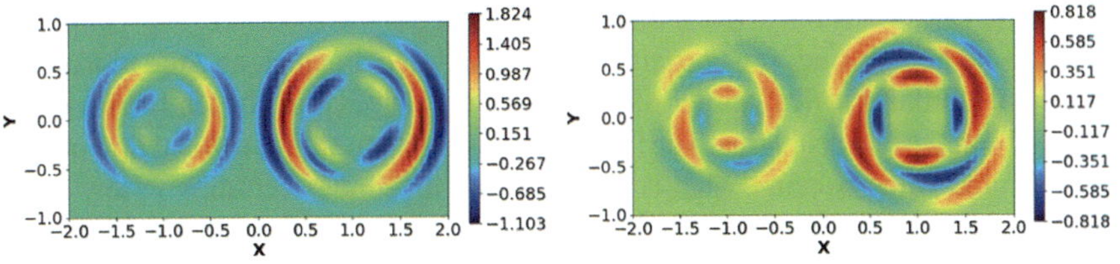

Fig. 23 Stresses at $t = 0.35$. (σ_{xx} in left diagram, σ_{xy} in right diagram). All waves are still in their "starting boxes"

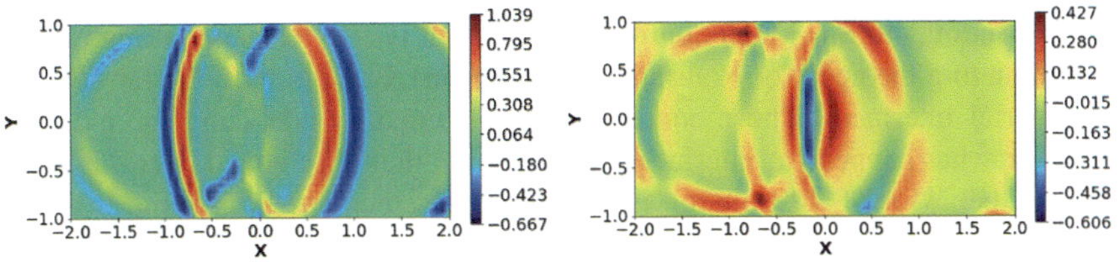

Fig. 24 Stresses at $t = 0.95$. (σ_{xx} in left diagram, σ_{xy} in right diagram). S-waves are still in their boxes, but P-waves have left without visible reflection both from the domain boundary and into the neighboring subdomain. Vorticity has been created at the interface by passage of the S-waves

6 Final Musings and Future Directions

In my role as a computational philosopher, I have tried to focus primarily on objectives rather than techniques. I take these to be much more than than the achievement of formal accuracy that is the topic of much contemporary research. Accuracy as commonly defined is a low frequency phenomenon. As such it is amenable to the precise probing of Taylor series and Fourier analysis. However, the things that go wrong with practical calculations happen at high frequencies. They include phase error, spurious oscillations, loss of resolution, and instability. These problems are often attacked by tuning free parameters of the scheme, but I hope to trace them to their causes even though I am unsure how rigorously that can be done.

After long thought, I feel that the key to understanding these issues lies in the flow of information. Both in a hyperbolic partial differential equation and in a digital computer there is a well-defined flow of information and I feel that the mapping between these flows should be as clean as possible. For example, the pathways should correspond. For the p.d.e.s the pathways are very sharply defined. However, in the computer they are necessarily diffused and should be cleaned up by making the stencil as compact as possible. They should also take the most direct routes from where the information is available to where it is needed. I can express this by saying that the stencil should always be centered, but centered on the place where the information that we wish to convey is located. Often, of course, this will be "upwind".

Not only should the pathways be clean, but they should carry the most useful information. Determining what is useful requires careful analysis of the differential operators. Doing this only in one dimension can be very misleading. For example, it provides no help in studying vorticity which is, in three dimensions, "the muscle and sinews of fluid dynamics" [56]. "Model problems" should be chosen for what they reveal about the real problems. The large literature on "multidimensional upwinding" is of limited value because what is at play is not a "wind" according to any useful analogy. Neither is it confined to a linear path. A fanciful analogy that works for me personally is that of grazing buffalo spreading over a prairie! The effect is described by integral equations such as (19). The herd may seem to move in one direction, but individual buffalo can move in many directions. The pasture is vector-valued. The value of doing an authentic multidimensional analysis is felt by extending the benefits of "upwinding" to low Mach number flows [57]. It is also felt, just as importantly, in eliminating the need for Non-reflecting Boundary Conditions at the far field and at material interfaces. Those conditions are obtained by analyzing the p.d.e.s to discover what happens at interfaces, and they are then discretized. Here, it seems that the behavior of the solution is so well represented by the scheme that such analysis is almost unnecessary.

A great deal of work remains to bring these ideas to practical use. Many small details like the exact form of an average may enable small, cumulative, improvements. One elusive but important detail is the handling of no-flow boundary conditions at solid walls. The major extension required is to three dimensions but I expect this to be straightforward. After all, the Active Flux method presented here is just the two-dimensional projection of a hypothetical three-dimensional scheme. The improvements may be greater in three dimensions, both because one more dimension of the grid can be coarsened, and because of the heightened influence of vorticity. A start has been made on the Navier-Stokes equations by He [17]. These equations have been expressed in hyperbolic form by Nishikawa and collaborators [58] and reformulated as hyperbolic moment equations by Levermore [59]. He [17] finds compressive and torsional waves in these equations, damped by bulk and shear viscosity, as well as "heat waves" damped by conductivity. It may be imagined that in some situations the calculations might be considerably improved by including this level of detail, but this does become expensive, and should probably be restricted to those locations where it would make a difference.

I have only discussed limiting in one context and that is because limiting is only needed to remedy some defect in the underlying method. Although overshoots are inevitable, they are greatly reduced by making a good choice of stencil. For many purposes, the unlimited results can be perfectly acceptable, as in Fig. 21. However, in case of Sect. 4.3.1 the use of some limiter proved essential. A semi-empirical limiter was found to work well enough. It can probably be improved, but I leave that until I am clearer about the causes.

I am confident that all of these details will be clarified in due time. I am also confident that CFD methods will eventually become available that are at least one order of magnitude more effective than those that are in current use. They may or may not be those that I have discussed here, but I am fairly sure they will make use of the same ideas.

References

1. Žižek, S.: How Philosophy Got Lost, online interview retrieved Jan 1, (2024). https://youtu.be/06KiOj6gjbs?si=dEMzQ_5fQRerWJgg
2. Lipman, M., et. al.: Philosophy in the Classroom, Temple University Press (1980)
3. Iserles, A., Strang, G.: The optimal accuracy of difference schemes. Trans. Am. Math. Soc. **277**(2), 779–803 (1983)
4. Hedstrom, G.W.: Models of difference schemes for $u_t + u_x = 0$ by partial difference equations. Math. Comp. **62**, 497–530 (1975)
5. Bouche, D., Bonnaud, G., Ramos, D.: Comparison of numerical schemes for solving the advection equation. App. Math. Letters **16**(2), 171–190 (2003)
6. Roe, P.L.: Designing CFD methods for bandwidth-a physical approach. Comput. Fluids **214**, 104774 (2021)
7. Tam, C.K.W., Webb, J.C.: Dispersion-relation-preserving finite difference schemes for computational acoustics. J. Comp. Phys. **107**, 262–281 (1993)
8. Cockburn, B., Shu, C.W.: Runge-Kutta discontinuous Galerkin methods for convection-dominated problems. J. Sci. Comput. **16**, 173–261 (2001)
9. Eymann, T., Roe, P.: Active flux schemes. In: 49th AIAA Aerospace Sciences Meeting including the New Horizons Forum and Aerospace Exposition p. 382 (2011)
10. Roe, P.L.: A simple explanation of superconvergence for discontinuous Galerkin solutions to ut+ ux= 0. Commun. Comput. Phys. **21**(4), 905–912 (2017)
11. Van Leer, B.: Towards the ultimate conservative difference scheme IV A new approach to numerical convection. J. Comput. Phys. **23**(3), 276–299 (1977)
12. Popov, M.V., Ustyugov, S.D.: Piecewise parabolic method on local stencil for gasdynamic simulations. Comput. Math. Math. Phys. **47**, 1970–1989 (2007)
13. Ii, S., Xiao, F.: CIP/multi-moment finite volume method for Euler equations: a semi-Lagrangian characteristic formulation. J. Comput. Phys. **222**(2), 849–871 (2007)
14. Eymann, T.A., Roe, P.L.: Multidimensional active flux schemes. In: 21st AIAA computational fluid dynamics conference. p. 2940 (2013)
15. Whitehead, A.N.: The Concept of Nature, reprinted 2007, Cosimo Books (1920)
16. Roe, P.L., Turkel, E.: The quest for diagonalization of differential systems. In: Barriers and Challenges in Computational Fluid Dynamics, pp. 351–369. Springer, Netherlands, Dordrecht (1998)
17. He, F.: Toward a new-generation numerical scheme for the compressible navier-stokes equations with the active flux method. PhD dissertation (Applied and Interdisciplinary Mathematics, University of Michigan) available at https://deepblue.lib.umich.edu (2021)
18. Fan, D.: On the acoustic component of the Active Flux schemes for nonlinear hyperbolic conservation laws. PhD dissertation (Aerospace Engineering, University of Michigan) available at https://deepblue.lib.umich.edu (2017)
19. Maeng, J.B.: On the advective component of the Active Flux schemes for nonlinear hyperbolic conservation laws. PhD dissertation (Aerospace Engineering, University of Michigan) available at https://deepblue.lib.umich.edu (2017)
20. Bai, Y.: Shock capturing for high-order computational fluid dynamics. PhD dissertation (Aerospace Engineering, University of Michigan) available at https://deepblue.lib.umich.edu (2023)
21. Morton, K.W., Roe, P.L.: Vorticity-preserving lax-wendroff-type schemes for the system wave equation. SIAM J. Sci. Comput. **23**(1), 170–192 (2001)
22. Lung, T.B., Roe, P.L.: Toward a reduction of mesh imprinting. Int. J. Numer. Meth. Fluids **76**(7), 450–470 (2014)
23. Poisson, S.D.: Mémoire sur la théorie do son. J. Ecole Polytech. Paris **7**, 319–370 (1808)
24. John, F.: Plane waves and spherical means applied to partial differential equations. Wiley Intersciences. Reprinted by Springer, 2013 (1955)
25. Whitham, G. B.: *Linear and Nonlinear Waves*, Wiley Interscience, reprinted 2011 (1974)

26. Evans, L.: Partial Differential Equations. American Mathematical Society, Providence (1988)
27. Calhoun, D., Chudzik, E., Helzel, C.: The cartesian grid active flux method with adaptive mesh refinement. J. Sci. Comput. **94**(3), 54 (2023)
28. Abgrall, R.: A combination of residual distribution and the active flux formulations or a new class of schemes that can combine several writings of the same hyperbolic problem: application to the 1d Euler equations. Communications on Applied Mathematics and Computation **5**(1), 370–402 (2023)
29. Samani, I.: Active Flux methods with Gradient Degrees of Freedom, PhD dissertation (Aerospace Engineering, University of Michigan) available at https://deepblue.lib.umich.edu (2024)
30. Olivera, B.O., Azevedo, F.F.A.: Stability and order-of accuracy analysis of high-order schemes formulated using the flux reconstruction approach. arXiv:2308617v1 [physics.fluid.dyn] (2023)
31. Takewaki, H., Nishiguchi, A., Yabe, T.: Cubic interpolated pseudo-particle method (CIP) for solving hyperbolic-type equations. J. Comput. Phys. **61**, 261–268 (1985)
32. Goodrich, J., Hagstrom, T., Lorenz, J.: Hermite methods for hyperbolic initial-boundary value problems. Math. Comput. **75**, 595–630 (2006)
33. Appelo, D., Hagstrom, T., Vargas, A.: Hermite methods for the scalar wave equation. SIAM J. Sci. Comput. **40**(6), A3902–A3927 (2018)
34. Pickering, B.P., Jackson, C.W., Scogland, T.R., Feng, W.C., Roy, C.J.: Directive-based GPU programming for computational fluid dynamics. Comput. Fluids **114**, 242–253 (2015)
35. Roe, P.L.: Fluctuations and signals-a framework for numerical evolution problems. Numerical methods for fluid dynamics, **11** (1982)
36. Levecque, R.J.: The dynamics of pressureless clouds and delta waves. J. Hyperbolic Diff. Eqns. **01**(2), 315–327 (2004)
37. Chavanis, P.H.: BEC dark matter, Zeldovich approximation, and generalized Burgers equation. Phys. Rev. D **84**(6), 063518 (2011)
38. Smoller, J.: Shock waves and reaction-diffusion equations (Vol. 258). Springer Science & Business Media (2012)
39. Givoli, D.: High-order local non-reflecting boundary conditions: a review. Wave Motion **39**(4), 319–326 (2004)
40. Mueller, T.R., Vogt, D.M., Fischer, M., Phillipsen, B.A.: On the far-field boundary condition treatment in the framework of aeromechanical computations using ANSYS CFX. Proceedings of the Institution of Mechanical Engineers, Part A: Journal of Power and Energy,2 **35**(5), pp. 1103–1118 (2021)
41. Ralph, J., Ross, J., Zylstra, A., et al.: The impact of low-mode symmetry on inertial fusion energy output in the burning plasma state. Preprint available at ResearchSquare (2024)
42. Barsukow, W.: Stationarity preserving schemes for multi-dimensional linear systems. Math. Comput. **88**(318), 1621–1645 (2019)
43. Ramsey, S.D., Kamm, J.R., Bolstad, J.H.: The Guderley problem revisited. Int. J. Comput. Fluid Dyn. **26**(2), 79–99 (2012)
44. Kuzmin, D., Löhner, R., Turek, S.: Flux-corrected transport: principles, algorithms, and applications. Springer (2012)
45. Lovelace, G.: Computational challenges in numerical relativity in the gravitational-wave era. Nature Comput. Sci. **1**, 450–452 (2021)
46. Nyquist, H.: Certain topics in telegraph transmission theory. Trans. AIEE. **47**(2) pp. 617-644. Reprinted in: *Proc. IEEE*, **90(2)**, 2002 (1928)
47. Shannon, C.E.: Communication in the presence of noise. Proceedings of the Institute of Radio Engineers. **37** (1), pp.10-21. Reprinted in: *Proc. IEEE,***86** (2), 1998 (1949)
48. Wikipedia contributors. Nyquist-Shannon sampling theorem. Wikipedia, The Free Encyclopedia. February 1, 2024, 11:33 UTC. Accessed February 5, (2024)
49. Papoulis, A.: Generalized sampling expansion. IEEE Trans. Circ. Syst. **24**(11), 652–654 (1977)
50. Mengaldo, G., Moura, R.C., Giralda, B., Peiró, J., Sherwin, S.J., et al.: Spatial eigensolution analysis of discontinuous Galerkin schemes with practical insights for under-resolved computations and implicit LES. Comput. Fluids **169**, 349–364 (2018)

51. Lesieur, M., Métais, O., Comte, P.: Large-eddy simulations of turbulence. Cambridge University Press (2005)
52. Grinstein, F., Margolin, L., Rider, W.: Implicit large eddy simulation. Cambridge University Press (2007)
53. Vargas, A., Chan, J., Hagstrom, T., Warburton, T.: Variations on Hermite methods for wave propagation. Commun. Comput. Phys. **22**(2), 303–337 (2017)
54. Samani, I., Roe, P.L.: Acoustics on a Coarse Grid. In: AIAA SCITECH 2023 Forum(p. 1156) (2023)
55. Nakazawa, S.: The application of a two-dimensional upwind leapfrog scheme to linear elastodynamics PhD dissertation (Aerospace Engineering, University of Michigan) available at https://deepblue.lib.umich.edu (2001)
56. Küchemann, D.: The Aerodynamic Design of Aircraft, p. 2012. Pergamon Press. reprinted by AIAA, Oxford (1978)
57. Barsukow, W., Hohm, J., Klingenberg, C., Roe, P.L.: The active flux scheme on Cartesian grids and its low Mach number limit. J. Sci. Comput. **81**, 594–622 (2019)
58. Li, L., Lou, J., Nishikawa, H., Luo, H.: Reconstructed discontinuous Galerkin methods for compressible flows based on a new hyperbolic Navier-Stokes system. J. Comput. Phys. **427**, 110058 (2021)
59. Levermore, C.D., Morokoff, W.J.: The Gaussian moment closure for gas dynamics, SIAM. J. Appl. Math. **59**, 72–96 (1998)
60. Bourgeois, R., Tremblina,P., Kokh, S., Padiolea, T.: Recasting an operator splitting solver into a standard finite volume flux-based algorithm. The case of a Lagrange-Projection-type method for gas dynamics. preprint posted on arXiv:2203.14596v3 [math.NA] 27 Oct 2023 (2023)

Using Neural Network for Improving an Explicit Algebraic Stress Model in 2D Flow

Lars Davidson

Abstract Neural Network (NN) is used to improve an Explicit Algebraic Reynolds Stress Model (EARSM). The turbulent kinetic energy and its dissipation are predicted using the standard $k - \omega$ model. The NN model is trained in channel flow of $Re_\tau =$ 10,000. The NN model is stored to disk and subsequently loaded into the CFD code. The NN model is called every iteration to compute the β coefficients in the EARSM, i.e. the CFD solver and the NN model are fully coupled. The Reynolds stresses are used in the momentum equations and the production term in the k and ω equations. It is found that when training the NN model, the target data cannot only be taken from DNS. The reason is that the stress-strain relation and the turbulent kinetic energy of the DNS data are different from those of the $k - \omega$ model. Hence, the target data are taken both from DNS and a $k - \omega$ simulation. The new EARSM-NN model is used for predicting channel flow at $Re_\tau = 2000$, 5200 and 10,000 and flat-plate boundary layer at $2500 \leq Re_\theta \leq 8000$. The EARSM-NN model gives much better results than the standard EARSM.

1 Introduction

The relation between strain rate and Reynolds stresses in two-equation eddy-viscosity turbulence models is usually linear (the Boussinesq assumption). Non-linear turbulence models were developed in the 1990s, see e.g. [1]. These non-linear models are based on a subset of the generalized nonlinear form given by Pope [2] which is expressed in ten tensors, T_{ij}^n and five invariants, i.e.

$$a_{ij} = \sum_{n=1}^{10} G^{(n)} T_{ij}^n, \quad a_{ij} = \overline{v_i' v_j'} - \frac{2}{3} k \delta_{ij} \tag{1}$$

CUFS-2024-PAPER ID, Cambridge, United Kingdom, 4-5 March 2024.

L. Davidson (✉)
Division of Fluid Dynamics, Department of Mechanics and Maritime Sciences,
Chalmers University of Technology, 412 96 Gothenburg, Sweden
e-mail: lada@chalmers.se

J. C. Tyacke and N. R. Vadlamani (eds.), *Proceedings of the Cambridge Unsteady Flow Symposium 2024*, https://doi.org/10.1007/978-3-031-69035-8_2

where a_{ij} is the anisotropy tensor. Ling *et al.* [3] seem to be the first who used neural networks (NN) and the generalized nonlinear for predicting the turbulent Reynolds stress. They use a general non-linear Reynolds stress model in which the stresses are expressed by Eq. (1). They trained their neural network using the five invariants as input and the 10 tensors as output. Six cases were used for training, namely duct flow, channel flow, a jet in cross-flow, an inclined jet in cross-flow, flow around a square cylinder and flow through a converging-diverging channel. However, when they include their NN model into their CFD solver, they apply the NN model only once and inject the NN stresses and then keep them frozen (sometimes called the frozen substitution [4]).

The objective of [5] was also to develop a model for the turbulent stress tensor, $\overline{v_i'v_j'}$. Contrary to the work by [3], they include more influence parameters. They include also the gradients of pressure and turbulent kinetic energy, i.e.

$$b_{ij} = f\left(|\bar{s}|, |\bar{\Omega}|, \partial\bar{p}/\partial x_i, \partial k/\partial x_i\right)$$

where $|\bar{s}| = (\bar{s}_{ij}\bar{s}_{ij})^{1/2}$ and $|\bar{\Omega}| = (\bar{\Omega}_{ij}\bar{\Omega}_{ij})^{1/2}$. They propose another three influence variables to account for viscous (low-Re number) effects near walls; they use local Reynolds number, $k^{1/2}y/\nu$ (y denotes wall-distance), turbulent kinetic energy and the ratio of turbulent to mean flow time scales. They use a machine learning method based on random forest regression.

Duraisamy *et al.* [6] gives a comprehensive review on Machine Learning and turbulence modeling. They survey recent developments in bounding uncertainties in RANS models using physical constraints and they present methods how to use machine learning to improve turbulence models.

Weatheritt and Sandberg [7] employed Genetic Expression Programming (GEP) to train an EARSM. The improvement of the Reynolds stress is chosen as the target application. They train the model in backward facing step and periodic hill flow. Then they do *à posteriori* simulations of the hill flow. However, when applying the GPA model they use a frozen velocity field predicted by a SST $k-\omega$ model.

Akoleka *et al.* [8] use the methodology developed in [7]. They train an EARSM using data of a DNS of a low-pressure turbine blade. Then they used the EARSM-GEP model to carry out 2D URANS simulations of the same turbine blade. The underlying turbulence model was the $k-\overline{v'^2}-\omega$ transition model. However, they used the frozen concept, solving only $\overline{v'^2}$ and ω and taking the remaining variables from DNS.

Neural Network (NN) is used in the present work to improve an EARSM. The turbulent kinetic energy and its dissipation are predicted by the standard $k-\omega$ model. The NN model is trained in channel flow of $Re_\tau = 10{,}000$. The NN model is stored to disk and subsequently loaded into the CFD code. The NN model is called every iteration to compute the β coefficients in the EARSM. Then the Reynolds stresses are computed which are used in the momentum equations and the production term in the k and ω equation.

It is found that when training the NN model, the target data cannot only be taken from DNS. The reason is that stress-strain relation and the turbulent kinetic energy of the DNS data are different from those of the $k - \omega$ model. Hence, the target data are taken both from DNS and a $k - \omega$ simulation.

2 Flow Equations

The momentum equations read

$$\frac{\partial \bar{v}_j \bar{v}_i}{\partial x_j} = -\frac{1}{\rho}\frac{\partial \bar{p}}{\partial x_i} + \frac{\partial}{\partial x_j}\left[\left(\nu + \nu_{eff}\right)\left(\frac{\partial \bar{v}_i}{\partial x_j} + \frac{\partial \bar{v}_j}{\partial x_i}\right)\right] - \frac{\partial (\overline{v_i' v_j'})_r}{\partial x_j} \tag{2}$$

When EARSM is used, ν_{eff} is the effective viscosity including the EARSM coefficient β_1, see Eq. (16). The last term includes $(\overline{v_i' v_j'})_r$ which is the residual stress tensor in EARSM. The total stress tensor is

$$\overline{v_i' v_j'} = -\nu_{eff}\left(\frac{\partial \bar{v}_i}{\partial x_j} + \frac{\partial \bar{v}_j}{\partial x_i}\right) + (\overline{v_i' v_j'})_r \tag{3}$$

When the standard $k - \omega$ is used without EARSM the residual stress vanishes and $\nu_{eff} = \nu_t$.

2.1 The Numerical Solver

The **pyCALC-RANS** code is used [9]. It is an incompressible, finite volume code written in Python. It is fully vectorized (i.e. no for loops). The convective terms in all equations are discretized using the Hybrid central/upwind scheme. The numerical procedure is based on the pressure-correction method, SIMPLEC, and a collocated grid arrangement using Rhie-Chow interpolation [10].

2.2 The $k - \omega$ Model

The Wilcox $k - \omega$ turbulence model reads [11]

$$\begin{aligned}
\frac{\partial \bar{v}_j k}{\partial x_j} &= P^k + \frac{\partial}{\partial x_j}\left[\left(\nu + \frac{\nu_t}{\sigma_k}\right)\frac{\partial k}{\partial x_j}\right] - C_\mu k\omega \\
\frac{\partial \bar{v}_j \omega}{\partial x_j} &= C_{\omega_1}\frac{\omega}{k}P^k + \frac{\partial}{\partial x_j}\left[\left(\nu + \frac{\nu_t}{\sigma_\omega}\right)\frac{\partial \omega}{\partial x_j}\right] - C_{\omega 2}\omega^2 \\
\nu_t &= \frac{k}{\omega}
\end{aligned} \tag{4}$$

The standard coefficients are used, i.e. $C_{\omega 1} = 5/9$, $C_{\omega 2} = 3/40$, $\sigma_k = \sigma_\omega = 2$ and $C_\mu = 0.09$. When EARSM is used, the production term is computed as

$$P^k = -\overline{v_i' v_j'}\frac{\partial \bar{v}_i}{\partial x_j} \tag{5}$$

and the dissipation in the EARSM reads

$$\varepsilon = C_\mu k\omega \tag{6}$$

In the $k - \omega$ model (without EARSM), the production term is computed as

$$P^k = \nu_t\left(\frac{\partial \bar{v}_i}{\partial x_j} + \frac{\partial \bar{v}_j}{\partial x_i}\right)\frac{\partial \bar{v}_i}{\partial x_j} \tag{7}$$

3 The EARSM

The Algebraic Stress Model (ASM) [12] with the LRR pressure-strain model [13] reads [14]

$$\begin{aligned}
\left(c_1 - 1 + P^k/\varepsilon\right) a_{ij} &= -\frac{8}{15}\bar{s}_{ij} + \frac{7c_2+1}{11}(a_{ik}\bar{\Omega}_{kj} - \bar{\Omega}_{ik}a_{kj}) \\
&\quad - \frac{5-9c_2}{11}\left(a_{ik}\bar{s}_{kj} + \bar{s}_{ik}a_{kj} - \frac{2}{3}a_{mn}\bar{s}_{nm}\delta_{ij}\right) \\
a_{ij} = \frac{\overline{v_i' v_j'}}{k} - \frac{2}{3}\delta_{ij}, \quad \bar{s}_{ij} &= \frac{1}{2}\left(\frac{\partial \bar{v}_i}{\partial x_j} + \frac{\partial \bar{v}_j}{\partial x_i}\right), \quad \bar{\Omega}_{ij} = \frac{1}{2}\left(\frac{\partial \bar{v}_i}{\partial x_j} - \frac{\partial \bar{v}_j}{\partial x_i}\right)
\end{aligned} \tag{8}$$

Note that the last term in Eq. (8) is zero if c_2 is set to 5/9 [15]. Equation (8) can be written as [16]

$$\begin{aligned}
Na_{ij} &= -A_1\bar{s}^*_{ij} + (a_{ik}\bar{\Omega}^*_{kj} - \bar{\Omega}^*_{ik}a_{kj}) - A_2\left(\bar{s}^*_{ik}a_{kj} + a_{ik}\bar{s}^*_{kj} - \frac{2}{3}\delta_{ij}\bar{s}^*_{mn}a_{nm}\right) \\
\bar{s}^*_{ij} &= \frac{k}{\varepsilon}\bar{s}_{ij}, \quad \bar{\Omega}^*_{ij} = \frac{k}{\varepsilon}\bar{\Omega}_{ij}
\end{aligned} \tag{9}$$

where

$$A_1 = 1.54, \quad A_2 = 0.37, \quad A_3 = 1.45, \quad A_4 = 2.89 \tag{10}$$

In order to get an explicit form of Eq. 9, Girimaji [17, 18] and Wallin & Johansson [16, 19], formulated a_{ij} in terms of the strain-rate tensor ($\bar{s}_{ij}$) and the vorticity tensor ($\bar{\Omega}_{ij}$). In 2D, it reads [2]

$$a_{ij} = \beta_1 \bar{s}^*_{ij} + \beta_2 \left(\bar{s}^*_{ik} \bar{s}^*_{kj} - \frac{1}{3} \bar{s}^*_{mn} \bar{s}^*_{nm} \delta_{ij} \right) + \beta_4 (\bar{s}^*_{ik} \bar{\Omega}^*_{kj} - \bar{\Omega}^*_{ik} \bar{s}^*_{kj}) \tag{11}$$

By inserting Eq. (11) in Eq. (9), Girimaji [17, 18] and Wallin & Johansson [16] derived an explicit form which in 2D reads [16] (a detailed derivation is given in [14])

$$\beta_1 = -\frac{A_1 N}{Q}, \quad \beta_2 = 2\frac{A_1 A_2}{Q}, \quad \beta_4 = -\frac{A_1}{Q}, \quad Q = N^2 - 2II_\Omega - \frac{2}{3}A_2^2 II_S \tag{12}$$

where N is given by the cubic equation

$$N^3 - A_3 N^2 - \left(\left(A_1 A_4 + \frac{2}{3} A_2^2 \right) II_S + 2II_\Omega \right) N + 2A_3 \left(\frac{1}{3} A_2^2 II_S + II_\Omega \right) = 0$$
$$II_S = \bar{s}^*_{mn} \bar{s}^*_{nm}, \quad II_\Omega = \bar{\Omega}^*_{mn} \bar{\Omega}^*_{nm}. \tag{13}$$

Equation (13) can be solved analytically. The analytical solution for the positive root reads [16]

$$N = \begin{cases} \frac{A_3}{3} + \left(P_1 + \sqrt{P_2}\right)^{1/3} + \text{sign}\left(P_1 - \sqrt{P_2}\right) \left|P_1 - \sqrt{P_2}\right|^{1/3}, \; P_2 \geq 0 \\ \frac{A_3}{3} + 2\left(P_1^2 - P_2\right)^{1/6} \cos\left[\frac{1}{3} \arccos\left(\frac{P_1}{\sqrt{P_1^2 - P_2}}\right)\right], \; P_2 < 0 \end{cases} \tag{14}$$

where

$$P_1 = \left(\frac{A_3}{27} + \left(\frac{A_1 A_4}{6} - \frac{2}{9} A_2^2 \right) II_S - \frac{2}{3} II_\Omega \right) A_3$$
$$P_2 = P_1^2 - \left(\frac{A_3}{9} + \left(\frac{A_1 A_4}{3} + \frac{2}{9} A_2^2 \right) II_S + \frac{2}{3} II_\Omega \right)^3$$

The Reynolds stress tensor including only the first term in Eq. (11) reads (see Eq. 8)

$$\overline{v'_i v'_j} = \beta_1 k \bar{s}^*_{ij} + \frac{2}{3} k \delta_{ij} = \beta_1 \frac{k^2}{\varepsilon} \bar{s}_{ij} + \frac{2}{3} k \delta_{ij} = -\nu_{eff} \bar{s}_{ij} + \frac{2}{3} k \delta_{ij} \tag{15}$$

where

$$\nu_{eff} = -0.5 \beta_1 \frac{k^2}{\varepsilon} \tag{16}$$

is the effective viscosity and $\varepsilon = C_\mu k\omega$, see Eq. (6). Equation (15) corresponds to the Boussinesq assumption with $\beta_1 = -2C_\mu$. The discretized momentum equation on matrix form reads

$$AW = b \tag{17}$$

where W is $\bar{v}_1$ or $\bar{v}_2$. The term including the effective viscosity, ν_{eff}, in Eq. (15) is included in A which greatly improves the numerical stability of the CFD code.

3.1 The Neural Network Model

Instead of computing β_1, β_2 and β_4 from Eqs. (12) and (13), I will in the present work make them functions of some input parameter(s) (to be determined) using Neural Network (NN). The process can be depicted as:

1. Choose input parameter(s) involving e.g. the velocity gradient, the shear stress, the dissipation, the wall distance which should all be non-dimensional.
2. The output (target) parameters are β_1, β_2, β_4.
3. Train the NN model in fully-developed channel flow.
4. Use the NN model to compute β_1, β_2, β_4 in the EARSM (k and ω predicted with the $k-\omega$ model) in the **pyCALC-RANS** CFD code.

In fully-developed channel flow the EARSM (Eq. 11) reads:

$$\begin{aligned} a_{11} &= \frac{1}{12}\left(\frac{\partial \bar{v}_1^*}{\partial y}\right)^2 (\beta_2 - 6\beta_4), \quad a_{22} = \frac{1}{12}\left(\frac{\partial \bar{v}_1^*}{\partial y}\right)^2 (\beta_2 + 6\beta_4) \\ a_{33} &= -\frac{2\beta_2}{12}\left(\frac{\partial \bar{v}_1^*}{\partial y}\right)^2, \quad a_{12} = \frac{\beta_1}{2}\frac{\partial \bar{v}_1^*}{\partial y}, \quad \frac{\partial \bar{v}_1^*}{\partial y} = \frac{k}{\varepsilon}\frac{\partial \bar{v}_1}{\partial y} \end{aligned} \tag{18}$$

From the relations above, I get the targets for the NN model

$$\beta_1 = \frac{2a_{12}}{\frac{\partial \bar{v}_1^*}{\partial y}}, \quad \beta_2 = \frac{6(a_{11}+a_{22})}{\left(\frac{\partial \bar{v}_1^*}{\partial y}\right)^2}, \quad \beta_4 = \frac{a_{22}-a_{11}}{\left(\frac{\partial \bar{v}_1^*}{\partial y}\right)^2} \tag{19}$$

$a_{ij}, k/\varepsilon \equiv (\omega C_\mu)^{-1}$ and $\frac{\partial \bar{v}_1}{\partial y}$ are computed from DNS data of channel flow. Note that a_{33} is defined by $a_{ii} = 0$.

The output parameters of the NN model are β_1, β_2 and β_4. What input parameters should be used? The NN model should be applicable at difference Reynolds numbers so it should be a good idea to choose input parameters which also are Reynolds number independent. The NN model will be used in the CFD code and validated against DNS data in channel flow at $Re_\tau = 2000$ [20], $Re_\tau = 5200$ [21] $Re_\tau = 10{,}000$ [22] and flat-plate boundary-layer flow $Re_\tau = 5\,500$ [23].

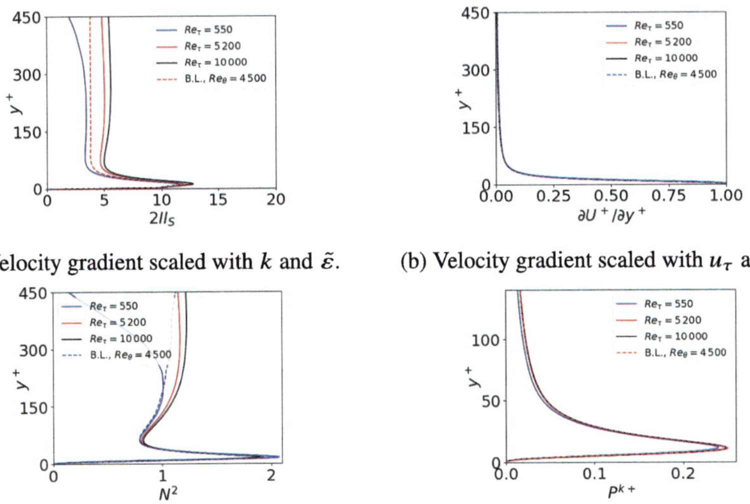

(a) Velocity gradient scaled with k and $\tilde{\varepsilon}$.

(b) Velocity gradient scaled with u_τ and ν.

(c) Ratio of P^k to $\tilde{\varepsilon}$.

(d) Production scaled with u_τ and ν.

Fig. 1 DNS data $\tilde{\varepsilon} = \varepsilon - \nu\partial^2 k/\partial y^2$

Figure 1 presents four possible dimensionless input parameters. $II_S = \frac{1}{2}\left(\frac{k}{\tilde{\varepsilon}}\frac{\partial \bar{v}_1}{\partial y}\right)^2$, $\frac{\partial U^+}{\partial y^+}$, $N = P^k/\tilde{\varepsilon}$ and P^{k+}. Note that $\tilde{\varepsilon} = \varepsilon - \nu\partial^2 k/\partial y^2$ is used because its near-wall behaviour is similar to that of the dissipation tern in the $k - \omega$ model. The velocity gradient seems to be best (i.e. least Re number dependent), but it turns out that it is very difficult to get the NN model to converge with this input variable. The reason is probably its large gradient near the wall, see Fig. 1b. I choose the second best, i.e. the production term together with y^+ as input parameters. I scale the two input parameters using `MinMaxScaler()` so that they are in the range [0, 1]

I use the NN in Python's `pytorch`. Figure 2 shows the NN model schematically. The optimizer is set as

```
optimizer = torch.optim.SGD(neural_net.parameters(), lr=l_rate)
```

with learning rate `l_rate = 0.07`; the number of epochs is 5000. When predicting the β coefficientsusing the 20% of DNS data the maximum error, e_i, defined as ($i = 1, 2$, or 4)

$$e_i = \max\left(\frac{|\beta_i - \beta_{i,DNS}|}{\beta_{i,DNS}}\right)$$

is less than 2.5%. More detail on the use of `pytorch` can be found in [24].

Since I use y^+ as input variable, I must train the NN model at the largest Reynolds number (which has the largest y^+ value), i.e. channel flow at $Re_\tau = 10{,}000$. An option could be to train the model at all Reynolds numbers. I exclude data in the viscous sublayer ($y^+ \leq 5$) because the gradient of β_2 and β_4 (see Eq. 19) are very large near the wall. I also exclude data near the center ($y^+ > 9\,800$) where $\frac{\partial \bar{v}_1^*}{\partial y}$ is very

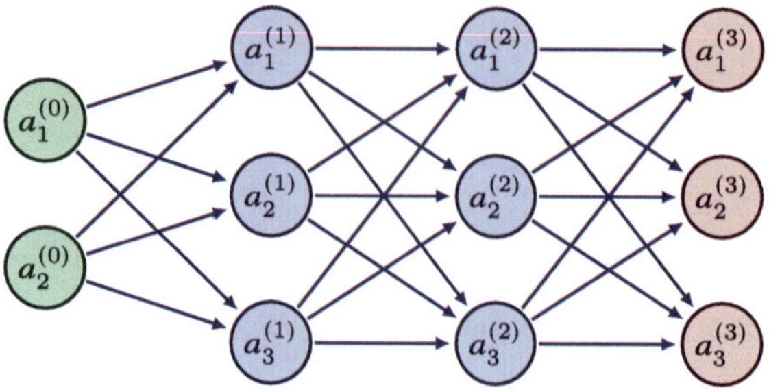

Fig. 2 The Neural Network with two inputs variables, $a_1^{(0)} = y^+$ and $a_2^{(0)} = P^+$ and three output variables, $a_1^{(3)} = \beta_1$, $a_2^{(3)} = \beta_2$ and $a_3^{(3)} = \beta_4$. There are three neurons and two hidden layers in this figure; in the simulations I use 50 neurons

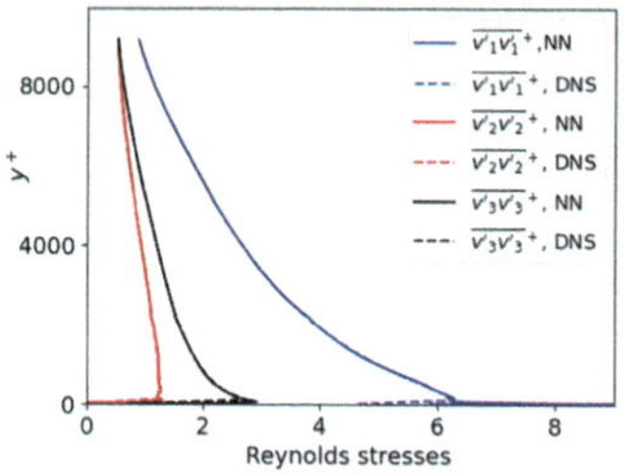

(a) Reynolds normal stresses.

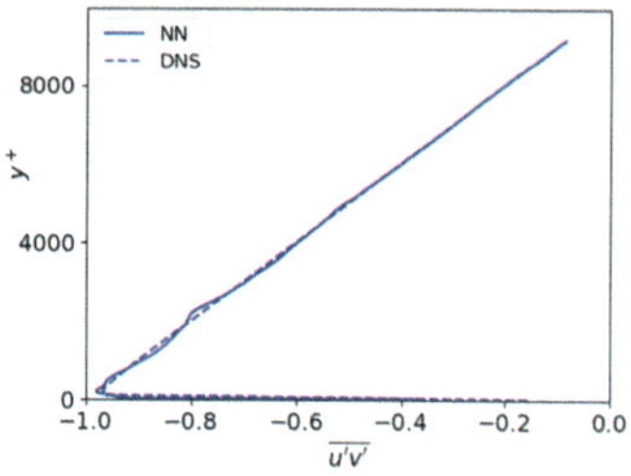

(b) Reynolds shear stress.

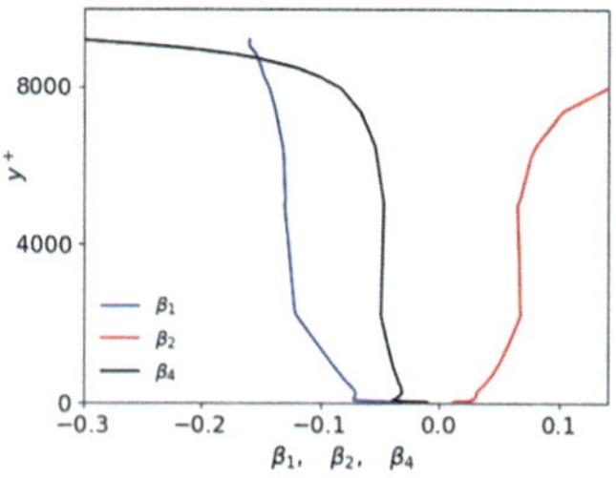

(c) Predicted EARSM coefficients.

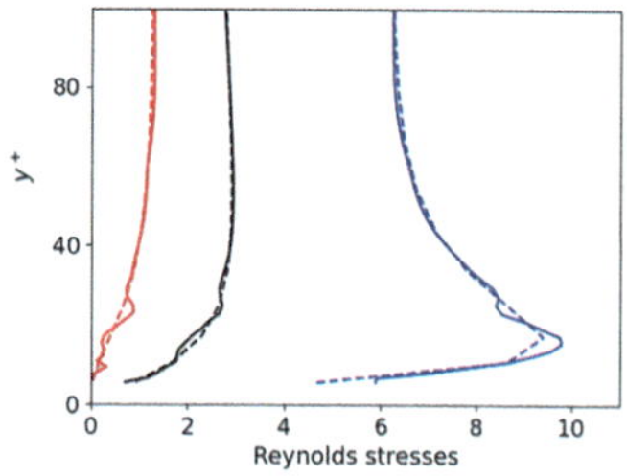

(d) Reynolds normal stresses near the wall.

Fig. 3 Predicted with NN model (no CFD). $Re_\tau = 10{,}000$. Markers: DNS data [22]

small. The turbulence is negligible in both these regions. I train on 80% of the data (approximately 800 randomly chosen data points) and test on (predict) the remaining 20%.

Figure 3 presents the predicted stresses. The agreement is—as it should—almost perfect. However, Fig. 3d shows that there are some small discrepancies near the wall. Figure 3c presents the EARSM coefficients predicted by the NN model.

4 The NN Model Incorporated in the CFD Solver

The prediction of Reynolds stresses presented in the previous section is *à priori* study. In the literature, these studies are very common. But they are of little use unless the NN model is coupled to a CFD code because in channel flow the velocity profile is entirely determined by the turbulent shear stress (see discussion in the next section).

Hence, in the next step I save the NN model to disk and then a load it into the CFD code. I include the NN model in the CFD code as follows:

1. Load the NN model
2. Solve $\bar{v}_1$, $\bar{v}_2$ and P' equations. The Reynolds stresses $\overline{v_1'^2}$, $\overline{v_2'^2}$, $\overline{v_1'v_2'}$ in the $\bar{v}_1$ and $\bar{v}_2$ equations (see Eq. 2) are taken from the previous iteration.
3. Compute $\beta_1, \beta_2, \beta_4$ using the NN model. Limits are set on both input and output parameters corresponding to min and max values during the training process.
4. Compute the anisotropic Reynolds stresses (a_{11}, a_{22}, a_{12}) using the β coefficients, see Eq. (18).
5. Compute the Reynolds stresses $\overline{v_1'^2} = ka_{11} + \frac{2}{3}k$, $\overline{v_2'^2} = ka_{22} + \frac{2}{3}k$, $\overline{v_1'v_2'} = ka_{12}$.
6. Solve the k and ω equations. The Reynolds stresses are used in the production term, see Eq. (5). In fully-developed channel flow $\overline{v_1'^2}$ and $\overline{v_2'^2}$ have no effect since $\partial\bar{v}_1/\partial x_1 = \partial\bar{v}_2/\partial x_2 = 0$, but in flat-plate boundary layer flow they have a small effect.
7. End of iteration. Repeat from Item 2 until convergence (1000 s of iterations).

4.1 Channel Flow with the NN Model Trained on DNS Data

Channel flow simulations are made. The grid has 110 cells in the y direction and is stretched by 10% from the walls. The first cell center is at $y^+ \simeq 0.5$. Figure 4 presents the predicted mean flow and the Reynolds stresses. The agreement of the predicted mean flow with the DNS data is poor. The shear stress agrees well DNS data but that does not signify anything. When using a finite volume CFD method, the shear stress will always agree with DNS data. Integrating the momentum equation the total shear stress, $\tau_{12,tot}$ (viscous plus turbulent), reads [14]

$$\tau_{12,tot} = \tau_w + \frac{1}{\rho}\frac{\partial \bar{p}}{\partial x_1}x_2 = \tau_w\left(1 - \frac{x_2}{\delta}\right) \tag{20}$$

where δ is the half-channel width and the total shear stress is defined as

$$\tau_{12,tot} = \nu\frac{\partial \bar{v}_1}{\partial x_2} - \overline{v_1'v_2'}. \tag{21}$$

At the last step in Eq. (20), I used the fact that the pressure gradient balances the wall shear stress. The velocity gradient will adapt so that Eq. (20) is satisfied.

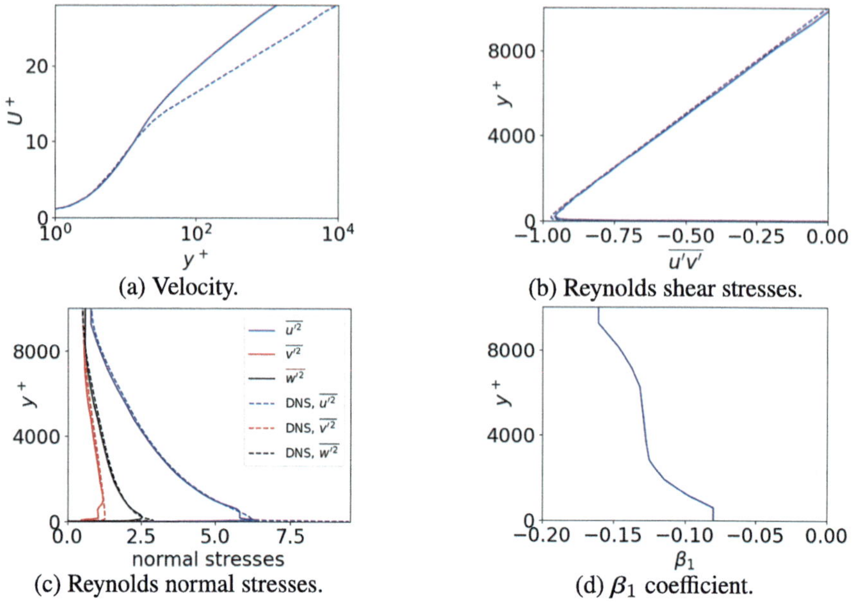

(a) Velocity.

(b) Reynolds shear stresses.

(c) Reynolds normal stresses.

(d) β_1 coefficient.

Fig. 4 The NN model incorporated in the CFD code. $Re_\tau = 10{,}000$. Dashed lines: DNS data [22]

Why is the velocity so poorly predicted? To find out, I make a CFD simulation using the $k - \omega$ model (no EARSM, no NN model). Figure 5a, b show the predicted velocity and shear stress and the agreement with DNS data is (or seems to be) excellent. The relation between the velocity gradient and the shear stress in the $k - \omega$ model is given by the Boussinesq assumption, see Eq. (15)

$$\overline{v_1' v_2'} = -2\nu_t \bar{s}_{12} = -\frac{2k}{\omega}\bar{s}_{12} = -\frac{2C_\mu k^2}{\varepsilon}\bar{s}_{12} \tag{22}$$

using $\varepsilon = C_\mu k\omega$, see Eq. (6). Comparing Eqs. (22) and (15) gives that

$$\beta_1 = -2C_\mu \tag{23}$$

The poor agreement in Fig. 4a can now be understood. The $k - \omega$ models predicts an (seemingly) excellent velocity profile (Fig. 5a) where the relation in Eq. (22) is used with $C_\mu = 0.09$. If the CFD and NN should give the same results, then the relation in Eq. (23) must be satisfied; Fig. 4d shows that this is not the case. The reason is that although the total shear stresses in the $k - \omega$ predictions and the DNS data both follow the linear law in Eq. (20), the velocity gradients are different. Figure 5c shows that the discrepancy is more than 20%.

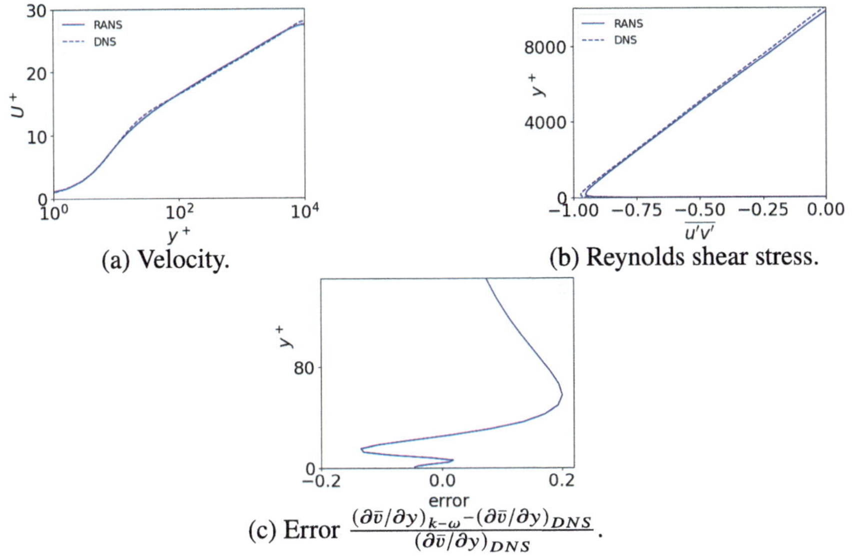

(a) Velocity.

(b) Reynolds shear stress.

(c) Error $\frac{(\partial\bar{v}/\partial y)_{k-\omega}-(\partial\bar{v}/\partial y)_{DNS}}{(\partial\bar{v}/\partial y)_{DNS}}$.

Fig. 5 CFD using the $k-\omega$ model. Channel flow, $Re_\tau = 10{,}000$. Markers: DNS data [22]

4.2 Channel Flow with the NN Model Trained on $k-\omega$ and DNS Data

Instead of training the NN model on DNS data as in Sect. 4.1, I will train on data taken both from DNS and the $k-\omega$ simulation shown in Fig. 5. The fact that this may be necessary was noted in [25]. I will use the following data:

- Input: P^k and y^+ from $k-\omega$ prediction
- Target: β_1, β_2 and β_4 computed from $\underbrace{\overline{v_1'^2}, \overline{v_2'^2}}_{DNS}$ and $\underbrace{\overline{v_1'v_2'}, k, \varepsilon}_{k-\omega}$

This will ensure that the relation between the shear stress and the velocity gradient as well as the turbulent kinetic energy are the same in the training process as in the CFD simulation. Note that $k_{DNS} = 0.5(\overline{v_1'^2}_{DNS} + \overline{v_2'^2}_{DNS} + \overline{v_3'^2}_{DNS})$ is not equal to the turbulent kinetic energy, $k_{k-\omega}$, predicted by the $k-\omega$ model. The spanwise normal stress, $\overline{v_3'^2}$, predicted by the NN model will adapt in order to satisfy $k_{k-\omega} = 0.5(\overline{v_1'^2}_{DNS} + \overline{v_2'^2}_{DNS} + \overline{v_3'^2}_{DNS})$. Hence, $\overline{v_3'^2}$ will be incorrectly predicted by the NN model (it even goes negative near the wall). Thus, the proposed EARSM-NN model is applicable only in two-dimension flows. In order to make the model applicable in three dimension, a new $k-\omega$ (or $k-\varepsilon$) model must be developed which satisfies $k_{DNS} = k_{k-\omega}$. This is out of the scope of the present work.

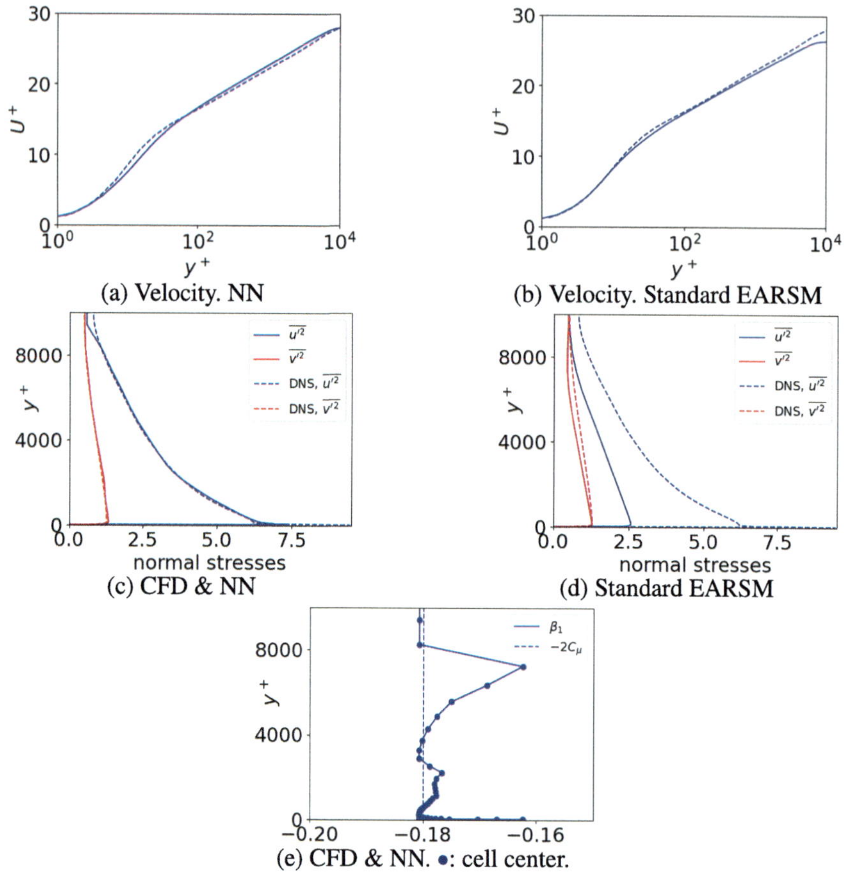

Fig. 6 CFD predictions with EARSM-NN model and standard EARSM. The NN model is trained using DNS and $k - \omega$ data. Channel flow, $Re_\tau = 10{,}000$. Dashed lines: DNS

Figure 6 presents CFD prediction using the EARSM-NN and the standard EARSM (see Eqs. (10), (12) and (14)). Both models predict the velocity well although the standard EARSM gives a slight overprediction near the center and the EARSM-NN profile exhibits somewhat too low values in the buffer layer. However, the EARSM-NN predicts much better normal stresses than the standard EARSM. Figure 6e presents the β_1 coefficient and its values are close to $-2C_\mu$ as it should. Setting $\beta_1 = -2C_\mu$ in EARSM-NN gives very similar results (not shown) as in Fig. 8a, c.

At $Re_\tau = 5200$ (Fig. 7) and $Re_\tau = 2000$ (Fig. 8) the EARSM-NN model predicts much better velocity profiles and normal Reynolds stresses than the standard EARSM. At the lower Reynolds number, The EARSM-NN model (Fig. 8c) overpredicts the streamwise normal stress by some 15%, whereas the standard EARSM (Fig. 8d) model predicts up to 50% too small values.

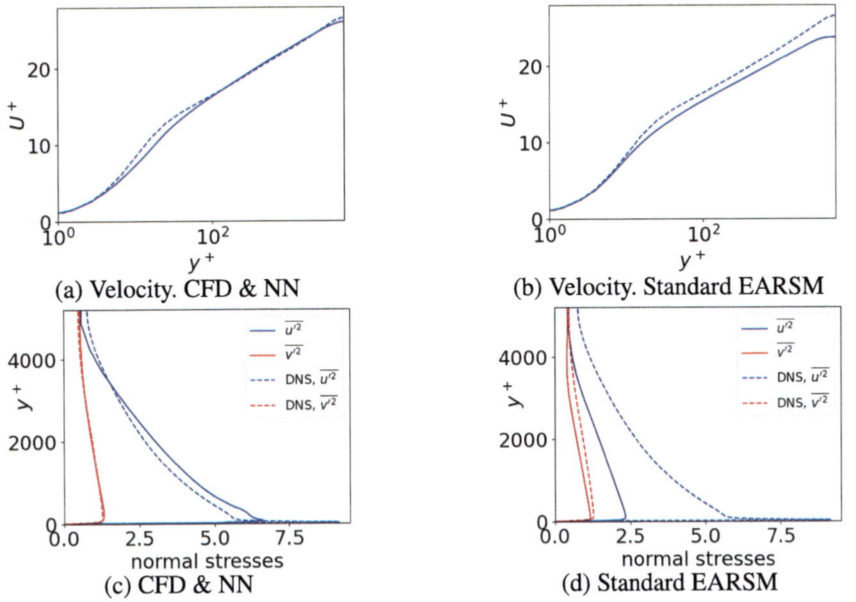

Fig. 7 CFD predictions with EARSM-NN model and standard EARSM. The NN model is trained using DNS and $k - \omega$ data. $Re_\tau = 5200$. Dashed lines: DNS

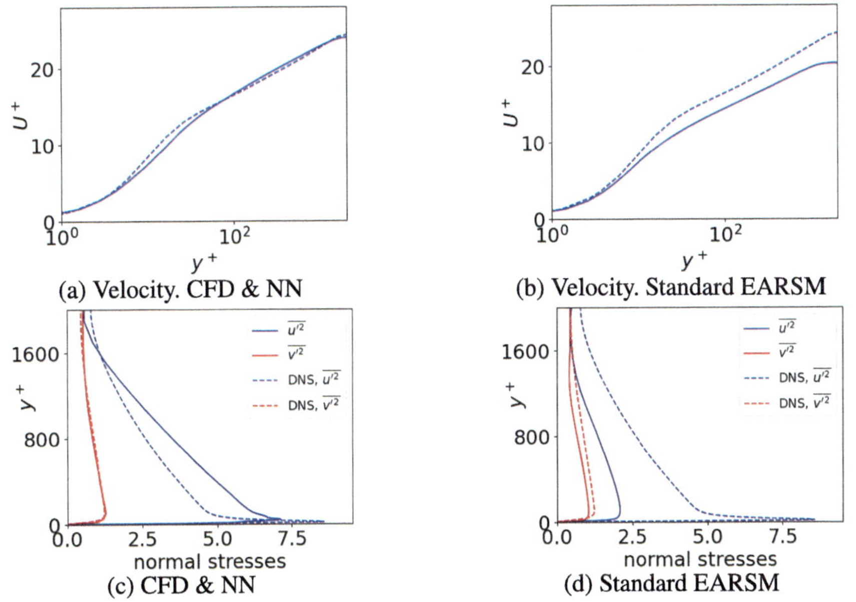

Fig. 8 CFD predictions with EARSM-NN model and standard EARSM. The NN model is trained using DNS and $k - \omega$ data. Channel flow, $Re_\tau = 2000$. Dashed lines: DNS

4.3 Flat-Plate Boundary-Layer Flow

In this section I present simulations of a flat-plate boundary layer. The grid has 300×90 (x, y) cells and is stretched by 10% from the wall but limiting the cell size to $\Delta y_{max} = 0.5$. The first cell center is at $y^+ \simeq 0.5$. The streamwise grid spacing is constant, $\Delta x = 0.5$. The inlet boundary-layer thickness is $\delta_{in} \simeq 0.8$. A refined mesh (600×180) gives identical results as the coarse mesh.

A pre-cursor $k - \omega$ simulation of a flat-plate boundary layer is carried out and $\bar{v}_1$, $\bar{v}_1$, k and ω are stored at $Re_\theta = 2\,500$ where θ denotes the boundary-layer momentum thickness. These stored data are used as inlet boundary condition in the subsequent flat-plate boundary-layer simulations using the two EARSM models.

Figure 9 presents the predicted skin friction along with velocity and normal stresses at $Re_\theta = 2\,500$. Again, the EARSM-NN model performs much better than the standard EARSM although the normal stresses are less well predicted than for the channel flow. But the EARSM-NN model predicts the skin friction within 5% of experimental data whereas the standard EARSM gives some 14% too large values.

The channel flow simulations at $Re_\tau = 10{,}000$ were also carried out setting $\beta_1 = -2C_\mu$ and it was noted that the results were similar to those in Figs. 8a, c. This was also found for the other two Reynolds numbers, $Re_\tau = 5\,200$ and $Re_\tau = 2\,000$ (not shown). However, in the boundary layer flow the results are somewhat worse setting $\beta_1 = -2C_\mu$; the skin friction is then over-predicted by 6%, see Fig. 10a. The skin friction predicted with the Wilcox $k - \omega$ model is also 6% too large (see Fig. 10b), but the trend is much worse. The β_1 coefficient predicted with EARSM-NN is shown in Fig. 10c and it can be seen that near the wall it is larger than $-2C_\mu$.

5 Conclusions

An Explicit Algebraic Reynolds Stress Model (together with Wilcox $k - \omega$ model) has been improved using Neural Network (NN). The NN model is trained in channel flow at $Re_\tau = 10{,}000$. It is found that target data cannot be taken only from DNS because the stress-strain relation and the turbulent kinetic energy are not the same in DNS and $k - \omega$ predictions. Hence the target data are taken both from DNS ($\overline{v_1'^2}$ and $\overline{v_2'^2}$) and a $k - \omega$ simulation ($\frac{\partial \bar{v}_1}{\partial y}$, $\overline{v_1' v_2'}$, k, $\varepsilon = C_\mu k \omega$). In this way the strain-stress relation and the turbulent kinetic energy are the same in the training process as in the CFD-NN predictions. Since k in the training process is taken from the $k - \omega$ results it means that $k \neq 0.5(\overline{v_1'^2}_{DNS} + \overline{v_2'^2}_{DNS} + \overline{v_3'^2}_{DNS})$. In the NN model, the spanwise Reynolds stress adapts to satisfy $a_{ii} = 0$ which means that $\overline{v_3'^2}$ is not correctly predicted (it even goes negative in the near-wall region). Hence, the EARSM-NN model is applicable only to two-dimensional flow where $\overline{v_3'^2}$ is not used. One way to make the model applicable in three-dimensional flow is to develop a $k - \omega$ (or $k - \varepsilon$ model) which accurately predicts the turbulent kinetic energy.

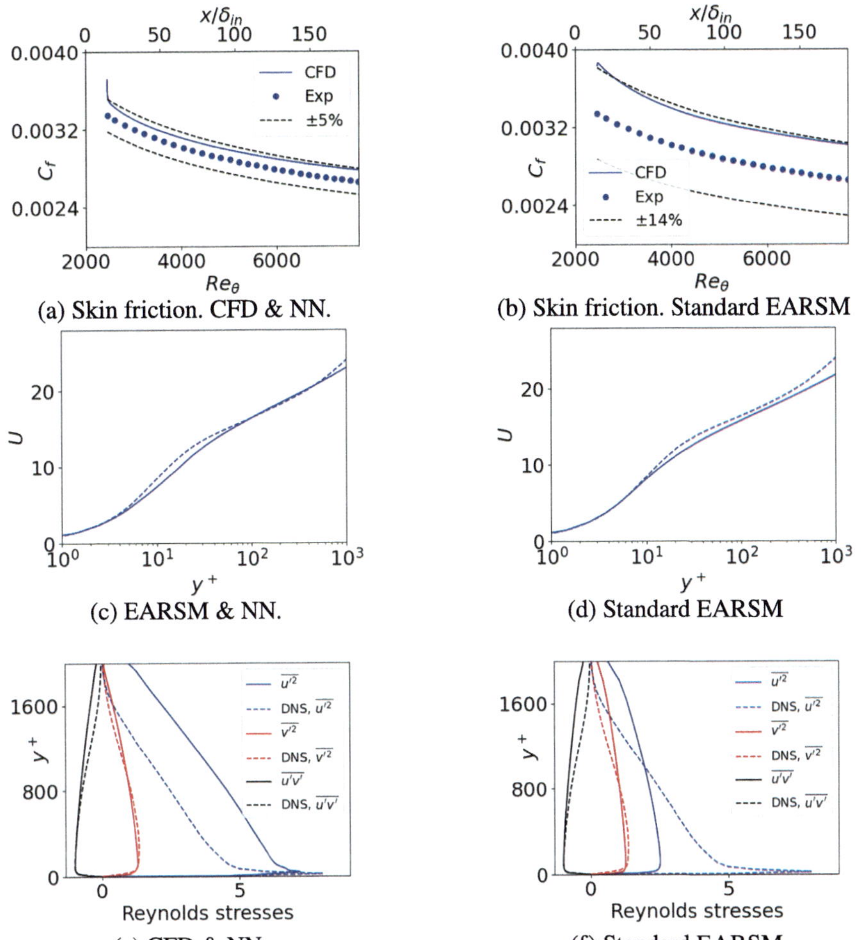

(a) Skin friction. CFD & NN.

(b) Skin friction. Standard EARSM

(c) EARSM & NN.

(d) Standard EARSM

(e) CFD & NN

(f) Standard EARSM

Fig. 9 CFD predictions with EARSM-NN and standard EARSM. The NN model is trained using DNS and $k - \omega$ data. Flat-plate boundary layer, $Re_\theta = 5\,500$. Dashed lines: DNS

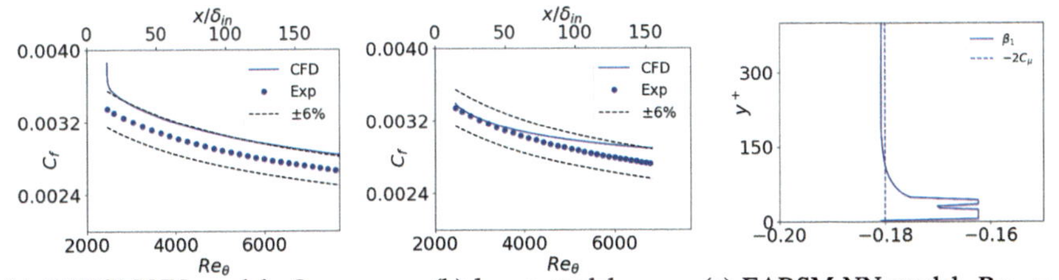

(a) EARSM-NN model. $\beta_1 = -2C_\mu$.

(b) $k - \omega$ model.

(c) EARSM-NN model. $Re_\theta = 5\,500$.

Fig. 10 CFD predictions. Flat-plate boundary layer

Acknowledgement This study was partly financed by the *Strategic research project on Chalmers on hydro- and aerodynamics* and *Chalmers Transport Area of Advance*, Grant No. C 2023-0125-19.

The computations were enabled by resources provided by the National Academic Infrastructure for Supercomputing in Sweden (NAISS), partially funded by the Swedish Research Council through grant agreement no. 2022-06725.

References

1. Craft, T.J., Launder, B.E., Suga, K.: Prediction of turbulent transitional phenomena with a nonlinear eddy-viscosity model. Int. J. Heat Fluid Flow **18**, 15–28 (1997)
2. Pope, S.B.: A more general effective-viscosity hypothesis. J. Fluid Mech. **472**, 331–340 (1975)
3. Ling, J., Kurzawski, A., Templeton, J.: Reynolds averaged turbulence modelling using deep neural networks with embedded invariance. J. Fluid Mech. **807**, 155–166 (2016). https://doi.org/10.1017/jfm.2016.615
4. Yin, Y., Shen, Z., Zhang, Y., Chen, H., Fu, S.: An iterative data-driven turbulence modeling framework based on reynolds stress representation. Theor. Appl. Mech. Lett. **12**(5), 100381 (2022). https://doi.org/10.1016/j.taml.2022.100381
5. Wu, J.-L., Xiao, H., Paterson, E.: Physics-informed machine learning approach for augmenting turbulence models: a comprehensive framework. Phys. Rev. Fluids **3**, 074602 (2018). https://doi.org/10.1103/PhysRevFluids.3.074602
6. Duraisamy, K., Iaccarino, G., Xiao, H.: Turbulence modeling in the age of data. Annu. Rev. Fluid Mech. **51**(1), 357–377 (2019). https://doi.org/10.1146/annurev-fluid-010518-040547
7. Weatheritt, J., Sandberg, R.: A novel evolutionary algorithm applied to algebraic modifications of the rans stress-strain relationship. J. Comput. Phys. **325**, 22–37 (2016). https://doi.org/10.1016/j.jcp.2016.08.015
8. Akolekar, H.D., Weatheritt, J., Hutchins, N., Sandberg, R.D., Laskowski, G., Michelassi, V.: Development and use of machine-learnt algebraic reynolds stress models for enhanced prediction of wake mixing in low-pressure turbines. J. Turbomach. **141**(4), 041010 (2019). https://doi.org/10.1115/1.4041753
9. Davidson, L.: pyCALC-RANS: a 2D python code for RANS, Division of Fluid Dynamics, Dept. of Mechanics and Maritime Sciences, Chalmers University of Technology, Gothenburg (2021)
10. Rhie, C.M., Chow, W.L.: Numerical study of the turbulent flow past an airfoil with trailing edge separation. AIAA J. **21**, 1525–1532 (1983)
11. Wilcox, D.C.: Reassessment of the scale-determining equation. AIAA J. **26**(11), 1299–1310 (1988)
12. Rodi, W.: A new algebraic relation for calculating the Reynolds stresses. ZAMM **56**, T219–T221 (1976)
13. Launder, B.E., Reece, G.J., Rodi, W.: Progress in the development of a Reynolds-stress turbulence closure. J. Fluid Mech. **68**(3), 537–566 (1975)
14. Davidson, L.: Fluid mechanics, turbulent flow and turbulence modeling, eBook, Division of Fluid Dynamics. Chalmers University of Technology, Gothenburg, Dept. of Mechanics and Maritime Sciences (2021)
15. Taulbee, D.B.: An improved algebraic Reynolds stress model and corresponding nonlinear stress model. Phys. Fluids A **4**, 2555–2561 (1992)
16. Wallin, S., Johansson, A.V.: A new explicit algebraic Reynolds stress model for incompressible and compressible turbulent flows. J. Fluid Mech. **403**, 89–132 (2000)
17. Girimaji, S.S.: Fully-explicit and self-consistent algebraic Reynolds stress model. Theoret. Comput. Fluid Dyn. **8**, 387–402 (1996)

18. Girimaji, S.: Fully-explicit and self-consistent algebraic Reynolds stress model, ICASE Rep. 95-82, Institute for Computer Applications in Science and Engineering NASA Langley Research Center, Hampton, VA, USA (1995)
19. Wallin, S., Johansson, A.: A new explicit algebraic Reynolds stress turbulence model including an improved near-wall treatment, in: C.-J. Chen, C. Shih, J. Lienau, R. J. Kung (Eds.), Proc. Flow Modeling and Turbulence Measurements VI, Tallahassee F. L., Balkema, pp. 399–406 (1996)
20. Hoyas, S., Jimenez, J.: Reynolds number effects on the reynolds-stress budgets in turbulent channels, Physics of Fluids A 20 (101511) (2008)
21. Lee, M., Moser, R.D.: Direct numerical simulation of turbulent channel flow up to $Re_\tau \approx 5200$. J. Fluid Mech. **774**, 395–415 (2015). https://doi.org/10.1017/jfm.2015.268
22. Hoyas, S., Oberlack, M., Alcántara-Ávila, F., Kraheberger, S.V., Laux, J.: Wall turbulence at high friction Reynolds numbers. Phys. Rev. Fluids **7**, 014602 (2022). https://doi.org/10.1103/PhysRevFluids.7.014602
23. Sillero, J., Jimenez, J., Moser, R.: One-point statistics for turbulent wall-bounded flows at Reynolds numbers up to $\delta^+ \simeq 2000$, Phys. Fluids **25** (105102) (2014)
24. Davidson, L.: Using machine learning for improving a non-linear $k - \varepsilon$ model: a first attempt, Tech. rep., Division of Fluid Dynamics, Dept. of Mechanics and Maritime Sciences, Chalmers University of Technology, Gothenburg (2023)
25. Duraisamy, K.: Perspectives on machine learning-augmented reynolds-averaged and large eddy simulation models of turbulence. Phys. Rev. Fluids (2021). https://doi.org/10.1103/PhysRevFluids.6.050504

Methods in Turbulence

Advanced Post-processing Techniques for Scale-Resolved Databases

Spencer L. Stahl and Datta V. Gaitonde

Abstract There remains a persistent need for advanced physics-informed and data-driven methods and strategies to address challenges in extracting insights from massive, scale-resolved databases. Physics-informed methods sift primitive variable fluctuations to yield physically more revealing components. Data driven methods are agnostic to the governing equations and generate modes with spatial structure, spectral content and amplitude. Two distinct new strategies are highlighted for multi-scale flow phenomena: (i) application of physics-informed and data-driven methods in sequence and (ii) recovering space-time localization and causal perspectives for transient or intermittent events that may otherwise be lost in data-driven spectral methods. For illustration, we choose a pair of complex propulsion-related problems comprised of impinging jets and multi-stream nozzles, respectively, at ground-test Reynolds numbers. Together, these contain instabilities, feedback, shock interactions and separation, thus forming challenging test beds for advanced methodologies. We show that sequential application of Doak's physics-informed decomposition into acoustic, hydrodynamic (vortical), and thermal (entropic) components followed by spectral proper orthogonal decomposition, elicits dominant as well as nuanced mechanisms underlying different tonal observations. The second strategy applies conditional space-time proper orthogonal decomposition to highlight transient flow events, distilling causal mechanisms that depend on synchronization of multi-channel information pathways and phase speeds. The results stress the immense scope and utility of advanced post-processing methods.

S. L. Stahl (✉) · D. V. Gaitonde
Department of Mechanical and Aerospace Engineering, The Ohio State University, Columbus, OH 43210, USA
e-mail: stahl.174@osu.edu

D. V. Gaitonde
e-mail: gaitonde.3@osu.edu

J. C. Tyacke and N. R. Vadlamani (eds.), *Proceedings of the Cambridge Unsteady Flow Symposium 2024*, https://doi.org/10.1007/978-3-031-69035-8_3

1 Introduction

Spatiotemporally scale-resolved simulations using $O(1)$ billion or more mesh points, are becoming increasingly commonplace [9]. The resulting databases, many on very complex configurations, have come to be viewed as truth models, suited for accurate determination of not just integrated quantities of practical interest, such as lift or drag, but also containing critical information on the underlying non-linear mechanisms. Large experimental databases are also being compiled with ever more capable high-bandwidth, wide-windowed, diagnostic techniques. Extracting the insights encapsulated in such data for fundamental research, or to inform design modifications or scalable control strategies, has proven to be a daunting challenge.

For reference, the different components of modern CFD analyses are shown in Fig. 1. Fluctuation data obtained from DNS, LES or WMLES (acronyms are provided in the figure itself), or through (mostly linearized) stability or perturbation techniques, constitute massive 4D truth models. To harness the potential of these databases, recent research has developed, rediscovered or modified post-processing techniques, many with sub-variants, that identify distinctive mechanisms and features that are not immediately apparent in the raw data. We may distinguish two broad sets of procedures. Data-driven methods (DDMs) are agnostic to the governing equations that obtained the data and generally trace their roots in system identification; they operate on snapshot sequences (black arrows in Fig. 1) to obtain *modes* [20]). Several examples are listed in Fig. 1; those pertinent to this paper will be elaborated in context below. Physics-informed methods (PIMs) (blue arrows) use the properties of the Navier-Stokes equations to essentially sift or filter the data into *components* that better highlight features of interest than the usual conserved, primitive or trivially derived variables such as vorticity or Mach number. For the purposes of this paper, key among PIMs is Doak's Momentum Potential Theory [3], MPT, which precisely i.e., without linearization or approximation, extracts Kovásznay-type (acoustic, thermal

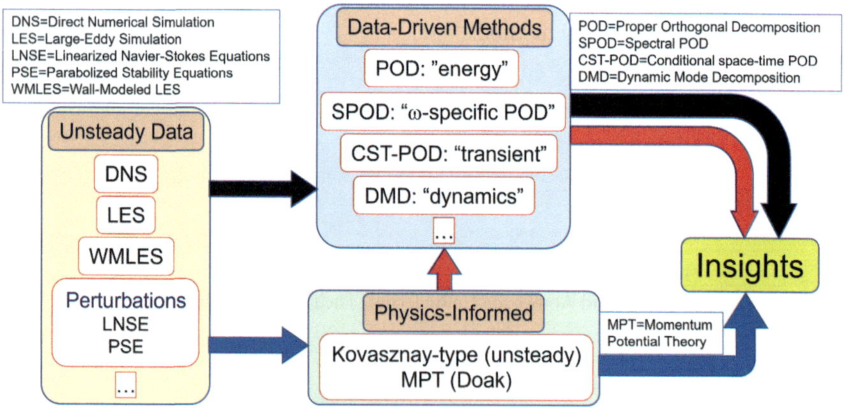

Fig. 1 Relationship of fluctuation data generation, data-driven and physics-informed techniques

and entropic) components for enhanced understanding. The modes or components from DDMs or PIMs methods, respectively, are then typically subjected to traditional approaches such as space-time Fourier decompositions that convert the data into wavenumber-frequency space to further refine the analysis.

Although DDMs yield substantial distillation of information i.e., perform model order reduction, they are most helpful for systems that are low-ranked, where only one or a few modes are sufficient to describe the critical dynamics. Most problems of interest at high Reynolds numbers do not exhibit the low-ranked property and further refinement of post-processing procedures becomes necessary. Furthermore, application of PIMs or DDMs in isolation can still obscure some of the more complicated phenomena whose final manifestation is a result of competing mechanisms. The first objective of this paper is to show the advantages, even necessity, of sequential application of PIMs and DDMs (red arrows of Fig. 1).

The results of DDMs are generally displayed in the frequency-domain, in which transient, intermittent or rare effects of critical importance to the flow dynamics can remain concealed. Our second objective is to demonstrate suitable post-processing steps to analyze such circumstances. Specifically, we employ conditional space-time proper orthogonal decomposition (CST-POD), with a strategy to identify and sample key events associated with extreme fluctuations, then decompose the ensemble of events to isolate transient dynamics. Unlike spectral POD (SPOD), the resulting CST-POD modes consist of multiple frequencies and better preserve cause and effect relations around the event.

In selecting problems to highlight these procedures, we seek those that have complex features that go well beyond isolated free or bounded shear layers. Specifically, we adopt problems that include instabilities (global or local, inviscid or viscous), intermittent behavior, unsteady separation, shock oscillations and flow-acoustic dependencies including tonal feedback. The non-linear mutual interactions of these features further obfuscate the underlying mechanisms at play. Although we restrict attention to statistically stationary problems here, we note that non-stationary situations such as scramjet unstart or transition to turbulence can also be addressed with similar strategies [19].

We select two complex multi-scale propulsion-related problems that are rich in the above described features with implication to thrust and noise penalties that dominate design. Both have readily available published comprehensive well-validated LES databases. The first (Fig. 2) considers a jet impinging on a ground plane of interest in rockets and short takeoff and vertical landing (STOVL) aircraft. The dominant mechanism is established by a feedback loop in which unstable Kelvin-Helmholtz (KH) vortices in the jet shear layer impinge on the ground, generating acoustic waves which travel back towards the nozzle exit and influence the KH process until an equilibrium is established, that includes tonal content. In the present work, we show by combining PIMs and DDMs that different tones in the flow are actually not mere harmonics, but are rather distinct mechanisms.

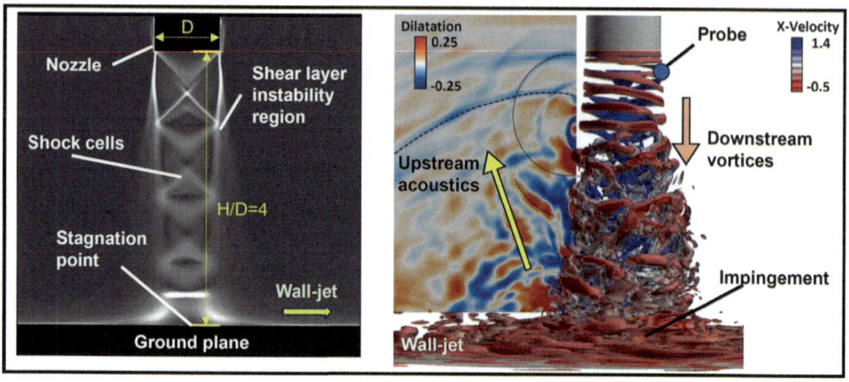

Fig. 2 Round impinging jet problem with various flow features marked. Left: Schlieren image, Right: Visualization using Q-criteria isolevels of 1 colored by axial velocity, and a radial plane showing dilatation

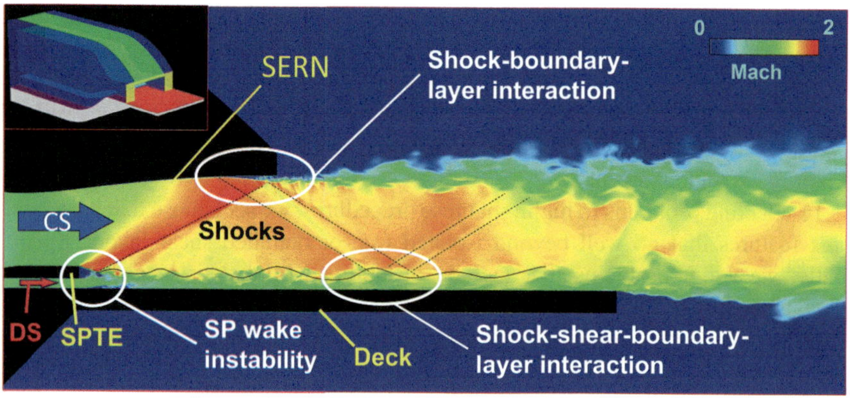

Fig. 3 Multistream problem with Single Expansion Ramp Nozzle (SERN). Left: Simulated configuration, Right: centerplane Mach contours with key features marked. SPTE=Splitter Plate Trailing Edge, CS=Core Stream (Mach 1.6), DS=Deck Stream (Mach 1)

The second problem (Fig. 3 inset) considers a multi-stream propulsion system of the type envisaged for above-wing nozzles. For ease of airframe integration, the lower surface (wing or "aft-deck") is flat, while the upper surface is a Single Expansion Ramp Nozzle (SERN). In such designs, a separate "deck stream" is inserted between the hot core propulsive plume and the aft-deck to avoid damage. For the current purpose, we consider a spanwise homogeneous case (no sidewalls) to focus on the symmetry plane dynamics. The flow on a typical plane, shown in Fig. 3, indicates that the resulting shear layer between the core stream (CS) and deck stream (DS) experiences shedding as well as KH instabilities, which in turn affect the shock system and subsequent separated shock/boundary layer interactions (SBLI). In this problem, the acoustic tones do not resonate through feedback, but nonetheless contribute to shock interactions downstream on the SERN as well as the deck. The focus of the analysis in Sect. 4 is to compare frequency and temporal perspectives through independent SPOD and CST-POD analyses to highlight cause and effect relationships

2 Summary of Methodology

2.1 *Fluid-Thermal Decomposition*

The physics based decomposition comprises two steps: 1) decomposition of momentum via Momentum Potential Theory (MPT) [3] and 2) decomposition of enthalpy. First, the "momentum density" ($\rho\mathbf{u}$) is decomposed into acoustic, thermal, and hydrodynamic components using Helmholtz decomposition exploiting the fact that the continuity equation ($\frac{\partial \rho}{\partial t} + \nabla \cdot (\rho\mathbf{u}) = 0$) is linear in $\rho\mathbf{u}$.

$$\rho\mathbf{u} = \overline{\mathbf{B}} + \mathbf{B}^{'} - \nabla\psi^{'}\,, \qquad \nabla \cdot \overline{\mathbf{B}} = 0\,, \quad \nabla \cdot \mathbf{B}^{'} = 0. \tag{1}$$

Here, $\mathbf{B}$ and ψ are the (solenoidal) hydrodynamic (or vortical) component and (curl-free) irrotational potential, respectively. Primes denote fluctuating quantities; for a time-stationary flow, the mean scalar potential gradient $\nabla\overline{\psi}$ is both irrotational and solenoidal, and assumed to be zero without loss of generality. Substitution into the compressible continuity equation directly yields a Poisson equation for ψ' i.e., $\nabla^2\psi^{'} = \frac{\partial \rho^{'}}{\partial t}$. ψ' may be decomposed into acoustic (irrotational-isentropic) $\psi^{'}_A$, and thermal (irrotational-isobaric) $\psi^{'}_T$ components, which, when combined with thermodynamic identities yields [3]:

$$\psi^{'} = \psi^{'}_A + \psi^{'}_T, \qquad \nabla^2\psi^{'}_A = \frac{1}{c^2}\frac{\partial p^{'}}{\partial t}, \qquad \nabla^2\psi^{'}_T = \frac{\partial \rho}{\partial S}\frac{\partial S^{'}}{\partial t}. \tag{2}$$

where c is the instantaneous local speed of sound, p is the thermodynamic pressure, and S is the entropy calculated as $S = S_j + C_p[\log(T/T_j)] - R[\log(p_j/p)]$. C_p is the specific heat of air at constant pressure, R is the ideal gas constant, and j denotes reference properties.

In the second step, the fluctuating stagnation enthalpy (FSE) is decomposed. FSE of the fluid per unit mass is defined as $H' = h' + \frac{1}{2}\left(\frac{\mathbf{m}_i\mathbf{m}_i}{\rho^2}\right)^{'}$, where $h = C_vT + p/\rho$ is the static enthalpy, C_v is the specific heat at constant volume, and $m_i = \rho\mathbf{u}$ is the momentum density vector. Following Jenvey [6], H' is split into its hydrodynamic ($H^{'}_B$), acoustic ($H^{'}_A$) and thermal ($H^{'}_T$) components, and after linearization as detailed in Ref. [6], may be written as:

$$H_B^{'} = \left(\frac{c}{\rho}\right)\overline{\mathbf{M}}.\mathbf{B}'; \qquad H_T^{'} = \left(\frac{\overline{c}^2}{\gamma - 1}\right)\left(\frac{S'}{C_p}\right)\left[1 + (\gamma - 1)\overline{\mathbf{M}}.\overline{\mathbf{M}}\right] \tag{3}$$

and

$$H_A^{'} = \left(\frac{p'}{\overline{\rho}}\right)(1 - \overline{\mathbf{M}}.\overline{\mathbf{M}}) - \left(\frac{\overline{c}}{\overline{\rho}}\right)(\overline{\mathbf{M}}.\nabla\psi_a^{'}) \tag{4}$$

where $\overline{\mathbf{M}}$ is the local mean Mach number vector. Equations governing the evolution of the different FSE components have also been derived [10] but not required for the present analysis.

2.2 Spectral Proper Orthogonal Decomposition

The SPOD algorithm implemented in this work uses the Welch block method [14], which forms Cross-Spectral Density (CSD) matrices from blocks of snapshots. First, the entire sequence of N snapshots contained in the matrix $\mathbf{Q}(\mathbf{X}, t)$ is partitioned into N_{BLK} overlapping blocks. Each snapshot block $\mathbf{Q}_{n,BLK}$ captures a realization of the flow over window size ΔT and contains approximately N/N_{BLK} snapshots. Second, the discrete Fourier transform is applied to each snapshot block $\mathbf{Q}(\mathbf{X}, t)_{n,BLK} \rightarrow \tilde{\mathbf{Q}}(\mathbf{X}, f)_{n,BLK}$. Third, new matrices are assembled, $\tilde{\mathbf{Q}}(\mathbf{X}, f_i)$, that collect shared frequencies f_i from each transformed block; i.e., there are now N/N_{BLK} new matrices that contain Fourier coefficients belonging to a single frequency.

The next step performs POD by constructing a cross-spectral density (CSD) matrix

$$\tilde{C}_i = \frac{1}{N - 1}\tilde{Q}_i\tilde{Q}_i^H \tag{5}$$

where H is the Hermitian transpose. Finally, SPOD modes are obtained from the eigenvalue decomposition of the CSD matrix. N_{BLK} mutually orthogonal SPOD modes are obtained for each frequency. The CSD eigenvalue decomposition is repeated for each frequency. Consequently, the relative energy contained within each SPOD mode (eigenvalue magnitude) can be compared across all frequencies, acting as power spectra for the entire flow field.

2.3 Conditional Space-Time POD

Under conditions of an asymptotically long time horizon ($t \rightarrow \infty$), the generalized space-time decomposition recovers SPOD modes described in Refs. [4, 7, 22]. However, for transient analysis with focus on shorter time-horizons $t \rightarrow \Delta T$ containing events with multi-frequency behavior, the conditional space-time decomposition

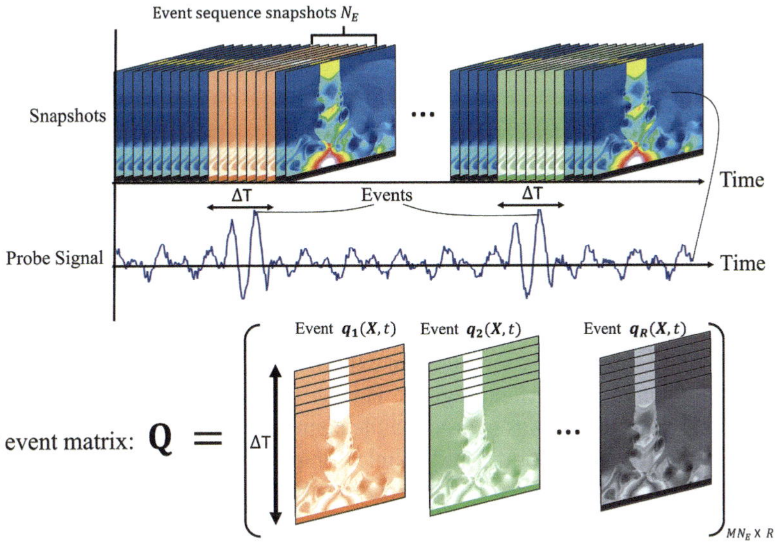

Fig. 4 Construction of the CPOD event matrix **Q**. Sequences of flow snapshots that correspond to the time of an event are flattened into each column of the matrix **Q**. Every column is a unique event

(CST-POD) is appropriate [5, 15]. Although related to simpler ensemble averages, the singular value decomposition (SVD) step highlights the underlying dynamics or event types. CST-POD is typically applied to statistically stationary data-sets but rather than continuous snapshot sequences, it isolates specific events in the flow; that is, particular snapshot sequences are sampled from the original snapshot matrix, as illustrated in Fig. 4. Here the "events" are arbitrary, but usually correspond to intermittent large fluctuations in a local region of the flow, as detected from a probe signal for example.

Regarding the structure of the input matrix **Q** for R events, CST-POD flattens event realizations from a consecutive sequence of snapshots, $\mathbf{q}_r(\mathbf{x}, t)$, into each column of **Q** for every event realization $r = 1, 2, ..., R$, where $R << N$. The CPOD sequence window ΔT (defined by the time preceding t^- and following t^+ each event at moment t_o) is an adjustable parameter based on the number of snapshots spanning each event, N_E, and thus included in each column of **Q**. Once the conditional filtering has selected R realizations of each event, the data matrix **Q** is populated.

CST-POD modes are obtained by an eigenvalue decomposition formulated on the two-point correlation tensor $\mathbf{Q}\mathbf{Q}^*\mathbf{W}\mathbf{\Phi} = \mathbf{\Phi}\mathbf{\Lambda}$, where $\mathbf{\Phi}$ contains eigenvectors $\phi_i(\mathbf{x}, t)$, and $\mathbf{\Lambda}$ is a diagonal matrix containing eigenvalues λ_i. The eigenvectors are the CPOD modes and the eigenvalues are the corresponding energies. In practice, CPOD modes may be calculated by a cheaper and faster singular value decomposition, $\mathbf{Q} = \mathbf{\Phi}\mathbf{\Sigma}\boldsymbol{\psi}^T$, yielding R CST-POD modes $\phi_i(\mathbf{x}, t)$ with corresponding singular values $(\sigma_1, \sigma_2, ... \sigma_{N_r})$ in the diagonal matrix $\mathbf{\Sigma}$. Similar to POD, the CPOD modes are ranked by variance and may be normalized by singular values; in addition to being functions of both space and time, CPOD modes are also orthogonal in the space-time inner product.

3 Impinging Jet: Combined Modal Decompositions

The signatures of the principal KH (hydrodynamic) instabilities and acoustic feedback of impinging jets (Fig. 2) are contained in all primitive and conserved variables but the underlying mechanisms are not immediately apparent. For example, pressure is most conveniently measured in experiments and directly correlates with the acoustic signal in the farfield regions. However, in the turbulent part of the jet, the pressure field also contains hydrodynamic content [21]; since an exact separation of pressure into acoustic and hydrodynamic components is not feasible (cf. Ref. [23] for a linearized development), interpretations based on pressure can be misleading. Finer, non-dominant features, which influence the major mechanisms are even more obscure in the primitive variables.

Doak's MPT (Sect. 2.1) provides an elegant solution to this problem and is applied to a Large-Eddy Simulation (LES) of an impinging jet [11, 18] mimicing available experimental data [2]. The Reynolds number, $Re = \rho_{ref} U_{ref} L_{ref} / \mu_{ref}$ based on nozzle exit conditions is 5.8×10^5 and the ground plate is located at a distance of $h/D = 4$ from the nozzle exit. The full 3D compressible Navier-Stokes equations are solved using a third-order upwind-biased Roe-scheme [13], in conjunction with the van Leer harmonic limiter [24], for the spatial discretization of the inviscid fluxes. The viscous fluxes are computed using a second-order central differencing scheme, and an implicit time integration method is used [1, 12]. The simulations have been verified and validated, and additional descriptions of the established flow may be found in Refs. [11, 18]

Here we build on the analysis to highlight tonal genesis by considering H, the raw value of which, at a representative instant in time, is shown in Fig. 5a. The three split components are displayed in Fig. 5b, c and d, respectively. Some of the main anticipated features are apparent. H_B and H_T are generally confined to the turbulent portion of the jet, with the former highlighting features in the shear layer and shock interactions in the core, while the latter displays a gradual spread into the core region. H_A is the most interesting, in that it not only highlights coherent structures in the KH region (acoustic wavepacket) but also clearly delineates wavefronts in the outer, more quiescent, region to establish a feedback pathway, where the other two components are negligible.

The patterns of Fig. 5 contain all spatio-temporal components in the unsteady solution. The axisymmetric nature of round jets results in mechanisms that are dominated by azimuthal modes; it then becomes advantageous and insightful to perform azimuthal mode decomposition, which has the added advantage of converting the 3D flow into an effectively 2D system. The lower modes are the most impactful

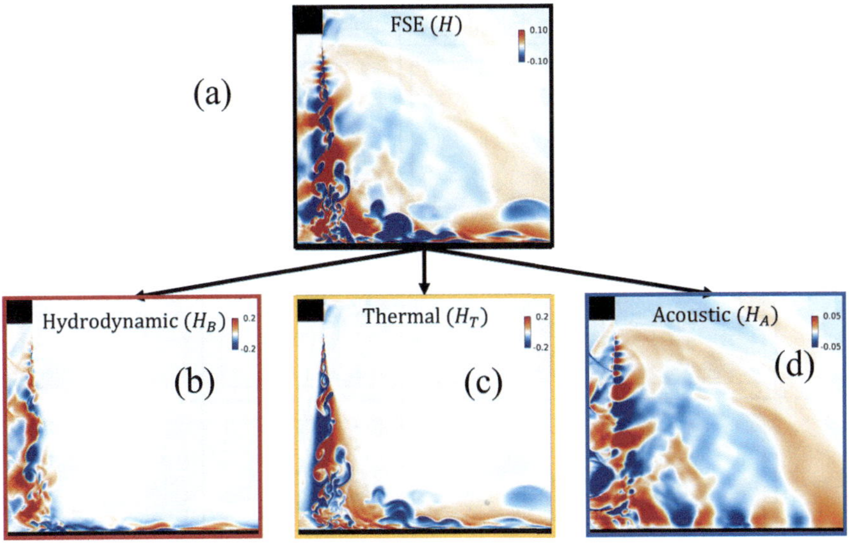

Fig. 5 Decomposition of **a** total FSE into **b** hydrodynamic, **c** thermal, and **d** acoustic components

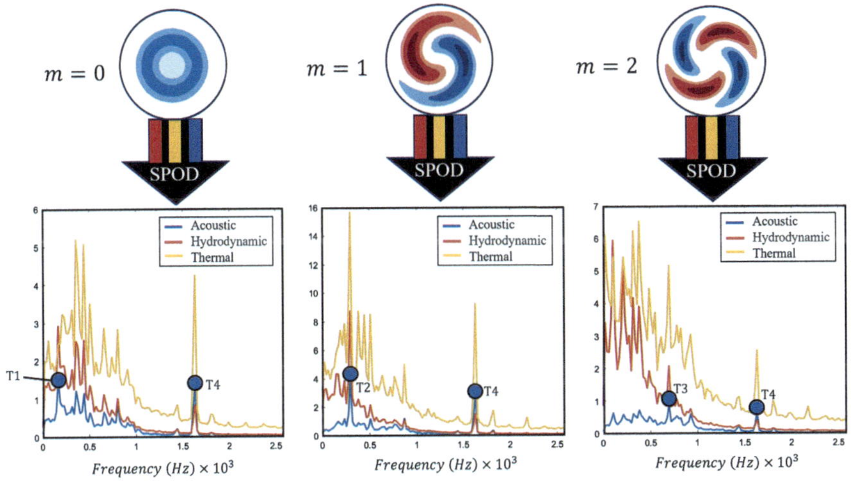

Fig. 6 Decomposition of spectral content in H_A, H_B and H_T with SPOD for azimuthal modes $m = 0$, $m = 1$ and $m = 2$

because of their larger energy content; here we consider $m = 0$ (axisymmetric), $m = 1$ (helical) and $m = 2$ (double-helical) azimuthal modes. The general form of these is depicted in the top part of Fig. 6. Each of the three FSE components is decomposed into the three azimuthal modes, yielding a total of nine fields of interest.

The relative importance of the nine components is determined by considering their frequency-specific energy content with SPOD. Details of the SPOD parameters

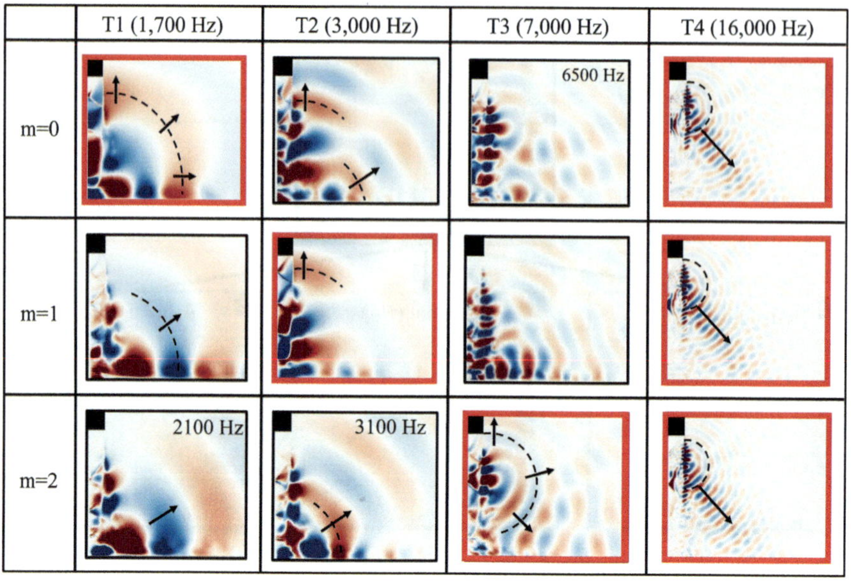

Fig. 7 SPOD modes of the peak acoustic FSE tones at each azimuthal mode

such as number of snapshots and sampling frequency are not presented for brevity; they may be found in a prior SPOD analysis of the same jet but using pressure as the variable of interest [17], which yields different information from the H considered here.

The nine spectral amplitudes of all FSE components and azimuthal mode numbers are displayed in the lower part of Fig. 6. All curves display tones (sharp peaks) whose implication is weighty, since they can cause significant damage to proximal life and structures. The thermal and hydrodynamic components are much larger in amplitude for all azimuthal modes and frequencies. Although contributing the least energy, the acoustic component is key to the feedback loop; this highlights the need for PIMs since they successfully isolate subtler pathways in the flow. Therefore, the acoustic component is analyzed further to discern the different competing feedback mechanisms. We do this by considering the four prominent tones (blue circles), designated T1 through T4 respectively, at $1{,}700Hz$ (fundamental feedback tone), $3{,}000Hz$, $7{,}000Hz$ and $16{,}000Hz$. The three lower tones, T1, T2, T3 appear, respectively, in modes $m = 0$, $m = 1$ and $m = 2$, while the highest frequency tone, T4, appears in all modes.

The spatial structures of the acoustic FSE component at each of the four frequencies and across all azimuthal modes are presented in Fig. 7. The azimuthal modes that correspond to the peak tone in Fig. 6 are highlighted with red borders i.e., $m = 0$, $m = 1$ and $m = 2$ are important respectively for the T1, T2 and T3, while all modes contribute to T4. Consider the azimuthal modes of the T1 fundamental feedback tone. The dominant $m = 0$ mode displays a more circular arc-like wavefront (dashed lines) that reaches the nozzle exit, thus closing the feedback loop. On the other hand, $m = 1$ and 2 modes better propagate acoustic radiation radially along the wall-jet. In general, low frequencies between T1 and T2 display radiation directed towards the nozzle receptivity region for the $m = 0$ and $m = 1$ azimuthal modes, while the $m = 2$ mode propagates acoustic waves from the wall-jet region away from the nozzle.

In contrast, the results indicate that the T3 and T4 SPOD modes are correlated to shock associated noise. The former is a result of screech feedback originating from the first shock cell and is observed to be synchronized with the impingement feedback mechanism through the upstream region of the wavefront [11]. This tone is strongest in the $m = 2$ azimuthal mode, radiating acoustics both upstream and downstream, while the other azimuthal modes are correlated to wavepackets throughout the entire jet shear-layer. The highest frequency shock-associated tone, T4, shares nearly identical mode shapes representing noise radiated in the downstream direction. The analysis thus facilitates a clear distinction in the mechanisms associated with the different tones, which is a first step towards formulating any control strategy.

We now consider the energetic content of the three FT components. First, the cumulative energy contained across all FSE and azimuthal modes is calculated from the sum of the SPOD spectra and shown in Fig. 8a. The thermal, hydrodynamic, and acoustic energy contribute approximately 65%, 25%, and 10%, respectively. These relative proportions are consistent across azimuthal modes. The cumulative energy also indicates that the helical $m = 1$ mode contains the most energy in the jet, peaking at T2 in the spectra of Fig. 6. Therefore, this mode may be referred to as the "optimal" impingement feedback mode. The FSE contributions for the different modes at each tone are shown in Fig. 8b, and from the prior analysis may be attributed to the fundamental impingement feedback cycle (T1, $m = 0$), optimal impingment (T2, $m = 1$), screech feedback (T3, $m = 2$), and downstream shock associated noise (T4, $m = 0, 1, 2$).

Finally, the forms of the largest energy modes are displayed in Fig. 8c for each of the four tones; this procedure essentially reverses the azimuthal decomposition procedure to yield the corresponding three-dimensional fields. The thermal energy of the T2, $m = 1$ mode contains the largest energy content, attributed to the entropy increase associated with the swirling motion component of the irreversible viscous shear-stress related effects, oriented in the azimuthal direction. The axisymmetric $m = 0$ mode lacks this azimuthal motion and is thus a relatively milder source of shear related dissipation because its motion is limited to the axial direction; in this case, the vortical (or hydrodynamic) energy is the largest contributor. With regard to the T3 and T4 shock-associated tones, the most prominent difference in mode shapes is the reduction in the wavelength of coherent structures at the higher frequency. The downstream shock-associated tone at T4 exhibits all targeted azimuthal wave numbers, and therefore is the only SPOD mode necessarily requiring calculation using effectively the full three-dimensional field. In summary, the combination of PIM based on MPT, DDM based on SPOD, and traditional wavenumber-frequency transformations, isolate the underlying mechanisms that generate different tones, as well as the detailed energy contributions of resonance modes.

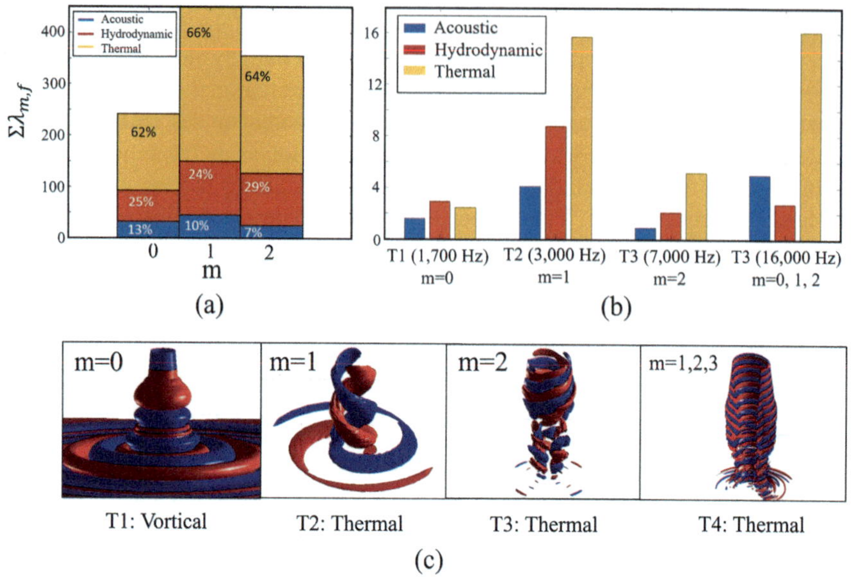

Fig. 8 **a** Cumulative SPOD energy across all frequencies represented by their FSE and azimuthal contributions. **b** FSE and azimuthal energy contained in the modes corresponding to the four peak acoustic tones. **c** Three-dimensional expansion of the largest energy contribution to each azimuthal mode

4 Multi-stream Jet: Spectral and Transient Modes

The need to use DDMs that retain space-time properties is illustrated with the multistream nozzle, introduced previously in Fig. 3. The basic flowfield has been extensively examined experimentally [8] and computationally [16]. The current work leverages the LES databases developed in the latter with the numerical techniques similar to that of the impinging jet described earlier. Further details of the mesh, methodology and validation may be found there and are suppressed here for brevity. Although the flow contains multiple shear-layer, boundary-layer and shock interactions, a primary feature is the shedding instability at the splitter plate trailing edge (SPTE), which substantially influences the flowfield and manifests as a prominent tone in the farfield. Unlike the impinging jet of Sect. 3, this acoustic component lacks feedback resonance mechanisms; rather the waves propagate across the channel to influence the shock system and affect the different SBLI on the SERN as well as on the aft-deck, in addition to direct interactions with the aft-deck boundary layer, all of which result in a rich range of spatio-temporal coherent structures.

We once again initiate the analysis with SPOD, which is a logical tool to explore the detailed mechanisms by which the shedding, with its dominant frequency dynamics, influences the rest of the flow. For this, eight blocks of 1,024 density fluctuation snapshots are used with 50% overlap, to resolve a non-dimensional frequency range

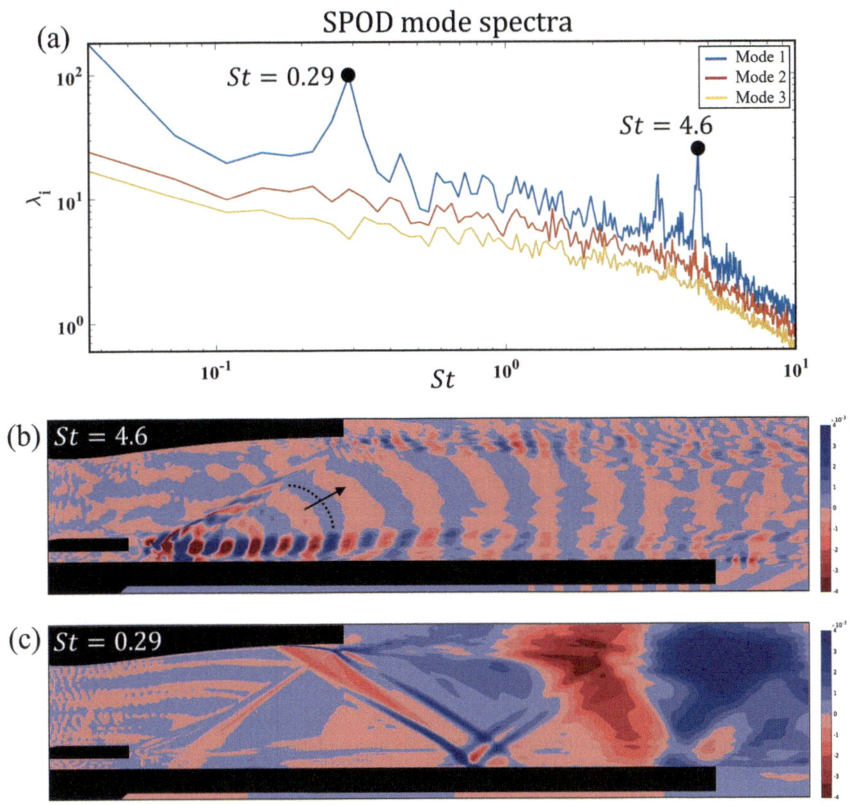

Fig. 9 **a** SPOD spectra of the first three modes. Leading SPOD mode shapes of the **b** $St = 4.6$ acoustic tone and **c** $St = 0.29$ low frequency tone

from $St = 0.036$ to 18.5, where $St = f L_{ref}/U_{ref}$, L_{ref}, and U_{ref} are the hydraulic diameter (4.45×10^{-2} m) and the averaged combined core and bypass flow (500 m/s), respectively. The SPOD spectra of the first three modes are shown in Fig. 9a, indicating two dominant peaks at $St = 0.29$ and $St = 4.6$. The features at the latter frequency are very similar to those of the primary tone reported in Refs. [8, 16], with minor quantitative differences potentially attributable to the spanwise homogeneous nature of the present flow. The corresponding most energetic tonal mode shapes are shown in frames (b) and (c). The higher frequency mode ($St = 4.6$) displays the shedding pattern downstream of the SPTE, and clearly corresponds to the aforementioned tone. The lower frequency mode ($St = 0.29$) highlights shocks that are reflected off the trailing edge of the top surface, then from the aft-deck, followed by interaction with the shear layer in the SERN wake.

The distinct nature of the two SPOD mode shapes suggest two independent phenomena; however, SPOD by itself does not guide understanding of causation, particularly when dynamics are contributed from multiple frequencies. To address this, we use CST-POD (Sect. 2.3) for a transient perspective that exposes the downstream

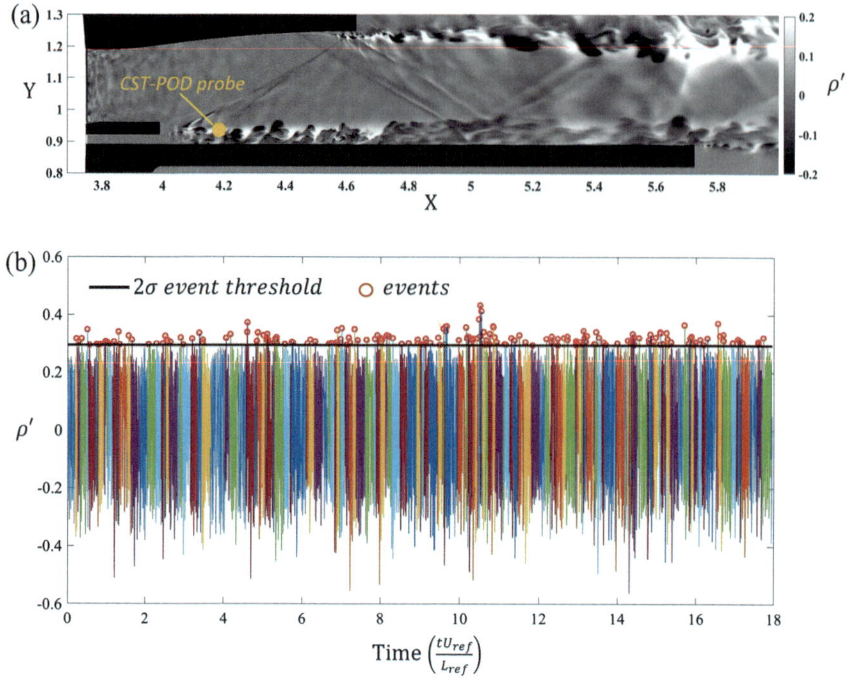

Fig. 10 **a** Density fluctuations and CST-POD probe location. **b** CST-POD probe signal identifies events by a 2 standard deviation threshold. Colored segments of the signal correspond to the duration of each event

effects of the acoustic waves on the reflected shocks. Since the source of the acoustic tone is immediately downstream of the SPTE, the first step is to identify large fluctuation events in that region using a numerical probe placed in the flow as shown in Fig. 10a. Density fluctuation measurements from the probe are plotted in Fig. 10b. If the events are defined as those where density fluctuations are greater than two standard deviations, the trace evidences a total of 192 event samples. Scrutiny of these events indicates that a non-dimensional event time-window of $\Delta T = 0.5$, or 50 snapshots, is sufficient to capture coherent effects, which occur at much slower time scales than the $St = 4.6$ frequency associated with the acoustics.

Once the events are defined, the snapshots from those events are processed with CST-POD as in Ref. [15] to yield ranked modes that are localized in time. The first CST-POD mode here is equivalent to the ensemble average, and is not of much interest for the causal dynamics. Thus, the results shown in Fig. 11, progressing in time from top to bottom, correspond to the second mode, which better educes the underlying unsteadiness. The figures display several markers, including two arrows to visually trace coherent structures in time, curved black dashed lines to designate the acoustic wavefront and three straight lines for shock oscillation structures (alternating black and white, except at $t = 0$, which only shows two since the structure is

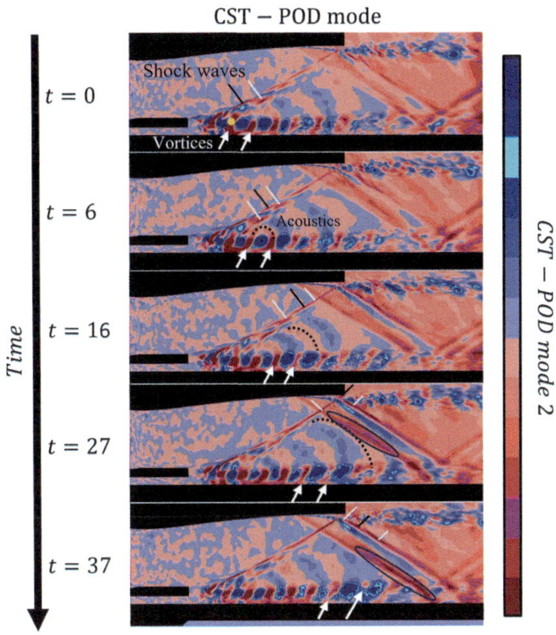

Fig. 11 CST-POD mode correlated to large fluctuations from vorticies in the splitter-plate wake (arrows). An emmited acoustic wave (dashed line) and shock waves (lines) are tracked throughout the CST-POD mode sequence, highlighting the acoustic-shock interactions (circle)

being initiated). At the moment of the spike in density, $t = 0$, a large vortex structure passes through the probe location (marked by the arrow on the left). An acoustic wavefront (curved dashed line) is generated at $t = 6$ between the two arrow-marked vortex structures. With time, this wavefront follows the propagation pattern previously observed in the $St = 4.6$ SPOD mode. The structures marked by the lines follow the progression of shock wave disturbances emanating from the splitter plate trailing edge, which are then reflected from the SERN surface. The mode at time $t = 27$ captures the acoustic wavefront intersecting the first reflected shock (marked with a black oval), which creates a ripple in the shock impinging on the aft-deck. Simultaneously, the three shock disturbances along the splitter plate shock reach the top surface, and are redirected along the second reflected shock.

Although the wave motion in the two reflected shocks is evident in the $St = 0.29$ SPOD mode of Fig. 9c, it alone does not indicate the role of the acoustic interactions. The CST-POD mode on the other hand captures the interactions of acoustics and shocks, as well as the relative phase speeds of the signals that connect these features. For example, all the disturbances identified in Fig. 11 are initiated in the SPTE wake at approximately the same time, but propagate along different paths and speeds. Acoustic and shock wave disturbances essentially outrun the vortex structures to trigger coherent events that account for these differences. This example demonstrates how a transient CST-POD analysis complements SPOD analysis to provide a more complete picture of jet dynamics that includes time-local features.

5 Conclusions

Novel integration of physics-informed and data-driven methods are shown to provide unparalleled insights into the finer details of complex turbulent flows. The procedure builds on classical, commonly used data-driven methods such as azimuthal and spectral proper orthogonal decomposition (SPOD), which are generally employed on measured or computed primitive variables such as velocity, pressure or those depending on their derivatives. It is shown that introducing a physics-informed sifting procedure, separating acoustic, hydrodynamic and thermal components, as well as their fluctuating stagnation enthalpies, prior to application of data-driven methods has significant benefits in isolating different competing mechanisms. The example case of a supersonic jet impinging on a ground plane crystallizes the benefits. The main tones are identified using SPOD in the filtered acoustic component, and the underlying azimuthal structures for each tone in the resonating jet are systematically exposed in a manner not feasible with primitive variables. The results show that though thermal and hydrodynamic energy comprise the major portion of the total energy – the helical thermal mode has the most resonance contribution – it is the smaller acoustic component that ultimately influences the mechanisms of feedback and shock-associated noise.

The second procedure presented in this work addresses the fact that transient dynamics are obscured by naive application of common modal decomposition methods such as SPOD. For this, a conditional space-time proper orthogonal decomposition (CST-POD) is used to complement the frequency correlations of SPOD. The benefits are exemplified by application to a complex multi-stream nozzle flow containing a tonal instability, whose effects propagate and convect along different pathways formed by multiple shock waves interacting with shear and boundary-layers. Transient CST-POD, in combination with spectral analysis, identifies and isolates the propagating coherent structures and their phase speeds, thus successfully correlating different previously perplexing temporal effects. In particular, the CST-POD mode correlated to the acoustic source location traces the propagation of acoustic disturbances along three shock wave interaction paths, connecting the multi-scale features by extracting the different phase speeds of the disturbances along each path.

References

1. Beam, R.M., Warming, R.F.: An implicit factored scheme for the compressible navier-stokes equations. AIAA J. **16**(4), 393–402 (1978)
2. Bhargav, V.N., Song, M., Sellappan, P., Alvi, F.S., Kumar, R.: Unsteady characteristics of resonant supersonic dual impinging jets. In: AIAA Scitech 2020 Forum (2020). https://doi.org/10.2514/6.2020-0498
3. Doak, P.E.: Momentum potential theory of energy flux carried by momentum fluctuations. J. Sound Vib. **131**(1), 67–90 (1989). https://doi.org/10.1016/0022-460X(89)90824-9
4. Frame, P., Towne, A.: Space-time POD and the Hankel matrix. arXiv e-prints:2206.08995 (39), 1–28 (2022). http://arxiv.org/abs/2206.08995

5. Hack, M.P., Schmidt, O.T.: Extreme events in wall turbulence. J. Fluid Mech.**907** (2021)
6. Jenvey, P.L.: The sound power from turbulence: a theory of the exchange of energy between the acoustic and non-acoustic field. J. Sound Vib. **131**, 37–66 (1989)
7. Lumley, J.L.: Stochastic tools in turbulence. Academic Press, New York (1970)
8. Magstadt, A.S., Berry, M.G., Berger, Z.P., Shea, P.R., Ruscher, C.J., Gogineni, S.P., Glauser, M.N.: Flow structures associated with turbulent mixing noise and screech tones in axisymmetric jets. Flow Turbul. Combust. **98**, 725–750 (2017)
9. Pirozzoli, S., Modesti, D.: Direct numerical simulation of one-sided forced thermal convection in plane channels. J. Fluid Mech. **957**, A31 (2023)
10. Prasad, C., Gaitonde, D.: Assessment of a generalized acoustic variable for supersonic jet noise. In: AIAA AVIATION 2021 FORUM, Paper 2021-2101 (2021)
11. Prasad, C., Stahl, S., Gaitonde, D.: Exchange mechanisms between hydrodynamic and acoustic components of an under-expanded supersonic impinging jet. AIAA AVIATION 2021 FORUM (2021). https://doi.org/10.2514/6.2021-2118
12. Pulliam, T.H., Chaussee, D.: A diagonal form of an implicit approximate-factorization algorithm. J. Comput. Phys. **39**(2), 347–363 (1981)
13. Roe, P.L.: Approximate riemann solvers, parameter vectors, and difference schemes. J. Comput. Phys. **43**(2), 357–372 (1981)
14. Schmidt, O.T.: Guide to spectral proper orthogonal decomposition. AIAA J. **58**(3), (2020)
15. Schmidt, O.T., Schmid, P.J.: A conditional space-Time POD formalism for intermittent and rare events: example of acoustic bursts in turbulent jets. J. Fluid Mech. **867**, 1–12 (2019)
16. Stack, C.M., Gaitonde, D.V.: Shear layer dynamics in a supersonic rectangular multistream nozzle with an aft-deck. AIAA J. **56**(11), 4348–4360 (2018)
17. Stahl, S.L., Prasad, C., Gaitonde, D.V.: Distinctions between single and twin impinging jet dynamics. J. Acoust. Soc. Am. **150**(2), 734–744 (2021)
18. Stahl, S.L., Prasad, C., Gaitonde, D.V.: Feedback mechanisms in a supersonic impinging jet using conditional proper orthogonal decomposition. AIAA SCITECH 2022 Forum (2022)
19. Sullivan, J., Gaitonde, D.: Model order reduction of scramjet isolator shock dynamics during unstart. In: ASME International Mechanical Engineering Congress and Exposition, vol. 86656, p. V003T04A026. American Society of Mechanical Engineers (2022)
20. Taira, K., Hemati, M.S., Brunton, S.L., Sun, Y., Duraisamy, K., Bagheri, S., Dawson, S.T., Yeh, C.A.: Modal analysis of fluid flows: applications and outlook. AIAA J. **58**(3), 998–1022 (2020)
21. Tinney, C., Jordan, P.: The near pressure field of co-axial subsonic jets. J. Fluid Mech. **611**, 175–204 (2008)
22. Towne, A., Schmidt, O.T., Colonius, T.: Spectral proper orthogonal decomposition and its relationship to dynamic mode decomposition and resolvent analysis. J. Fluid Mech. **847**, 821–867 (2018)
23. Unnikrishnan, S., Gaitonde, D.V.: A pressure decomposition framework for aeroacoustic analysis of turbulent jets. Euro. J. Mech. B/Fluids **81**(January), 41–61 (2020). https://doi.org/10.1016/j.euromechflu.2020.01.006
24. Van Leer, B.: Towards the ultimate conservative difference scheme. V. A second-order sequel to godunov's method. J. Comput. Phys. **32**(1), 101–136 (1979)

Implications of Under-Resolved DNS Data in the Buffer Zone of Wall-Bounded Turbulent Flows for Turbulence Modeling

Svetlana V. Poroseva

Abstract In the paper, data from direct numerical simulation of an incompressible planar turbulent flow in a strained channel (*Jeyapaul et al., 2015*) are used to investigate relations between derivatives of the streamwise mean velocity in the wall-normal direction, turbulence production by the shear, the shape of probability density functions of velocity fluctuations in the streamwise and wall-normal directions, and balance errors in the DNS budgets of velocity moments up to the fourth order with the focus on the flow near-wall area. As time progresses, the flow undergoes deformation leading to its separation. A new criterion based on the streamwise mean velocity second derivative in the wall-normal direction is proposed to rigorously define the buffer zone boundaries. It is shown that the balance errors have the systematic contribution in this flow area. The errors are also the highest in comparison with the turbulence production in the buffer zone at any time. The most affected are odd- and even-order velocity moments with odd exponents of the velocity fluctuations. The systematic error in the data is suggested as a likely explanation of the deformed shape of the probability density function of the velocity fluctuation in the streamwise direction observed in the buffer zone at early times of this flow development. Unresolved reference data are of concern for modeling as they may mislead the modeling effort and the model quality assessment.

1 Introduction

A standard approach to validating a turbulence model is to compare the model results in a flow of interest with reference data from experiments and direct numerical simulations (DNS). Discrepancies between the model and the reference data revealed by their comparison are attributed to the model deficiencies. However, the reference data have their own sources of errors and uncertainties. When the errors in reference

CUFS-2024-12, Cambridge, United Kingdom, 4-5 March 2024

S. V. Poroseva (✉)
University of New Mexico, Albuquerque, NM 87131, USA
e-mail: poroseva@unm.edu

J. C. Tyacke and N. R. Vadlamani (eds.), *Proceedings of the Cambridge Unsteady Flow Symposium 2024*, https://doi.org/10.1007/978-3-031-69035-8_4

data are on the order of magnitude or larger than the data, meaningful conclusions about the model performance are questionable.

This situation is not hypothetical as one may think. A case in study is DNS data for incompressible planar wall-bounded turbulent flows. Several datasets for such flows [1–3] were examined in our previous studies [4–6]. These datasets are considered to be of high quality by the state-of-the-art standards and indeed, balance errors in the Reynolds stress transport (RST) equation budgets obtained in these simulations are small when compared to the leading terms. However, when compared with the minor budget terms representing viscous (molecular) diffusion, the balance errors can be up to two orders of the magnitude larger in the equations for all Reynolds stresses. Even though the viscous diffusion does not require modeling, uncertainty in its values is a source of uncertainty in values of other terms in the RST equations and, thus, has an unknown impact on assessing the model quality for those terms that have to be modeled.

In [5], it was demonstrated that the balance errors in the budgets of RST equations from DNS are systematic rather than statistical in their nature, and that the DNS data convergence does not necessarily imply that a converged solution is the right one. In both studies [4, 5], the grid resolution and particularly, in the near-wall flow area, was suggested as a possible leading source of errors in DNS statistics.

Similar conclusions about deficiency of the standard grid generation procedure based on the Kolmogorov's length scale were made in [6] from the analysis of discrepancies between the model and DNS data for velocity moments of third- and higher-order moments. Statistics affected the most were found to be odd-order velocity moments and the turbulence production by the shear in the transport equations of such moments. Another revelation from the study was the close alignment of the maximum values of these statistics with the maximum balance errors in the DNS budgets of odd-order velocity moments, with all being located in the buffer zone of the considered flows.

To summarize, results from [4–6] indicate a possibility of DNS being under-resolved in the wall-bounded flow buffer zone where the wall-normal gradient of the mean velocity component in the streamwise direction (hereafter, streamwise mean velocity) and turbulence effects tend to be at their maximum (when no other forcing is applied), and this is due to the current approach to the grid generation based on the Kolmogorov's length scale used in DNS. However, when A. N. Kolmogorov introduced the theory of local isotropy (and the characteristic turbulence scales), he did it with the explicit warning that the theory is not applicable near the flow boundaries [7]. A new length scale was proposed in [6] to address the grid resolution issue in DNS of wall-bounded flows. The current study investigates further possible relations between the derivatives in the wall-normal direction of the streamwise mean velocity, turbulence production by the shear and balance errors in the DNS budgets of velocity moments up to the fourth order, as well as the probability density function (PDF) of velocity fluctuations in the streamwise and wall-normal directions in the near-wall area of an incompressible planar turbulent flow in a strained channel. The DNS dataset from [2] is used in the study.

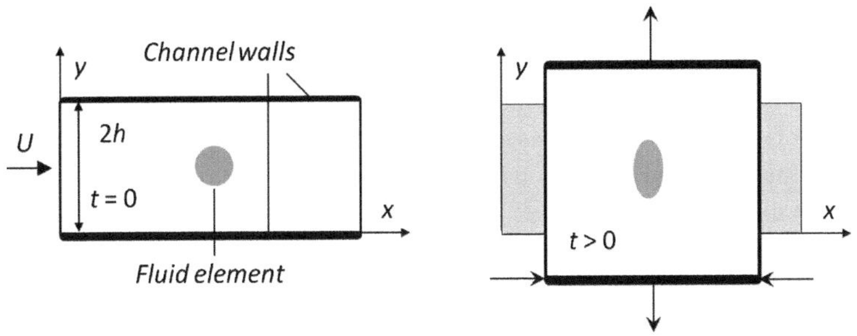

Fig. 1 Time-evolving channel flow under imposed strain: initial at $t = 0$ and deformed at $t > 0$

2 DNS Data Description

In [2], simulations were first conducted in a planar fully-developed incompressible turbulent channel flow at $Re_{\tau 0} = 392$, where $Re_{\tau 0}$ is based on the friction velocity $u_{\tau 0}$ and the channel half-height h. The unstrained channel flow data were used then as the initial conditions in simulations of a strained channel flow.

In a strained channel, the initial channel flow domain was deformed by applying simultaneously streamwise compression and wall-normal divergence to the domain boundaries in the plane (x, y) as shown in Fig. 1, where x and y are streamwise and wall-normal directions, respectively. The imposed deformation field varied linearly in space: $U_i^*(\overrightarrow{x}, t) = A_{ij}(t)x_j, i = 1, 2, 3$ with $\overrightarrow{U^*} = (U^*, V^*, 0)$. The strain, $A_{ij} = \partial U_i^*/\partial x_j$, imposed at $t = 0$ was spatially uniform and constant in time: $\partial A_{ij}/\partial t = 0$ with two non-zero components: $A_{11} = \partial U^*/\partial x < 0$ (streamwise compression) and $A_{22} = \partial V^*/\partial y > 0$ (wall-normal divergence) such that $|A_{11}| = A_{22} = 0.0118$. This value constitutes 31% of the ratio of $u_{\tau 0}$ to the initial channel half-width.

The flow in a strained channel is a time-developing approximation of a spatially-developing adverse pressure-gradient turbulent boundary layer. In the DNS dataset, PDFs are available at three non-dimensional times: $t_s^{'} = 0,\ 0.365,$ and 0.772 corresponding to $s = 0,\ 5,\ 7$ when the non-dimensional time is defined as $t_s^{'} = A_{22}\, t_s$. These are times considered in the current study. At $t_0^{'} = 0$, the flow is unstrained. At $t_7^{'} = 0.772$, the flow is separated.

3 Results

3.1 Definition of the Buffer Zone

The buffer zone in wall-bounded turbulent flows is usually identified in relation to the normalized mean velocity profile $u^+ = U/u_\tau$ as the distance between the viscous sublayer and the log-law region where linear and logarithmic approximations in terms of $y^+ = yu_\tau/\nu$ are applicable, respectively [8]. Hereafter, U is the streamwise mean velocity , u_τ is the friction velocity, and ν is the fluid kinematic viscosity. In [8], the buffer zone is specified to be within the interval $5 < y^+ < 30$. In other studies, the upper boundary of this region was expanded farther.

The definition of the buffer zone as the distance between two other flow regions depends on the boundaries of those regions and approximations used to define them. This makes it difficult to use such a definition when a flow undergoes separation causing the mean velocity profile to deform. A more beneficial definition of the buffer zone would be the one independent from other flow regions and their approximations. The approach proposed in this paper is to define the buffer zone in terms relevant to the turbulence production, which reaches its maximum in this flow area when no other forcing is applied.

In the strained-channel flow considered in the current study, the turbulence production is dominated by the flow shear to compare with that by the strain at any time [6]. The expression for this term in the transport equation of a statistical velocity moment $\langle u^n v^m w^k \rangle$ in planar wall-bounded flows is the following

$$P^{nmk} = -n\langle u^{n-1} v^{m+1} w^k \rangle \frac{\partial U}{\partial y}, \tag{1}$$

[4, 6], where u, v, and w are components of the turbulent velocity, $\vec{\mathbf{u}}$, in the streamwise, wall-normal, and spanwise directions, respectively.

The term in (1) common for all velocity moments is the wall-normal gradient of the streamwise mean velocity. Figure 2a, c and e show how this gradient profile plotted against y^+ evolves with time. Three slopes can be recognized in the profile when moving away from the channel wall towards the channel center at all considered times. At t_0', the first slope next to the wall is a slope of slow decay. At later times, the gradient grows right from the wall and reaches its maximum at $y^+ = 1.85$ at t_5' and 7.12 at t_7'. Notice that traditionally, the mean flow velocity in this flow area called the viscous sublayer is approximated by $U^+ = U/u_\tau \sim y^+$ (and thus, $U \sim y$) [8]. That is, one would expect to see a constant value of the wall-normal streamwise mean velocity gradient instead of a slope next to the wall. However, the data does not support this approximation.

Away from the wall, the region of rapid linear decay can be recognized in all three profiles of the wall-normal gradient of the streamwise mean velocity followed by the region of slow decay. The three slopes observed in the streamwise mean velocity gradient profile are time-dependent.

Less variability is observed in the profiles of the second derivative of streamwise mean velocity in the wall-normal direction presented in Fig. 2b, e, and f. In all three profiles, two regions can be recognized. In Region I next to the wall, the second derivative rapidly decays until it reaches its minimum and then, grows approaching values close to the zero. Region I is followed by Region II where the second derivative values are close to zero. In this study, "zero" values are those that are two orders of the magnitude less than the absolute value of the profile minimum value.

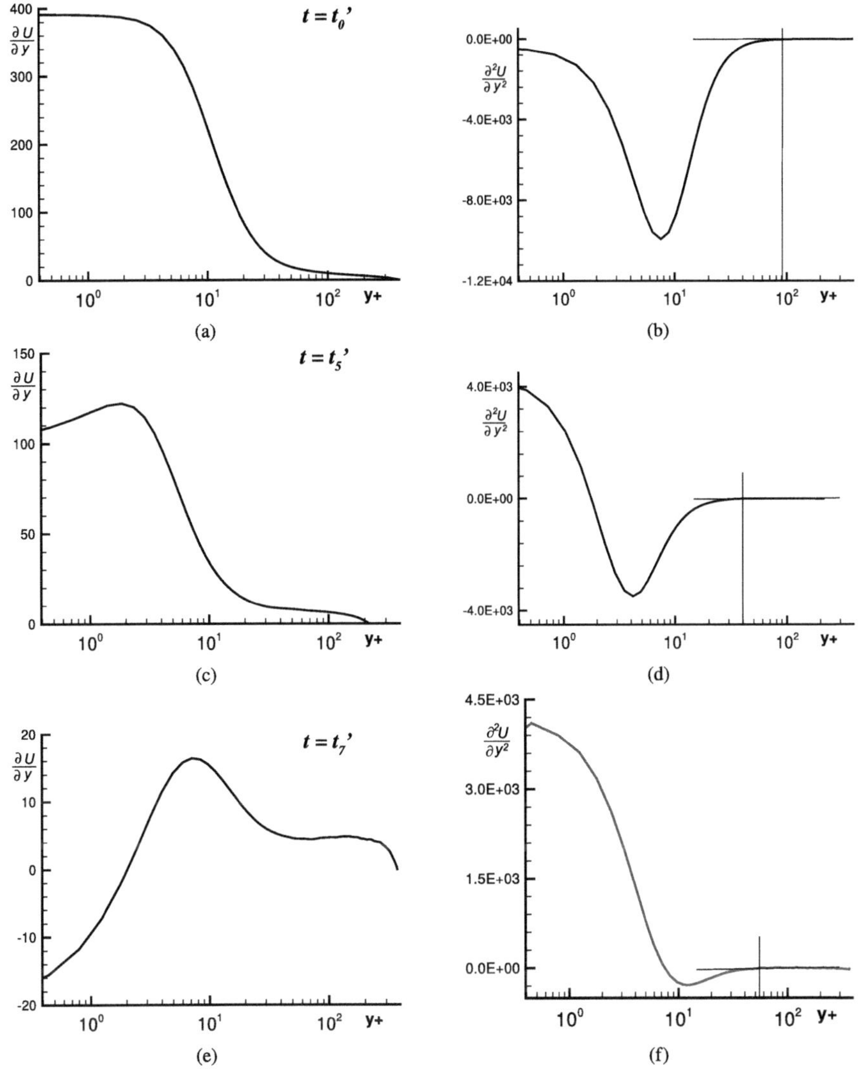

Fig. 2 Time evolution of the profiles of first (**a, c, e**) and second (**b, d, f**) derivatives in the wall-normal direction of the streamwise mean velocity: **a, b** $t_0^{'}$, **c, d** $t_5^{'}$, and **e, f** $t_7^{'}$

Both boundaries of Region I can be rigorously determined using the criterion: $\frac{\partial^2 U}{\partial y^2} \rightarrow 0$ (to the degree of accuracy the reference data). They are provided in Table 1 rounded to the nearest integer. Whether there is a practical reason to distinguish the lower boundary of Region I from the wall, remains to be investigated further. Currently, no particular benefits for that are identified.

Table 1 Boundaries of Regions I and II in the strained channel identified from the second derivative of the streamwise mean velocity in the wall-normal direction

Time	Region I	Region II
t_0'	$0 \le y^+ < 60$	$60 < y^+$
t_5'	$2 \le y^+ < 31$	$31 < y^+$
t_7'	$8 \le y^+ < 53$	$53 < y^+$

3.2 Production by the Shear and the Balance Errors in the Buffer Zone

Figure 3 shows the turbulence production by the shear in the transport equations for velocity moments up to the fourth order along with the balance errors, Err, in the budgets of these equations obtained from DNS data [2] in the unstrained channel ($t = t_0'$). For the transport equations of the Reynolds stresses and the third-order velocity moments, the balance errors can be calculated as

$$Err_{ij} = D_{ij}^M + D_{ij}^T + P_{ij} + \Pi_{ij} + \epsilon_{ij}, \tag{2}$$

$$Err_{ijk} = D_{ijk}^M + D_{ijk}^T + P_{ijk} + P_{ijk}^{(T)} + \Pi_{ijk} + \epsilon_{ijk}. \tag{3}$$

For the transport equations of higher-order velocity moments, expressions similar to (3) are used to calculate the balance errors. In (2) and (3), terms D^M, D^T, P, P^T, Π, and ϵ represent viscous diffusion, turbulent diffusion, production (by the shear and by the strain), production-by-turbulence, velocity/pressure-gradient correlations, and viscous dissipation, respectively, defined in the standard manner for the Reynolds decomposition of the turbulent flow field applied to an incompressible turbulent flow. For the flow considered in the current study, the equations and the terms can be found in [2, 6].

In the figure, the terms are normalized by $u_{\tau 0}^l/\nu$, where $l =$ 4, 5, and 6 for the terms in the transport equations for the second-, third-, and fourth-order velocity moments, respectively. Velocity moments shown in every sub-figure of Fig. 3 identify the transport equation the profiles correspond to. Region I is highlighted by the filled rectangular shape in each sub-figure. For every velocity moment, Region I covers extremes in both profiles: turbulence production by the shear and the balance errors. This is also the region of where the production contributes the most in the budgets of all relevant velocity moments at this time. The largest extreme in all production profiles is closely located to the location of the minimum value of the second derivative of the streamwise mean velocity, $y^+ = 8$, and is aligned with such a location in the turbulence production profile for $\langle u^3 \rangle$. The farthest from this location is the extreme in the turbulence production profile for $\langle uv^3 \rangle$.

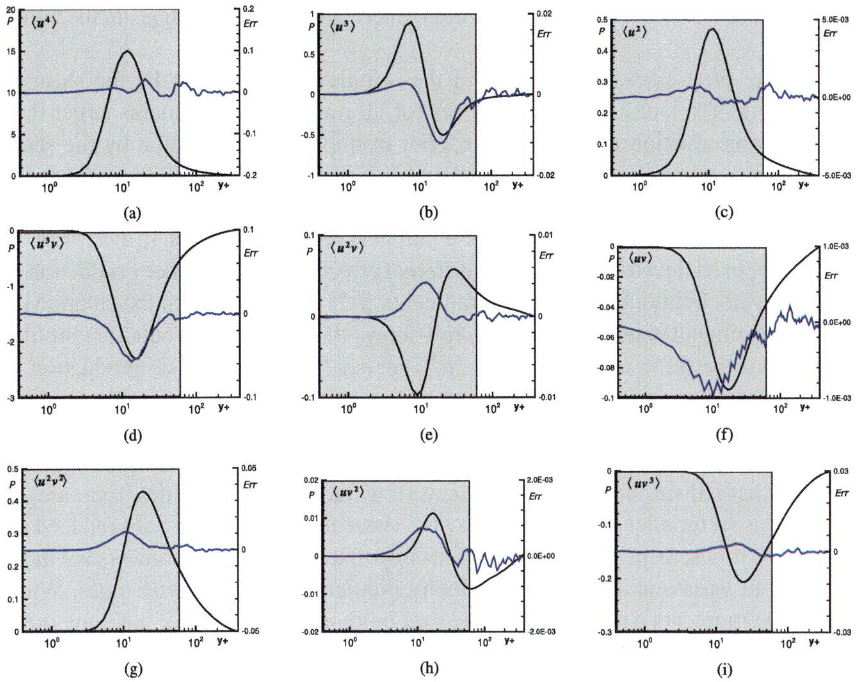

Fig. 3 Turbulence production by the shear (black lines) and the balance errors (blue lines) in the DNS budgets of velocity moments of various orders at $t = t_0'$ (in the unstrained channel). Region I from Table 1 is shown by the grey rectangular shapes

As seen in the figure, the balance errors are also the highest in the buffer zone for all velocity moments but $\langle u^2 \rangle$ and $\langle u^4 \rangle$. For these two moments, no particular tendency can be recognized at this time. This was previously noticed in [6], where the close alignment between the extremes in the profiles of turbulence production by the shear and the balance errors in the buffer zone was reported for the odd velocity moments. The phenomenon was attributed to the sensitivity of odd velocity moments to the direction of velocity fluctuations. However, any velocity moment $\langle u^m v^n \rangle$ with odd exponents m and n will be sensitive to the velocity fluctuation direction, and Fig. 3 supports this conclusion by showing a good alignment between the extremes in the turbulence production and the balance errors for all such moments. It also shows that the alignment is more pronounced when $m > n$ and for larger m and n.

When comparing the balance errors with the turbulence production values, one can find the errors to be the highest in the budgets of the odd velocity moments with $m + n = 3$ and for the fourth-order moments $\langle u^3 v \rangle$ and $\langle u^2 v^2 \rangle$, followed by those in the budgets of $\langle u v^3 \rangle$ and $\langle u v \rangle$. The lowest errors (in comparison with the turbulence production) are in the budgets of $\langle u^2 \rangle$ and $\langle u^4 \rangle$. That is, again, one can see the relation between the balance errors and the velocity moment order, the exponent of a velocity fluctuation, and the velocity fluctuation direction: odd moments, moments with larger

m and n, and with $m > n$ tend to have more uncertainty in their budgets in the buffer zone.

As time progresses, contribution of the turbulence production by the shear in the budgets of velocity moments reduces for all moments, but remains larger than that of the production by the strain [6]. For example, the production by the shear in the transport equations of $\langle u^2 \rangle$, $\langle u^3 \rangle$, and $\langle u^4 \rangle$ is more than 39, 72, and 34 times larger than the production by the strain, respectively, at $t_1^{'} = 0.02$, when the flow deformation has just started. At $t_7^{'}$, these numbers are: 4, 6, and 3.8, respectively.

The imposed flow deformation has different effect on the turbulence production of different velocity moments, with the moments $\langle u^m \rangle$ associated with the streamwise direction being affected the most. Figure 4 shows the turbulence production profiles at $t_7^{'}$. From the three moments: $\langle u^2 \rangle$, $\langle u^3 \rangle$, and $\langle u^4 \rangle$, the turbulence production in the transport equation of the odd moment $\langle u^3 \rangle$ is reduced the most, close to the two orders of the magnitude, when comparing the profile extremes at $t_0^{'}$ and $t_7^{'}$. This demonstrates a particular relation between this moment and the mean velocity gradient that reduces drastically near the wall when the flow separates. Extremes in the profiles of turbulence production by the shear for $\langle u^4 \rangle$ and $\langle u^2 \rangle$ are reduced by 70 and 30 times, respectively, by $t_7^{'}$ confirming that higher-order moments are more sensitive to variations in the mean velocity gradient than lower-order ones. When $n > 0$, the moments with $m > n$ are reduced more than the others of the same order $(m + n)$ pointing once again towards a relation between the mean velocity gradient in the wall-normal direction and velocity fluctuations in the streamwise direction.

The flow deformation affects not only the level of the turbulence production by the shear, but also the locations of the profile extremes. The largest extreme shifts from the buffer zone into Region II for all moments but $\langle u^3 \rangle$ and $\langle u^2 v \rangle$. For these two moments, the maximum turbulence production remains in the buffer zone with its location shifting from $y^+ = 8$ to 10 for $\langle u^3 \rangle$ and from $y^+ = 9$ to 16 for $\langle u^2 v \rangle$. Notice that the minimum value of the second derivative of the streamwise mean velocity also remains in the buffer zone under the flow deformation (Fig. 2d and f), with its location shifting from $y^+ = 8$ at $t_0^{'}$ to $y^+ = 13$ at $t_7^{'}$. This may explain the secondary peak in the production profiles for the velocity moments other than $\langle u^3 \rangle$ and $\langle u^2 v \rangle$, observable in Region I and this extreme location shifts away from the wall by $t_7^{'}$. Although the new location for this peak varies for different moments, it is within the interval $10 \le y^+ \le 20$ for all of them.

The balance errors reduce in the budgets of all velocity moments as time progresses, but less dramatically than it was observed for the turbulence production by the shear. What does not change with time is that when compared with the turbulence production, the balance errors are the highest in the budgets of the odd moments: $\langle u^3 \rangle$, $\langle uv^2 \rangle$, and $\langle u^2 v \rangle$. The errors also remain high and systematic in the buffer zone for all moments at considered times. When at later time, the mean velocity gradient is reduced, the systematic nature of the balance errors in the buffer zone becomes obvious even for the even moments such as $\langle u^2 \rangle$, $\langle u^4 \rangle$, and $\langle uv^3 \rangle$ (Fig. 4c, a, and i), which was difficult to recognize in the unstrained channel.

To summarize the discussion in this sub-section, the data analysis demonstrates the existence of a relation between balance errors, turbulence production by the shear,

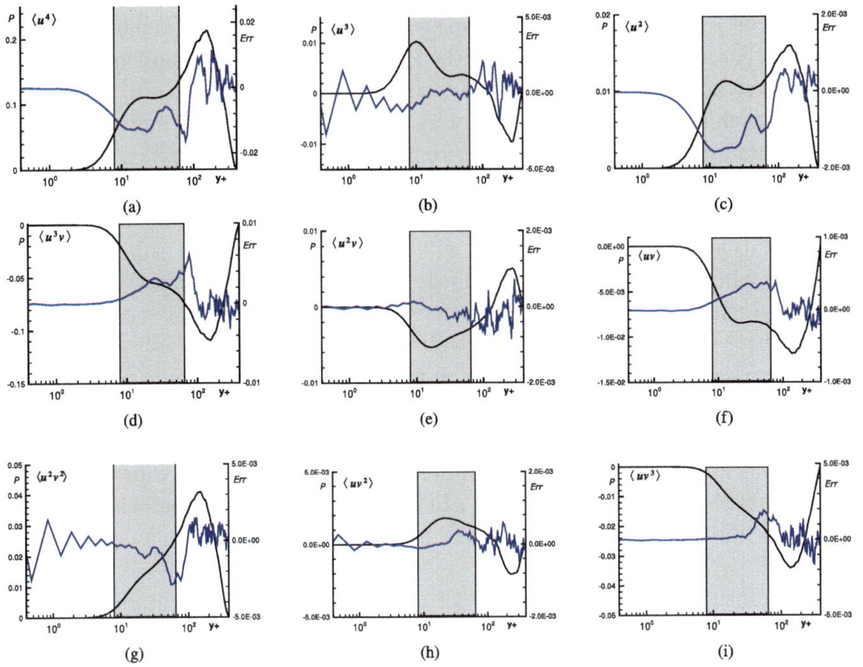

Fig. 4 Turbulence production by the shear (black lines) and the balance errors (blue lines) in the DNS budgets of velocity moments of various orders in the strained channel after the flow separates at $t = t_7^{'}$. Region I from Table 1 is shown by the grey rectangular shapes

and second derivative of the streamwise mean velocity in the wall-normal direction in the DNS budgets of velocity moments in the buffer zone of the considered flow at any time. In this flow area, the profiles of balance errors exhibit a presence of systematic contribution. Velocity moments with the highest level of balance errors to compare with the turbulence production are the odd-order ones and the even-order ones with odd exponents of velocity fluctuations. The higher the exponent of the velocity fluctuation in the streamwise direction in such moments, the higher the errors.

3.3 Probability Density Functions

Systematic contribution to the balance errors in the budgets of velocity moments should also affect PDFs of velocity fluctuations. In turbulence modeling, PDFs are used in various ways. Of interest for the current study are the PDF skewness and kurtosis used to determine the PDF deviation from the Gaussian one. The skewness and kurtosis of the Gaussian PDF have specific values: zero and three, respectively.

Inaccuracies in reproducing a PDF shape will affect these values and thus, modeling strategies based on the conclusions of the PDF proximity to the Gaussian shape.

The skewness, $S(u_i) = \langle u_i^3 \rangle / \langle u_i^2 \rangle^{3/2}$, and the kurtosis, $K(u_i) = \langle u_i^4 \rangle / \langle u_i^2 \rangle^2$, are normalized third- and fourth-order velocity moments, and thus, are more susceptible to high balance errors in the flow buffer zone than the Reynolds stresses based on the discussion in the sub-section above. This is why one may expect PDF to be deformed more in the buffer zone than outside, with deformation affecting the PDF maximum value and its location. PDFs associated with the velocity fluctuation in the streamwise direction will be affected more than those for the velocity fluctuation in the wall-normal direction. This is also based on the results from the previous sub-section and in [6], where velocity moments in the wall-normal direction are discussed at length. Since the balance errors are reduced with time, the PDF shape should improve with time.

All of that is demonstrated in Fig. 5, where transformation of PDFs for velocity fluctuations in the streamwise and wall-normal directions is shown with the distance from the wall at two times: $t_0^{'}$ and $t_7^{'}$. In the figure, three locations closest to the wall are within the buffer zone at both times, and the farthest one in the figures is in Region II.

In the unstrained channel (at $t_0^{'}$), PDF for the velocity component in the streamwise direction at $y^+ = 10$ (Fig. 5a) has unresolved area with no extremes where the maximum values are expected. Farther from the wall at $y^+ \geq 30$, the single maximum emerges. One may look at this as the flow physical feature at this location at the given time, but as demonstrated in Fig. 3, the largest balance errors are also occur at $y^+ < 30$ for the most of velocity moments including the velocity fluctuation in the streamwise direction. Until these errors are reduced, there is uncertainty about the actual PDF shape for this velocity fluctuation in the flow buffer zone.

As time progresses, changes in the shape of PDF of u are significant and particularly, in the buffer zone. At $t_7^{'}$, all PDFs have the single maximum. Reduction of the balance errors that occurs between $t_0^{'}$ and $t_7^{'}$ is a likely cause of that.

At both times shown in the figure, when PDF of u has the clear maximum, the PDF peak location appears to be affected by the streamwise mean velocity direction at the given y^+.

PDF for the velocity component in the wall-normal direction, v, exhibits the single maximum at any location from the wall (Fig. 5c) at both shown times. It also varies little in Region II at any time. As time progresses, changes in this PDF can mainly be seen in its maximum value in the close proximity to the wall ($y^+ = 10$ in the figure).

Overall, the presence of systematic balance errors in the flow buffer zone where the PDFs of velocity fluctuations transform the most, cannot be ignored. If collecting data for PDFs from DNS is not the option, collecting the odd moments, their budgets, and the balance errors, must be a mandatory requirement as a proof of the data quality. From all odd moments, the velocity moment $\langle u^3 \rangle$ in the streamwise direction appears to be the best indicator of simulation inaccuracies in the buffer zone of wall-bounded flows.

Notice that the data used in this study are well converged in statistical sense as discussed in [2, 6].

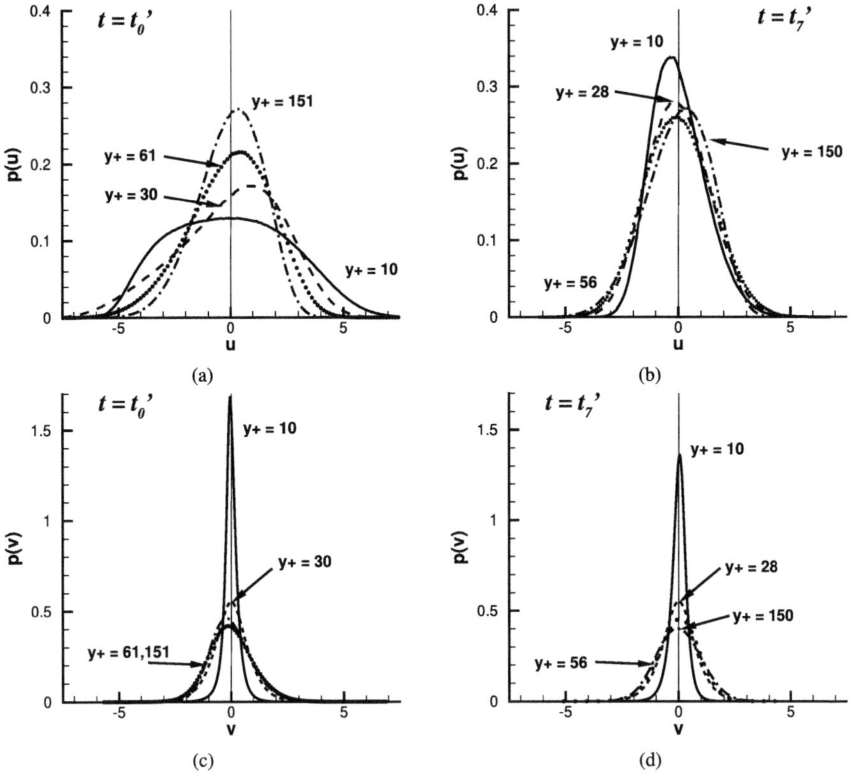

Fig. 5 Transformation of PDFs of velocity fluctuations u and v in the wall-normal direction and with time. Notations: —— $y^+ = 10$, - - - $y^+ = 28, 30$, ... $y^+ = 56, 61$, and _ . _ . _ $y^+ = 150, 151$

4 Conclusions

In the paper, statistical data from the direct numerical simulation of a strained channel flow are analyzed with the focus on the near-wall area and on the data quality for modelling purposes. Conclusions from *Poroseva & Murman (2021)* about the relation existing between the extremes in the profiles of the turbulence production by the shear and the balance errors in the DNS budgets of odd velocity moments are enhanced in the current study by demonstrating that this phenomenon is also observed in the budget of even velocity moments with odd exponents of velocity fluctuations. This is because all such moments are sensitive to the velocity fluctuations direction. It is shown that the balance errors are the highest to compare with the turbulence production in the budgets of these moments. The errors are higher for the moments of higher orders and for those with larger exponents of the velocity fluctuation in the streamwise direction.

The existence of a relation between the second derivative of the streamwise mean velocity in the wall-normal direction, the turbulence production by the shear, and the balance errors from the DNS budgets of velocity moments is established in the buffer zone of the considered flow at any time. The distance between the near-wall locations of the zeros in the profile of the second derivative of the streamwise mean velocity in the wall-normal direction zeros is proposed as a more rigorous definition of the buffer zone, independent of approximations of other flow regions in the streamwise mean velocity profile. In the considered flow, the buffer zone upper boundary is found to be at $y^+ < 65$ at any time. The lower boundary is proposed to be at the wall at any time.

The data analysis showed that profiles of the balance errors in the budgets of all considered velocity moments exhibit the systematic contribution in the buffer zone. This contribution is suggested as a possible cause of deformation of PDF of the velocity fluctuation in the streamwise direction observed in this flow area.

Collecting the odd moments (and particularly, the velocity moment $\langle u_i^3 \rangle$ in the streamwise direction), their budgets, and the balance errors, is proposed as the best practice requirement to demonstrate the DNS data quality.

Acknowledgements The author would like to thank Dr. Mohamed Abuhegazy, a recent graduate from the Department of Mechanical Engineering, University of New Mexico, for assistance with preparing the manuscript to comply with the production requirements.

References

1. Lee, M., Moser, R.D.: Direct numerical simulation of turbulent channel flow up to $Re_\tau \approx 5200$. J. Fluid Mech. **774**, 395–415 (2015)
2. Jeyapaul, E., Coleman, G.N., Rumsey, C.L.: Higher-order and length-scale statistics from DNS of a decelerated planar wall-bounded turbulent flow. Int. J. Heat Fluid Flow **54**, 14–27 (2015)
3. Sillero, J.A., Jiménez, J., Moser, R.D.: One-point statistics for turbulent wall-bounded flows at Reynolds numbers up to $\delta^+ \approx 2000$. Phys. Fluids **25**(10), 1–15 (2013)
4. Poroseva, S.V., Colmenares F, J.D., Murman, S.M.: On the accuracy of RANS simulations with DNS data. Phys. Fluids **28**(11), (2016)
5. Poroseva, S.V., Jeyapaul, E., Murman, S.M., Colmenares F, J.D.: The effect of the dns data averaging time on the accuracy of RANS-DNS simulations. AIAA, **2016-3940**, (2016)
6. Poroseva, S.V., Murman, S.M.: Analysis of high-order velocity moments in a strained channel flow. Int. J. Heat Fluid Flow **89**, (2021)
7. A. N. Kolmogorov: A The local structure of turbulence in incompressible viscous fluid for very large Reynolds numbers. Doklady Akad. Nauk SSSR, **30**(4), 299–303 (2021). English translation by V. Levin: Turbulence and stochastic process: Kolmogorov's ideas 50 Years on. Proc. R. Soc. Lond. A, **434**, 9–13 (1890)
8. Pope, S. B.: Turbulent flows, (Cambridge university press, 2000)

A Practical Approach for Computing Sensitivities in Chaotic Turbulent Flows

Krzysztof Fidkowski

Abstract We present a data-driven approach for calculating adjoint sensitivities in unsteady turbulent flows, with application to shape optimization and output-based adaptation. Such simulations are much more expensive than steady models, so that each function evaluation, usually a statistic such as a time-averaged output, requires orders of magnitude more computational time and resources than in steady-state. This cost impedes the application of unsteady simulations to many-query studies, such as optimization, particularly via gradient-free methods that rely on numerous function evaluations. Furthermore, the chaotic nature of unsteady turbulent simulations prevents the calculation of gradients via adjoint-based methods, which become unstable and require expensive regularization techniques to provide meaningful answers. Our approach does not use unsteady adjoint equations but instead relies on unsteady data to train a corrected turbulence model, which then yields the required adjoint solutions. It is non-intrusive and inexpensive, requiring only a small number of unsteady forward simulations, but sufficiently powerful to capture unsteady effects in the sensitivities. Results for high-order discretizations of the unsteady Navier-Stokes equations, augmented by a corrected Spalart-Allmaras turbulence closure with and without a transition model, demonstrate the ability of the approach to accurately compute sensitivities in problems with unsteady turbulence.

1 Introduction

Advances in computational power have enabled scale-resolving $\tilde{\omega}$ simulations of turbulent flow, yet the high cost of these simulations prohibits their use in a multi-query setting, such as design optimization. In addition, the chaotic nature of unsteady turbulence prevents the direct application of adjoint methods to computing the local sensitivities that are required for efficient optimization, uncertainty quantification, and mesh adaptation.

CUFS-2024-PAPER ID, Cambridge, United Kingdom, 4-5 March 2024

K. Fidkowski (✉)
University of Michigan, Ann Arbor, USA
e-mail: kfid@umich.edu

J. C. Tyacke and N. R. Vadlamani (eds.), *Proceedings of the Cambridge Unsteady Flow Symposium 2024*, https://doi.org/10.1007/978-3-031-69035-8_5

A much less expensive turbulence modeling technique is Reynolds-averaging of the Navier-Stokes (RANS) equations, and this approach remains of high value in many practical simulations. Reynolds averaging regularizes the chaotic nature of turbulent flow and enables efficient adjoint-based sensitivities. However, RANS models suffer from lower accuracy due to the problem of closure, wherein the effect of unresolved scales on the relatively coarse resolved ones must be modeled.

Thus far, scale-resolving simulations and RANS have been treated largely separately. When sensitivities are required in design, particularly in preliminary design where model accuracy is perhaps not critical, RANS is the method of choice, and the turnaround of analyses is quick. On the other hand, when high accuracy is needed or when off-design performance must be assessed, scale-resolving methods become necessary. These higher-fidelity analyses may identify new physical phenomena, but presently it is not possible to close the design loop due to the lack of efficient sensitivity calculations in scale-resolving simulations.

This work addresses the problem of high-fidelity sensitivity calculations in turbulent flow by combining RANS and scale-resolving approaches. It reviews and extends our previous work, in which we have shown how RANS can be augmented with a correction, here called a dynamic closure, driven by unsteady simulations and implemented in a data-driven model [1, 2]. The dynamic closure allows for stable adjoint-based sensitivities, which are calculated from the RANS model. It also keeps the costs of optimization and adaptation simulations low because multiple iterations of optimization/adaptation are done using RANS, whereas unsteady simulations are only needed a few times to rebase the correction.

Much of our previous work has focused on fully-turbulent flow, where RANS models without transition are suitable for correction. However, for moderate Reynolds numbers in the transitional regime, fully-turbulent RANS models become less suitable, requiring larger corrections that cannot always be accurately modeled from the state. Hence, in this work, we extend the dynamic correction approach to RANS with a transition model.

2 Models and Discretization

We consider the Reynolds-averaged form of the compressible Navier-Stokes equations, discretized using a discontinuous finite-element method. This section describes the specific models and numerical method used.

2.1 *Baseline Turbulence Model*

The one-equation Spalart-Allmaras (SA) model [3, 4] serves as the baseline RANS closure for our work. The RANS-SA equations written in index notation with implied summation are

$$
\begin{aligned}
&\partial_t \rho \quad + \partial_j(\rho u_j) && = 0 \\
&\partial_t(\rho u_i) + \partial_j(\rho u_j u_i + p\delta_{ij}) - \partial_j \tau_{ij} && = 0 \\
&\partial_t(\rho E) + \partial_j(\rho u_j H) \qquad - \partial_j(u_i \tau_{ij} - q_j) && = 0 \\
&\partial_t(\rho \tilde{\nu}) + \partial_j(\rho u_j \tilde{\nu}) \qquad - \partial_j\left[\tfrac{1}{\sigma}\rho(\nu + \tilde{\nu} f_n)\partial_j \tilde{\nu}\right] && = S_{\tilde{\nu}}
\end{aligned}
\tag{1}
$$

where the source term for the additional turbulence model $\tilde{\nu}$ equation is

$$
S_{\tilde{\nu}} = \tilde{\gamma}\beta P - D - \frac{1}{\sigma}(\nu + \tilde{\nu} f_n)\partial_j \rho \partial_j \tilde{\nu} + \frac{c_{b2}\rho}{\sigma}\partial_j \tilde{\nu}\partial_j \tilde{\nu}. \tag{2}
$$

ρ is the density, ρu_j is the momentum, E is the total energy, $H = E + p/\rho$ is the total enthalpy, $p = (\gamma - 1)\left(\rho E - \frac{1}{2}\rho u_i u_i\right)$ is the pressure, γ is the ratio of specific heats, P is the turbulence production, D is the turbulence destruction, $\tilde{\gamma}$ is an intermittency function, β is a production correction term, and i, j index the spatial dimension. Details of the production/destruction terms, viscous stresses, and closure coefficients can be found in previous work [5]. In particular, the eddy viscosity $\mu_t = \rho \nu_t$ is computed as $\mu_t = \rho \tilde{\nu} f_{v1}$ when $\tilde{\nu} > 0$, where $f_{v1} = \chi^3/(\chi^3 + c_{v1}^3)$ and $\chi = \tilde{\nu}/\nu$.

2.2 *Transition Model*

The intermittency function $\tilde{\gamma}$ in Eq. (2) attenuates turbulent production to maintain laminar flow until transition to turbulence. Various methods exist for modeling $\tilde{\gamma}$, many relying on additional equations. Presently, we use an algebraic model, BCM [6, 7], named after the initial authors. This model is heavily empirical and contains tunable factors, but it is inexpensive and simple to implement. The results will show that it is suitable for predicting sensitivities after correction.

The intermittency function of the BCM model is

$$
\tilde{\gamma} = 1 - \exp\left(-\sqrt{T_1} - \sqrt{T_2}\right), \tag{3}
$$

where

$$
T_1 = \frac{\max(Re_\theta - Re_{\theta c}, 0)}{c_{t1} Re_{\theta c}}, \qquad T_2 = \max\left(\frac{\nu_t}{c_{t2}\nu}, 0\right) \tag{4}
$$

The momentum thickness Reynolds number is correlated to the vorticity Reynolds number, Re_S through [7],

$$
Re_\theta = \frac{Re_S}{2.193}, \qquad Re_S = \frac{d^2 S}{\nu}, \tag{5}
$$

and hence can be measured locally without any wall-normal integration. The critical momentum thickness Reynolds number is based on the empirical formula

$$Re_{\theta c} = 803.73\,(Tu_\infty + 0.6067)^{-1.027}\,, \tag{6}$$

where Tu_∞ is the freestream turbulence level that need not correlate directly to $\tilde{\nu}_\infty$ [8].

The closure coefficients used in this work are

$$c_{t1} = 0.002,\;\; c_{t2} = 50,\;\; Tu_\infty = .01. \tag{7}$$

Finally, β that appears in Eq. (2) is the dynamic-closure production correction factor that is applied independently of the intermittency.

2.3 Finite-Element Discretization

We use a discontinuous Galerkin (DG) finite-element spatial discretization [9], with the Roe [10] convective flux and the second form of Bassi and Rebay (BR2) [11] for the viscous treatment. The state is approximated on an unstructured mesh of non-overlapping elements using polynomials of order p. The semi-discretized form of the equations is

$$\mathbf{M}\frac{d\mathbf{U}}{dt} + \mathbf{R}(\mathbf{U}) = \mathbf{0}, \tag{8}$$

where $\mathbf{U} \in \mathbb{R}^N$ is the discrete state vector, N is the total number of unknowns including the state rank, $\mathbf{R}(\cdot) \in \mathbb{R}^{N\times N}$ is the nonlinear spatial residual, and $\mathbf{M} \in \mathbb{R}^{N\times N}$ is the block-element sparse mass matrix. For steady simulations, the time derivative term drops out, but pseudo-time continuation is used to drive the steady residual to zero [12]. The solver consists of a Newton-Raphson method with the generalized minimum residual (GMRES) [13] linear solver, preconditioned by an element-line Jacobi smoother with a coarse-level ($p = 1$) correction [14, 15]. For unsteady simulations, we use a third-order modified extended backward difference formula [16] applied to the the semi-discrete form.

Efficiently solving for the correction field requires an adjoint, which here is calculated in discrete form. For a scalar output $J(\mathbf{U})$, the discrete steady adjoint vector, $\mathbf{\Psi} \in \mathbb{R}^N$, is the local sensitivity of J to perturbations in the steady residual, $\mathbf{R}$ [17]. Linearization of the residual and output shows that the adjoint satisfies the following linear equation,

$$\left(\frac{\partial \mathbf{R}}{\partial \mathbf{U}}\right)^T \mathbf{\Psi} + \left(\frac{\partial J}{\partial \mathbf{U}}\right)^T = \mathbf{0}. \tag{9}$$

This equation is solved using the same preconditioned-GMRES method used in the primal solver.

The adjoint vector can be used to efficiently compute sensitivities of the output with respect to shape or operational parameters, $\mathbf{x}$, of the problem,

$$\frac{dJ}{d\mathbf{x}} = \frac{\partial J}{\partial \mathbf{x}} + \mathbf{\Psi}^T \frac{\partial \mathbf{R}}{\partial \mathbf{x}}. \tag{10}$$

3 Correction Field Inversion and Modeling

We perform field inversion to obtain the correction field $\beta(\vec{x})$ that appears in Eq. (2). The objective for the field inversion arises from matching higher-fidelity data, experimental or numerical, in the form of integrated scalars, surface distributions, or entire flow-fields.

3.1 *High-Fidelity Data*

We obtain high-fidelity numerical data through unsteady, scale-resolving, simulations, which are averaged for training steady RANS models [1, 2]. Let $\bar{\mathbf{U}}$ be a statistically-steady flow state computed from the time-average of the unsteady simulation after initial transients,

$$\bar{\mathbf{U}} = \frac{1}{T_f - T_i} \int_{T_i}^{T_f} \mathbf{U}(t)\, dt, \tag{11}$$

where T_i is the start time, taken sufficiently large to minimize the effect of the transients, and T_f is the final time, taken sufficiently large to yield an adequate statistical mean. The goal of field inversion is to determine the $\beta(\vec{x})$ that makes the corrected RANS solution match $\bar{\mathbf{U}}$ or a functional of it.

The field-inversion error, $\mathcal{F}$, measures the accuracy of this match, and presently we consider both scalar and distribution measures,

$$\mathcal{F}^{\text{output}}(\mathbf{U}(\beta), \bar{\mathbf{U}}) = \frac{1}{2}(J(\mathbf{U}(\beta)) - J(\bar{\mathbf{U}}))^2, \tag{12}$$

$$\mathcal{F}^{\text{dist}}(\mathbf{U}(\beta), \bar{\mathbf{U}}) = \frac{1}{2} \int_{\text{airfoil}} \|\boldsymbol{\sigma}(\mathbf{U}(\beta)) \cdot \vec{n} - \boldsymbol{\sigma}(\bar{\mathbf{U}}) \cdot \vec{n}\|^2\, ds, \tag{13}$$

where $J(\mathbf{U})$ is a scalar output, such as a drag coefficient, $\boldsymbol{\sigma}$ is the stress tensor, and $\vec{n}$ is the unit normal vector on the surface, which is an airfoil in the problems considered here. Instead of the time-averaged unsteady solution, experimental data can also be used in the above error measures.

3.2 Field Inversion

We represent the correction field using a $p = 1$ Lagrange DG basis on each element, with a nominal value of 1. Denoting by $\boldsymbol{\beta}$ the vector of basis coefficients in the approximation, the field inversion becomes a discrete optimization problem,

$$\begin{aligned} \min_{\boldsymbol{\beta}} \quad & J^{\text{inv}} \equiv \mathcal{F}(\mathbf{U}(\boldsymbol{\beta}), \bar{\mathbf{U}}) + \frac{\gamma^{\beta}}{2}(\boldsymbol{\beta} - 1)^T \mathbf{M}(\boldsymbol{\beta} - 1) \\ \text{s.t.} \quad & \mathbf{R}(\mathbf{U}, \boldsymbol{\beta}) = \mathbf{0} \end{aligned} \tag{14}$$

where $\mathcal{F}$ is either the output or distribution measure, and the additional term in the objective is a continuous Tikhonov regularization that makes the inverse problem well-posed by penalizing large deviations of the correction field from the nominal value of one. $\mathbf{M}$ is the mass matrix from the spatial discretization, and γ^{β} is a small user-prescribed parameter set to 10^{-5}, although a wide range of values was found to yield nearly identical results.

The field inversion problem is solved using 20 iterations of a limited-memory Broyden-Fletcher-Goldfarb-Shanno (L-BFGS) algorithm [18], with a history of 10 updates for approximating the inverse Hessian. Each iteration amounts to a steady RANS solution, which is not expensive as the solution from each iteration provides the initial guess for the state at the next iteration. An adjoint is used to evaluate the objective gradient, and the residual linearization with respect to $\boldsymbol{\beta}$ is calculated efficiently using finite differences and only a few residual evaluations [1].

3.3 Machine-Learning Model

The correction field obtained from field inversion is static and specific to the geometry and flow conditions used in the inversion. We write it as $\beta(\vec{x})$ to denote that it only depends on the spatial coordinates. In applications such as design optimization and mesh adaptation, we are interested in changes to the geometry and discretization, and we therefore seek a model for β in terms of local flow information, similarly to other terms in the turbulence model.

We model the mapping between local state information and the correction field using an artificial neural network [1, 19–22]. The inputs are the local state ($\mathbf{u}$), its gradient ($\nabla\mathbf{u}$), and the wall distance function (d). The lack of scale invariance in the inputs is not problematic as we do not seek a general-purpose model; instead, we retrain the model as the optimization or adaptation progresses. The utility of the model comes from its ability to yield accurate gradients, a property that we currently assess for transitional flows.

The neural network used in this work consists of a single hidden layer of 30 neurons and a sigmoid activation function between the input and hidden layer. The adaptive moment (Adam) algorithm in TensorFlow [23] yields the weights and biases. The true β values for training come from the field inversion correction, sampled at element quadrature points, for one simulation. Once a network is trained, it is implemented as a physics model in the RANS turbulence source calculation. No changes to the inputs to this calculation are required, since the turbulent source already uses the state, its gradient, and the wall distance function. The network is differentiated with respect to the state and its gradient for the Newton solver and the discrete adjoint solver. Due to the nonlinear nature of the network, higher-order accurate quadrature rules are used for the corrected RANS model: $2p + 3$ for solution order p.

4 Results

This section presents two sensitivity studies using the proposed dynamic closure method, one for a low Reynolds number flow and one for transitional flow. Discrete solutions are obtained using $p = 2$ spatial approximation. A third-order modified backwards difference scheme (MEBDF) [16] is used to advance the unsteady cases in time, using a time step of $\Delta t = .05c/U_\infty$, where U_∞ is the free-stream speed. No turbulence model is used in the unsteady runs. After a transient, a time horizon of $T_f - T_i = 30c/U_\infty$ is used to calculate the average state according to Eq. (11). The problems are two-dimensional but share features of more complex, and expensive, three-dimensional turbulent simulations.

4.1 Camber Sensitivity at $Re = 10^4$

To demonstrate the sensitivity calculation, we consider a family NACA 4-digit airfoils, X412, at $M = 0.2$, $Re = 10^4$, and angle of attack $\alpha = 0$. The mesh consists of 1978 $q = 3$ quadrilateral elements. At this Reynolds-number, the flow exhibits unsteady behavior that prevents the use of direct unsteady adjoints for time-averaged outputs [24].

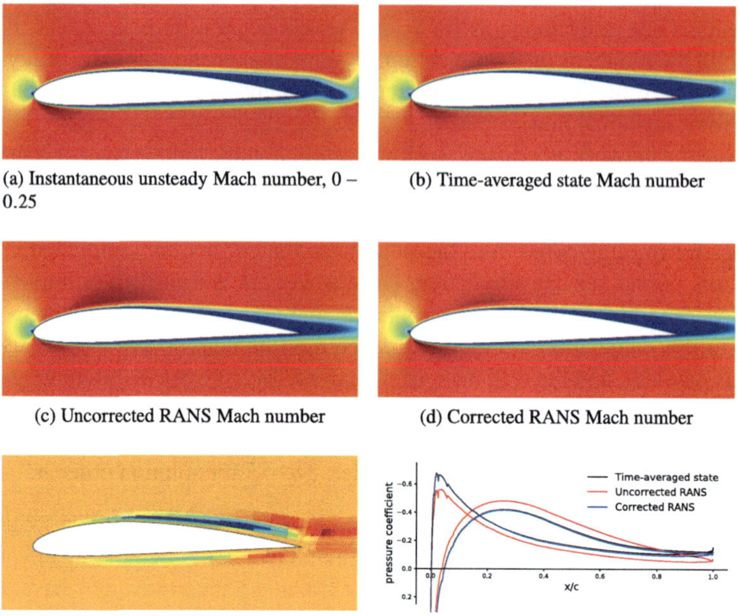

(a) Instantaneous unsteady Mach number, 0 – 0.25

(b) Time-averaged state Mach number

(c) Uncorrected RANS Mach number

(d) Corrected RANS Mach number

(e) Correction field, $\beta(\vec{x})$, 0.3 – 1.3

(f) Surface pressure coefficient

Fig. 1 Field-inversion results for unsteady flow around a NACA 3412 airfoil at $M = 0.2$, $Re = 10^4$, $\alpha = 0$

Figure 1 compares an instantaneous snapshot and various time-averaged solutions for one airfoil in the family, a NACA 3412. The baseline (uncorrected) RANS solution exhibits a smaller wake and a different pressure distribution compared to the averaged unsteady result. With a correction to the production term, the RANS model matches the averaged unsteady result well. Figure 1 shows that the Mach number profile and pressure distribution of the corrected RANS result are nearly identical to the time-averaged unsteady result. The field inversion in this case is done using the stress distribution error function, Eq. (13). The $\beta(\vec{x})$ plot indicates that reduced turbulent production on the boundary layer edges, particularly over the upper surface, and increased production towards the trailing edge and wake are required to match the averaged unsteady stress distribution.

The correction field, $\beta(\vec{x})$, gives physical insight into where turbulent production needs to be enhanced or suppressed in order to match a higher-fidelity result. Its existence derives from the availability of the higher-fidelity data, in this case, a time-averaged unsteady flowfield. With such data available, the corrected RANS simulation does not appear to add much value, since the unsteady data could be used instead. Furthermore, when the mesh or geometry changes, the correction field must be re-mapped, possibly in spatial regions where the correction did not exist. There is also the question of whether a static correction field is appropriate for sensitivity

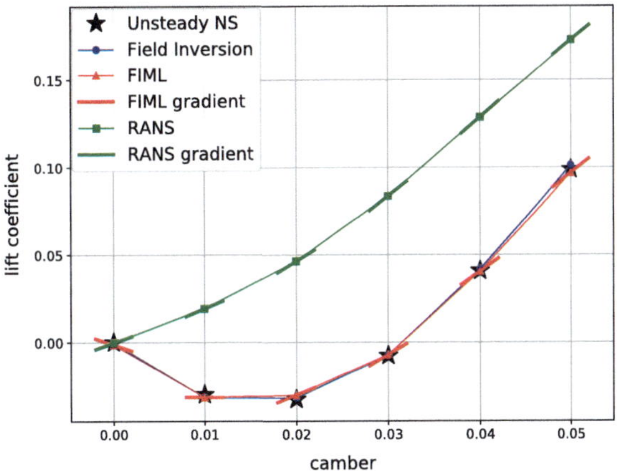

Fig. 2 Field-inversion and machine learning results for lift-coefficient calculations in unsteady flow around NACA X412 airfoils at $M = 0.2$, $Re = 10^4$, $\alpha = 0$

calculations, or whether it should change when the design or operating conditions change. These are the questions that we address in this example by studying the neural network model of β.

To demonstrate the ability of the neural-network model to produce an accurate correction field, we now consider multiple airfoils of varying camber. For six airfoils of cambers 0 to 5% of the chord, unsteady simulations, field inversions, and neural network trainings are separately performed. The output of interest is the lift coefficient and its sensitivity to camber changes.

Figure 2 shows the lift-coefficient results of the various simulations. The uncorrected RANS results show an increasing lift coefficient with camber, and adjoint-based gradients that match the trend from the multiple simulations. However, this trend disagrees with the time-averaged unsteady results, which show an initial decrease in the averaged lift coefficient for small cambers, followed by an increase. The result is not surprising, as the RANS model is not designed for such low Reynolds numbers.

Applying field inversion to each case and using the resulting correction field $\beta(\vec{x})$ in the turbulence model yields data that track the time-averaged results well. More importantly, the neural-network model for $\beta(\vec{x})$ also performs very well, as indicated by the "FIML" data points in Fig. 2. These data correspond to simulations performed by using the neural network model and not the inverted $\beta(\vec{x})$ field, which was used only for training. Furthermore, when differentiated as part of the adjoint solver, the FIML result yields a reasonably-accurate PDE-constrained gradient of the lift coefficient with respect to the camber. The ability to accurately predict the lift coefficient value and its gradient with respect to shape parameters is important when using the FIML correction model in optimization.

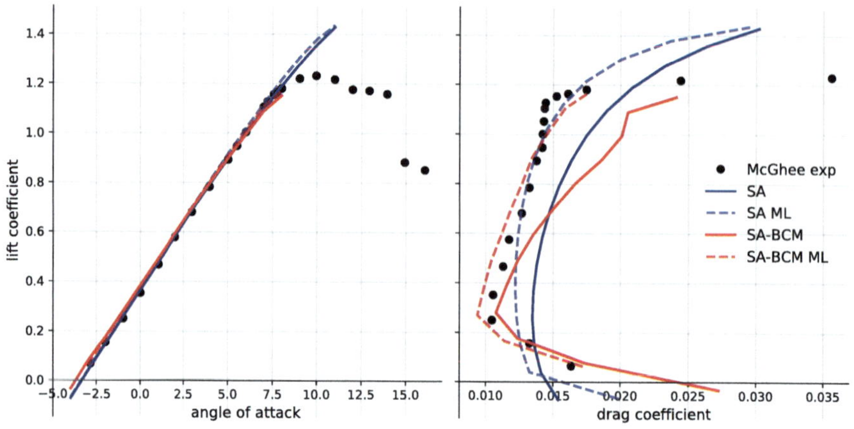

Fig. 3 Eppler airfoil lift and drag polar at $Re = 200{,}000$

4.2 *Transitional Flow*

We apply the SA-BCM model in an FIML setting for an Eppler 387 airfoil, for which experimental data are available at various Reynolds numbers [25]. Both experimental and unsteady numerical data are used for the field inversion objective. We use a Mach number of 0.1, a free-stream SA working variable of $\tilde{\nu}_\infty = 3\nu$ for the fully-turbulent SA model, and $\tilde{\nu}_\infty = 0.1\nu$ for the SA-BCM model [8]. The computational mesh consists of 3584 curved $Q = 3$ quadrilateral elements and extends approximately 100 chord lengths away from the airfoil.

4.2.1 Eppler Airfoil, $Re = 200\,k$

Running the baseline RANS model (SA) and the transitional model (SA-BCM) at different angles of attack for $Re = 200k$ produces the lift and drag polars shown in Fig. 3. We focus on moderate angles of attack, but not stall, in which neither model performs well. The results indicate that both models are accurate at predicting the lift, which is not overly difficult when the flow is attached. On the other hand, they generally over-predict the drag, albeit for different reasons: the fully-turbulent SA model over-predicts skin friction drag as it misses the laminar flow before transition, while the SA-BCM model over-predicts pressure drag due to the flow not transitioning to turbulence sufficiently quickly, resulting in more separated flow.

For both the SA and SA-BCM models, we run field inversion at $\alpha = 6°$, using the experimental drag coefficient as the objective. We then generate machine-learning correction models, SA ML and SA-BCM ML, using the baseline network and use them in the same angle of attack study. As shown in Fig. 3, the lift coefficient does not change much, whereas the drag coefficient decreases as expected given the lower

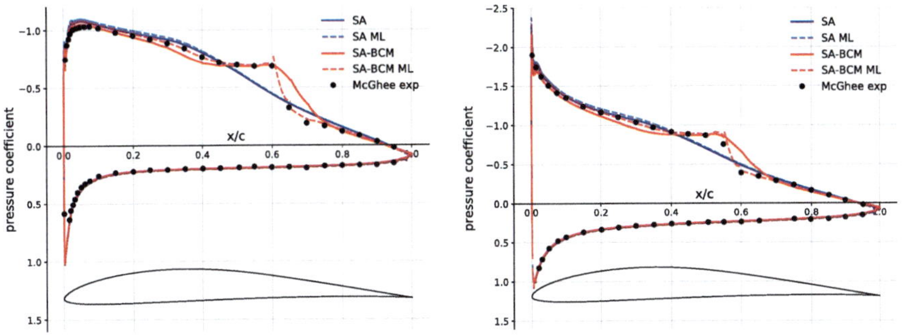

(a) $\alpha = 4°$ (b) $\alpha = 6°$

Fig. 4 Eppler airfoil pressure coefficient distributions at $Re = 200{,}000$. The corrected results all use the ML model obtained from field inversion at $\alpha = 6°$

objective drag for the field inversion. Both models predict the experimental/target c_d at $\alpha = 6°$ well. At larger angles of attack, neither model predicts the drag dip, which is caused in the experiments by a decrease in pressure drag due to the disappearance of the laminar separation bubble (LSB). However, at lower angles of attack, the SA-BCM results follow the experimental drag polar better, in value and particularly the slope, which is of interest in sensitivity calculations, than SA alone.

To further compare the results, we look at the pressure coefficient distributions, shown in Fig. 4 for $\alpha = 4°, 6°$. The SA (fully turbulent) model does not predict transition, and its correction barely changes the c_p – it does affect the skin friction, however. The SA-BCM model without correction transitions to turbulence slightly too late, and too slowly at both of the angles of attack. Correction moves this transition earlier, virtually on top of the experimental data, through higher β values near the transition location, and this results in lower pressure drag values for SA-BCM ML. We observe improvement at $\alpha = 4°$ even though the inversion and ML training were performed at $\alpha = 6°$, and this illustrates that the correction remains valid locally, which is of interest for accurate sensitivity predictions.

The skin friction results in Fig. 5 indicate a large drop in c_f for SA ML compared to SA alone, caused by suppression of turbulent production (low β) in what is supposed to be the laminar region. The resulting profiles still differ from the transition model, as they are without an LSB or transition. Only the average c_f, together with the pressure distribution, are consistent with the drag objective, as enforcing matching of an integrated quantity does not guarantee matching of the distribution. For SA-BCM ML, the shift in the transition location on the upper surface is apparent and consistent with the c_p result.

Finally, Fig. 6 shows the Mach number and β field plots. The fully-turbulent SA Mach number contours indicate a thin boundary layer without transition. The correction in SA ML reduces the skin friction through low β values, particularly on the front upper portion of the airfoil. The effect on the boundary layer thickness is minimal, and hence the negligible change observed in the c_p distribution. On the

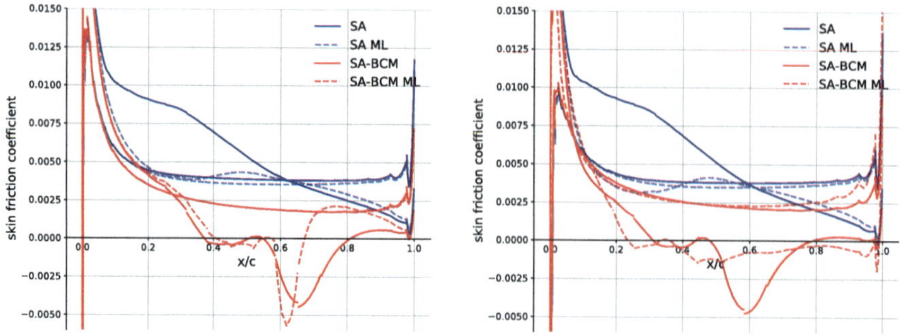

(a) $\alpha = 4°$ (b) $\alpha = 6°$

Fig. 5 Eppler airfoil skin friction coefficient distributions at $Re = 200{,}000$. The corrected results all use the ML model obtained from field inversion at $\alpha = 6°$

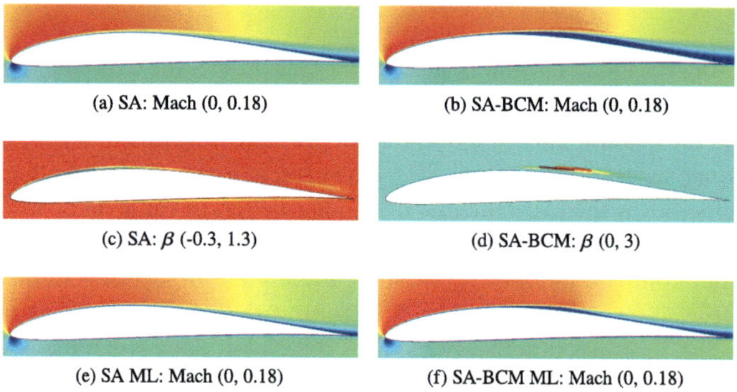

Fig. 6 Eppler airfoil Mach and correction-field contours at $Re = 200{,}000$ and $\alpha = 6°$

other hand, the SA-BCM boundary layer is thicker after the laminar separation. The correction consists of a high β at transition to accelerate the reattachment. This leads to a smaller LSB in SA-BCM ML, and hence less pressure drag.

4.2.2 Eppler Airfoil, $Re = 60\ k$

As a final test case, we consider the same Eppler 387 airfoil at a lower Reynolds number, $Re = 60k$, and one angle of attack, $\alpha = 8°$. The flow is now more likely to separate earlier, and transition is important in keeping the flow attached. In addition to experimental data [25], we also generate higher-fidelity numerical data by running an unsteady two-dimensional simulation using the Navier–Stokes equations without any additional models. The pressure distribution from the time-average of the unsteady state agrees well with the experimental data, as shown in Fig. 7, suggesting that the dominant transition mechanism is well-approximated by 2D assumptions. Figure 7 also shows the results of the SA and SA-BCM models: SA predicts the average

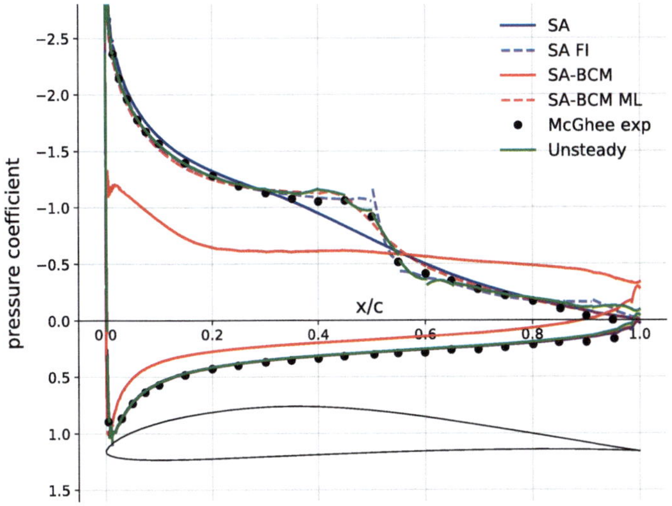

Fig. 7 Eppler airfoil pressure coefficient distributions at $Re = 60{,}000$, $\alpha = 8°$. Note, FI indicates field inversion, and ML indicates machine learning

pressure distribution well but misses the LSB and transition, as it assumes fully-turbulent flow. SA-BCM does quite poorly: the flow separates in the laminar portion and never re-attaches.

Figure 8 shows the Mach number and correction factor field plots for the various simulations. The unsteady plots illustrate the breakdown of the laminar flow and transition at approximately the half-chord location. The SA boundary layer remains thin due to the fully-turbulent assumption, whereas the SA-BCM flowfield exhibits massive separation on the upper surface due to late transition.

In this case, due to the availability of the time-averaged unsteady flowfield data, the airfoil surface stress distribution is used for the field inversion objective. The pressure distributions for the field inversion results are nearly identical to the unsteady data, which was the target. For SA-BCM, we observe a massive difference in the flowfield, as the correction causes reattachment and a large change in the pressure distribution. Note the large values of β near the transition location that are responsible for the earlier/faster transition and associated re-attachment. For SA, the change in the flowfield is also noticeable, as the LSB appears. This is caused by increased turbulent production (higher β) at the desired transition location.

Whereas field inversion produces good results with appropriate corrections for both SA and SA-BCM, the machine-learning model is only successful in the SA-BCM case. For SA, the model cannot learn the correction from local state information. The difficulty lies in predicting the sudden low-to-high swing in β at transition, where the state and wall distance do not change much over a small chord-wise distance. For SA-BCM, the β values are much smaller and possible to learn from local information. The machine learning loss function history confirms this, with the

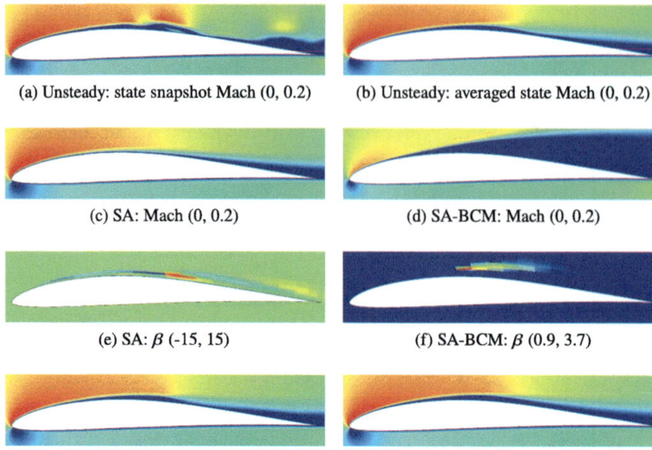

(a) Unsteady: state snapshot Mach (0, 0.2) (b) Unsteady: averaged state Mach (0, 0.2)

(c) SA: Mach (0, 0.2) (d) SA-BCM: Mach (0, 0.2)

(e) SA: β (-15, 15) (f) SA-BCM: β (0.9, 3.7)

(g) SA FI: Mach (0, 0.2) (h) SA-BCM ML: Mach (0, 0.2)

Fig. 8 Eppler airfoil Mach and correction-field contours at $Re = 60{,}000$ and $\alpha = 8^\circ$

loss function decreasing less than an order of magnitude in the SA case, compared to four orders of magnitude in the SA-BCM case. A caveat for the SA-BCM model is that the result depends on the parameters of the model, in particular c_{t2} in Eq. (7), which was set to 1 for these runs. The baseline value of 50 causes delayed transition and requires much larger swings in β values that the ML model cannot learn. We are currently working on a combined field and parameter inversion method to address this sensitivity.

5 Conclusions

This paper presents a method for computing sensitivities in unsteady turbulent flow using dynamic corrections obtained from field inversion and machine learning. This method uses adjoints of steady forward runs that contain dynamic models. The truth data for the corrections can come from experiments or higher-fidelity data, such as unsteady scale-resolving simulations, for which adjoints can be impractical to compute. The method consists of two steps: first, field inversion yields the correction as a function of domain position; second, a neural network learns this correction as a function of local state, gradient, and wall distance. The resulting dynamically-corrected model can be linearized to compute adjoints and gradients. The results show that the gradients in this manner are accurate when compared to unsteady data. The present paper also extends our previous work in this area to transitional flows, using an algebraic transition model, SA-BCM. Results for the Eppler 387 airfoil indicate that the corrected SA-BCM results are more accurate than the corrected SA results when inverting on an integrated quantity, such as the drag coefficient, and that they yield a more readily learned correction field when inverting on distribution data. Additional work remains to mitigate sensitivity to parameters of the models, in particular BCM.

References

1. Fidkowski, K.J.: Gradient-based shape optimization for unsteady turbulent simulations using field inversion and machine learning. Aerosp. Sci. Technol. **129**, 107843 (2022). https://doi.org/10.1016/j.ast.2022.107843
2. Fidkowski, K.J.: Output-based error estimation and mesh adaptation for unsteady turbulent flow simulations. Comput. Method. Appl. Mech. Eng. **399**, 115322 (2022). https://doi.org/10.1016/j.cma.2022.115322
3. Spalart, P.R., Allmaras, S.R.: A one-equation turbulence model for aerodynamic flows. La Recherche Aérospatiale (1), 5–21 (1994)
4. Allmaras, S.R., Johnson, F.T., Spalart, P.R.: Modifications and clarifications for the implementation of the Spalart-Allmaras turbulence model. Seventh international conference on computational fluid dynamics (ICCFD7) **1902**, (2012)
5. Fidkowski, K.J.: Three-dimensional benchmark RANS computations using discontinuous finite elements on solution-adapted meshes. AIAA Paper 2018–1104 (2018). https://doi.org/10.2514/6.2018-1104
6. Cakmakcioglu, S.C., Bas, O., Kaynak, U.: A correlation-based algebraic transition model. J. Mech. Eng. Sci. **232**(21), 3915–3929 (2018). https://doi.org/10.1177/0954406217743537
7. Cakmakcioglu, S.C., Bas, O., Mura, R., Kaynak, U.: A revised one-equation transitional model for external aerodynamics. AIAA Paper 2020-2706 (2009). https://doi.org/10.2514/6.2020-2706
8. Crivellini, A., Ghidoni, A., Noventa, G.: Algebraic modifications of the $k - \tilde{\omega}$ and Spalart-Allmaras turbulence models to predict bypass and separation-induced transition. Comput. Fluid **253**, 105791 (2023). https://doi.org/10.1016/j.compfluid.2023.105791
9. Fidkowski, K.J.: Output error estimation strategies for discontinuous Galerkin discretizations of unsteady convection-dominated flows. Int. J. Numer. Method. Eng. **88**(12), 1297–1322 (2011). https://doi.org/10.1002/nme.3224
10. Roe, P.L.: Approximate Riemann solvers, parameter vectors, and difference schemes. J. Comput. Phys. **43**, 357–372 (1981). https://doi.org/10.1016/0021-9991(81)90128-5
11. Bassi, F., Rebay, S.: Numerical evaluation of two discontinuous Galerkin methods for the compressible Navier-Stokes equations. Int. J. Numer. Method. Fluid **40**, 197–207 (2002). https://doi.org/10.1002/fld.338
12. Ceze, M.A., Fidkowski, K.J.: Constrained pseudo-transient continuation. Int. J. Numer. Method. Eng. **102**, 1683–1703 (2015). https://doi.org/10.1002/nme.4858
13. Saad, Y., Schultz, M.H.: GMRES: A generalized minimal residual algorithm for solving nonsymmetric linear systems. SIAM J. Sci. Comput. **7**(3), 856–869 (1986). https://doi.org/10.1137/0907058
14. Fidkowski, K.J., Oliver, T.A., Lu, J., Darmofal, D.L.: p-Multigrid solution of high-order discontinuous Galerkin discretizations of the compressible Navier-Stokes equations. J. Comput. Phys. **207**, 92–113 (2005). https://doi.org/10.1016/j.jcp.2005.01.005
15. Persson, P.-O., Peraire, J.: Newton-GMRES preconditioning for discontinuous Galerkin discretizations of the Navier-Stokes equations. SIAM J. Sci. Comput. **30**(6), 2709–2733 (2008). https://doi.org/10.1137/070692108
16. Cash, J.R.: The integration of stiff initial value problems in ODEs using modified extended backward differentiation formulae. Comput. Math. Appl. **9**(5), 645–657 (1983). https://doi.org/10.1016/0898-1221(83)90122-0

17. Fidkowski, K.J., Darmofal, D.L.: Review of output-based error estimation and mesh adaptation in computational fluid dynamics. AIAA J. **49**(4), 673–694 (2011). https://doi.org/10.2514/1.J050073
18. Liu, D.C., Nocedal, J.: On the limited memory BFGS method for large scale optimization. Math. Prog. **45**, 503–528 (1989). https://doi.org/10.1007/BF01589116
19. Singh, A.P., Medida, S., Duraisamy, K.: Machine-learning-augmented predictive modeling of turbulent separated flows over airfoils. AIAA J. **55**(7), 2215–2227 (2017). https://doi.org/10.2514/1.j055595
20. Holland, J.R., Baeder, J.D., Duraisamy, K.: Towards integrated field inversion and machine learning with embedded neural networks for RANS modeling. AIAA Paper 2019-1884 (2019). https://doi.org/10.2514/6.2019-1884
21. Ho, J., West, A.: Field inversion and machine learning for turbulence modelling applied to three-dimensional separated flows. AIAA Paper 2021–2903 (2021). https://doi.org/10.2514/6.2021-2903
22. Jäckel, F.: A closed-form correction for the Spalart-Allmaras turbulence model for separated flow. AIAA Paper 2022–0462 (2022). https://doi.org/10.2514/6.2022-0462
23. Abadi, M., Agarwal, A., Barham, P., Brevdo, E., Chen, Z., Citro, C., Corrado, G.S., Davis, A., Dean, J., Devin, M., Ghemawat, S., Goodfellow, I., Harp, A., Irving, G., Isard, M., Jia, Y., Jozefowicz, R., Kaiser, L., Kudlur, M., Levenberg, J., Mané, D., Monga, R., Moore, S., Murray, D., Olah, C., Schuster, M., Shlens, J., Steiner, B., Sutskever, I., Talwar, K., Tucker, P., Vanhoucke, V., Vasudevan, V., Viégas, F., Vinyals, O., Warden, P., Wattenberg, M., Wicke, M., Yu, Y., Zheng, X.: TensorFlow: Large-scale machine learning on heterogeneous systems (2015). Software available from http://tensorflow.org/
24. Shimizu, Y.S., Fidkowski, K.J.: Output error estimation for chaotic flows. AIAA Paper 2016-3806 (2016). https://doi.org/10.2514/6.2016-3806
25. McGhee, R.J., Walker, B.S., Millard, B.F.: Experimental results for the Eppler 387 airfoil at low Reynolds numbers in the Langley low-turbulence pressure tunnel. NASA Tech. Memorandum 4062 (1988)

Modal Analysis of High-Fidelity Simulations in Turbomachinery

Christian Morsbach, Bjoern F. Klose, Michael Bergmann, and Felix M. Möller

Abstract We revisit recently published high-fidelity implicit large eddy simulation data sets obtained with a high-order discontinuous Galerkin spectral element method and analyse them using Proper Orthogonal Decomposition (POD) as well as Spectral Proper Orthogonal Decomposition (SPOD). The first configuration is the MTU T161 low-pressure turbine cascade with resolved end wall boundary layers in a clean version and one with incoming turbulent wakes. We focus on the behaviour of the laminar separation bubble and the secondary flow system and how these phenomena are affected by incoming wakes. The second configuration is a transonic compressor cascade in which we analyse the unsteady behaviour of the shock wave boundary layer interaction. Throughout the paper, we try to discuss not only the flow physics but also insights into how the modal decomposition techniques can help facilitate understanding and where their limitations are. POD can, e.g. lead to misconceptions due to its missing frequency resolution capability. These can be avoided by using SPOD, which, on the other hand, does need more data to achieve a similar signal-to-noise ratio in the computed modes. We conclude that these techniques are most effective, when used in combination with other unsteady analyses.

CUFS-2024-4, Cambridge, United Kingdom, 4-5 March 2024

C. Morsbach (✉) · M. Bergmann
Institute of Propulsion Technology, German Aerospace Center (DLR), Cologne, Germany
e-mail: christian.morsbach@dlr.de

M. Bergmann
e-mail: michael.bergmann@dlr.de

B. F. Klose · F. M. Möller
Institute of Test and Simulation for Gas Turbines, German Aerospace Center (DLR), Cologne, Germany
e-mail: bjoern.klose@dlr.de

F. M. Möller
e-mail: felix.moeller@dlr.de

J. C. Tyacke and N. R. Vadlamani (eds.), *Proceedings of the Cambridge Unsteady Flow Symposium 2024*, https://doi.org/10.1007/978-3-031-69035-8_6

1 Introduction

Computational Fluid Dynamics (CFD) simulations have revolutionized our ability to predict and understand complex flow phenomena. With increasing computing resources and the development of high-order methods, simulations produce ever more high-fidelity data sets, which are difficult to analyse. While these data sets have the potential to improve our understanding of flow physics, the sheer amount of data, and its temporal and spatial resolution, make gaining insights a challenging task. Hence, most Large Eddy Simulation (LES) studies only discuss time-averaged quantities or features based on snapshots.

To facilitate the analysis of these data sets, researchers have recently developed data analysis techniques, such as Proper Orthogonal Decomposition (POD) [4, 21], Spectral Proper Orthogonal Decomposition (SPOD) [17], and Dynamic Mode Decomposition (DMD) [16], which have gained significant attention in the computational fluid dynamics community. These methods allow for the extraction of the most energetic modes (POD) and sorting by frequency content (SPOD).

For turbomachinery flows, these modal decomposition techniques have traditionally been applied to experimental data. Some recent examples include POD of Particle Image Velocimetry (PIV) data from a centrifugal compressor inlet [2], POD and DMD applied to analyse PIV data of a laminar separation bubble undergoing transition to turbulence [1, 7, 12], POD used to investigate the interaction of high- and low-pressure turbine based on unsteady total pressure data in planes between the blade rows [6], or an investigation of transonic buffet using DMD of PIV data [8]. Very early numerical studies were based on unsteady Reynolds-averaged Navier-Stokes (RANS) data, e.g. [5], but more recently, turbulence resolving approaches such as hybrid RANS/LES or LES have become feasible. Fiore *et al.* [9] recently applied the SPOD method to a low-pressure turbine flow with laminar inflow, as well as with wakes generated by a row of upstream cylinders. They found that the cylinder wakes amplify certain modes over the blade and increase turbulence downstream of the trailing edge over a reduced frequency bandwidth. Mechanisms of trailing edge (TE) cutback film cooling flows at different blowing rations have been investigated using SPOD on Delayed Detached Eddy Simulation (DDES) data by Wang *et al.* [20]. Here, the authors analysed how the structure of vortex shedding changes with blowing ratio. He *et al.* [10] recently presented an extensive analysis of the compressor tip leakage flow using SPOD on data obtained with a DDES. They used the complete 3D flow field in the tip region of the blade as input for the SPOD and identified low-rank behaviour over a range of frequencies associated with an oscillation of the tip leakage vortex and vortex shedding after its breakdown.

At the German Aerospace Center (DLR), we generate LES data with a numerical test rig for turbomachinery based on the high-order Discontinuous Galerkin Spectral Element Method (DGSEM). In this study, we will reconsider previously published high-fidelity data sets of turbomachinery flows [11, 14] generated with our DGSEM solver TRACE [3] and extract the dominant dynamic features using POD and SPOD. The data sets include the interaction of wakes with downstream low-pressure turbine blades and the shock boundary layer interaction in a compressor cascade. The study will discuss the advantages and limitations of each method and provide insights into how these techniques can be used to gain a deeper understanding of complex flow phenomena.

2 Modal Decomposition Techniques

In this section, we will very briefly introduce POD and SPOD to clarify the terminology.

2.1 Proper Orthogonal Decomposition (POD)

POD by itself is not a new method. Its theoretical background and formulation have been reviewed decades ago [4]. The formulation here is inspired by the review of Schmidt and Colonius [17] as well as the tutorial by Weiss [21]. For a consistent derivation of how a snapshot POD can be conveniently computed on non-equidistant grids using a Singular Value Decomposition (SVD), we refer the interested reader to the appendix.

We define the matrix of M snapshots of fluctuating quantities (mean subtracted) where each snapshot is a row and spatial dimensions as well as multiple variables are flattened to size N:

$$\mathbf{X} = \begin{pmatrix} — \mathbf{x}_1^H — \\ \vdots \\ — \mathbf{x}_M^H — \end{pmatrix} \in \mathbb{C}^{M \times N}. \tag{1}$$

Generally, the snapshots can contain complex numbers as required for SPOD, while they will be real numbers in our case when we apply POD to flow data directly. The POD is based on the sample covariance matrix

$$\mathbf{C} = \frac{1}{M-1}\mathbf{X}^H\mathbf{X} \in \mathbb{C}^{N \times N} \tag{2}$$

through an eigenvalue problem with a weight matrix $\mathbf{W}$ and inner product $\langle \mathbf{x}_1, \mathbf{x}_2 \rangle = \mathbf{x}_1^H \mathbf{W} \mathbf{x}_2$ [17]:

$$\mathbf{C}\mathbf{W}\Phi = \Phi\Lambda, \quad \mathbf{W} = \mathrm{diag}\,(w_1, \ldots, w_N)\,. \tag{3}$$

Here, $\Phi \in \mathbb{C}^{N \times M}$ is the matrix of POD modes or eigenvectors and $\Lambda = \mathrm{diag}$ $(\lambda_1, \ldots, \lambda_M)$ is the matrix of eigenvalues. The modes are orthonormal under the scalar product defined above:

$$\Phi^H \mathbf{W} \Phi = \mathbf{1}. \tag{4}$$

Because the weight matrix is problem specific, we will discuss its form for each application separately. A linear combination of the modes recovers the original snapshots:

$$\mathbf{X} = \mathbf{A}\Phi^T = \sum_j a_{ij}\Phi_{jk} = \sum_j \tilde{X}_j \in \mathbb{C}^{M\times N}. \tag{5}$$

This can be interpreted as sum of modal contributions with the time coefficients determined by

$$\mathbf{A} = \mathbf{X}\mathbf{W}\Phi \in \mathbb{C}^{M\times N}. \tag{6}$$

2.2 Spectral Proper Orthogonal Decomposition (SPOD)

In this section, we briefly review the implementation of the SPOD algorithm based on Welch's periodogram method, which is used in this work. For a detailed description of the theoretical background, we refer the reader to Towne, Schmidt and collaborators [17, 19], as well as to Mengaldo & Maulik [13] for the implementation.

Just as done for the POD, we assemble the matrix of snapshots $\mathbf{X}$ (1). Next, we apply Welch's spectral estimation method by decomposing the matrix along its time axis into N_{blk} overlapping blocks and, assuming that each realisation in the snapshot matrix occurs periodically in time, apply a discrete Fourier transform on each block after multiplication of the data by a Hamming window. The result is a set of N_{blk} Fourier transformed blocks $\widetilde{\mathbf{x}}_k$ with N_{fft} frequencies each.

For each frequency, we assemble the matrix $\widetilde{\mathbf{X}}$

$$\widetilde{\mathbf{X}} = \begin{pmatrix} — & \widetilde{\mathbf{x}}_1^H & — \\ & \vdots & \\ — & \widetilde{\mathbf{x}}_{N_{\text{blk}}}^H & — \end{pmatrix} \in \mathbb{C}^{N_{\text{blk}}\times N}, \tag{7}$$

where, again, N is the number of spatial points times the number of variables and N_{blk} is the number of blocks. For each frequency, we compute the covariance matrix

$$\widetilde{\mathbf{C}} = \frac{1}{N_{\text{blk}}}\widetilde{\mathbf{X}}^H\widetilde{\mathbf{X}} \in \mathbb{C}^{N\times N}, \tag{8}$$

and obtain the SPOD modes by solving the eigenvalue decomposition of the covariance matrix for each frequency separately [17]:

$$\widetilde{\mathbf{C}}\mathbf{W}\Phi = \Lambda\Phi, \quad \mathbf{W} = \text{diag}\,(w_1, \ldots, w_N)\,. \tag{9}$$

Here, $\mathbf{W}$ is a weight matrix, Φ are the eigenvectors and Λ the eigenvalues.

3 Application to Turbomachinery Flows

In the following, we will apply both POD and SPOD to three different data sets with different flow phenomena. We will begin by investigating the transitional flow through the MTU T161 low-pressure turbine (LPT) with focus on separation-induced transition around midspan and large-scale unsteadiness of the secondary flow system. Additionally, we will analyse the influence of wakes from an upstream row. Finally, we will study shock wave boundary layer interaction in a linear compressor cascade.

3.1 Separation-Induced Transition and Secondary Flow Effects

We analyse the turbulent flow through an LPT cascade at an off-design Reynolds number of 90,000 and a Mach number of 0.65 computed as implicit LES with a high-order DGSEM scheme [14]. We use a data set containing the in-plane and spanwise components of the time-resolved wall shear stress vector (τ_{xy}, τ_z) on the blade suction side spanning 100 convective time units t_c based on chord length and outlet velocity with 3668 samples. The original sampling rate was 366 samples per t_c but we use only a tenth of that for the modal analysis, since that already covers the frequency range of interest. Figure 1 shows the configuration with a view of the blade suction side of the configuration with instantaneous vortex structures visualised by the Q-criterion to illustrate the separation induced transition around midspan and the secondary flow regions close to the hub and tip walls. This secondary flow system is dominated by the

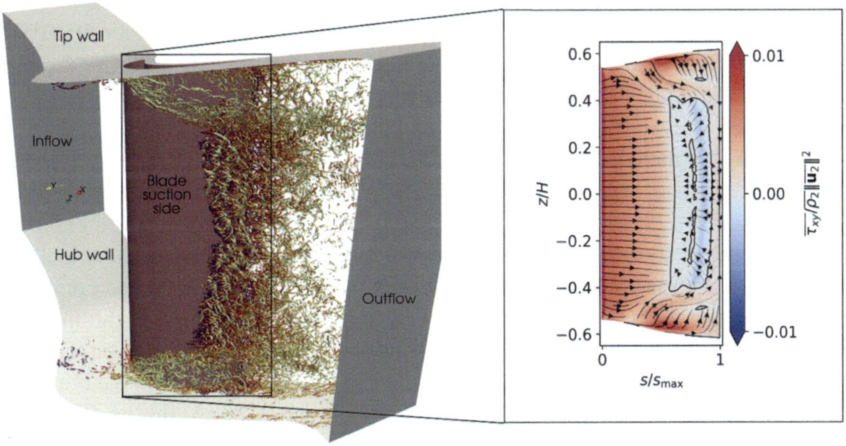

Fig. 1 Illustration of MTU T161 simulation domain with instantaneous vortex structures and mean in-plane wall shear stress τ_{xy} on the blade suction side with surface streamlines and $\tau_{xy} = 0$ as black contour line

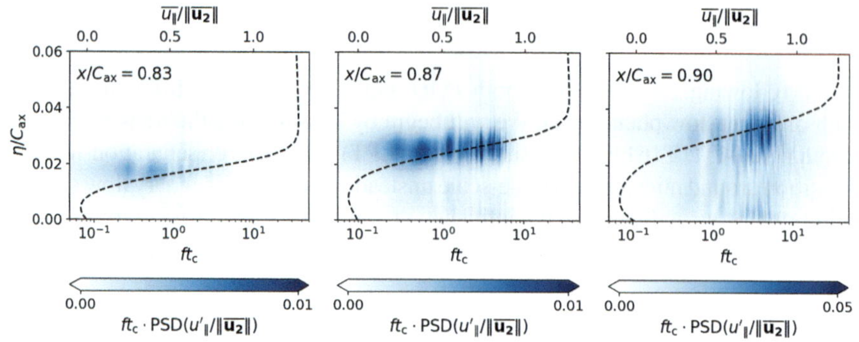

Fig. 2 Midspan premultiplied spectrum of $u'_\parallel$ for different distances η from the solid wall in boundary layer cuts of the MTU T161 blade suction side at $x/C_{\text{ax}} = 0.83$ (pre-transitional), $x/C_{\text{ax}} = 0.87$ (transition onset) and $x/C_{\text{ax}} = 0.90$ (half way to fully turbulent)

passage vortex (PV), which originates at the blade leading edge (LE) and is amplified by the passage cross flow due to the pressure difference between pressure and suction side. For a more detailed discussion, we refer the interested reader to Morsbach *et al.* [15]. On the right-hand side, the mean in-plane wall shear stress τ_{xy} on the blade suction side surface over the arc length from the LE at $s = 0$ to the TE at $s/s_{\text{max}} = 1$ is presented in the way we will display the mode shapes throughout this section. The z-coordinate is normalised by the upstream channel height H allowing to see the diverging end walls. We will employ classical Fourier analysis, space-only POD and SPOD to investigate a potential connection between the Kelvin-Helmholtz (KH) based transition mechanism around midspan and oscillations of the corner flow.

In addition to the surface data set, we use three wall-normal cuts through the suction side boundary layer at midspan to be able to relate frequencies to physical mechanisms in the flow. Figure 2 shows premultiplied Power Spectral Density (PSD) of the wall-parallel fluctuating velocity $u'_\parallel$ for these cuts varying with the distance from the solid wall. For orientation, the mean velocity profiles $\overline{u_\parallel}$ are plotted as black dashed lines. Both velocities are normalised with the time- and area-averaged velocity $\|\overline{\mathbf{u}_2}\|$ in the outflow plane of the domain. In the cut at $x/C_{\text{ax}} = 0.83$, where the separated shear layer is still laminar, only non-dimensional frequencies lower than 1 (time scales larger than a convective time unit) show noticeable content in the region of strongest shearing. Further downstream at $x/C_{\text{ax}} = 0.87$, where the flow begins its transition to turbulence, non-dimensional frequencies in the range 3–5 become excited due to the KH instability and have already surpassed the intensity found in the lower frequencies. These instabilities can be found to grow even further at $x/C_{\text{ax}} = 0.90$, where the maximum turbulent kinetic energy (TKE) in the boundary layer is at 50% of the maximum, reached slightly downstream. At this position, they dominate the whole spectrum and peaks at these frequencies can not only be found in the shear layer but also close to the solid wall. The scale of the colour bar was adapted to highlight both the low frequency and high frequency effects, which will be discussed in the modal analysis below.

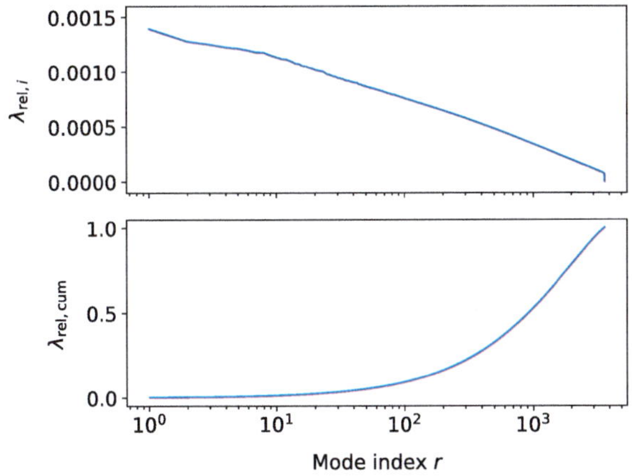

Fig. 3 POD relative eigenvalues $\lambda_{\text{rel},i} = \sigma_i^2 / \sum_{j=1}^{N} \sigma_j^2$ of MTU T161 suction side wall shear stress (*top*) and cumulative eigenvalues $\lambda_{\text{rel,cum}} = \sum_{i=1}^{r} \lambda_{\text{rel},i}$ (*bottom*)

Since our data set has points clustered towards the leading and trailing edge, as well as towards the end walls, we use the element area as weights to perform a POD using an SVD with (14), (17) and (18). The input variables are the two independent components of the fluctuating wall shear stress vector τ'_{xy} and τ'_z, equally weighted using the weighted 2-norm [17]:

$$\mathbf{X} = \left[\tau'_{xy},\ \tau'_z\right]^{\mathrm{T}}, \quad \mathbf{W} = \int_V \operatorname{diag}\left(\mathrm{d}V,\ \mathrm{d}V\right). \tag{10}$$

The resulting eigenvalues $\lambda_{\text{rel},i}$ (*top*) and the cumulative energy $\lambda_{\text{rel,cum}}$ contained in the first r modes (*bottom*) are shown in Fig. 3. It can be instantly noted that, in contrast to many textbook examples of POD, there is no large-scale organisation in the flow field responsible for a great portion the fluctuating energy. On the contrary, the most energetic mode only accounts for 0.14% while the tenth mode still accounts for 0.11%. Yet, the modes stand out above the noise threshold and the mode shapes offer some insights into the flow physics.

Figure 4 shows the first eight POD modes. Note that the modes are orthonormal by definition, so the corresponding eigenvalues are given in the title to indicate their relative strengths. The first two modes exhibit structures close to the end walls while the area around midspan shows relatively low amplitudes and rather noisy data. Mode 1 has larger amplitudes at the tip wall (positive z) in both components of the wall shear stress while mode 2 shows a very similar structure basically mirrored at the $z = 0$ plane. This corresponds to a periodic displacement of flow from the end walls into the channel and can be associated with an oscillation of the PV. The third mode is the first one to be dominated by structures associated with the KH instability in the separated shear layer around midspan. It is notable that the large

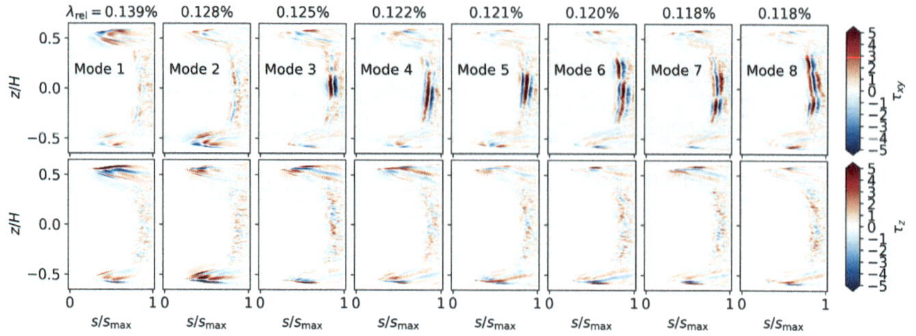

Fig. 4 MTU T161: First eight POD modes for in-plane (*top*) and spanwise (*bottom*) wall shear stress τ_{xy}, τ_z

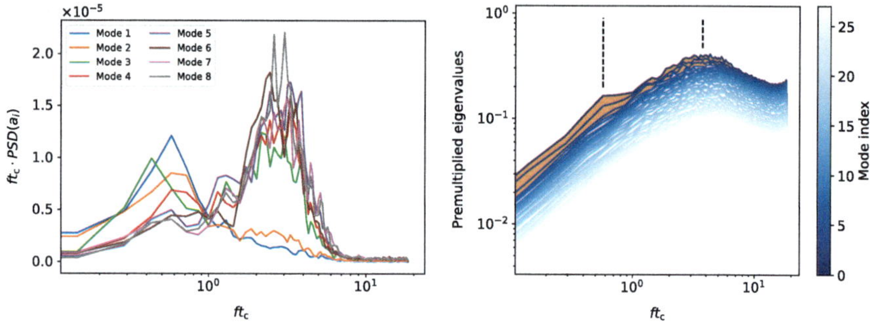

(a) Premultiplied PSD of POD time coefficients (b) Premultiplied SPOD eigenvalues

Fig. 5 Spectra for the MTU T161 without incoming wakes

amplitudes are only found close to midspan and are near perfectly aligned with the z axis. Larger structures close to the end walls can still be seen here, yet at a lower amplitude compared to modes 1 and 2. A spanwise variation of the KH structures can be found in the following modes be it either in the form of a tilt (mode 4, 5, 8) or an inversion of the phase at some spanwise location (mode 6 and 7). All the modes featuring KH structures still show some amplitude at the end walls suggesting that these phenomena might be connected.

A look at the POD time coefficients, especially at their spectral content, can reveal additional information on the modes. Figure 5a shows the premultiplied PSD of the time coefficients for the first eight modes, obtained via (6), computed with Welch's method with 256 samples per segment. Two distinct features can be identified: a low-frequency peak around 0.5 and a high frequency peak around 3. The first two modes, which do not show any sign of KH structures, clearly peak at the low frequency but show no contribution in the high frequency range. On the other hand, all modes with KH structures have the most significant contribution to the spectrum in this high frequency peak. While most of these modes do not show contributions above the noise

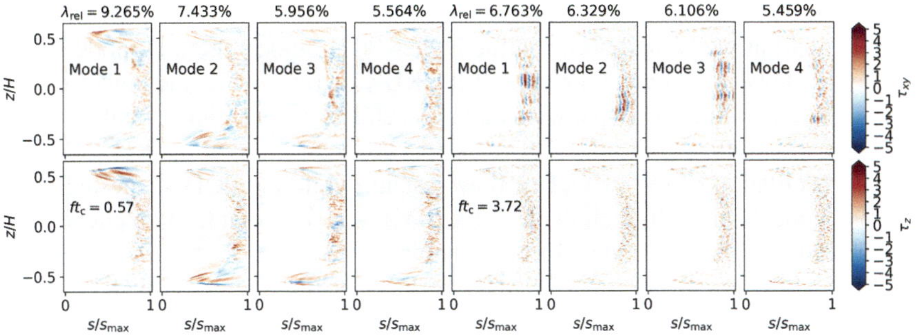

Fig. 6 MTU T161: First 4 SPOD modes for in-plane (*top*) and spanwise (*bottom*) wall shear stress τ_{xy}, τ_z at $ft_c = 0.57$ and 3.71

level in the low frequency range, mode 3 is an exception: it features a low frequency peak in the same order of magnitude as modes 1 and 2. This suggests some connection of the oscillation of the secondary flow system with the KH shedding through this mode shape.

At this point, it can be helpful to resort to SPOD to be able to obtain optimal spatial modes corresponding to a single frequency. We use exactly the same data set with weights chosen to represent a weighted 2-norm [17], a bin size of 256 and an overlap of 50%. Figure 5b shows the premultiplied SPOD spectrum with the area between mode 1 and 4 shaded in orange to indicate the energy contained within the first three modes. The spectrum peaks in the same high frequency range as the POD time coefficient spectra discussed above. As expected from the above discussion of the POD, the separation between the first and second mode is not as clear as in many textbook examples. Nevertheless, the first three to four modes stand out in terms of separation between the modes and suggest that some kind of low-rank behaviour can possibly be identified [10]. We, therefore, examine the first four modes at the frequencies 0.57 and 3.72 in Fig. 6. The choice of former is due to the subtle low-frequency peak while the latter is roughly the maximum premultiplied eigenvalue over all frequencies. Note, that slight changes to the investigated frequency do not qualitatively change the mode shapes in this case.

The first four columns show the first four modes of $ft_c = 0.57$. To present an indication for the energy of the respective mode, the mode eigenvalue λ_{rel} normalised with the sum over all eigenvalues at that frequency is given in the column header. The first two SPOD modes at $ft_c = 0.57$ show a great resemblance with the first two POD modes, which is a consistent picture since the frequency analysis of the latter revealed that most of their energy is found at this low frequency. Looking at the higher modes, there is again some indication that the oscillation of the secondary flow system has an effect on the midspan flow, although the data in the separation bubble is rather noisy. In contrast to our argument above using the POD analysis, however, we see no direct connection with KH structures but a larger scale oscillation in the third mode. KH structures in the laminar separation bubble (LSB), on the other hand can be clearly identified in the in-plane wall shear stress for all modes at $f t_c = 3.71$. Similar to what was found above in the higher POD modes, the first mode at this frequency shows a vertical alignment of the KH structures while the higher modes introduce some tilting. The spanwise wall shear stress shows mainly turbulent noise. An interesting difference to the POD concerns the connection of the KH structures and the oscillation of the secondary flow system: while the POD suggested some combined mechanism through mode 3, the SPOD suggests that the two phenomena are, indeed, not directly coupled as no significant structures in the secondary flow region can be found at $f t_c = 3.71$. This underlines the importance of SPOD to avoid drawing false conclusions due to the missing frequency resolution of POD. It has to be mentioned, though, that the SPOD modes are rather noisy and for a more quantitative analysis, more samples in excess of the available 100 convective time units would be beneficial. Very recently, a sub-sampling method was suggested to overcome the problems associated with short time signals by sacrificing the high-frequency part of the spectrum to be able to obtain more realisations of low-frequency phenomena [18].

3.2 Blade Row Interaction

Besides the clean inflow configuration with freestream turbulence only, a data set with incoming periodic disturbances generated by cylindrical wake generators was presented in [14]. POD was only used to obtain less noisy phase averages in the wake plane in this paper. In the following, we perform an SPOD analysis consistent with the one above. The data set consists of 3009 samples, spanning 82 convective time units corresponding to 50 bar passes. Again, we use only every tenth sample of the original data set sampled at 368 per t_c. The periodic wakes lead to an intermittently separated flow around midspan, which does not show any backflow on average [14].

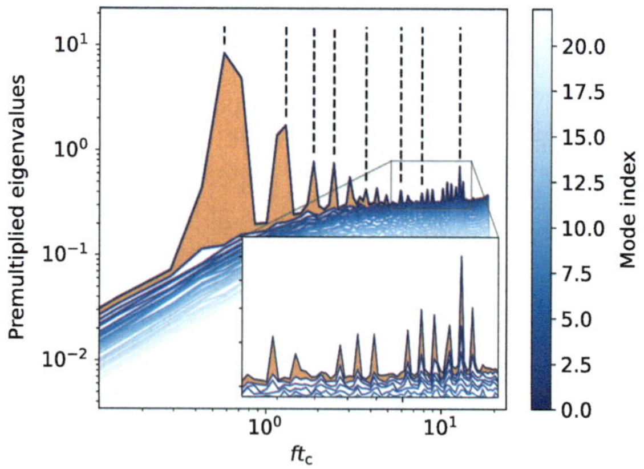

Fig. 7 MTU T161: SPOD spectrum for the case with incoming wakes; frequencies for mode analysis marked with dashed lines

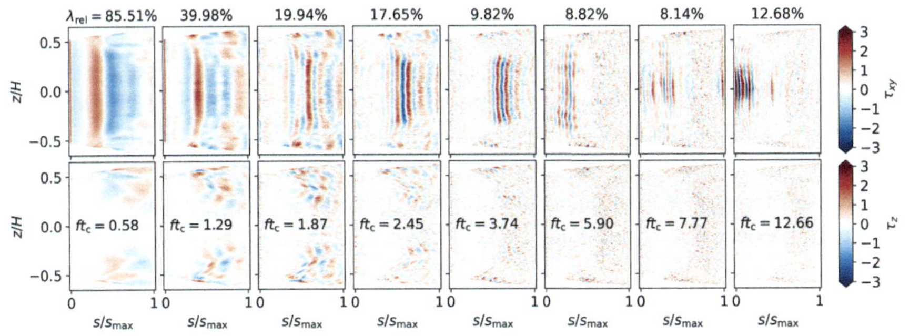

Fig. 8 MTU T161: First SPOD modes for in-plane (*top*) and spanwise (*bottom*) wall shear stress τ_{xy}, τ_z at dominant frequencies

The SPOD spectrum in Fig. 7 is dominated by the wake passing frequency $f t_c = 0.61$ and its harmonics. Coincidentally, this frequency is very close to the one of the large-scale secondary flow oscillations found in the case without wakes. The higher harmonics are in a frequency range corresponding to the KH shedding. In the high frequency range $f t_c > 5$, another set of peaks standing out of the turbulent noise can be found which does not occur in the case without wakes. In the following, we will analyse the first modes at the frequencies marked with dashed lines. An inspection of the spectrum and the second modes revealed that these do not generally show larger scale structures with a significant signal to noise ratio.

Figure 8 shows these modes, again along with their relative eigenvalue normalised with the sum of all eigenvalues at that frequency. Note that the ratio of the eigenvalues between different frequencies cannot be derived from these but must be obtained from Fig. 7. The first mode at $f t_c = 0.58$ is clearly associated with the large-scale wake passing and is responsible for most of the energy at that frequency with mode 2 at $\lambda_{\mathrm{rel}} = 1.26\%$ only (not shown in the figure). Note that due to the choice of blocks and the time step size, the available frequencies in the SPOD only approximately match the bar passing frequency and its harmonics. The large structures in the in-plane wall shear stress describe the periodic separation and reattachment, although this conclusion can only be drawn in combination with knowledge about the local mean values. In contrast to the case without wakes, the oscillation of the corner separation appears to be dominated by the wake passing as seen in large coherent structures in the both wall shear stress components. The first three harmonics quickly decrease in eigenvalue and show refined structures corresponding to the same phenomenon. At frequencies in the range of the KH instability of the case without wakes, i.e. $f t_c$ around 3, the characteristic vertical bands appear in the aft section of the blade. Again, from this frequency on, at best, a very subtle interaction with the secondary flow region can be identified.

When the frequency is further increased, another set of vertical structures with shorter wavelength can be found upstream, starting near the LE at $f t_c = 12.66$. The instantaneous flow field in a midspan plane (not shown) revealed, that these are the footprints of upstream traveling acoustic waves originating from the strong vortex shedding of the moving cylinder and their reflections from the adjacent blade pressure side. The high frequency structures near the LE could only be identified using the SPOD and cannot be observed in any of the first 30 POD modes. This, again, highlights the usefulness of the SPOD in distinguishing between different features of the flow which occur at different frequencies and are possibly hidden within high-energy low-frequency behaviour.

3.3 Shock Wave Boundary Layer Interaction

In this section, we analyse the modal content of the flow over a compressor cascade at transonic operating condition. The flow is simulated using a 4th order accurate LES based on a high-order DGSEM scheme. For a detailed description of the scheme, we refer to Bergmann *et al.* [3] and references therein. The linear cascade is computed based on a spanwise domain extrusion of 5% chord length, where periodic boundary conditions are applied along the pitchwise and the spanwise faces to approximate the flow over an infinite cascade of infinite blades. The computational domain consists of 2 276 829 hexahedral elements clustered around the blade and the shock location, adding to a total 145.7 million degrees of freedom per equation. The simulation is evaluated over 21 convective time units (t_c) based on the chord length c and mean inflow velocity $\|\overline{\mathbf{u}_0}\|$ and sampled at a non-dimensional frequency of 28 samples per time unit. We refer the interested reader to Klose *et al.* [11] for a more detailed description of the setup and discussion of the flow field for this test case.

Given that the compressor cascade is at transonic operating conditions, the SPOD is computed based on the compressible energy norm [17] with input vector

$$\mathbf{X} = \left[\rho',\ u'_{\parallel},\ u'_{\perp},\ u'_{z},\ T'\right]^{\mathrm{T}}, \tag{11}$$

and weights

$$\mathbf{W} = \int_V \operatorname{diag}\left(\frac{\overline{T}}{\gamma\overline{\rho}M_0^2},\ \overline{\rho},\ \overline{\rho},\ \overline{\rho},\ \frac{\overline{\rho}}{\gamma(\gamma-1)\overline{T}M_0^2}\right) \mathrm{d}V, \tag{12}$$

where ρ is the density, T the temperature, $(u_{\parallel},\ u_{\perp},\ u_z)$ the velocity components in wall-parallel, wall-normal and spanwise direction and M the Mach number. All physical quantities here are non-dimensionalised with respect to reference quantities at the inlet measurement plane (indicated by subscript 0) and the bar $\overline{\square}$ refers to the time-averaged field. A total of 568 snapshots are evaluated for the modal analysis for this test case, where the bin size for the SPOD is 142 snapshots with 50% overlap.

The flow is characterized by an upstream laminar boundary layer, shock-induced separation and transition to turbulence. Features of the instantaneous flow field are shown in Fig. 9a, where contours of the density are shown at the mid-chord section

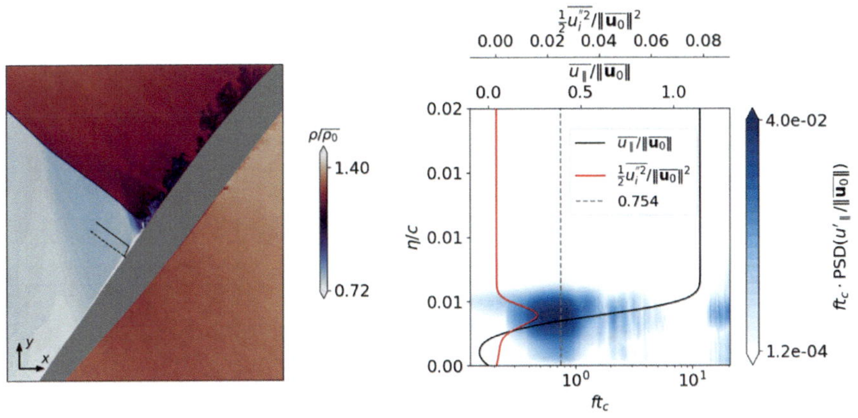

(a) Density contours (b) PSD along boundary layer probe

Fig. 9 DLR compressor cascade: **a** instantaneous density contours with boundary layer profile indicator in black; **b** contours of the frequency spectrum of $u_{\|}$ along the boundary layer with mean velocity and TKE profiles

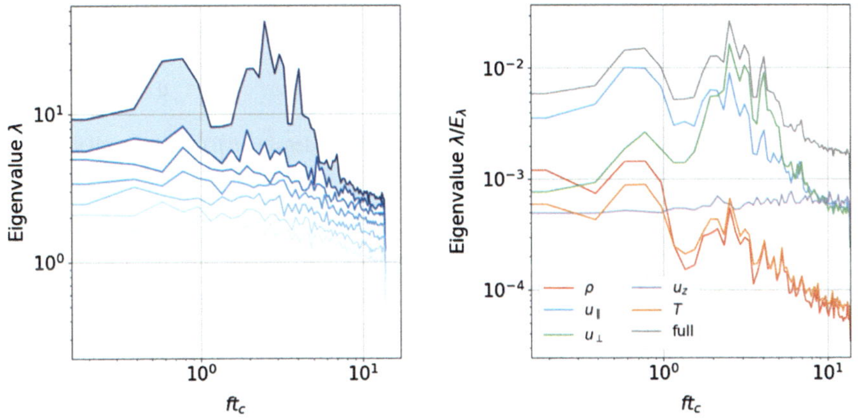

(a) Eigenvalue spectrum of q (b) Relative contributions

Fig. 10 DLR compressor cascade: **a** eigenvalue spectrum of $q = [\rho, u_{\|}, u_{\perp}, u_z, T]$; **b** relative contributions to the first mode of the individual components of q

of the blade around the shockwave and the premultiplied frequency spectrum of the wall-parallel velocity component $u_{\|}$ along a wall-normal probe is given in Fig. 9b. Additionally, the mean velocity and TKE profiles are indicated, as well as the peak in the spectrum at $f t_c = 0.75$. As shown by Klose *et al.* [11], low-amplitude oscillations of the normal shock occur in the range $0.6 < f t_c < 0.8$, as well as high-frequency fluctuations at $f t_c = 2.6$, which were associated with a wake mode.

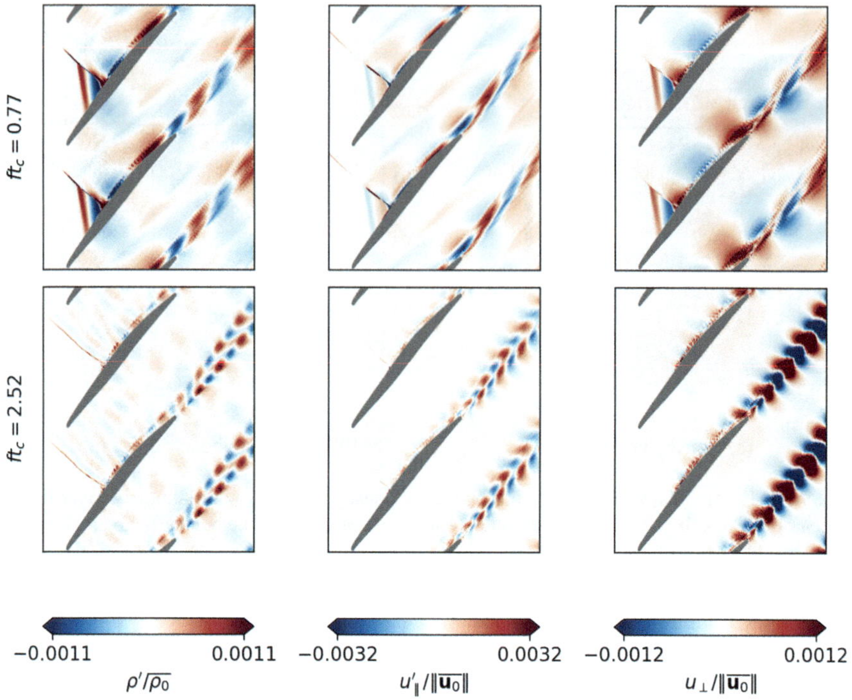

Fig. 11 DLR compressor cascade: First SPOD modes of the density (*left*), blade-parallel velocity (*middle*) and blade-normal velocity (*right*)

The eigenvalue spectrum in Fig. 10a is obtained by computing the SPOD of the input vector (11) and the weights (12) as a whole, where the shaded area highlights the separation between the first and second SPOD modes. A distinctly separated peak occurs at $ft_c = 0.77$ and a series of separated peaks at around $ft_c = 2.52$, closely matching the frequency peak along the boundary layer probe. To extract the contributions of each term to the overall fluctuating energy, the SPOD eigenvalues of the first mode for the individual components (with their respective weights) are shown in Fig. 10b, together with the result of the *full* input vector. All components are normalised by the integrated eigenvalues of the *full* input vector, E_λ. While the wall-parallel velocity component $u'_{\parallel}$ dominates the energy in the lower frequency regime, the wall-normal component $u'_{\perp}$ is the main contributor for the higher frequency peaks. The components measuring the compressibility effects (ρ', T') also predominantly contribute energy to the low-frequency modes, where they still only make up around 15% of the total energy.

The shape of the first SPOD mode for $ft_c = 0.77$ and $ft_c = 2.52$ for ρ', $u'_{\parallel}$ and $u'_{\perp}$ is given in Fig. 11. At $ft_c = 0.77$, the compression area upstream of the normal shock is distinctly highlighted by the density, where the thickening of the boundary layer is accompanied by a mild oblique compression wave. The movement of the normal shock itself is also tracked by the $u'_{\parallel}$ velocity component and shows as a distinct, sharp mode structure normal to the blade surface. At the higher frequency of $ft_c = 2.52$, the shock wave is no longer discernible in the velocity components. Here, $u'_{\perp}$ is dominant and its mode shape proves that the oscillating energy is almost exclusively contained in the wake.

4 Conclusion

We have presented the application of two modal decomposition techniques, namely POD and SPOD, to previously published LES data sets of turbomachinery flows computed with our high-order DGSEM solver TRACE. While POD could be used to draw first conclusions even when large-scale effects did not dominate the turbulent spectrum in energy, the frequency resolution capabilities of SPOD allowed for a clearer analysis and separation of effects. In the LPT flow without incoming wakes, the weak but dominant feature was a low-frequency oscillation of the secondary flow system at time scales longer than a convective time unit. The SPOD showed that these oscillations were coupled with a large-scale motion of the LSB. We found, that the KH instability, occurring at higher frequencies was easy to detect and did not directly interact with the secondary flow system. An SPOD of the flow through the LPT with incoming wakes revealed the wakes' influence on the blade including the secondary flow system while the method was also able to detect the local KH mechanism in the intermittent separation bubble. It also revealed the footprint of very high frequency acoustic perturbations, which originate from the vortex shedding of the upstream moving cylinder. In this case, the frequency decomposition benefitted the analysis to reveal low-energy effects otherwise obscured by the dominant large-scale motion. Applied to the flow around a transonic compressor cascade, SPOD was able to discern two effects: first, low-frequency oscillations are associated with shock-wave boundary layer interaction and show higher contributions of compressible components to the total energy. Second, high-frequency oscillations are associated with vortex shedding in the wake in which compressibility does not play a significant role. It has to be noted, that many statements made in this paper depended on a combined analysis of modal decomposition techniques with other methods such as averages, frequency analysis, space-time diagrams or visualisation of the instantaneous flow field. In this respect, modal decomposition techniques are a valuable addition in the toolbox of unsteady flow analysis in turbomachinery flows.

Appendix

In the following, we will re-arrange (3) so that we can determine the POD modes and eigenvalues using an SVD of a weighted snapshot matrix. Inserting (2) into (3) yields

$$\left(\frac{\mathbf{X}\sqrt{\mathbf{W}}}{\sqrt{M-1}}\right)^H \frac{\mathbf{X}\sqrt{\mathbf{W}}}{\sqrt{M-1}}\Phi = \Phi\Lambda \tag{13}$$

with $\sqrt{\mathbf{W}} = \operatorname{diag}\left(\sqrt{w_1}, \ldots, \sqrt{w_N}\right)$. The weighted snapshot matrix can be written as compact SVD

$$\frac{\mathbf{X}\sqrt{\mathbf{W}}}{\sqrt{M-1}} = \mathbf{U}\Sigma\mathbf{V}^H, \quad \mathbf{U} \in \mathbb{C}^{R\times R},\ \Sigma \in \mathbb{C}^{R\times R},\ \mathbf{V} \in \mathbb{C}^{N\times R} \tag{14}$$

with $R = \min(M, N)$ and $\mathbf{U}^H\mathbf{U} = \mathbf{V}^H\mathbf{V} = \mathbf{1}$. This yields

$$\mathbf{V}\Sigma^H\Sigma\mathbf{V}^H\Phi = \Phi\Lambda. \tag{15}$$

With (4), it can be written as

$$\mathbf{V}\Sigma^H\Sigma\mathbf{V}^H = \sqrt{\mathbf{W}}\Phi\Lambda^H\left(\sqrt{\mathbf{W}}\Phi\right)^H. \tag{16}$$

So the eigenvalues Λ and modes Φ are obtained from the weighted snapshot SVD by

$$\Lambda = \Lambda^H = \Sigma^H\Sigma \tag{17}$$

$$\Phi = \sqrt{\mathbf{W}^{-1}}\mathbf{V} \tag{18}$$

with $\sqrt{\mathbf{W}^{-1}} = \operatorname{diag}\left(\sqrt{w_1^{-1}}, \ldots, \sqrt{w_N^{-1}}\right)$.

References

1. Alessandri, A., Bagnerini, P., Gaggero, M., Lengani, D., Simoni, D.: Dynamic mode decomposition for the inspection of three-regime separated transitional boundary layers using a least squares method. Phys. Fluids **31**(4), 044103 (2019). https://doi.org/10.1063/1.5086225
2. Banerjee, D., Selamet, A., Dehner, R.: Proper orthogonal decomposition analysis of particle image velocimetry data at the inlet of a centrifugal compressor. In: ASME international mechanical engineering congress and exposition, vol. 8: Fluids Engineering; Heat Transfer and Thermal Engineering, p. V008T10A016 (2022). https://doi.org/10.1115/IMECE2022-94308
3. Bergmann, M., Morsbach, C., Klose, B.F., Ashcroft, G., Kügeler, E.: A numerical test rig for turbomachinery flows based on large eddy simulations with a high-order discontinuous Galerkin scheme-Part I: sliding interfaces and unsteady row interactions. J. Turbomach. **146**(2), 021005 (2023). https://doi.org/10.1115/1.4063734
4. Berkooz, G., Holmes, P., Lumley, J.L.: The proper orthogonal decomposition in the analysis of turbulent flows. Annu. Rev. Fluid Mech. **25**(1), 539–575 (1993). https://doi.org/10.1146/annurev.fl.25.010193.002543
5. Cizmas, P.G., Palacios, A.: Proper orthogonal decomposition of turbine rotor-stator interaction. J. Propul. Power **19**(2), 268–281 (2003)

6. Dellacasagrande, M., Sterzinger, P.Z., Zerobin, S., Merli, F., Wiesinger, L., Peters, A., Maini, G., Heitmeir, F., Göttlich, E.: Unsteady flow interactions between a high- and low-pressure turbine: Part 2—rotor-synchronic averaging and proper orthogonal decomposition of the unsteady flow fields. In: Turbo Expo: Power for Land, Sea, and Air, vol. 2A: Turbomachinery, p. V02AT45A011. The American Society of Mechanical Engineers (2019). https://doi.org/10.1115/GT2019-90824
7. Dotto, A., Lengani, D., Simoni, D., Tacchella, A.: Dynamic mode decomposition and Koopman spectral analysis of boundary layer separation-induced transition. Phys. Fluids **33**(10), 104104 (2021). https://doi.org/10.1063/5.0065554
8. Feldhusen-Hoffmann, A., Lagemann, C., Loosen, S., Meysonnat, P., Klaas, M., Schröder, W.: Analysis of transonic buffet using dynamic mode decomposition. Exp. Fluids **62**(4), 66 (2021). https://doi.org/10.1007/s00348-020-03111-5
9. Fiore, M., Gojon, R., Sáez-Mischlich, G., Gressier, J.: LES of the T106 low-pressure turbine: spectral proper orthogonal decomposition of the flow based on a fluctuating energy norm. Comput. Fluids **252**, 105761 (2023). https://doi.org/10.1016/j.compfluid.2022.105761
10. He, X., Fang, Z., Rigas, G., Vahdati, M.: Spectral proper orthogonal decomposition of compressor tip leakage flow. Phys. Fluids **33**(10), 105105 (2021). https://doi.org/10.1063/5.0065929
11. Klose, B.F., Morsbach, C., Bergmann, M., Hergt, A., Klinner, J., Grund, S., Kügeler, E.: A numerical test rig for turbomachinery flows based on large eddy simulations with a high-order discontinuous Galerkin scheme-Part II: shock capturing and transonic flows. J. Turbomach. **146**(2), 021006 (2023). https://doi.org/10.1115/1.4063827
12. Lengani, D., Simoni, D., Ubaldi, M., Zunino, P., Bertini, F.: Experimental investigation on the time-space evolution of a laminar separation bubble by proper orthogonal decomposition and dynamic mode decomposition. J. Turbomach. **139**(3), 031006 (2016). https://doi.org/10.1115/1.4034917
13. Mengaldo, G., Maulik, R.: PySPOD: A python package for spectral proper orthogonal decomposition (SPOD). J. Open Source Softw. **6**(60), 2862 (2021). https://doi.org/10.21105/joss.02862 https://doi.org/10.21105/joss.02862
14. Morsbach, C., Bergmann, M., Tosun, A., Klose, B.F., Bechlars, P., Kügeler, E.: A numerical test rig for turbomachinery flows based on large eddy simulations with a high-order discontinuous Galerkin scheme-Part III: secondary flow effects. J. Turbomach. **146**(2), 021007 (2023). https://doi.org/10.1115/1.4063511
15. Morsbach, C., Bergmann, M., Tosun, A., Klose, B.F., Kügeler, E., Franke, M.: Large eddy simulation of a low-pressure turbine cascade with turbulent end wall boundary layers. Flow, Turbulence and Combustion (2023). https://doi.org/10.1007/s10494-023-00502-6. Published online. In press
16. Schmid, P.J.: Dynamic mode decomposition of numerical and experimental data. J. Fluid Mech. **656**, 5–28 (2010). https://doi.org/10.1017/S0022112010001217
17. Schmidt, O.T., Colonius, T.: Guide to spectral proper orthogonal decomposition. AIAA J. **58**(3), 1023–1033 (2020). https://doi.org/10.2514/1.J058809
18. Schneider, N., Köhler, S., von Wolfersdorf, J.: Experimental detection of organised motion in complex flows with modified spectral proper orthogonal decomposition. Fluids **8**(6), 184 (2023). https://doi.org/10.3390/fluids8060184
19. Towne, A., Schmidt, O.T., Colonius, T.: Spectral proper orthogonal decomposition and its relationship to dynamic mode decomposition and resolvent analysis. J. Fluid Mech. **847**, 821–867 (2018). https://doi.org/10.1017/jfm.2018.283
20. Wang, R., He, X., Yan, X.: Spectral proper orthogonal decomposition analysis of trailing edge cutback film cooling flow. Phys. Fluids **34**(10), 105106 (2022). https://doi.org/10.1063/5.0098796
21. Weiss, J.: A tutorial on the proper orthogonal decomposition. In: AIAA aviation 2019 forum. Dallas, Texas, USA (2019). https://doi.org/10.2514/6.2019-3333

A Hybrid RANS-LES Dataset for Data-Driven Turbulent Mean Flow Reconstruction

Omid Bidar, Sean R Anderson, and Ning Qin

Abstract This paper presents results for a set of parameterised periodic hill geometries using the hybrid RANS-LES approach based on the Spalart-Allmaras improved delayed-detached eddy simulation (IDDES). These results are compared with data from direct numerical simulations. The dataset is intended to be used for data-driven Reynolds-averaged Navier-Stokes (RANS) closure modelling for aerodynamic shape optimisation. As a demonstration, the IDDES data for the baseline geometry is used for turbulent mean flow reconstruction with the $k-\omega$ shear stress transport model. Flow reconstruction results demonstrate considerable improvements compared to the baseline RANS model.

1 Introduction

Data-driven flow simulations and/or analyses are an active research pursuit in the fluid dynamics community. While the research breadth is very broad–e.g. enhanced reduced order modelling through dynamic mode decomposition [2], physics-informed neural networks [3], turbulence closure modelling [1] etc.–this paper presents the first step towards exploring data-driven turbulence closure modelling in the context of aerodynamic shape optimisation. Our final goal is to study the deep learning-enhancement of turbulence models in the contexts of Reynolds-averaged Navier-Stokes (RANS) simulations when performing aerodynamic shape

CUFS-2024-7, Cambridge, United Kingdom, 4-5 March 2024.

O. Bidar (✉)
University of Sheffield, Sheffield, England
e-mail: obidar1@sheffield.ac.uk

S. R. Anderson
School of Electrical and Electronic Engineering, University of Sheffield, Sheffield, England
e-mail: s.anderson@sheffield.ac.uk

N. Qin
School of Mechanical, Aerospace and Civil Engineering, University of Sheffield, Sheffield, England
e-mail: n.qin@sheffield.ac.uk

J. C. Tyacke and N. R. Vadlamani (eds.), *Proceedings of the Cambridge Unsteady Flow Symposium 2024*, https://doi.org/10.1007/978-3-031-69035-8_7

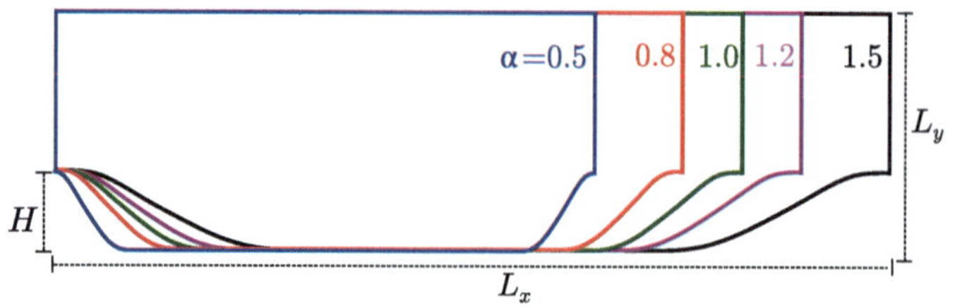

Fig. 1 Parameterised periodic hill geometries for generating the training dataset

optimisation in complex flow conditions–where existing turbulence closures are prone to inaccuracies. To that end, here we present a training dataset for a set of parameterised periodic hills, generated using hybrid RANS-LES simulations. This approach is especially appealing in high Reynolds number aerodynamic flows such as those encountered in the aerospace industry, since the computational resources requirement is much lower than large eddy or direct numerical simulations.

In aerodynamic shape optimisation, many flow simulations are required to iteratively find the optimal shape, which largely restricts the highest level of fidelity in terms of turbulence modelling to RANS-based simulations due computational resources required. While potentially adequate in on-design scenarios, e.g. wing optimisation in steady, level flight, the accuracy in flow predictions can become problematic in more complex settings, e.g. wing near stall, where complex flow phenomena such as separation due to adverse pressure gradients is poorly predicted by RANS-based turbulence models. Recent advances in machine-learning based augmentation of turbulence models for such complex flows have inspired us to investigate their application in aerodynamic shape optimisation to address the aforementioned challenges in existing RANS-based approaches.

In order to investigate the potential of machine learning in enhanced turbulence modelling for aerodynamic shape optimisation a dataset of high-fidelity data is required. As a proof-of-concept we selected the periodic hill case, with the parameterisation by Xiao *et al.* [4], shown in Fig. 1. The original periodic hill case ($\alpha = 1.0$ in Fig. 1) is a simple geometry that features flow separation due to the curved surface, and flow reattachment on a flat plate. Additionally, this case has well-defined boundary conditions, and relatively affordable computational cost. As the analysis of separated flows over curved surfaces is a common characteristic of many practical application, the periodic hill case is a popular benchmark case for high-order simulations (LES and DNS), and evaluation of RANS turbulence models. Breuer *et al.* [5] performed detailed numerical simulations using DNS and highly-resolved LES, with comparisons against PIV experimental measurements up to $Re = 10{,}595$. Rapp and Manhart performed experimental measurements up to $Re = 37{,}000$ [32]. Other examples of numerical simulations include: [6, 21] (DNS), [7, 21, 22] (LES), and [23, 29–31] (hybrid RANS-LES).

While mean DNS fields for the set of geometries in Fig. 1 are available, we choose to generate our own dataset using hybrid RANS-LES methods. The reasons for this are two-fold: aerodynamic shape optimisation is generally performed where the goal is to optimise wall-quantities (e.g. minimising surface drag coefficient/skin friction), however, the aforementioned dataset does not provide these to benchmark the RANS-based predictions; secondly, after shape optimisation is performed using the machine learning-enhanced turbulence model, the flow predictions for the optimised shape must be validated using the higher-fidelity simulations used to generate the training data. Therefore, in order to have full access to all quantities of interest, and a consistent high-fidelity simulation setup for the training and validation data–while avoiding excessive computational requirements in DNS–we generate the data using improved-delayed detached-eddy simulations (IDDES). The available quantities from the DNS dataset will be used to verify our IDDES results.

Many approaches to utilising data assimilation and machine learning techniques for RANS-based turbulence modelling have been introduced over the past decade. These include: generating data-driven Reynolds stress model closures using novel neural network architectures with embedded invariance properties [8]; formulating algebraic nonlinear closures using gene expression programming and symbolic regression [9, 10]; using data for uncertainty quantification, and many more, comprehensively reviewed in the following papers: [1, 11].

Our overall goals in this research is to explore field inversion and machine learning[12, 13]. In the paper the focus is the former step, i.e. field inversion. The field inversion step involves perturbation of the transport equation(s) for an existing turbulence model, and the solution of multiple inverse problems (for varying flow conditions and/or geometries) to reconstruct turbulent mean flow given the high-fidelity data. Field inversion-based flow reconstruction is case specific, i.e. it cannot be used for prediction of flows outside the training dataset. Therefore, machine learning approaches such as deep neural networks (e.g. [12]) or Gaussian processes (e.g. [13]) are used to generalise the model corrections for flow conditions/geometries beyond—but similar to—the training dataset. Benefits of the FIML approach are model-consistency, and the ability to work with relatively sparse data, e.g. [12, 14]. We stress, this paper will only present field inversion results, while the machine learning generalisation is the subject of ongoing work.

The rest of the paper is structured as follows: in Sect. 2 we introduce the IDDES methodology and compare the results against the DNS data; in Sect. 3 we introduce the field inversion approach for turbulent mean flow reconstruction and present the results for the baseline shape ($\alpha = 1.0$); and finally the paper is concluded with an outline of future work in Sect. 4.

2 IDDES Dataset

Hybrid methods utilise a closure model in the near-wall region of the flow, while performing eddy-resolving simulation away from the wall. The predictions with these methods can be significantly more accurate than RANS models, while requiring a coarser mesh than required for wall-resolved LES, or DNS [15, 16]. In this work we use the Spalart-Allmaras IDDES model [17], outlined in the Appendix, to generate the dataset for the parameterised periodic hills.

2.1 Periodic Hill Flow Simulation Overview

The periodic hill flow is driven by a constant pressure gradient which is established by adding a source term in the momentum equation. The Reynolds number, $Re_H = 5,600$ for all the parameterised geometries in Fig. 1, is defined as follows:

$$Re_H = \frac{U_b H}{\nu}, \qquad U_b = \frac{1}{2.035H} \int_{H}^{3.035H} U_x(y)\, dy, \tag{1}$$

where U_b is the bulk velocity, ν is the kinematic viscosity, and U_x is the streamwise velocity component.

Cyclic boundary conditions are applied at the inlet, outlet, and the spanwise directions, while no-slip is imposed at the hills and top wall. Two different mesh resolutions are used. For $\alpha = \{0.5, 0.8, 1.0\}$ the number of cells is $n_x \times n_y \times n_z = 200 \times 160 \times 80 = 2.56 \times 10^6$, while for $\alpha = \{1.2, 1.5\}$, the number of cells is $n_x \times n_y \times n_z = 400 \times 220 \times 80 = 7.04 \times 10^6$. The mesh are deliberately dense to reduce mesh related inaccuracies. The geometry in the spanwise direction is extended by half the streamwise length of the baseline geometry, i.e. $L_z = 9H/2$ following the DNS dataset. The time-step used is $\Delta t = 0.0004$, and the data is collected over a time period of $T = 350H/U_b$.

The simulations are run in OpenFOAM [26], based on the finite-volume method. The convective term in the momentum equation is discretised using the hybrid scheme of Travin *et al.* [27] blending a second-order central difference scheme for the LES regions, and a second-order upwind scheme for the RANS regions. All other terms were discretised using a second-order limited linear differencing scheme. The PIMPLE algorithm is used for the pressure-velocity coupling, which combines the Pressure-Implicit with Splitting of Operators (PISO) and the Semi-Implicit Method for Pressure Linked Equations (SIMPLE) algorithms [28].

2.2 Results and Comparison to DNS

The IDDES results—averaged over time and space in the spanwise (z-axis) direction—will be compared against the DNS data from Xiao *et al.* [4] and some commonly used RANS-based turbulence models implemented in OpenFOAM: $k - \epsilon$ [25], Menter $k - \omega$ [24], and $k - \omega$ SST [20]. For brevity, the focus will be on the baseline periodic hill geometry ($\alpha = 1.0$). The streamwise and wall-normal velocity

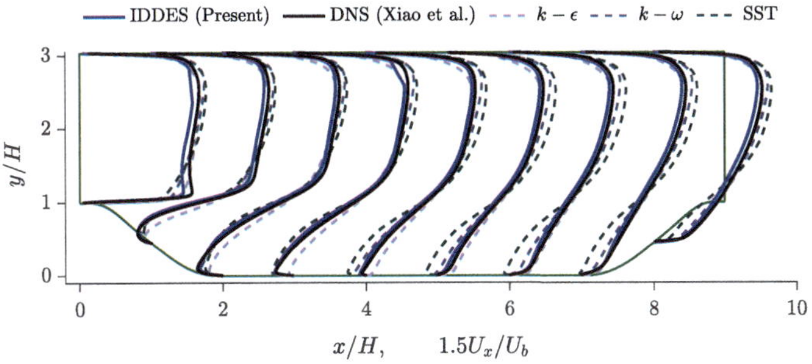

Fig. 2 Comparison of the streamwise velocity profiles

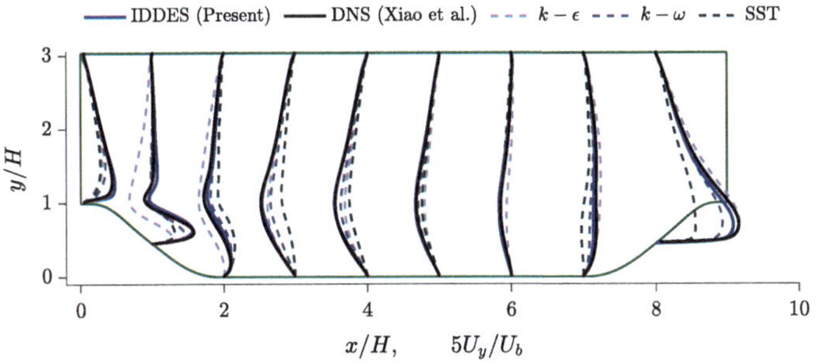

Fig. 3 Comparison of the wall-normal velocity profiles

profiles comparisons are shown in Figs. 2 and 3, respectively. The IDDES results agree well with the DNS reference, while there are discrepancies in all the RANS models considered. The most significant errors in the RANS models are present in separation zone after the first hill crest.

Similarly, profiles of normal and shear Reynolds stresses are shown in Figs. 4 and 5, respectively. It is clear that the IDDES is largely able to match the DNS data, while all the RANS models show significant inaccuracies, especially in the normal Reynolds stress predictions. Figure 6 shows the skin friction predictions. As Xiao *et al.* [4] dataset does not provide wall quantities, we use the DNS data by Krank *et al.* [6] as surrogate. Again, IDDES result closely match the reference data, while the RANS models fail to accurately capture the flow separation and re-attachment, especially, in the case of $k-\epsilon$ and $k-\omega$ SST models, which significantly under-predict or over-predict the size of the separation zone, respectively.

Figure 7 compares the drag coefficient prediction on the hill wall for the different geometries, with the different models. It is clear that the drag rises as the hill crests become steeper. This is due to the higher flow separation, as illustrated in the sizes of the recirculation zones for the steepest ($\alpha = 0.5$) and shallowest ($\alpha = 1.5$) hill geometries, in Fig. 8. The RANS-based drag predictions are higher than the IDDES results, considerably so in the case of $k-\epsilon$ and the $k-\omega$ model.

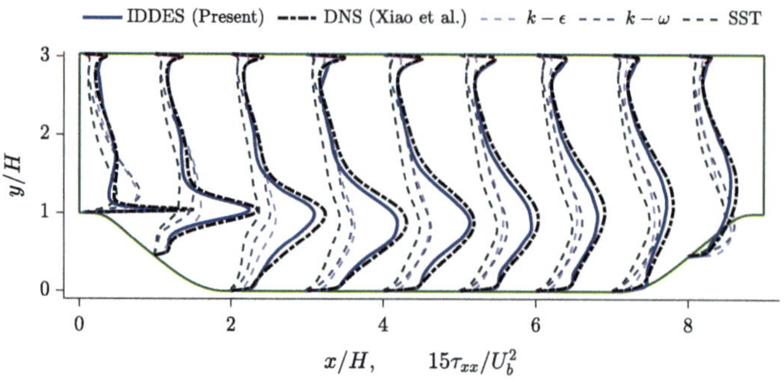

Fig. 4 Comparison of the normal Reynolds stress profiles

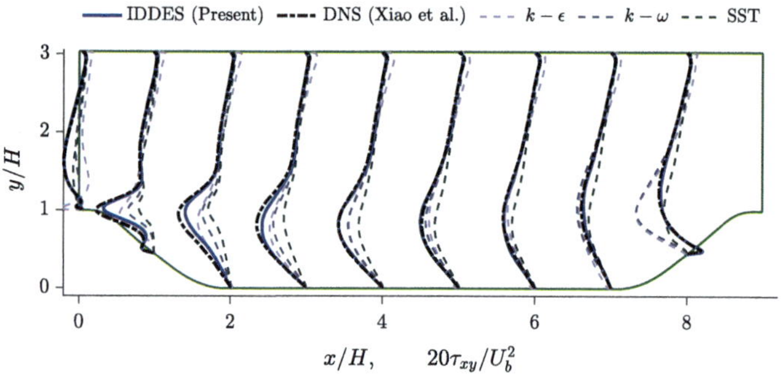

Fig. 5 Comparison of the Reynolds shear stress profiles

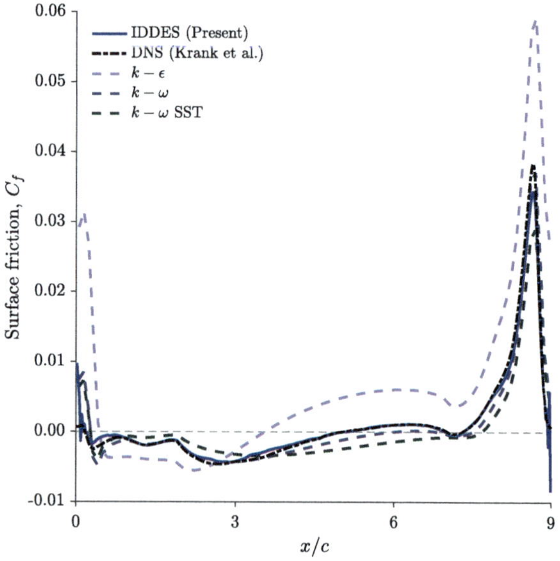

Fig. 6 Comparison of the skin friction on the hill wall

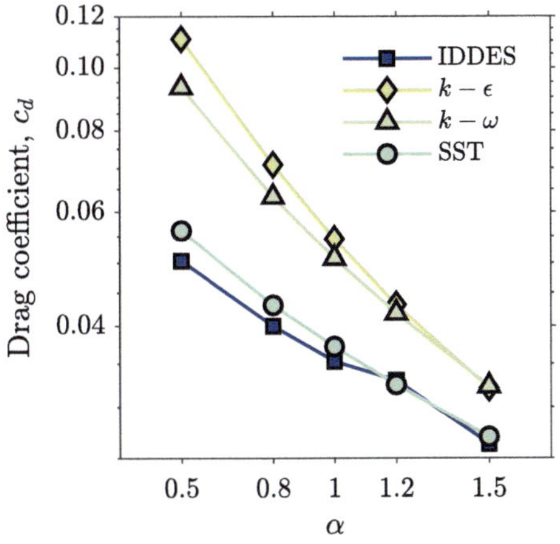

Fig. 7 Comparison of the drag coefficient predictions by the IDDES and various RANS models along periodic hill walls for the various geometries

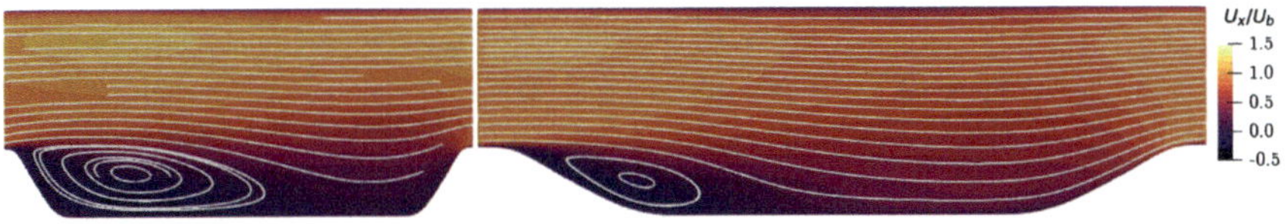

Fig. 8 Normalised streamwise velocity predictions overlaid by the streamlines from the IDDES simulations for the shape $\alpha = 0.5$ (left), and $\alpha = 1.5$ (right)

Finally, comparisons of the root-mean-squared errors for the IDDES and various RANS-based simulations against the DNS dataset are shown in Fig. 9. For the four quantities considered (streamwise and wall normal velocity component, and the normal and shear stresses), the IDDES results show considerably lower errors compared to the RANS-based predictions. The $k-\omega$ model has a marginally lower error for the streamwise velocity predictions compared to the IDDES results, however, is not consistent in the other three quantities considered.

3 Turbulent Mean Flow Reconstruction

In this second half of the paper we use the IDDES data for the baseline shape ($\alpha = 1.0$) to demonstrate RANS-based turbulent mean flow reconstruction using the field inversion approach.

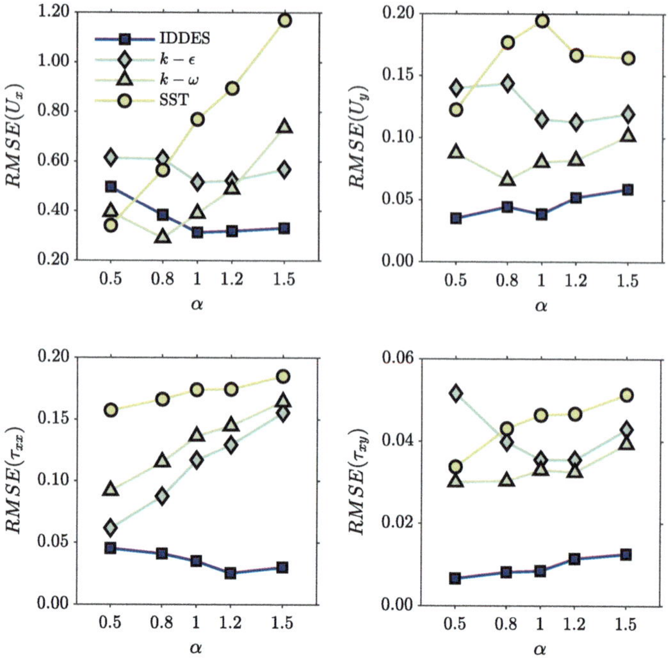

Fig. 9 Root-mean-squared error comparison of the IDDES results compared to three RANS-based results. α, U_x, U_y, τ_{xx}, τ_{xy} represent the five periodic hill geometries in Fig. 1, the streamwise and wall-normal velocity components, and the normal and shear Reynolds stresses, respectively

3.1 Field Inversion Formulation

In field inversion the turbulent flow reconstruction is performed by perturbing one of the transport equations in the turbulence model. We use the two-equation $k-\omega$ SST model as the baseline model in this case, and perturb the turbulence dissipation transport equation ω, by a scalar field β_{FI}. The transport equation for the turbulence dissipation is thus expressed as follows in the general form:

$$\frac{D\omega}{Dt} = \beta_{\mathrm{FI}}\left(\boldsymbol{x}\right)\mathcal{P}_{\omega} + \mathcal{T}_{\omega} - \mathcal{D}_{\omega}, \tag{2}$$

where $\mathcal{P}$, $\mathcal{T}$, and $\mathcal{D}$ are the production, transport, and dissipation terms of the transport equation, respectively. The baseline model is recovered with $\beta = 1$.

With the goal of reducing the error between the baseline $k-\omega$ SST model and the IDDES data, optimum values for β_{FI} is sought in all the mesh cells by minimising an objective function of the following form:

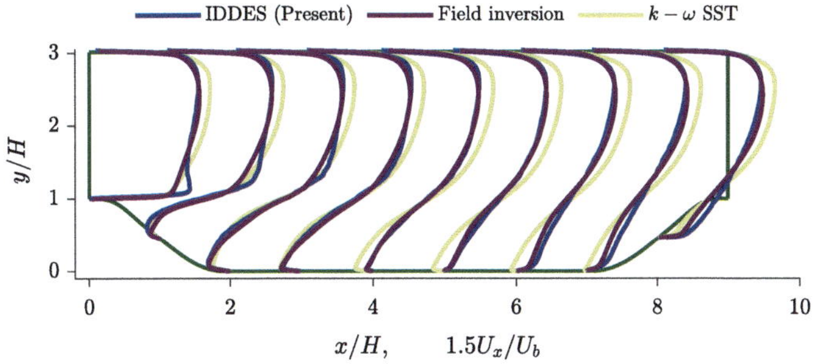

Fig. 10 Comparison of the streamwise velocity profiles

$$\min_{\beta} \quad \mathcal{J} = \left(\sum_{i=1}^{N_d} w_i \frac{1}{\mathcal{J}_{0,i}} \|\mathcal{G}_i(\beta) - \boldsymbol{d}_i\|_2^2 \right) + \lambda \|\beta - \beta_{\text{prior}}\|_2^2, \tag{3}$$

where $\|.\|_2$ is the L_2 norm; the index $i \in \mathbb{R}^{N_d}$ represents the different data quantities (e.g. velocity, and pressure); $\boldsymbol{d}_i$ represents the specific quantity from high-fidelity data, with $\mathcal{G}_i(\beta)$ representing the RANS model equivalent; w_i are the weights; $\mathcal{J}_{0,i}$ are the least-square errors between the baseline model and data; λ is a regularisation parameter to avoid an ill-posed optimisation formulation which is set to 10^{-10}, and β_{prior} is typically assumed to be 1, to bias the solution closer the baseline model. The high fidelity data we use are the streamwise and wall normal velocity fields, and the drag coefficient on the hill wall. This is a multi-objective optimisation problem. In [14] we investigated the effects of different weights w_i for the terms in Eq. (3), and generated a Pareto-optimal front, concluding that equal weighting is appropriate. Therefore, we choose equal weights for all quantities. Due to the high dimensionality of the problem, an adjoint-based optimisation problem is solved, with the implementation documented in [18].

3.2 Results

The streamwise velocity profiles and the lower wall skin friction predictions are shown in Figs. 10 and 11, respectively. The field inversion is able to significantly improve the predictions by the baseline $k-\omega$ SST model. Significant improvements are also achieved in the drag coefficient prediction on the hill wall, and the location of flow reattachment, as shown in Table 1.

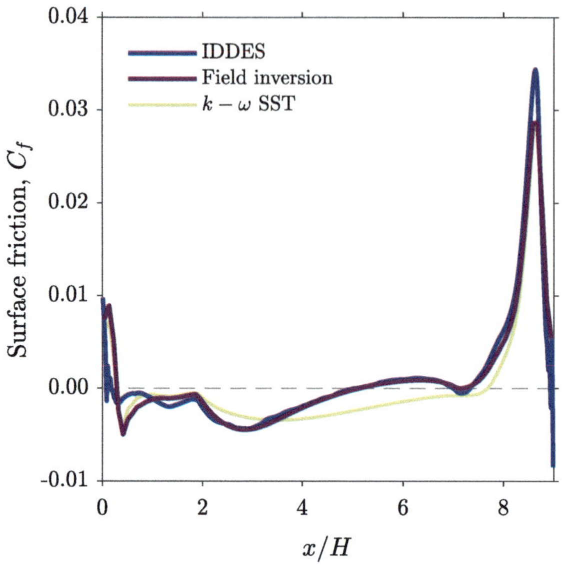

Fig. 11 Comparison of the skin friction prediction on the hill wall

Table 1 Drag coefficient and flow reattachment location comparison on the lower wall for the baseline geometry

Quantity	IDDES	Baseline	Field inversion
C_d	0.0353	0.0372	0.0354
$(x/H)_{\text{re-attach.}}$	5.02	7.68	5.12

The error reduction in the baseline model after mean flow reconstruction is summarised in Table 2. There are significant reductions in error in most quantities of interest (i.e. velocity prediction, and surface quantities), whereas, the errors in the Reynolds stress predictions are marginally reduced. The latter is in line with results previously reported in literature, where field inversion cannot achieve the same level of error reduction in the Reynolds stress predictions as the other quantities of interest (e.g. [19]).

4 Conclusion

We generated and presented turbulent flow simulation results for a set of parameterised periodic hill geometries which involve mild to massive flow separations. The simulations were performed using the improved delayed-detached eddy simulation (IDDES) approach. The IDDES results were compared to reference data from

Table 2 Root-mean-square comparison of the baseline $k-\omega$ SST and the field inversion predictions with respect to the IDDES data

Quantity	RMSE Baseline	RMSE field inversion	% error reduction
U_x	0.8290	0.3521	57.53
U_y	0.1765	0.1023	42.07
τ_{xx}	0.1485	0.1372	7.61
τ_{xy}	0.0418	0.0394	5.57
C_f	0.0021	0.0012	42.25
C_d	1.8749×10^{-3}	0.1190×10^{-3}	93.65

$U_x, U_y, \tau_{xx}, \tau_{xy}, C_f, C_d$ represent the streamwise and wall-normal velocity components, and the normal and shear Reynolds stresses, skin friction and drag coefficient on hill wall, respectively

direct numerical simulations (DNS), and three popular turbulence models—$k-\epsilon$, $k-\omega$, and $k-\omega$ SST—in the context of steady Reynolds-averaged Navier-Stokes (RANS) simulations. IDDES results matched well with the reference DNS data, while all the RANS-based results had varying degrees of discrepancies for the various quantities considered. We then demonstrated turbulent mean flow reconstruction through the adjoint-based field inversion approach to improve the $k-\omega$ SST model. Reconstructed flow results showed significant error reductions in all quantities of interest compared to the original model. Generating high -fidelity datasets using the hybrid RANS-LES approach seems particularly promising for high Reynolds number flows where generating LES or DNS data can be computationally demanding. As mentioned in the introduction, the final goal is to use the generated data for machine-learning-based enhancement of a turbulence model in the context of aerodynamic shape optimisation, which is the subject of our ongoing work.

Appendix: Spalart-Allmaras IDDES model

The surrogate eddy viscosity $\tilde{\nu}$ is defined as [17]:

$$\frac{\partial \tilde{\nu}}{\partial t} + U_i \frac{\partial \tilde{\nu}}{\partial x_j} = c_{b1} \tilde{S} \tilde{\nu} + \frac{1}{\sigma} \left[\nabla \cdot (\tilde{\nu} \nabla \tilde{\nu}) + c_{b2} (\nabla \tilde{\nu})^2 \right] - c_{w1} f_w \left(\tilde{r} \left(\frac{\tilde{\nu}}{l_{IDDES}} \right)^2 \right), \tag{4}$$

where the turbulent eddy viscosity is defined as $\nu_t = f_{v1} \tilde{\nu}$. The functions f_{v1} and f_w are for near-wall corrections, $\tilde{S}$ is the strain rate tensor, the non-dimensional term $\tilde{r}$ is defined as $\nu_t / (\tilde{S} \kappa^2 d_w^2)$, where κ is the von Kármán constant, d_w is the distance from the wall, and $\{\sigma, c_{b1}, c_{b2}, c_{w1}\}$ are model constants. The modified length scale l_{IDDES} is used to switch the transition from the unsteady RANS to scale-resolving LES, and along with the intermediate variables and functions, is defined as:

$$
\begin{aligned}
l_{IDDES} &= \tilde{f}_d\,(1+f_e)\,d_w + (1-\tilde{f}_d)l_{LES},\\
l_{LES} &= C_{DES}\psi\Delta, \quad C_{DES} = 0.65,\\
\Delta &= \min(\max([C_w d_w, C_w h_{max}, h_{wn}])),\\
\tilde{f}_d &= max(1-f_d, f_B),\\
f_d &= 1-\tanh\left[8\left(r_d^3\right)\right],\\
f_B &= \min\left[2\exp\left(-9\alpha^2\right), 1.0\right],\\
\alpha &= 0.25 - \frac{d_w}{h_{max}},\\
f_e &= max\left[(f_{e1}-1), 0\right]\psi f_{e2},
\end{aligned}
\tag{5}
$$

for detailed discussions refer to Shur *et al.* [17].

References

1. Duraisamy, K., Iaccarino, G., Xiao, H.: Turbulence modeling in the age of data. Annu. Rev. Fluid Mech. **51**, 357–377 (2019)
2. Rowley, C., Dawson, S.: Model reduction for flow analysis and control. Annu. Rev. Fluid Mech. **49**, 387–417 (2017)
3. Cai, S., Mao, Z., Wang, Z., Yin, M., Karniadakis, G.: Physics-informed neural networks (PINNs) for fluid mechanics: a review. Acta. Mech. Sin. **37**, 1727–1738 (2021)
4. Xiao, H., Wu, J., Laizet, S., Duan, L.: Flows over periodic hills of parameterized geometries: a dataset for data-driven turbulence modeling from direct simulations. Comput. Fluids. **200**, 104431 (2020)
5. Breuer, M., Peller, N., Rapp, C., Manhart, M.: Flow over periodic hills-numerical and experimental study in a wide range of Reynolds numbers. Comput. Fluids **38**, 433–457 (2009)
6. Krank, B., Kronbichler, M., Wall, W.: Direct numerical simulation of flow over periodic hills up to $Re_H = 10, 595$. Flow Turbul. Combust. **101**, 521–551 (2018)
7. Gloerfelt, X., Cinnella, P.: Large eddy simulation requirements for the flow over periodic hills. Flow Turbul. Combusti. **103**, 55–91 (2019)
8. Ling, J., Kurzawski, A., Templeton, J.: Reynolds averaged turbulence modelling using deep neural networks with embedded invariance. J. Fluid Mech. **807**, 155–166 (2016)
9. Weatheritt, J., Sandberg, R.: The development of algebraic stress models using a novel evolutionary algorithm. Int. J. Heat Fluid Flow **68**, 298–318 (2017)
10. Schmelzer, M., Dwight, R., Cinnella, P.: Discovery of algebraic Reynolds-stress models using sparse symbolic regression. Flow Turbul. Combust. **104**, 579–603 (2020)
11. Xiao, H., Cinnella, P.: Quantification of model uncertainty in RANS simulations: a review. Prog. Aerosp. Sci. **108**, 1–31 (2019)
12. Singh, A., Medida, S., Duraisamy, K.: Machine-learning-augmented predictive modeling of turbulent separated flows over airfoils. AIAA J. **55**, 2215–2227 (2017)
13. Parish, E., Duraisamy, K.: A paradigm for data-driven predictive modeling using field inversion and machine learning. J. Comput. Phys. **305**, 758–774 (2016)
14. Bidar, O., He, P., Anderson, S., Qin, N.: Turbulent mean flow reconstruction based on sparse multi-sensor data and adjoint-based field inversion. AIAA AVIATION 2022 Forum (2022)
15. Fröhlich, J., Terzi, D.: Hybrid LES/RANS methods for the simulation of turbulent flows. Prog. Aerosp. Sci. **44**, 349–377 (2008)
16. Chaouat, B.: The state of the art of hybrid RANS/LES modeling for the simulation of turbulent flows. Flow Turbul. Combust. **99**, 279–327 (2017)

17. Shur, M., Spalart, P., Strelets, M., Travin, A.: A hybrid RANS-LES approach with delayed-DES and wall-modelled LES capabilities. Int. J. Heat Fluid Flow **29**, 1638–1649 (2008)
18. Bidar, O., He, P., Anderson, S., Qin, N.: An open-source adjoint-based field inversion tool for data-driven RANS modelling. AIAA AVIATION 2022 Forum (2022)
19. Singh, A., Duraisamy, K., Zhang, Z.: Augmentation of turbulence models using field inversion and machine learning. 55th AIAA Aerosp. Sci. Meet. p. 0993 (2017)
20. Menter, F., Kuntz, M., Langtry, R.: Ten years of industrial experience with the SST turbulence model. Turbul. Heat Mass Transf. **4**, 625–632 (2003)
21. Balakumar, P., Park, G.: DNS/LES simulations of separated flows at high reynolds numbers. 45th AIAA fluid dynamics conference (2015)
22. Frohlich, J., Mellen, C.P., Rodi, W., Temmerman, L., Leschziner, M.A.: Highly resolved large-eddy simulation of separated flow in a channel with streamwise periodic constrictions. J. Fluid Mech. **526**, 19–66 (2005)
23. Heinz, S., Mokhtarpoor, R., Stoellinger, M.: Theory-based reynolds-averaged navier-stokes equations with large eddy simulation capability for separated turbulent flow simulations. Phys. Fluids. **32**, (2020)
24. Wilcox, D.: Turbulence modeling for CFD. DCW industries La Canada, CA (1998)
25. Launder, B., Spalding, D.: The numerical computation of turbulent flows, pp. 96–116. Numerical Prediction Of Flow, Heat Transfer, Turbulence And Combustion (1983)
26. Jasak, H., Jemcov, A., Tukovic, Z.: OpenFOAM: A C++ library for complex physics simulations. International workshop on coupled methods in numerical dynamics. pp. 1–20 (2007)
27. Travin, A., Shur, M., Strelets, M., Spalart, P.: Physical and numerical upgrades in the detached-eddy simulation of complex turbulent flows. Adv. LES Complex Flows. pp. 239–254 (2002)
28. Robertson, E., Choudhury, V., Bhushan, S., Walters, D.: Validation of OpenFOAM numerical methods and turbulence models for incompressible bluff body flows. Comput. Fluids **123**, 122–145 (2015)
29. Saini, R., Karimi, N., Duan, L., Sadiki, A., Mehdizadeh, A.: Effects of near wall modeling in the improved-delayed-detached-eddy-simulation (IDDES) methodology. Entropy **20**, 771 (2018)
30. Chaouat, B., Schiestel, R.: Hybrid RANS/LES simulations of the turbulent flow over periodic hills at high Reynolds number using the PITM method. Comput. Fluids **84**, 279–300 (2013)
31. Heinz, S., Mokhtarpoor, R., Stoellinger, M.: Hybrid RANS-LES periodic hill flow simulations up to extreme reynolds numbers. AIAA Scitech 2021 Forum (2021)
32. Rapp, C., Manhart, M.: Flow over periodic hills: an experimental study. Exp. Fluids **51**, 247–269 (2011)

Exploring Unsteady Flows

A Modular Multifidelity Approach for Multiphysics Oleo-Pneumatic Shock Absorber Simulations

Ahmed A. Sheikh Al-Shabab, Paulo A. S. F. Silva, Bojan Grenko, Panagiotis Tsoutsanis, and Martin Skote

Abstract A numerical framework is developed and tested to simulate the internal dynamics of oleo-pneumatic shock absorbers. A modular approach is devised to address the multiphysics nature of the problem, starting with three dimensional scale resolving turbulence simulations and two dimensional axisymmetric multiphase URANS simulations. These simulations capture the main dominant aspects of energy dissipation through turbulence, and the multiphase mixing which can affect the working fluid properties. Internal flow simulations are run on a representative shock absorber geometry based on dimensions provided in the validation study together with sizing guidelines from the literature. A lower fidelity two-equation dynamic system solver is used to scan the design space and test the sensitivity towards various design parameters, in addition to identifying parameter combinations that would be of interest to investigate using higher fidelity methods.

1 Introduction

Oleo-pneumatic shock absorbers have been a relatively stable component on aircraft for several decades without witnessing progress comparable to that made in other areas, such as aeroengines or aerospace materials. This is partly due to the combined complexity of the internal flow dynamics and geometries involved, making experimental and computational work challenging and expensive. The multiphysics nature of the internal flow makes it difficult to identify the significant features of the flow field and the conditions under which specific phenomena become important in a given design.

A main difficulty in the investigation of oleo-pneumatic shock absorbers is the lack of widely researched standard geometries with known properties, similar to the NACA-series in the case of aerofoils. Therefore, a baseline geometry was generated

CUFS-2024-11, Cambridge, United Kingdom, 4-5 March 2024

A. A. Sheikh Al-Shabab (✉) · P. A. S. F. Silva · B. Grenko · P. Tsoutsanis · M. Skote
School of Aerospace, Transport and Manufacturing, Cranfield University, College Road, Cranfield, Bedfordshire MK43 0AL, UK
e-mail: A.A.Sheikh-Al-Shabab@Cranfield.ac.uk

J. C. Tyacke and N. R. Vadlamani (eds.), *Proceedings of the Cambridge Unsteady Flow Symposium 2024*, https://doi.org/10.1007/978-3-031-69035-8_8

by using the available information from the drop down experiment used for validation [1], while the missing parameters were estimated based on the general guidelines and sizing data available in the literature [2, 3].

The study of Milwitzky and Cook [1] was chosen as the reference experimental validation case in the present investigation, because it provides detailed drop test experimental measurements that can be used for both high-fidelity and low-fidelity model validation. The shock absorber used is of a small military trainer aircraft landing gear, with a relatively simple geometry, for which some key parameters are available. Therefore, a version of the full three-dimensional geometry can be approximately reconstructed based on these key parameters, together with the sizing guidelines in the literature [2, 3], for use in high-fidelity simulations. In addition, a comprehensive development of a dynamic system model of landing gear shock absorbers is presented by Milwitzky and Cook [1], including tire forces, inclination effects, friction forces and ground forces. This provides a useful basis for a low-fidelity simulation of the system.

The present investigation focuses on the development and validation of high and medium fidelity branches of a multifidelity framework developed to enable efficient generation of shock absorber performance databases and to facilitate design space exploration.

2 Methodology and Numerical Methods

The prediction of shock absorber performance requires accounting for a wide range of physical phenomena that are expensive to compute directly using fully resolved three-dimensional simulations. The main dominant features include the multiphase interaction between hydraulic oil and gas inside the shock absorber, see Fig. 2a, in addition to the turbulent shear flow through the orifice. The multiphase interaction has an influence on the variation of working fluid properties as the stroke progresses, while the developing turbulent flow downstream of the orifice is central to the energy absorption within the shock absorber [2].

It is possible to conduct an integrated multi-physics shock absorber simulation, that accounts for most of the flow physics in a single simulation. However, it is more efficient (and computationally stable) to adopt a modular approach using specialised tools and concentrating on each of the flow physics in turn, particularly when a large number of simulations are planned to explore the design space. In this modular approach, understanding the coupling between the various aspects of flow physics requires careful consideration, and potentially a few integrated fully coupled multi-physics simulations. Nevertheless, the benefit of taking advantage of the most efficient and well developed tool in each area is attractive. The chosen numerical solvers have been validated on model problems that are representative of the flow physics of interest [4].

Adopting a multi-fidelity approach, Fig. 1, increases the efficiency of design space exploration by using more expensive high-fidelity scale resolving simulations to provide fundamental flow physics understanding and to calibrate lower fidelity models

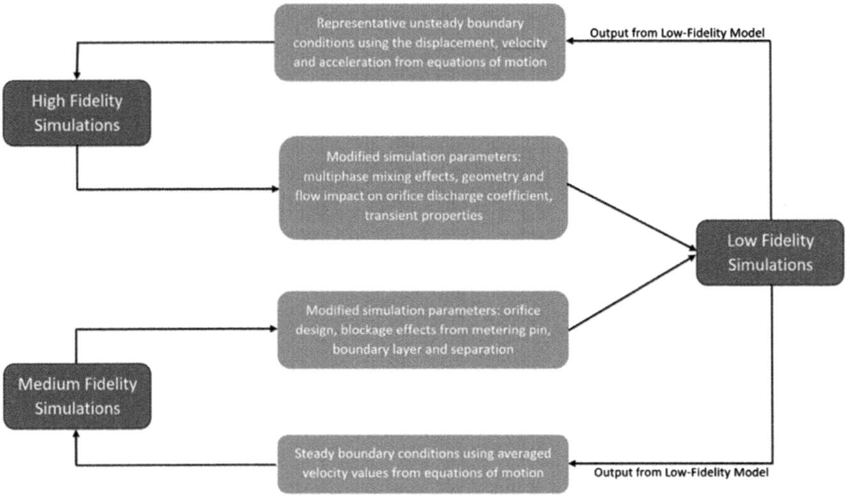

Fig. 1 Multi-fidelity interaction to refine discharge coefficient estimate and provide boundary conditions

in different flow regimes. Medium-fidelity models such as RANS simulations can also be utilised as an intermediate solution to conduct design refinement and fine tune low-fidelity models. RANS models have been shown to be effective at predicting free shear layer deflections in design studies [5, 6], which could be relevant here in the presence of different metering pin or valve flap designs. However, the two-equation dynamic system model low-fidelity solver [1, 4] will be tasked with carrying out the majority of simulations in design studies or database generation activities. It can also provide case specific boundary conditions for high-fidelity simulations from the relative piston displacement, velocity and acceleration. On the other hand, high-fidelity data can be used to calibrate critical low-fidelity solver parameters to improve prediction of the later stages of the shock absorber stroke.

2.1 Multiphase Simulation

Multiphase simulation cases are conducted using ANSYS Fluent 2019 R2 [7]. The cases are set up as 2D half sections with an axisymmetric boundary condition along the axis of symmetry. A Pressure based solver is used with the unsteady RANS $k - \omega \ SST$ [8] turbulence model. The ideal gas law is applied to account for gas compressibility. The domain is filled with two phases: hydraulic oil *MIL-H-5606*, and air, in accordance with the drop test [1]. A no-slip boundary condition is used on the side and top shock absorber walls, while an axisymmetric boundary condition is applied along the axis, and a variable inlet velocity is applied at the bottom wall. These boundary conditions are annotated on the computational domain and mesh visualization in Fig. 7. The time dependent inlet velocity is designed to replicate the shock absorber stroke rate, as an alternative to a moving mesh approach, because it reduces the computational effort and allows better control over mesh quality. The spatial discretization schemes are PRESTO (PREssure STaggering Option) scheme for pressure, modified HRIC for volume fraction, and second order upwind for the rest of the flow variables. The multiphase modelling method is an explicit formulation of the volume of fluid scheme with sharp interface modelling. A dual-time stepping approach is used with a maximum of 40 sub-iterations per physical time step. Variable time stepping method is applied with a CFL number constraint of 0.3, and a minimum allowable physical time step of $1e-8$ seconds.

2.2 *Turbulent Shear Layer Simulation*

OpenFoam (v2006) [9, 10] was used to conduct single phase time-dependent scale resolving turbulent shear layer simulations. OpenFoam uses an unstructured finite volume approach to simulate fluid flow, which facilitates the handling of complex geometries that are expected in industrial type applications, as in the case of the shock absorbers investigated here. Field properties are stored at the centroid of each cell and then calculated at each cell surface "face" by interpolating the values at the centroids of the two cells sharing that face. Conservation laws are then applied based on theses face values, in accordance with the chosen discretization scheme.

The PIMPLE algorithm [11] is selected in the present simulations, as it provides good stability and convergence control options. PIMPLE utilises an outer loop based on the PISO algorithm [12] to capture the field development in time, while an inner loop based on the SIMPLE algorithm [13] solves each field instance as a steady problem. The interpolation method used is a second order linear interpolation, which is a standard choice in many finite volume solvers. The convective term is discretized using a second order limited linear scheme, which provides a balance between accuracy and stability through limiting the nonlinear term. While the diffusive term is discretized using a second order linear scheme and temporal integration is conducted using a second order backward difference method. An adjustable time step is used to insure that the Courant–Friedrich–Lewy (CFL) number remains below 1.0 throughout, to aid in achieving a balance between efficiency and stability during these simulations.

After the discretization process, the resulting linear system for the pressure field is solved for using a Preconditioned Conjugate Gradient (PCG) method, while the velocity field system is found using a Preconditioned Bi-Conjugate Gradient method (PBiCG). The WALE subgrid scale model [14] is used to model the eddy viscosity of the subgrid scales in the Large Eddy Simulation, (LES) framework. The flow field is allowed to develop by allowing multiple through-flows across the entire domain, then time averaging is conducted until statistical convergence is observed for all turbulent properties of interest.

3 Computational Setup

The baseline orifice geometry, illustrated in Fig. 2, has an orifice with semi-circular lips. The rounded outer (upstream) edge of the orifice helps avoid any boundary layer separation on the orifice's inner surface, while the accelerating flow under the influence of the favourable pressure gradient through the orifice allows relatively strong flow turning in the flow direction so that the flow can be tangential to the axial (streamwise) direction at the orifice exit. The three dimensional simulation is conducted on a simplified geometry of the shock absorber that focuses on the orifice nearfield region and the developing shear layer in its vicinity.

Grids of varying resolution were generated and tested to establish grid independence and optimise the setup, where possible. The resolution in the shear layer region was tested using a preliminary simulation on a grid with 185,000 nodes. This included investigating the estimated mixing layer thickness near the orifice tip to determine the grid resolution required there. The boundary conditions are illustrated in Fig. 2b, while further information about the case is provided in Table 1.

The inlet boundary condition, in the single phase 3D simulation, is taken to be a constant vertical flow through the shock absorber $V = 1.0$ m/s as a representative figure of an approximately time-averaged value over the stroke period from the Milwitzky and Cook [1] drop test. The stroke shape and value, corresponding to the drop test, can be seen in Fig. 3b, where this figure was obtained using a dynamic

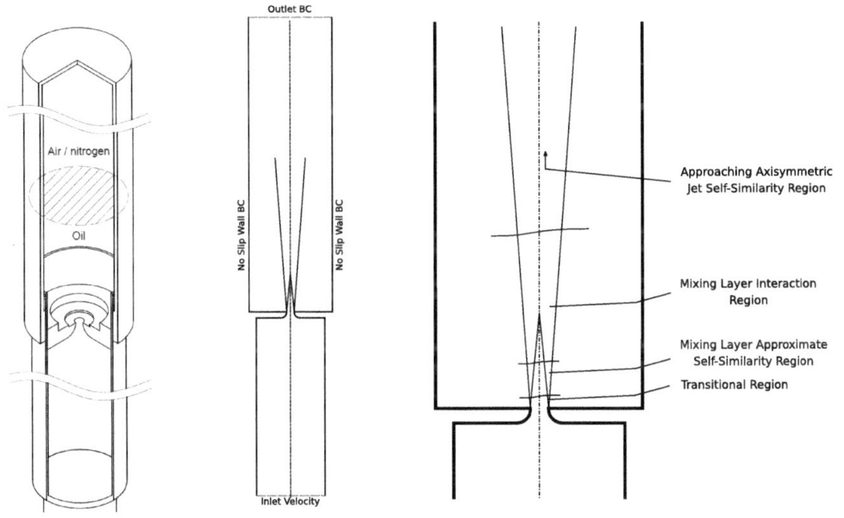

(a) Schematic (b) Setup (c) Free shear layer development

Fig. 2 Shock absorber geometry and setup details, taken from Fig. 1 in Sheikh Al-Shabab *et al.* [15]. **a** Schematic view of the simplified shock absorber geometry; **b** The computational domain with the applied boundary conditions; **c** Axisymmetric shear layer development regions, following Bradshaw et al. [16]

Table 1 Scale resolving 3D axisymmetric free shear layer simulation

Simulation parameter	Setting
Working fluid type	Hydraulic Oil (AN-VV-O-366B)
Hydraulic oil density, ρ_h (kg/m^3)	869
Hydraulic oil kinematic viscosity, ν_h (m^2/s)	1.35×10^{-5}
Time step	$\approx 8 \times 10^{-7}$
Orifice diameter, m	8.128×10^{-3}
Lower chamber diameter, m	7.462×10^{-2}
Upper chamber diameter, m	8.255×10^{-2}
Inlet velocity, m/s	1.0
$Re = \frac{U_{orifice} D_{orifice}}{\nu_h}$	3×10^4
Total grid nodes	1.8×10^6

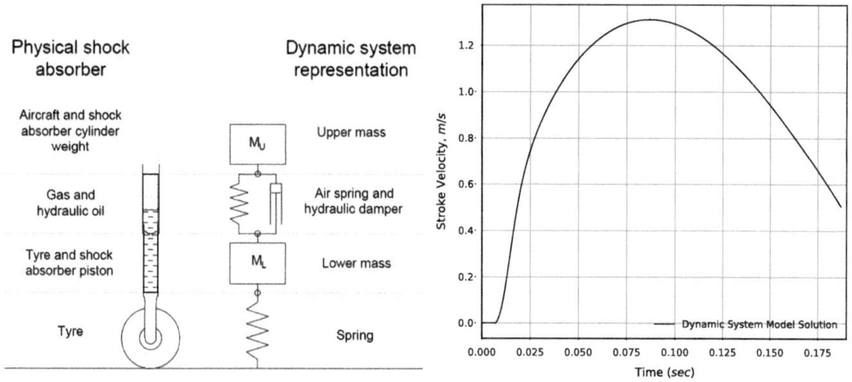

(a) 2DOF Dynamic system model (b) Stroke Rate

Fig. 3 Shock absorber stroke velocity profile from a dynamic system model simulating a drop test, which is used to estimate the inlet velocity value in the LES study

system model simulation of the drop test following the analysis in Milwitzky and Cook [1]. The initial delay in the response of the shock absorber, the flat region at time = 0.0 s, is due to the force required to deflect the tire and other internal effects (e.g., the gas pressure force), which leads to the shock absorber behaving as a rigid body during the early stages of impact; however, this is quickly overcome by the impact resulting in the subsequent stroke profile.

The flow in the present simulation is assumed to be incompressible, which is justified due to the low Mach number prevalent throughout the computational domain ($M < 0.3$), and, since the focus of the study is on the free shear layer under a constant inlet velocity, we ignore the influence of the transient compression pressure waves associated with the movement of the piston in a real shock absorber. This assumption helps to increase the efficiency of the simulation by allowing the numerical stability to be determined by the hydrodynamic time-scale rather than the more restrictive acoustic time-scale.

The edge of the free shear layer for each profile was found locally based on the vorticity magnitude and streamwise velocity value. Then the high and low velocity values across the profile were defined, and their difference was used for normalisation of the velocity values across that specific profile. The self-similarity distance variable η is defined according to Eq. (1) where the capital vertical coordinate Y has its origin at the orifice exit.

$$\eta = \frac{x - D_{orifice}/2}{Y}, \tag{1}$$

It should be noted that, in the 3D single phase simulation, the streamwise direction is along the vertical y-direction with the corresponding velocity component and turbulent (Reynolds) stress component, $\overline{v'v'}$. For the purposes of analysing the near-field free shear layer development, a two-dimensional (2D) section (x-y plane at $z = 0$) is extracted from the three-dimensional domain, where the mixing layer is quasi two-dimensional within a distance equivalent to a few orifice diameters downstream of the orifice exit. In this region, the mixing layer is expected to exhibit standard behaviour that can be used to validate the simulation against similar mixing layer characteristics reported in experiments, which is a substitute for the lack of internal shock absorber data available in the literature.

4 Results and Discussion

4.1 Single Phase Turbulent Shear Layer

Single phase simulations are used to focus on the orifice nearfield and the impact of modifications made in that region on the flow development. This includes the development of turbulent flow downstream, the pressure gradient around the orifice and its susceptibility to cavitation, as well as the overall hydraulic resistance and its likely impact on overall shock absorber performance.

The development of flow instabilities and turbulence in the free shear layers downstream of the orifice is visualised in Fig. 4, using the vorticity magnitude and Q-criterion, found by plotting iso-surfaces of the second invariant of the velocity strain rate tensor as shown in Fig. 4b. Driven by a strong primary instability, the development of increasingly complex flow structures in the mixing layer is observed as initially quasi two-dimensional instabilities, such as ring vortices, develop into more complex structures and breakdown to progressively smaller scales, until the full turbulent cascade is established further downstream.

Vorticity magnitude contours are more clearly shown in Fig. 4a, where high vorticity magnitude values in the turbulent free shear layers downstream of the orifice are expected as a consequence of the strong primary instabilities and the turbulent flow

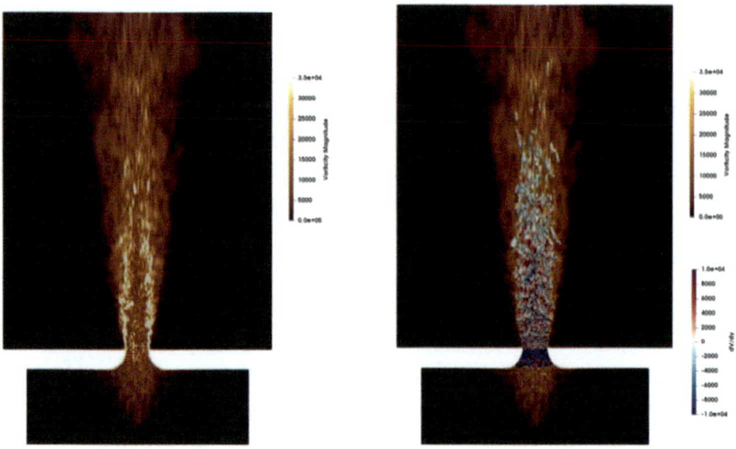

(a) Vorticity magnitude contours.

(b) Q Criterion iso-surfaces superposed on vorticity magnitude contours.

Fig. 4 Flow development in the vicinity of the orifice

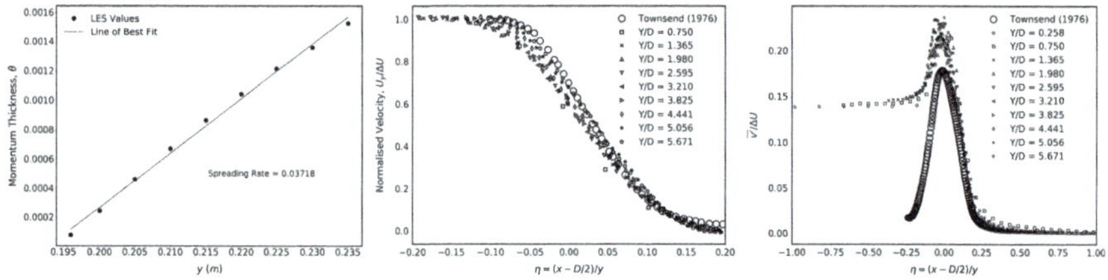

(a) Spreading rate

(b) Streamwise velocity

(c) Crossflow turbulent stress

Fig. 5 Free shear layer spreading rate and self-similarity validation. Plots modified from Sheikh Al-Shabab *et al.* [15]

in that region. However, it is also found that there is a region of significant vorticity magnitude upstream of the orifice. This is likely caused by the strong blockage effect of the orifice causing rapid flow radial convergence towards the axis of symmetry, resulting in a relatively *noisy* orifice jet core.

The development of the free shear layer downstream of the orifice can be validated against standard results for mixing layers from the literature [17, 18]. The spreading rate, shown in Fig. 5a, is calculated based on the growth rate of the shear layer's momentum thickness, where Jimenez [18] devised the following relation to fit the reported experimental measurements for a range of mixing layer stream ratios:

$$SR_{\theta} = \frac{d\theta}{dx} = 0.036\beta, \quad \beta = \frac{U_{High} - U_{Low}}{U_{High} + U_{Low}} \tag{2}$$

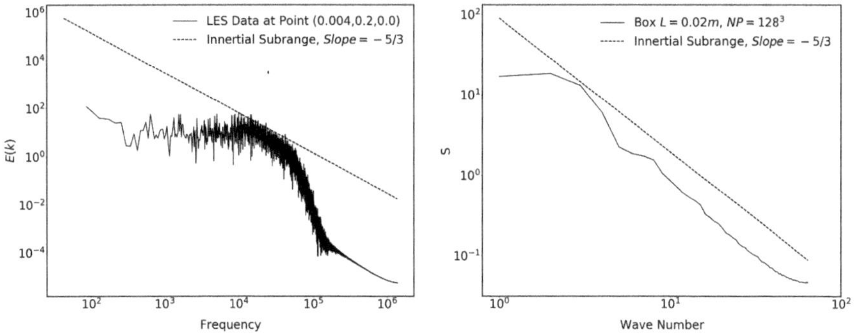

(a) Time spectrum at a point in mixing layer. (b) Spacial spectrum over a box.

Fig. 6 Spectra downstream of orifice

Noting that $\beta \approx 1$ for a single stream mixing layer, the calculated spreading rate value here of $SR_\theta = 0.037$ is within the expected range, particularly considering the spread in the mixing layer data [18] used to arrive at Eq. (2).

The tendency of the mixing layer towards self-similarity is also demonstrated in Fig. 5b, c for the streamwise velocity and the the cross flow Root Mean Squared (RMS) component $\sqrt{\overline{u'u'}}$ respectively. This state is reached once the free shear layer passes the initial transitional stage near the tip and enters the turbulent regime around one orifice diameter downstream of the exit, where the profiles of turbulent flow variables taken across the shear layer tend to collapse onto a single profile once they are normalised appropriately [17]. The first profile plotted at $Y/D_{orifice} = 0.75$ has a relatively larger deviation from the self-similar profile while further downstream profiles exhibit increasingly better agreement with the self-similar one. The higher turbulent stress values on the inside of the jet through the orifice are consistent with the high unsteadiness observed in the core of the jet in Fig. 4a.

From a design prospective, temporal and spatial energy spectra, Fig. 6, are important to assess the critical regions of energy dissipation in different orifice designs and configurations. Turbulent kinetic energy spectra are calculated from the unsteady velocity fluctuations time history at a point in Fig. 6a, while in Fig. 6b a box spanning the central region downstream of the orifice including the shear layers is used following Tsoutsanis [19]. Both spectra plots show the presence of an inertial subrange consistent with a developed turbulent cascade in their corresponding location. This will be further developed when applied to more complex designs, that could include metering pins, snubber chambers and other internal details.

4.2 Multiphase Simulation

Multiphase simulations are conducted on an unstructured two-dimensional axisymmetric domain, where the layout and mesh are shown in Fig. 7, with an unsteady velocity inlet profile corresponding to the piston stroke velocity given in Fig. 3b. The geometry used in these simulations, Fig. 2a, is simplified from the original to focus on the main multiphase interaction of interest, while reducing the geometrical complexity of orifice support tubes, cylinder telescoping bearings and snubber chamber details. Particularly since the details of these features are not publicly available in the literature.

The validation of the drop test case is performed against the experimental measurements of Milwitzky and Cook [1], which consist of the measured total force exerted by the shock absorber in the time span from impact at $0.0s$ to the end of the main stroke at $0.16s$. The total force calculated from the present simulation is compared with the experimental measurements in Fig. 8a, where the agreement is found to be strong in the first half of the stroke until roughly the maximum force point, after which there is a nearly constant underprediction of the total force by the computational simulaton relative to the experimental measurements. The observed discrepancy in the advanced stages of the stroke isn't surprising due to the increasing complexity of the flow physics and the likely impact of simplifying assumptions.

It is worth noting that the total force can't be extracted directly from the simulation in the manner an accelerometer can record the total force in the drop test. instead, the hydraulic and pneumatic forces are computed separately and then added to find the total force values. The individual curves of the pneumatic and hydraulic forces are also plotted in Fig. 8a to provide an insight into their contribution to the total

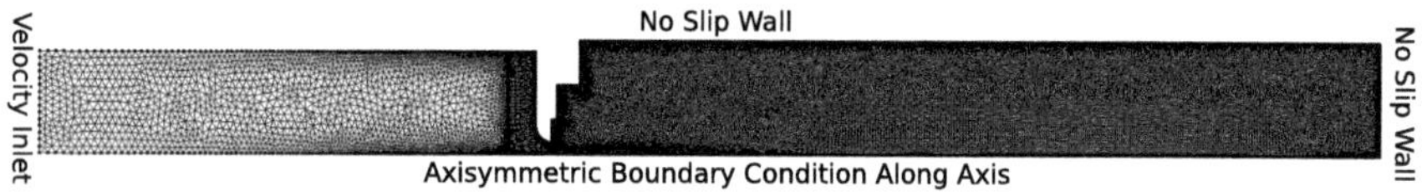

Fig. 7 Simplified oleo pneumatic shock absorber 2D half domain mesh for validation case of Milwitzky and Cook [1]

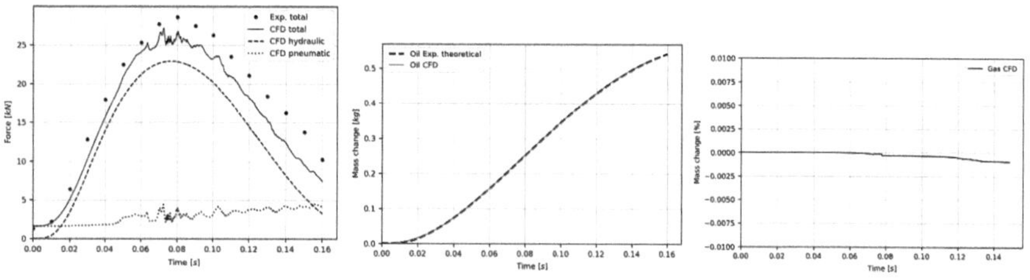

(a) Force validation plot (b) Hydraulic oil mass (c) Gas mass

Fig. 8 High-fidelity multiphase drop test case validation by comparison with force measurements and checking phase mass conservation

and ensure their overall shape and values are reasonable. The pneumatic force is calculated using averaged pressure values over the top wall of the shock absorber, accounting for the fact that the jet through the orifice impacts the central portion of the upper wall and also accounting for the strong unsteadiness in the circulating flow reaching the top of the shock absorber. The pneumatic pressure plot follows the approximate shape of the theoretical pressure increase expected of an ideal gas under compression.

The hydraulic force, on the other hand, is less straightforward to calculate [20], as it cannot be directly measured from the available parameters in simulations but is rather calculated using Eq. (3).

$$F_h = \rho \frac{A_h^3}{2C_D^2 A_0^2} \dot{s}|\dot{s}|, \tag{3}$$

where A_h is the hydraulic area, C_D is the discharge coefficient, A_0 is the orifice area, and $|\dot{s}|$ is the stroke rate or inlet velocity in the case of the static mesh in the present simulation.

In addition, the discharge coefficient C_D is an influential parameter in the hydraulic force calculation. It is calculated using Eq. (4).

$$C_D = \frac{Q_{experimental}}{Q_{theoretical}} = \frac{\dot{m}}{\rho A_0 u_{max}}, \tag{4}$$

Where $\dot{m}$ is the mass flow through the orifice, and u_{max} is the maximum velocity at the point of venna contracta, downstream of the orifice, The maximum velocity is taken as maximum value from probes placed downstream of the orifice as the location of the venna contracta can show some dependency on the stroke rate.

Further validation of high-fidelity multiphase simulation was performed by monitoring the mass conservation of the oil and gas phases, as shown in Fig. 8b, c for oil and gas respectively. This is important to ensure that the mesh is fine enough to resolve the smaller bubbles when the violent multiphase mixing process takes place. The oil phase is checked against the calculated rise in mass due to the inflow implemented to simulate the stroke rate, where it is shown in Fig. 8b that there is a nearly perfect match between the calculated and actual oil mass present in the simulation. On the other hand, the gas phase does show a slight loss of mass towards the end of the stroke in Fig. 8c, when the mixing is most vigorous. However, the percentage decrease of gas mass is less than 0.001 %, and thus insignificant for the purposes of this study.

The phase fraction contours in Fig. 9a showing the internal multiphase interaction at three different time values, clearly demonstrate the complexity of the interaction as the stroke progresses. The applied stroke rate causes the oil to rise through the orifice axis reaching the top wall. Then, The oil is deflected towards the side walls encapsulating the gas and creating a large recirculation zone extending along the full length of the upper chamber. The severity of the phase mixing increases as the oil breaks down the main gas mass into smaller bubbles and as those bubbles impinge on

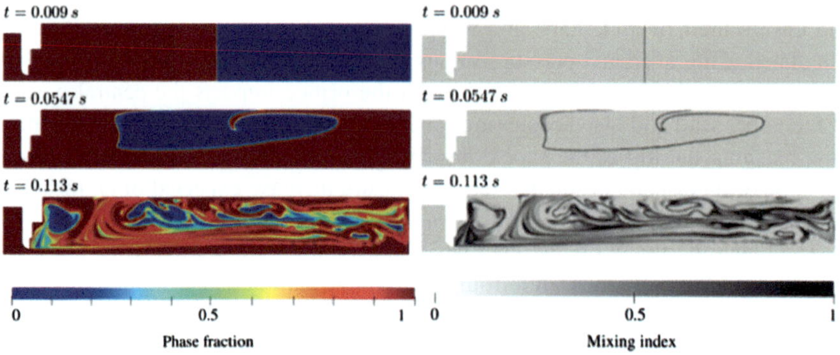

(a) Phase fraction (b) Mixing index

Fig. 9 Multiphase contour plots

the orifice jet and the shear layers on its sides, causing the gas bubble to breakdown and disperse further as parts of it get convected with the jet. Some of the gas is pushed into a smaller recirculation region on the wall side of the orifice plate. To facilitate quantitative assessment of multiphase mixing, different forms of a mixing quality index have been developed [21–23]. It is defined here according to Eq. (5).

$$MI = \Sigma(1 - \sqrt{(2\alpha - 1)^2}) \tag{5}$$

The mixing index, Eq. (5) allows the interface between the two phases to be visualised at a given α value, as seen in Fig. 9b for $\alpha = 0.5$. In addition, a total interface area or length can be calculated by multiplying the mixing index value by a representative local cell area or length, as in Fig. 10. Which enables the assessment of multiphase mixing severity in different shock absorber designs systematically.

Finally, it is also possible to monitor low pressure regions and their extent, as illustrated in Fig. 11 to evaluate the susceptibility to cavitation in various operating conditions. Detecting cavitation is important because of the excessive forces often associated with the collapse of cavitation bubbles, and their potential to reduce the lifespan of internal components. While cavitation modelling is not included in the present multiphase simulation, regions of low pressure or sever pressure gradients in the domain could indicate the need for a dedicated cavitation simulation under these conditions. These simulations can then be run on a dedicated high order multiphysics solver [25] to investigate the full extent of cavitation impact on cases of interest.

5 Conclusions

The internal dynamics of oleo-pneumatic shock absorber were simulated using a geometry developed from the literature. A modular approach was used to tackle the complex multiphysics problem, starting with the most critical aspects of shock

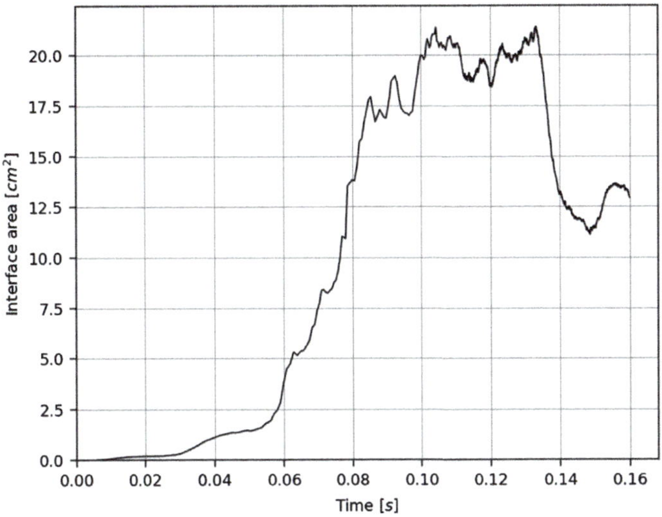

Fig. 10 Multiphase mixing index time history

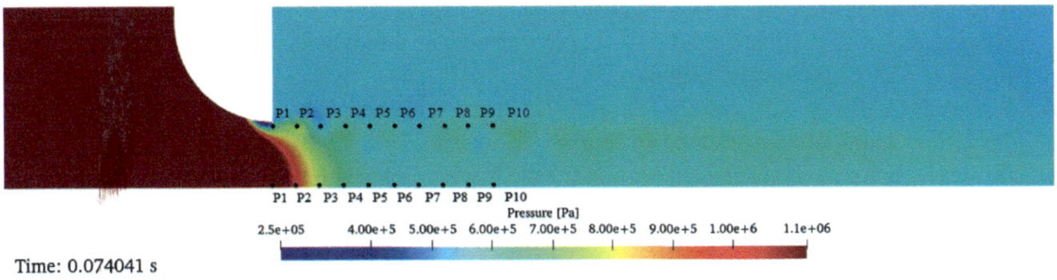

Fig. 11 Multiphase drop test simulation – pressure field around the orifice with data probes position displayed, Fig. 8 in Sheikh Al-Shabab *et al.* [24]

absorber flow physics; the turbulence development downstream of the main orifice and the multiphase mixing between oil and gas. The discussion focused on validating the simulations against drop test measurements and some standard physical results, in addition to assessing quantitative parameters that can be incorporated within an automated numerical framework. Low-fidelity simulations, based on a dynamic systems model of the shock absorber, were used to provide quick return simulations to explore the design space and provide case specific boundary conditions for higher fidelity simulations.

Future work will focus on adding additional areas of the relevant multiphysics affecting shock absorber operation, through simulations focusing on cavitation, heat transfer, compressibility etc. Furthermore, low-fidelity simulations will be calibrated to modify properties and parameters with stroke displacement, in order to optimise their performance over the entire stroke.

Acknowledgements This research was funded by Innovate UK grant number 10002411.
Data Access The underlying data can be accessed using the following DOI: 10.17862/cranfield.rd.24755505.

References

1. Milwitzky, B., Cook, F.E.: Analysis of landing-gear behavior, NACA Technical Note, no. TN-2755 (1952)
2. Currey, N.: Aircraft Landing Gear Design: Principles and Practices. AIAA education series, American Institute of Aeronautics and Astronautics (1988)
3. Schmidt, R.K.: The design of aircraft landing gear. SAE International (2021)
4. Sheikh Al-Shabab, A.A., Vitlaris, D., Lin, Z., Grenko, B., Tsoutsanis, P., Antoniadis, A.F., Skote, M.: Numerical investigation of oleo-pneumatic shock absorber: Setup and validation, WCCM-ECCOMAS2020
5. Al-Shabab, A.A.S., Tucker, P.G.: Toward active computational fluid dynamics role in open jet airfoil experiments design. AIAA J. **56**(8), 3205–3215 (2018)
6. Al-Shabab, A.S., Tucker, P.: Rans prediction of open jet aerofoil interaction and design metrics. Aeronauti. J. **123**(1266), 1275–1296 (2019)
7. ANSYS Fluent, Release 2019 R2, Theory guide (2019)
8. Menter, F.R.: Two-equation eddy-viscosity turbulence models for engineering applications. AIAA J. **32**(8), 1598–1605 (1994)
9. Weller, H.G., Tabor, G., Jasak, H., Fureby, C.: A tensorial approach to computational continuum mechanics using object-oriented techniques. Comput. Phys. **12**(6), 620–631 (1998)
10. Jasak, H.: Error analysis and estimation for the finite volume method with applications to fluid flows. PhD thesis, Imperial College London (1996)
11. Holzmann, T.: Mathematics, numerics, derivations and openfoam®. Holzmann CFD, Loeben, Germany (2016)
12. Issa, R.: Solution of the implicitly discretised fluid flow equations by operator-splitting. J. Comput. Phys. **62**(1), 40–65 (1986)
13. Patankar, S., Spalding, D.: A calculation procedure for heat, mass and momentum transfer in three-dimensional parabolic flows. Int. J. Heat Mass Transf. **15**(10), 1787–1806 (1972)
14. Nicoud, F., Ducros, F.: Subgrid-scale stress modelling based on the square of the velocity gradient tensor. Flow Turbul. Combust. **62**(3), 183–200 (1999)
15. Sheikh Al-Shabab, A.A., Grenko, B., Vitlaris, D., Tsoutsanis, P., Antoniadis, A.F., Skote, M.: Numerical investigation of orifice nearfield flow development in oleo-pneumatic shock absorbers. Fluids **7**(2), 54 (2022)
16. Bradshaw, P., Ferriss, D.H., Johnson, R.: Turbulence in the noise-producing region of a circular jet. J. Fluid Mech. **19**(4), 591–624 (1964)
17. Townsend, A.: The structure of turbulent shear flow. Cambridge university press (1980)
18. Jimenez, J.: Free shear flows. Universidad Politecnica de Madrid. Retrieved from torroja. dmt. upm. es, Madrid (2013)
19. Tsoutsanis, P., Nogueira, X., Fu, L.: A short note on a 3d spectral analysis for turbulent flows on unstructured meshes. J. Comput. Phys. **474**, 111804 (2023)
20. Ding, Y.W., Wei, X.H., Nie, H., Li, Y.P.: Discharge coefficient calculation method of landing gear shock absorber and its influence on drop dynamics. J. Vibroeng. **20**(7), 2550–2562 (2018)
21. Solehati, N., Bae, J., Sasmito, A.P.: Numerical investigation of mixing performance in microchannel t-junction with wavy structure. Comput. Fluids **96**, 10–19 (2014)
22. Wang, S., Huang, X., Yang, C.: Mixing enhancement for high viscous fluids in a microfluidic chamber. Lab Chip **11**(12), 2081–2087 (2011)

23. Nguyen, Q., Papavassiliou, D.V.: Quality measures of mixing in turbulent flow and effects of molecular diffusivity. Fluids **3**(3), 53 (2018)
24. Sheikh Al Shabab, A., Grenko, B., Vitlaris, D., Tsoutsanis, P., Antoniadis, A., Skote, M.: Numerical investigation of oleo-pneumatic shock absorber: a multi-fidelity approach, In: ECCOMAS Congress 2022 - 8th European Congress on Computational Methods in Applied Sciences and Engineering, ECCOMAS, (2022)
25. Antoniadis, A.F., Drikakis, D., Farmakis, P.S., Fu, L., Kokkinakis, I., Nogueira, X., Silva, P.A., Skote, M., Titarev, V., Tsoutsanis, P.: Ucns3d: an open-source high-order finite-volume unstructured cfd solver. Comput. Phys. Commun. **279**, 108453 (2022)

Large-Eddy Simulation of Structured Porous Coatings

C. J. Naicker, E. J. G. Arcondoulis, and James C. Tyacke

Abstract Porous coated cylinders have the potential to reduce the vortex shedding tonal noise as well as broadband noise compared to an uncoated cylinder, although the mechanism of noise reduction is not fully understood. Recent studies show that such benefits can be obtained for both randomised porous coatings as well as structured porous coatings. These coatings have been gaining interest due to their potential application in various fields, such as landing gear struts, wind turbine structures and heat exchangers. The current study presents the first fully resolved Large-Eddy Simulation (LES) of a Structured Porous Coated Cylinder (SPCC). High-fidelity LES data is presented modelling a spanwise section at $Re = 73,000$ (based on outer diameter). Wake profiles of streamwise mean and RMS velocity profiles are presented and compared with an uncoated cylinder and forms a basis to inform and improve Darcy-Forchheimer porous modelling. Boundary layer development on the inner cylinder and shear layer trajectory is also presented. Unsteady data is recorded using a novel solver-independent HDF5-based I/O library developed by Upstream CFD GmbH.

1 Introduction

The application of passive flow control techniques to bluff bodies has been gaining interest due to the need to reduce noise produced by such bodies, including landing gear struts, wind turbine structures and heat exchangers. Porous coatings are one

CUFS-2024-8, Cambridge, United Kingdom, 4-5 March 2024.

C. J. Naicker (✉) · J. C. Tyacke
Brunel University London, Kingston Lane, Uxbridge UB8 3PH, UK
e-mail: clinton.naicker@brunel.ac.uk

J. C. Tyacke
e-mail: james.tyacke@brunel.ac.uk

E. J. G. Arcondoulis
Bristol University, University Walk, Bristol BS8 1TR, UK
e-mail: elias.arcondoulis@bristol.ac.uk

J. C. Tyacke and N. R. Vadlamani (eds.), *Proceedings of the Cambridge Unsteady Flow Symposium 2024*, https://doi.org/10.1007/978-3-031-69035-8_9

such example of passive flow control. Boorsma et al. [1] tested perforated fairings as passive noise control mechanism for landing gear struts, for $2 \times 10^5 < Re < 4 \times 10^5$. The two shear layers were found to exhibit half the vorticity of the solid fairing, resulting in weaker, smaller scale vortices. Sueki et al. [2] tested various materials as porous coatings for high-speed pantographs, up to Re = 300,000 based on outer diameter. It was found that porous materials with approximately 13 Pores Per Inch (PPI) showed the greatest aerodynamic noise reduction, and that both porous metal and porous urethane had the same level of noise reduction. In addition to the application to bluff bodies, porous coatings have also been applied to airfoils. Geyer et al. [3] created 16 airfoils from different homogeneous materials and conducted aeroacoustics tests for $4 \times 10^5 < Re < 8 \times 10^5$. Low to medium frequency noise was found to decrease by up to 10 dB for the porous airfoils, with the reduction being strongly dependant on the flow resistivity of the porous material. However, the porous airfoils produced more noise at higher frequencies, which was attributed to the surface roughness. Porous coatings can therefore be seen to have significant potential industrial and commercial applications.

Porous coatings can be randomised (unstructured) in nature, such as porous polyurethane or metal foam, or can possess a regular structure (structured), such as a lattice. With respect to unstructured cylinder coatings, Sueki et al. [4] tested 3 types of porous coatings at $Re = 8.3 \times 10^4$. It was found that for the same surface roughness, open-cell type porous materials reduced sound more than the closed cell-type. In the wake, the region of zero velocity was found to spread wider than that of a bare cylinder, and sound reduction was attributed to the porous material depriving momentum and suppressing unsteady motion of vortices.

A numerical study was conducted by Bhattacharyya and Singh [5] on a porous-wrapped cylinder to determine laminar vortex shedding and wake flow, for $20 < Re < 250$. The Darcy-Brinkman-Forchheimer extended model was used to govern the flow in the porous zone. Similar to the work by Boorsma [1], the porous coating was found to reduce the strength of the separated shear layers which delayed vortex shedding. It was also found that a thinner porous layer with lower permeability has the equivalent effect of a thicker porous layer with higher permeability.

DNS and LES was used by Naito and Fukagata [6] to investigate flow around a cylinder having uniform thickness, permeability, and porosity. Simulations were carried out for $Re = 100, 1000, 3900$, and 1.0×10^5, where the porous coatings were modelled by the Darcy-Forchheimer model. At the higher Reynolds number, RMS lift coefficients were found to be negligible and the flow field was more symmetric compared to lower Reynolds numbers. It was conjectured that for higher Reynolds number flows, there will be an increase in the critical Reynolds number for laminar to turbulent transition, caused by the porous surface delaying the growth of instability in the boundary and shear layers. It was also conjectured that even for full turbulent flow, fluctuations in the flow field would be eliminated by the wide, low-energy region created by the porous material in the wake.

With respect to structured cylinder coatings, Arcondoulis et al. [7] designed an SPCC and compared it to an unstructured porous coated cylinder which was represented by the Darcy-Forchheimer model. Agreement between both cylinders was observed in the near and far wake.

Tomographic and 2D planar Particle Image Velocimetry (PIV) was conducted by Arcondoulis et al. [8] on two SPCCs, one with square-shaped pores and the other with elliptical-shaped pores. The water tunnel tests were conducted at $Re = 0.9 \times 10^4$ and $Re = 7.3 \times 10^4$. The SPCC with square pores exhibited greater re-circulation through the rear of the porous structure compared to the SPCC with elliptical pores at $Re = 0.9 \times 10^4$. Alternating negative to positive vorticity was noticed at the windward side of the elliptical SPCC, assumed to be caused by flow colliding with structural members and rotating clockwise and anti-clockwise around the members. The authors concluded that the significant difference between the internal and external flow fields of the two coatings show that subtle changes to the geometry can modify the near wake, and therefore the modify the strength of the vortex shedding.

Further PIV work done by Arcondoulis et al. [9] at $Re = 0.7 \times 10^4$ and $Re = 7.3 \times 10^4$, found that the vortex shedding process originated from the inner diameter of the SPCC since vortex shedding processes were found to emanate through the single pores at the outer diameter.

SPCC geometries with greater porosity on one half than the other was tested by Arcondoulis et al. [10] at Reynolds numbers of $Re = 10^5$, 1.53×10^5 and 1.66×10^5. When compared to an SPCC with uniform porosity, the SPCCs showed only small differences in the wake compared to each other. However, compared to the bare cylinder, the SPCCs exhibited a wider and deeper wake.

2 Methodology

Although the work done by Arcondoulis et al. [8] made use of transparent materials to create the SPCC, there are still some limitations in the field of view and mean flow development within the pores is not fully understood. Here, the same SPCC geometry and flow conditions as Arcondoulis et al. [8] are used to look deeply into the flow within the porous structure and develop porous modelling such that geometric detail can be approximated rather than resolved, enabling design based on predictive flow data.

2.1 Numerical Methods

The simulations were conducted using the rhoPimpleFoam solver in OpenFOAM v2212. For temporal discretisation, the Crank-Nicolson [11] scheme was used with a blending factor of 0.5. The simulation was run for 53 timescales, with temporal averaging conducted over the last 20 timescales, where one timescale, T^*, was defined as:

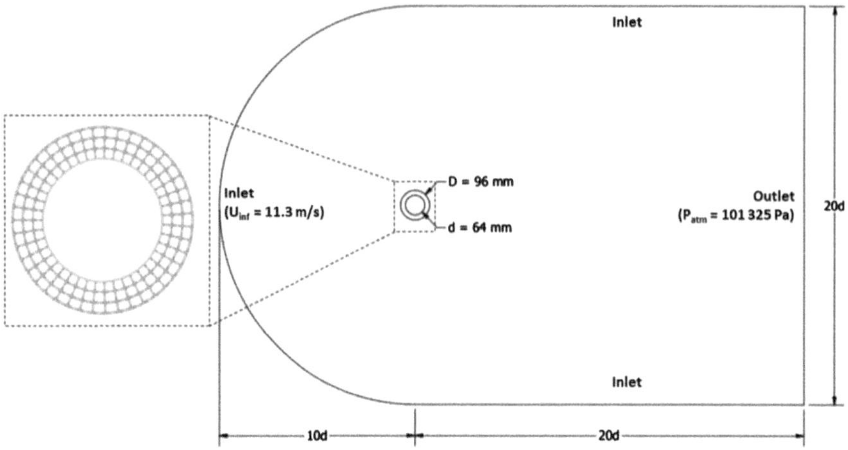

Fig. 1 Computational domain schematic

$$T^* = d/U, \tag{1}$$

Where d is the inner diameter of 0.064 m as shown in Fig. 1. In order to keep the Courant-Friedrichs-Lewy (CFL) number below 1, a timestep of $8.8 \times 10^{-5} T^*$ was used.

For spatial discretisation, the LUST (Linear-Upwind Stabilised Transport) divergence scheme was used. The LUST scheme is a fixed blended scheme with 75% central differencing and 25% second order upwind weights [12].

The Wall Adapting Local Eddy-viscosity (WALE) Subgrid-Scale (SGS) model [13] was used for the LES. The filter width was taken to be $\max(\Delta x, \Delta y, \Delta z)$, which is the distance between the cell centre and furthest face centre. Since the calculated delta using this method is equivalent to half the width of a regular hexagonal cell, a coefficient of 2 was used.

2.2 Case Setup

The velocity at the inlet was set to 11.3 m/s, resulting in a Reynolds number based on outer diameter of $Re = 7.3 \times 10^4$, matching measurements in [8]. The pressure at the outlet was set to be atmospheric pressure (101,325 Pa).

A no-slip condition was applied at the walls. The walls were also set to be adiabatic. The temperature was set to 291.15 K. Spanwise periodicity was used as shown in Fig. 4. A schematic of the domain is shown in Fig. 1.

2.3 SPCC Geometry and Mesh Generation

Figure 2a shows the SPCC geometry. The geometry is discussed in more detail in [14]. For the purposes of this simulation, a spanwise extent containing 3 pores was used, as shown in Fig. 2b. A study by Geyer [15] showed that the spanwise correlation length of a typical porous coated cylinder is greater than 7 outer diameters, which provides confidence that the simulation in this study will be representative of any section along the span. The inner diameter of the SPCC was 0.064 m and the outer diameter was 0.096 m. Cut planes through the geometry are shown in Fig. 3.

The mesh was generated using Pointwise 2022.2.2. To create the surface mesh, the geometry was decomposed into a section, as shown in Fig. 4. The surface mesh consists of triangles with an average height of 1.2×10^{-4} m, shown in Fig. 5. The side

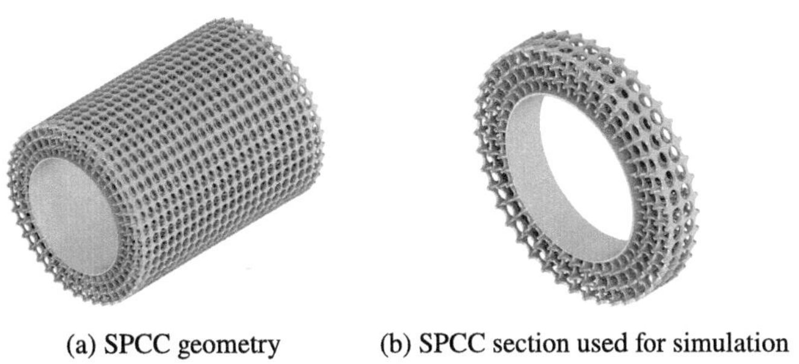

Fig. 2 SPCC geometry rendering generated in Autodesk Inventor Professional 2024

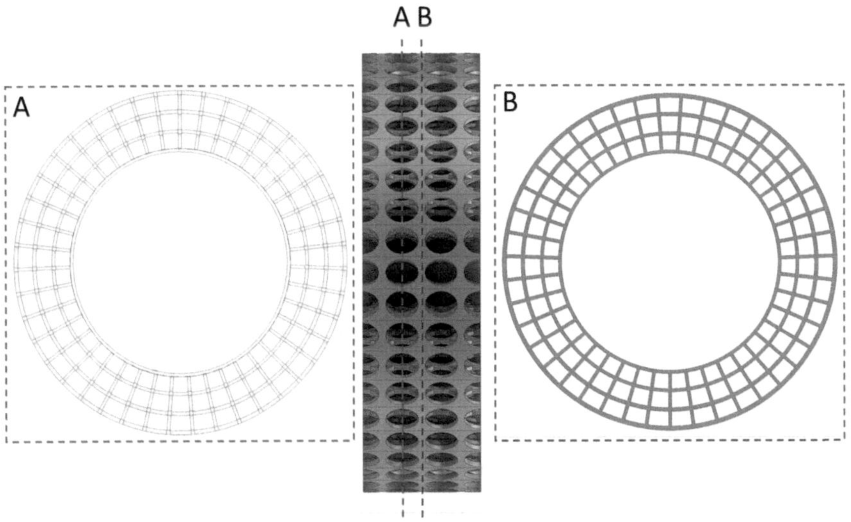

Fig. 3 Cut planes through a spanwise pore and in-between 2 pores

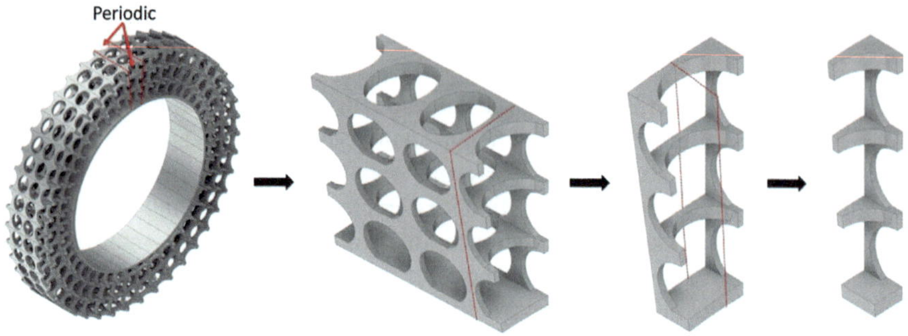

Fig. 4 Decomposition of geometry for meshing

Fig. 5 Surface mesh of SPCC section

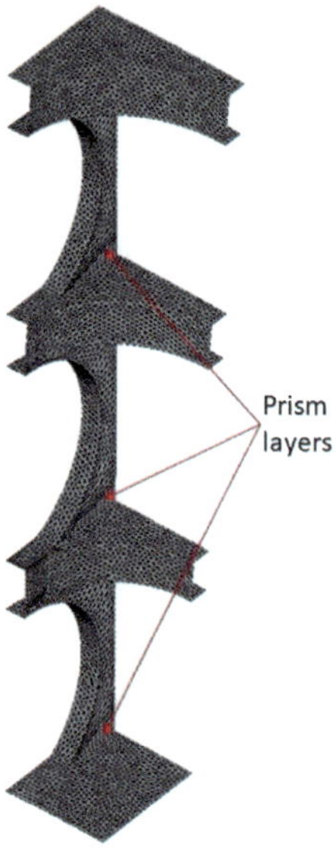

walls were set to be adaptive to allow prism layers to grow off the other walls. Pointwise's anisotropic tetrahedral extrusion (T-Rex) algorithm was used to grow prism layers off the walls. A maximum of 12 layers was requested from the solver, although the solver may produce less layers based on certain physical constraints and quality controls. The average wall $\Delta_{normal}+$ was kept below 2. Cell non-orthogonality was kept below 70°. A volume interface was used between the unstructured SPCC volume and the structured farfield volume. This was achieved by enclosing the SPCC with a structured cylindrical surface, as shown in Fig. 6. Mesh refinement on the leeward side is shown in Fig. 7. The farfield mesh consists of 66 spanwise layers, and the total cell count for the mesh is approximately 135 million cells.

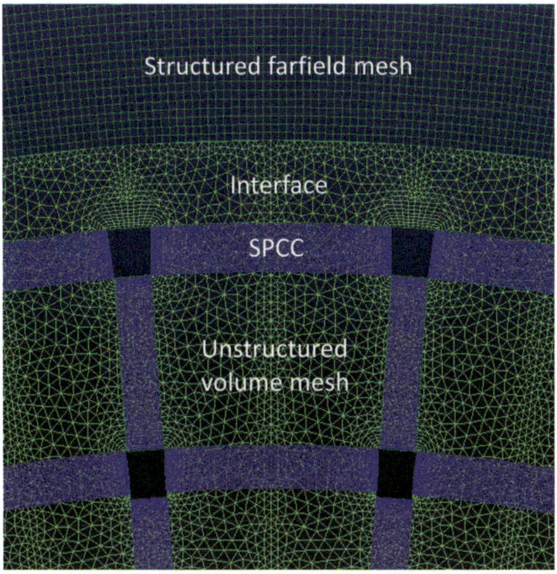

Fig. 6 Interface between unstructured and structured volume mesh

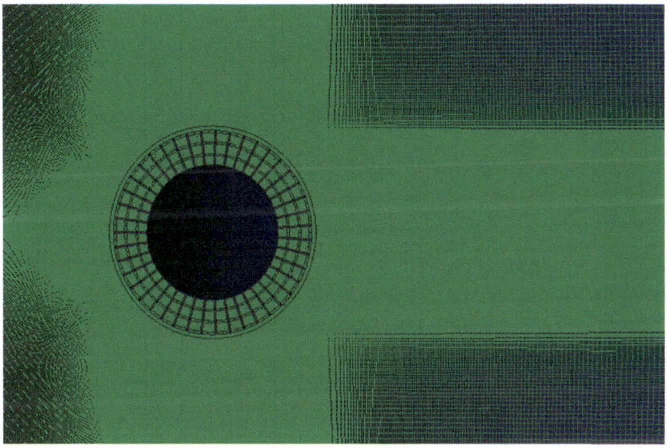

Fig. 7 Mesh refinement in the wake

2.4 Unsteady Data Collection

A novel parallel I/O solver-independent library developed by Upstream CFD GmbH was tested during this study. Depending on the required field outputs, OpenFOAM currently writes approximately 30 files per time step per processor used. For larger

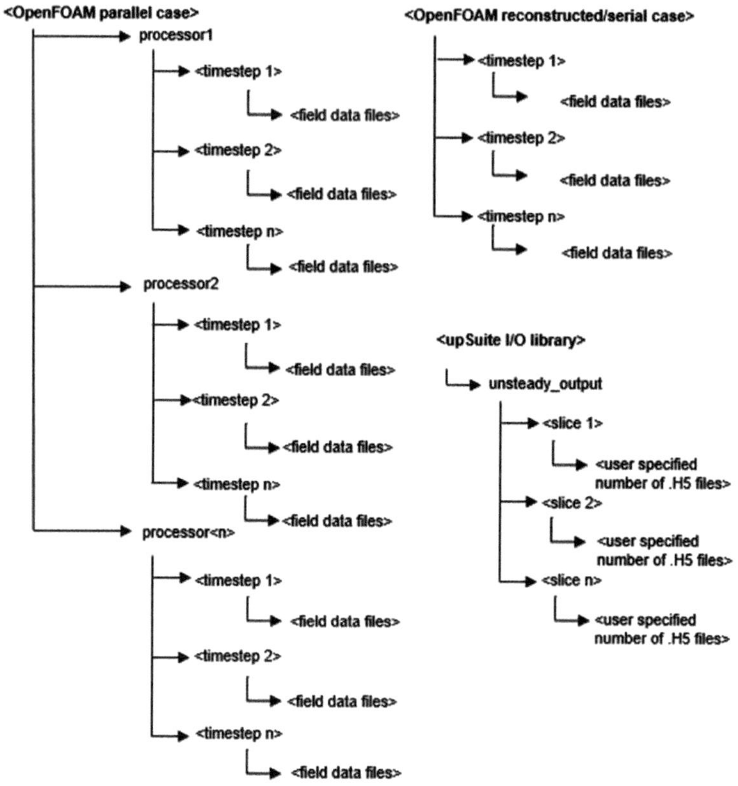

Fig. 8 Comparison between OpenFOAM and upSuite file structure

cases, this can result in millions of files per simulation. Writing unsteady data for the purposes of animations and time traces, for example, becomes increasingly difficult with high numbers of processing cores because of the way OpenFOAM writes data. The most efficient built-in way to record unsteady data in OpenFOAM is to write data to cut planes in VTK format, which still results in 1 file per timestep per cut plane. The HDF5-based I/O library allows the user to choose how many timesteps are written to a single HDF file in parallel, thus significantly reducing the number of files created during a simulation. A comparison between the OpenFOAM file structure and upSuite I/O library file structure is shown in Fig. 8.

3 Results

Figure 9 shows instantaneous Q-criterion iso-surfaces coloured by mean velocity magnitude at t = 0.24 s. Small scale turbulence within the porous layers interacting with the generated outer shear layer with subsequent delayed generation large-scale

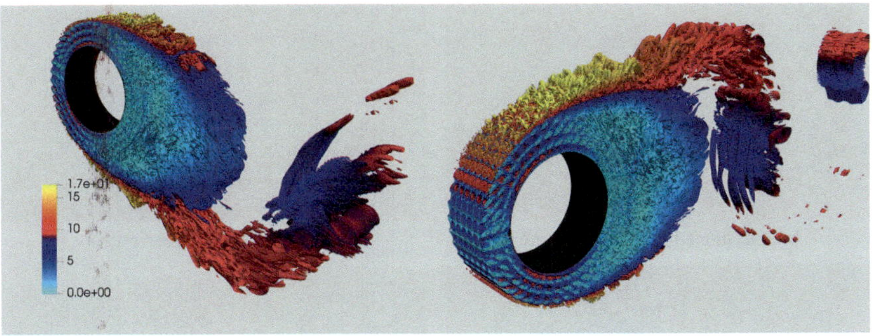

Fig. 9 Q-criterion iso-surfaces coloured by mean velocity magnitude, $Q = 150,000$

vortex shedding is noticeable. Similar to the work done by Arcondoulis et al. [9], vortex shedding emerging through the single pores at the outer diameter is also observed.

To form a basis for comparison, an LES was conducted for a bare cylinder with a diameter of 0.064 m, as per the SPCC inner diameter. The inflow velocity was set to be the same as for the SPCC case, i.e., 11.3 m/s. Contours of streamwise mean velocity for the SPCC and bare cylinder are shown in Fig. 10. By comparing the contours in Fig. 10a–c with Fig. 10d, the extended region of low velocity mentioned by Boorsma et al. [1] can be seen. Note that for all contour plots, flow is from left to right (positive x-direction), and the cut planes shown in Figs. 10 and 14 are located as follows:

- Back pore: $z/d = 0.086$
- Between pores: $z/d = 0.13$
- Front pore: $z/d = 0.17$

Figure 11 shows contours of mean velocity magnitude for the front cut-plane, where the boundary layer development on the inner cylinder of the SPCC can be seen. In water tunnel experiments conducted by Arcondoulis et al. [16] on the same geometry at $Re = 7,000$, a more prominent boundary layer was found to develop on the inner cylinder, whereas in the current study, the boundary layer separates earlier from the inner cylinder.

Figure 12 shows coalescence of multiple separated shear layers forming sub-channels of moderate velocity fluid. Near the upstream stagnation region and downstream separated region, flow within the porous layer returns to distinct wake-like formations. Flow is accelerated over the outer surface, as with usual cylinder flow.

The streamwise mean velocity distributions for 3 wake locations ($x/d = 1, 2, 3$) are presented in Fig. 13. To compensate for the shorter temporal statistics, spanwise averaging was conducted for all wake profiles of the SPCC. From Fig. 13, it can be seen that a larger wake deficit is achieved by the SPCC for $x/d = 1$. For $x/d = 2$, 3, the wake deficit of the SPCC is lower than that of the bare cylinder but the peak is broader.

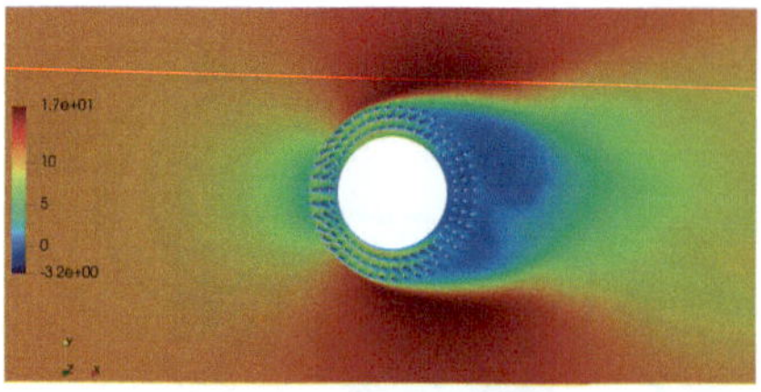

(a) Cut plane through back pore

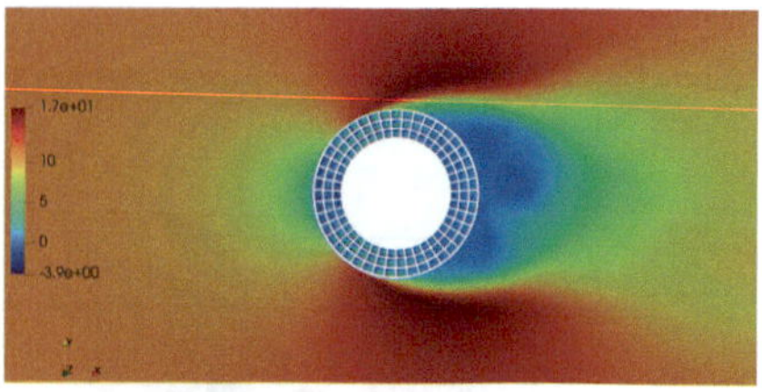

(b) Cut plane between pores

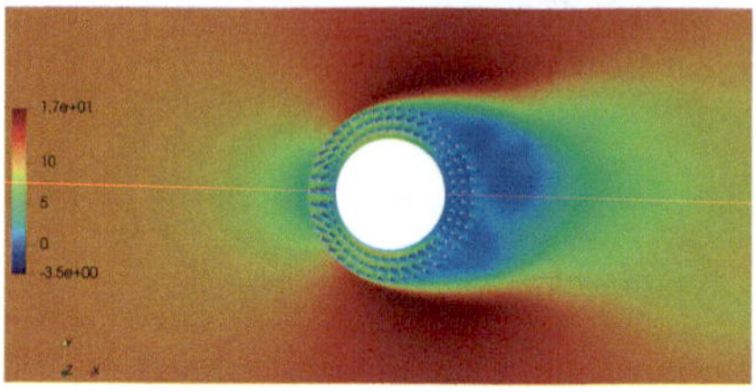

(c) Cut plane through front pore

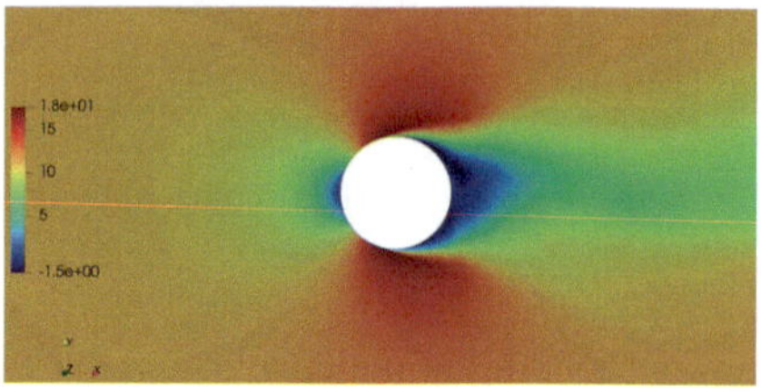

(d) Bare cylinder

Fig. 10 Streamwise mean velocity contours on 3 different cut planes through SPCC and bare cylinder

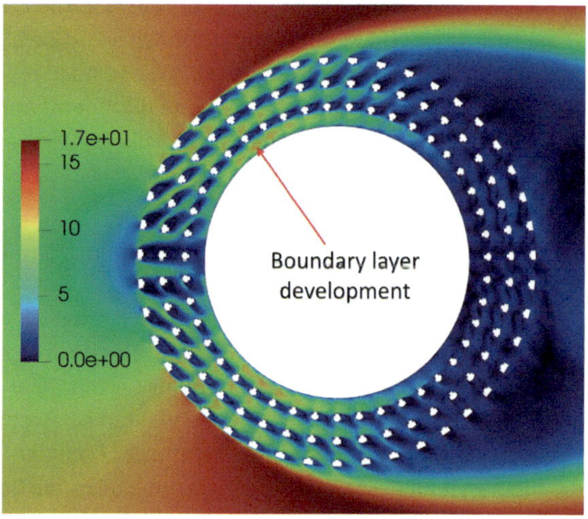

Fig. 11 Contours of mean velocity magnitude showing boundary layer development within the SPCC

Contours of streamwise RMS velocity for the SPCC and bare cylinder are shown in Fig. 14. The streamwise RMS velocity distributions for 3 wake locations are presented in Fig. 15. The locations are at $x/d = 1$, 2 and 3.

By comparing the streamwise RMS velocity profiles in Fig. 15, it can be seen that the peak for the bare cylinder is only greater than that of the SPCC for $x/d = 1$. For $x/d = 2$ and 3, the peaks are greater for the baseline cylinder, signifying a greater turbulence dissipation for the SPCC compared to the bare cylinder.

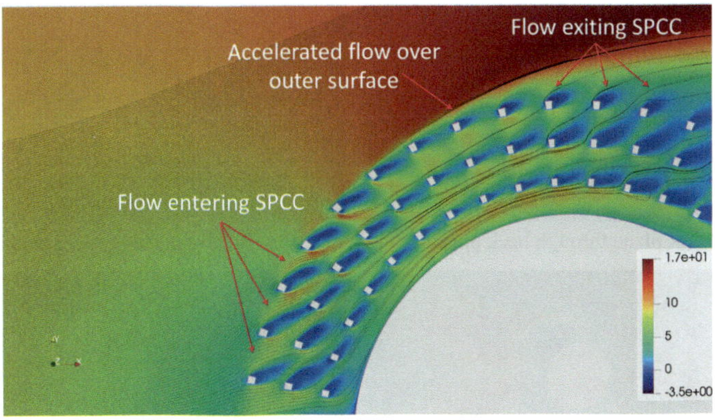

Fig. 12 Streamlines of streamwise mean velocity

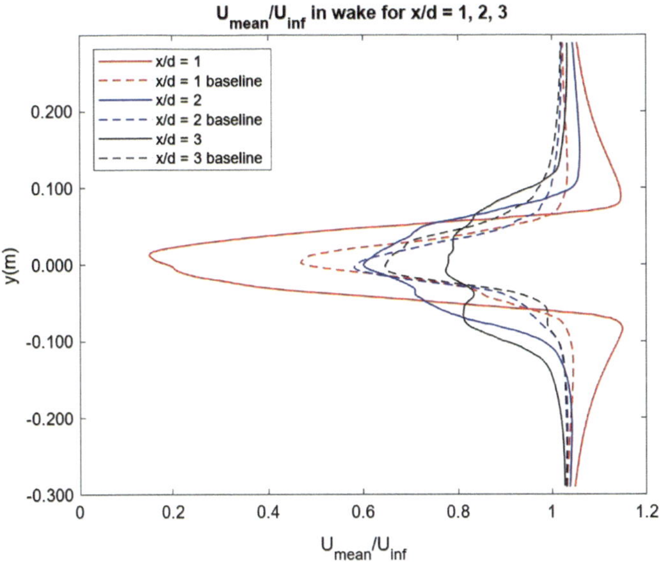

Fig. 13 Normalised streamwise mean velocity profiles at various wake locations for both SPCC and bare cylinder (baseline)

The shear layer trajectory for both the bare cylinder and the SPCC is shown using average spanwise vorticity contours in Fig. 16. From Fig. 16a, it can be seen that for the bare cylinder, the free shear layer is broad with intense regions of vorticity after separation. Due to the manner in which the fluid flows around the internal structures of the SPCC and consequently exits through the circumferential openings (shown in Fig. 12), with merging vorticity from the SPCC structure, higher vorticity is found towards the windward side of the SPCC at the inner and outer radii as shown in

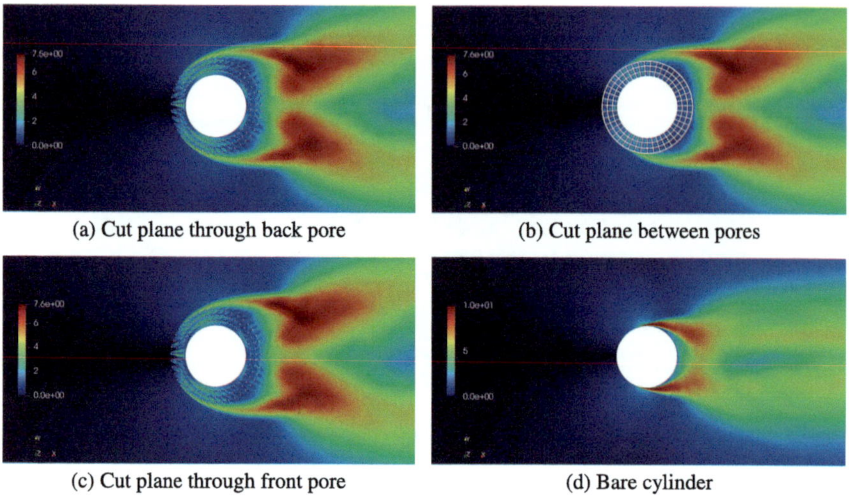
(a) Cut plane through back pore

(b) Cut plane between pores

(c) Cut plane through front pore

(d) Bare cylinder

Fig. 14 Streamwise RMS velocity contours on 3 different cut planes through SPCC and bare cylinder

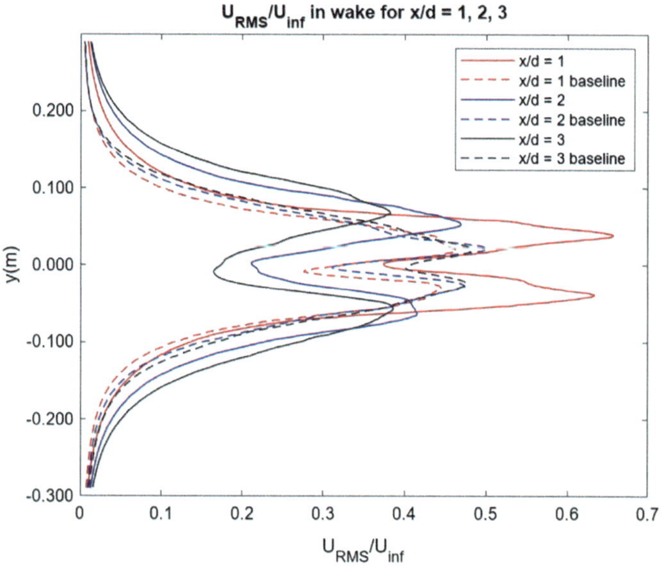

Fig. 15 Normalised streamwise RMS velocity profiles at various wake locations for both SPCC and bare cylinder (baseline)

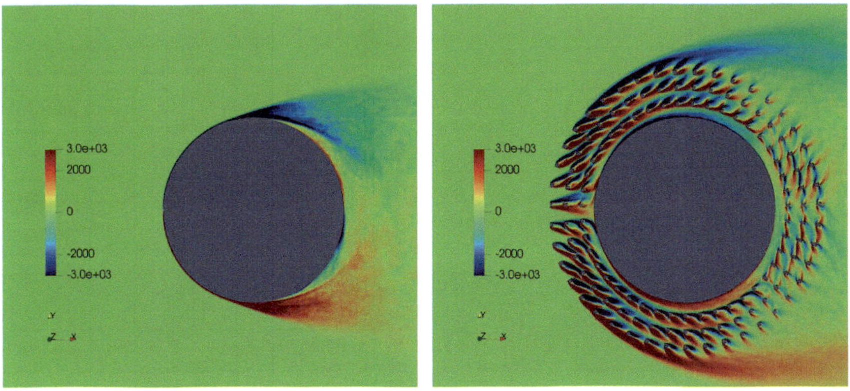

(a) Bare cylinder (b) SPCC

Fig. 16 Spanwise mean vorticity (ω_z) showing shear layer trajectory

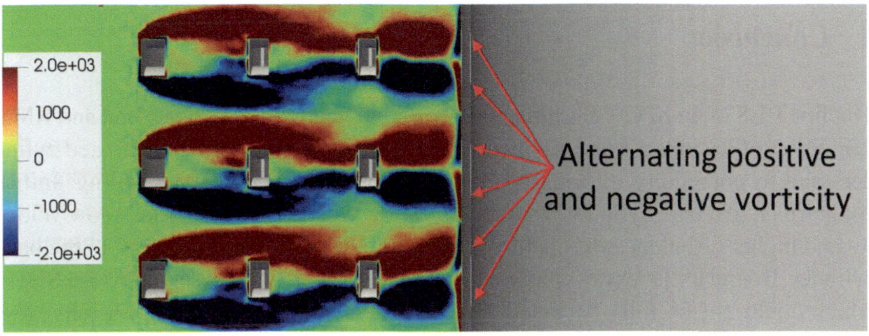

Fig. 17 Mean vorticity about the y-axis (ω_y) along the cylinder span

Fig. 16b. This interaction appears to diffuse the shear layer which also separates from the outer SPCC earlier.

Figure 17 is an x-z cut plane at y = 0 (i.e., through the centre if the SPCC, viewed from above), showing the mean vorticity, ω_y, along the span of the cylinder at the windward side. A similar trend of alternating positive and negative vorticity was found by [8] for the same geometry and is attributed to separation from the internal structures resulting in the alternating vorticity.

Near the aft stagnation region, the pressure differences induced by the vortex shedding in the wake creates a coupled mechanism, forcing the fluid back upstream into the SPCC. Instantaneous streamwise velocity contours (only showing reversed flow) for $T^* = 32$ s and 53 s is shown in Fig. 18. This coupling is under further investigation.

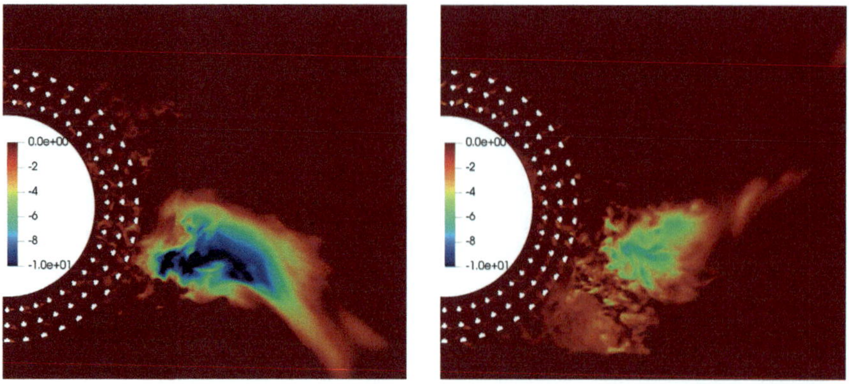

(a) $T^* = 32$ (b) $T^* = 53$

Fig. 18 Instantaneous streamwise velocity contours showing reversed flow

4 Conclusion

The first LES of an SPCC in uniform flow is conducted. Streamwise mean and RMS velocity profiles in the wake are presented for future comparison to uncoated cylinders and to potentially inform and improve Darcy-Forchheimer modelling. Initial comparison to a bare cylinder from an LES shows the extended region of low velocity and higher turbulence dissipation in the wake of the SPCC compared to the bare cylinder. Boundary layer development is observed on the inner cylinder of the SPCC which could not be fully accessed by experiments. Spanwise mean vorticity (ω_z) is used to determine the trajectory of the shear layer alongside streamwise RMS velocity profiles revealing a reduction in peak turbulence intensity in the shear layers. Streamlines are used to demonstrate the internal flow features of the SPCC and key regions of accelerated flow. Future work will involve further interrogation of the unsteady data collected to inform the Darcy-Forchheimer model to reduce mesh requirements, as well as a rigorous comparison with published experimental data.

Acknowledgements We are grateful to the UK Materials and Molecular Modelling Hub for computational resources, which is partially funded by EPSRC (EP/P020194/1 and EP/T022213/1). As a consortium partner, access to these resources was provided by Brunel University London. The authors would like to thank EPSRC for the computational time made available on the UK supercomputing facility ARCHER/ARCHER2 via the UK Turbulence Consortium (EP/R029326/1). This work used the ARCHER2 UK National Supercomputing Service (https://www.archer2.ac.uk). The upSuite software is currently in a non-public testing stage. Personal copies can be requested from Upstream CFD GmbH by E-mail. It is intended to release upSuite publicly in 2024. The authors thank Max Scholz for initial geometry support and many interesting discussions.

References

1. Boorsma, K., Zhang, X., Molin, N., Chow, L.C.: Bluff body noise control using perforated fairings. AIAA J. **47**, 33–43 (2009). https://doi.org/10.2514/1.32766
2. Sueki, T., Ikeda, M., Takaishi, T.: Aerodynamic noise reduction using porous materials and their application to high-speed pantographs. Q. Rep. RTRI **50**, 26–31 (2009)
3. Geyer, T., Sarradj, E., Fritzsche, C.: Measurement of the noise generation at the trailing edge of porous airfoils. Exp. Fluids **48**, 291–308 (2009)
4. Sueki, T., Takaishi, T., Ikeda, M., Arai, N.: Application of porous material to reduce aerodynamic sound from bluff bodies. Fluid Dyn. Res. **42**, 015004 (2010). https://doi.org/10.1088/0169-5983/42/1/015004
5. Bhattacharyya, S., Singh, A.K.: Reduction in drag and vortex shedding frequency through porous sheath around a circular cylinder. Int. J. Numer. Meth. Fluids **65**, 683–698 (2011)
6. Naito, H., Fukagata, K.: Numerical simulation of flow around a circular cylinder having porous surface. Phys. Fluids **24**, 117102 (2012). https://doi.org/10.1063/1.4767534
7. Arcondoulis, E.J.G., Liu, Y., Li, Z., Yang, Y., Wang, Y.: Structured porous material design for passive flow and noise control of cylinders in uniform flow. Materials **12** (2019). https://www.mdpi.com/1996-1944/12/18/2905
8. Arcondoulis, E., et al.: The internal and external flow fields of a structured porous coated cylinder and implications on flow-induced noise (2019). https://arc.aiaa.org/doi/abs/10.2514/6.2019-2648
9. Arcondoulis, E., et al.: Internal shear layer and vortex shedding development of a structured porous coated cylinder using tomographic particle image velocimetry. J. Fluid Mech. **967**, A17 (2023)
10. Arcondoulis, E., Geyer, T., Liu, Y.: An investigation of wake flows produced by asymmetrically structured porous coated cylinders. Phys. Fluids **33**, 037124 (2021)
11. Crank, J., Nicolson, P.: A practical method for numerical evaluation of solutions of partial differential equations of the heat-conduction type. Math. Proc. Cambridge Philos. Soc. **43**, 50–67 (1947)
12. OpenFOAM: User Guide v2112. Lust divergence scheme. howpublishedhttps://www.openfoam.com/documentation/guides/latest/doc/guide-schemes-divergence-lust.html. noteAccessed: (24 August 2023)
13. Franck, N., Ducros, F.: Subgrid-scale stress modelling based on the square of the velocity gradient tensor. Flow Turbul. Combust. **62**, 183–200 (1999)
14. Arcondoulis, E., et al.: The design and noise characteristics of a structured porous coated cylinder in uniform flow (2018). https://arc.aiaa.org/doi/abs/10.2514/6.2018-2963
15. Geyer, T.F.: Experimental evaluation of cylinder vortex shedding noise reduction using porous material. Exp. Fluids **61**, 1–21 (2020)
16. Arcondoulis, E., Liu, Y., Geyer, T.F., Sedaghatizadeh, N., Arjomandi, M.: Aeroacoustic performance of cylinders with a circumferential varying porous coating (2019). https://doi.org/10.2514/6.2020-2527

Reducing Jet Noise from an Underexpanded Biconical Nozzle

Daniel Joseph Bodony and Sandeep Ravikumar Murthy

Abstract The intense noise radiated by jet aircraft exhaust nozzles causes structural vibration, fatigue, carrier deck personnel operational difficulties, and community environmental concerns. Prior work into the physics and control of jet noise have identified several important sound sources, including wavepackets, screech, Mach wave radiation, and broadband shock associated noise, to name a few. Reducing the loudest sources of jet noise, without sacrificing propulsive performance, has thus-far relied on intuition, parametric survey, or optimal control techniques. With the aim of developing a more general and robust method of jet noise reduction (JNR), we seek a physics-based JNR approach that is built upon a linear resolvent analysis appropriate for mean flows with strong shocks. The effect of flow discontinuities on linear analysis, including the optimal forcing and response modes that arise from resolvents, is investigated.

1 Introduction

High speed jet noise from Naval tactical aircraft causes operational difficulties limiting communication between pilot and carrier deck crew, quickly damages deck crew hearing, and leads to sound-induced structural vibrations and fatigue. Several decades of experimental, theoretical, and computational investigations into the physics and control of jet noise have identified several important sound sources, including wavepackets [1], screech [2], Mach wave radiation [3], and broadband shock associated noise [4]. Reducing the loudest sources of jet noise, without sacrificing propulsive performance, has relied on intuition [5], parametric survey [6], or optimal control techniques [7].

CUFS-2024-23, Cambridge, United Kingdom, 4–5 March 2024.

D. J. Bodony (✉) · S. R. Murthy
Department of Aerospace Engineering,University of Illinois at Urbana-Champaign, Urbana, USA
e-mail: bodony@illinois.edu

S. R. Murthy
e-mail: srmurth2@illinois.edu

J. C. Tyacke and N. R. Vadlamani (eds.), *Proceedings of the Cambridge Unsteady Flow Symposium 2024*, https://doi.org/10.1007/978-3-031-69035-8_10

A general and robust physics-based approach to jet noise reduction (JNR), that relies on the observation that coherent structures in turbulent jets strongly resemble instability wavepackets and considers the possibility that these coherent structures are governed by linear dynamics of small-amplitude fluctuations on a turbulent base-flow, is the focus of the current investigation. Many articles in the past [8–10] have considered the possibility that a linear model can faithfully reproduce the coherent turbulence structures in jet-like flows. These articles have been summarized in [1] and are shown to faithfully reproduce the observed spatial growth of coherent turbulence structures near the nozzle over a significant but limited range of frequencies. However, for the underlying theoretical model behind the linear analysis to be justified, the linear perturbations developing in the base flow must be deterministic in nature. Therefore extending such a model to analyze chaotic nonlinear fluctuations around a statistical turbulent mean state has remained vague [11].

A promising framework to understand aerodynamic turbulent jet noise is formulated by constructing a linearised Navier-Stokes system subject to stochastic forcing such that the statistical moments of the turbulent flow are governed by model equations [12, 13]. Literature that investigates such a stochastic forcing applied to linearised Navier-Stokes equations predominantly focus on time-domain formulations of covariance dynamics; however, coherence in jet turbulence has often been analyzed in the frequency domain [11]. Specifically, spectral proper orthogonal decomposition (SPOD) has been used as a means to extract empirical coherent structures at a given frequency from numerical flow data [10, 11, 14]. Recently linear stability analyses of jets has been carried out in a frequency-domain framework using optimal forcing/response structures [14]. The forcing and response modes in this formalism are global in nature, and they are constructed as the singular modes of the global resolvent operator [15]. This analysis has been employed in the literature to model the stochastic dynamics of the Navier-Stokes equations [16]. The work by [17] demonstrates that the spatial structure of the optimal linear flow response agree with the leading SPOD modes, obtained from numerical simulations, and has established a formal justification for a direct comparison between optimal linear response structures and SPOD modes.

This paper describes current research towards extending linear-based JNR strategies to jets with strong shocks. Although such jets are common in high-performance aircraft, there is scant literature discussing forward and adjoint sensitivity analyses about flows with discontinuities in general, with even less prior work within the computational fluid dynamics realm. The question of how to address sensitivities with shock-laden flows follows a presentation of our large-eddy simulation framework and its application to underexpanded jets.

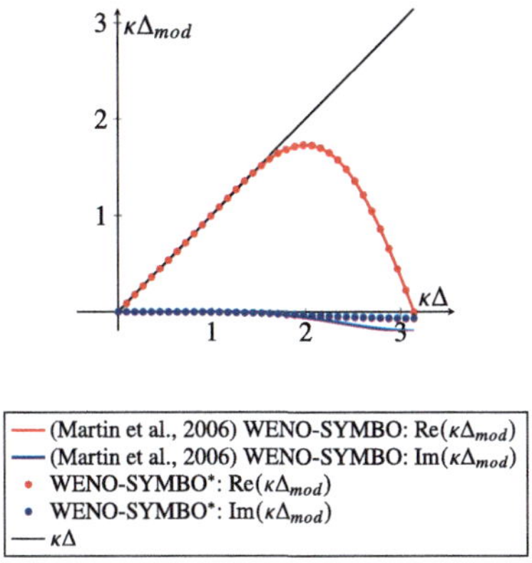

Fig. 1 Bandwidth efficiency for the WENO-SYMBO and WENO-SYMBO* schemes

2 Large-Eddy Simulation Predictions of Jet Noise

This section describes the numerical implementation and application of our flow solver for simulating a 3D supersonic flow through a bi-conical tactical aircraft nozzle, shown in Fig. 3. Additionally, it describes the numerical procedure used to educe the coherent large-scale structures responsible for the intense jet noise radiated by such flows using a linear analysis. Before describing how the LES and linear analysis capabilities are incorporated into our in-house compressible flow solver `PlasCom2`, we provide a brief summary of the solver's inherent capabilities.

2.1 Large-Eddy Simulation Methodology

`PlasCom2` solves the equations of mass, momentum, and energy conservation for a compressible, viscous fluid with an ideal gas equation of state, Fourier law of heat conduction and Newtonian viscous stresses $\tau = \mu[\nabla \boldsymbol{u} + (\nabla \boldsymbol{u})^{\mathrm{T}}] + \lambda(\nabla \cdot \boldsymbol{u})\boldsymbol{I}$, where $\boldsymbol{I}$ is the identity tensor and $\boldsymbol{u}$ is the velocity field. The shear viscosity is temperature dependent and is modelled as $\mu/\mu_0 = (T/T_0)^{2/3}$. The equations are expressed in computational coordinates $\boldsymbol{\xi} = \Xi(\boldsymbol{x})$, whose mapping Ξ to physical coordinates $\boldsymbol{x}$ of a mesh is one-to-one and onto [18]. The solver has the ability to handle overset meshes [19], whilst solving the non-dimensionalized compressible Navier-Stokes equations, in parallel, using a collocated finite difference approximation. The spatial derivatives are approximated by an author-modified WENO-SYMBO* scheme (to remove the long-time instability in [20]), that are sixth order accurate in the interior. This scheme incorporates free-stream preservation as described in [21] and bandwidth optimization as per a modified version of [20] (see Fig. 1). Repeated first derivatives are used to approximate the viscous terms and implementation of the boundary conditions uses the ghost cell approach.

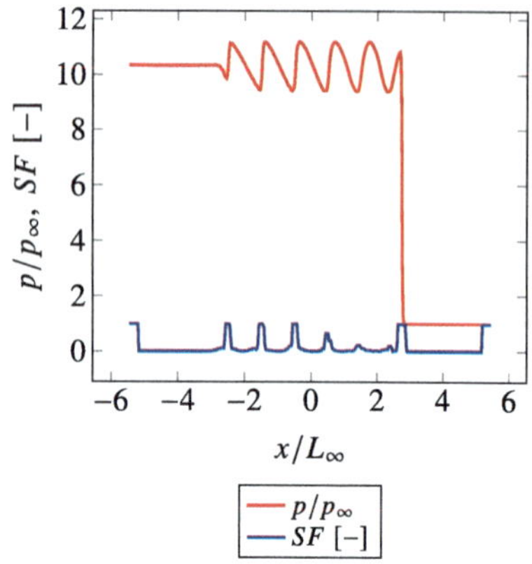

Fig. 2 Switch function (SF) identifying the regions where WENO-SYMBO* is needed

The choice of symmetric stencils in the WENO-SYMBO* scheme helps reduce dissipation far from any flow-discontinuities. However, it was observed that even with these improvements the large eddy simulations of the decay of homogenous isotropic turbulence did not match the experimental observations presented in [22]. Thus, `PlasCom2` uses a hybrid scheme that selectively turns on WENO-SYMBO* only in the presence of flow-discontinuities. The flow-discontinuities are sensed using a shock sensor (switch function 6 from [23]) as shown in Fig. 2. The resulting scheme is able to handle both shock laden flows and the cascade of energy from the large to the smallest scales of turbulence (Figs. 3, 4, 5, 6, 7, 8 and 9).

The dynamic SGS model for compressible turbulence presented in [24] is used to model the effect of the sub-grid scale dissipation on the unresolved yet represented scales of motion in the LES simulation. This model computes the strain rate fields at two different scales (test and selective filters help acquire this information) and utilizes the spectral information in the large-scale field to extrapolate the small-scale stresses. The model then uses an algebraic identity derived by [25] along with an eddy viscosity model to represent the SGS stresses at both scales. The expression derived for the eddy viscosity coefficient is both a function of space and time, and hence can be described as dynamic in nature. The coefficient can be negative in some regions and thus does not exclude backscatter, provides for the proper asymptotic behavior of the stresses near the wall without ad-hoc damping functions, and it vanishes in laminar flow without ad-hoc intermittency functions.

2.2 *Large Eddy Simulation Results*

Using the optimized WENO-SYMBO* based shock capturing scheme, with a Ducros shock sensor, the shock-laden turbulent flow issuing from a biconical nozzle, shown in Fig. 3 with nozzle pressure ratio (NPR = $p_0/p_\infty = 4$) have been simulated for a series three total temperature ratios (TTR = $T_0/T_\infty = 1, 3, 7$). Several simulation

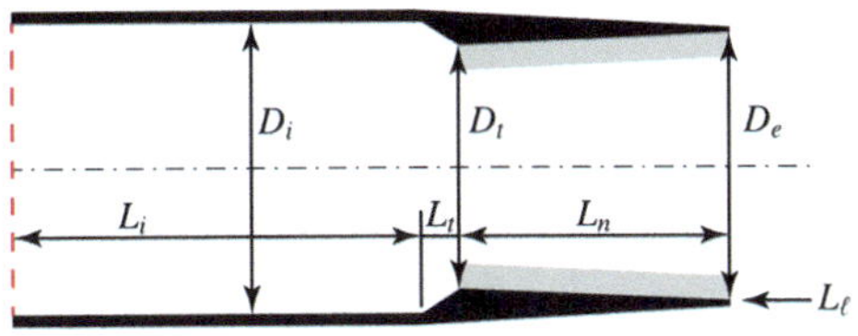

Parameter	Value
D_t/D_e	67.06/72.85
L_t/D_e	9.75/72.85
L_n/D_e	73.56/72.85
D_i/D_e	79.34/72.85
L_i/D_e	1.09
L_ℓ/D_e	0.508/72.85

Fig. 3 Biconical nozzle geometric parameters

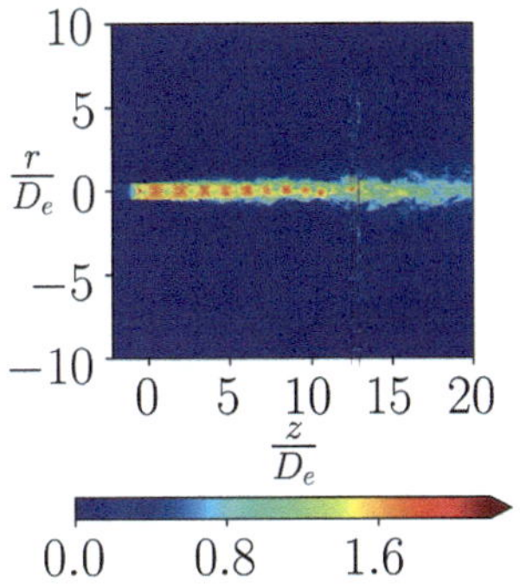

Fig. 4 Instantaneous Mach number for NPR = 4, TTR = 1 jet

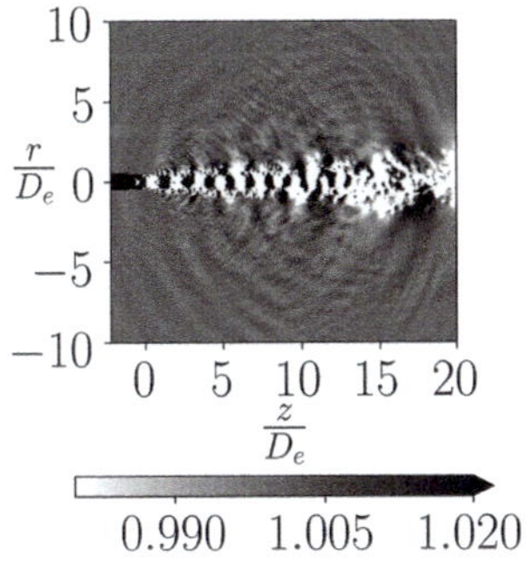

Fig. 5 Instantaneous pressure field for NPR = 4, TTR = 1 jet

parameters are given in Table 1. The flow fields corresponding to the TTR = 1 conditions are shown, along with their corresponding sound fields, in Figs. 4, 5. Results from the higher temperature jets are available in [26]. For TTR = 1, the LES predictions shown in Figs. 6, 7, 8 and 9 faithfully match the experimental data along the jet centerline and away from it. show The flow is characterized by a strong shock cell pattern near the nozzle exit that initiates with a normal shock. More detailed analysis of the jet and near-jet fields are provided in [26].

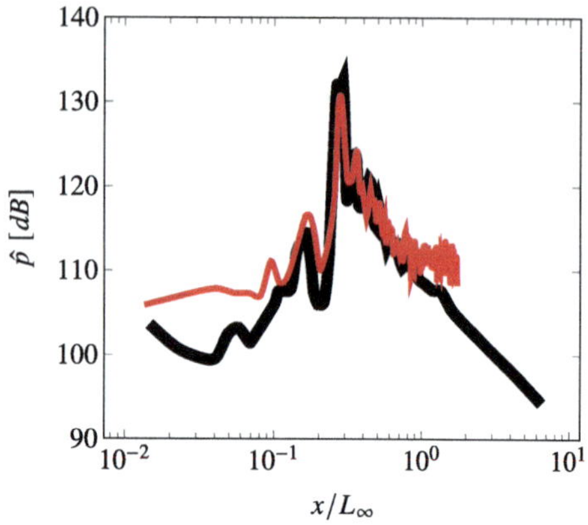

Fig. 6 TTR = 1, NPR = 4, LES-based SPL calculation at $(z/D_e, r/D_e) = (0, 1.5)$ (represented by red line) compared against experimental data (represented by black line) from [27]

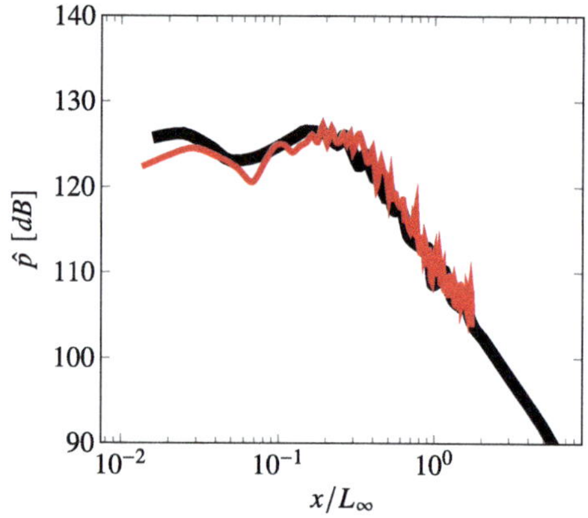

Fig. 7 TTR = 1, NPR = 4, LES-based SPL calculation at $(z/D_e, r/D_e) = (20, 4.8)$ (represented by red line) compared against experimental data (represented by black line) from [27]

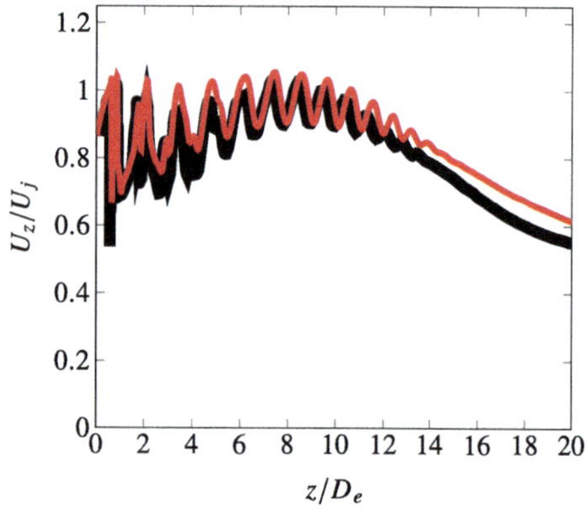

Fig. 8 TTR = 1, NPR = 4, LES-based centerline axial velocity (represented by red line) comparison against the slip wall (no boundary layer) LES simulation data (represented by black line) from [28]

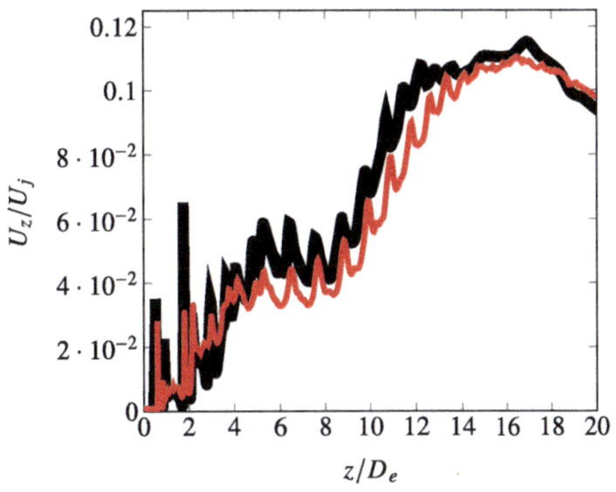

Fig. 9 TTR = 1, NPR = 4, LES-based centerline axial turbulence intensity (represented by red line) comparison against slip wall (no boundary layer) LES simulation data (represented by black line) from [28]

Table 1 Simulation details common to all simulations

Prandtl number (Pr)	0.71
γ	1.4
R_{gas}	287.05 $[m^2 s^{-2} K^{-1}]$
C_p	1004.675 $[m^2 s^{-2} K^{-1}]$
Wall boundary condition	Adiabatic slip wall

2.3 *Numerical Discretization for the Linearized Analysis of the Mean Flow*

The linear analysis about the jet mean flow to support JNR are also performed using our in-house code `PlasCom2`. The construction of the linearized Navier-Stokes operator ($\boldsymbol{L}$), used for the global linear stability analysis of the jets, is as shown below (in Matlab-like notation):

$$\boldsymbol{L}(:\,,\ i) = \frac{\tilde{\boldsymbol{R}}(\boldsymbol{Q} + \epsilon\ \boldsymbol{e}_i) - \tilde{\boldsymbol{R}}(\boldsymbol{Q})}{\epsilon}, \tag{1}$$

where, $\epsilon = 10^{-8}$, the vector $\boldsymbol{e}_i$ is a unit vector with a 1 in the ith position, $\boldsymbol{Q}$ is the vector of unknowns and $\tilde{\boldsymbol{R}}$ is the discretized right-hand side of the non-linear compressible Navier-Stokes equations, including the WENO-SYMBO* terms, sponges and ghost boundary points, which accepts the state ($\boldsymbol{Q}$) or its perturbed form ($\boldsymbol{Q} + \epsilon\ \boldsymbol{e}_i$) as input.

The presence of the strong shocks in the jet mean flow, particularly the "barrel shock" that forms along the centerline just downstream of the nozzle exit, implies that care must be taken when considering the linearized dynamics, especially when the linearized operator depends on the smooth definition of the derivative. The next section briefly describes several aspects for sensitivity analysis, adjoint analysis, and resolvent analysis with shock-laden mean flows.

2.4 Discretization Requirements for Convergent Adjoint Modes in the Presence of Shocks

The requirements for the construction of a physically-meaningful adjoint operator derived from the spatially-discretized non-linear operator were investigated by Bodony & Fikl [29] using the inviscid Burgers' equation. Since the partial differential equations being discretized are non-linear, an artificial dissipation was used to regularize the solution when shocks formed. The general form of the first derivative operator used in the discretization was given by $\boldsymbol{D}_1[\mathbf{U}] = \boldsymbol{P}^{-1}\boldsymbol{Q}[\mathbf{U}]$ (with $\mathbf{U}$ is the base flow state). Additionally, the artificial dissipation was expressed in the following form $\boldsymbol{A}_D = -\boldsymbol{P}^{-1}\tilde{\boldsymbol{D}}_p^T\boldsymbol{\phi}[\mathbf{U}]\tilde{\boldsymbol{D}}_p$, where $\tilde{\boldsymbol{D}}_p$ is a finite difference approximation of the unscaled pth derivative. Using these operators, the following conditions on the derivative operator $\boldsymbol{Q}[\mathbf{U}]$ and the artificial dissipation term $\boldsymbol{\phi}[\mathbf{U}]$ were derived that ensured that adjoint solutions of the inviscid Burgers' equation would converge: (1) $\boldsymbol{\phi}[\mathbf{U}] > 0$, (2) $\boldsymbol{\phi}[\mathbf{U}]$ is differentiable with respect to $\mathbf{U}$, and (3) $\boldsymbol{Q}[\mathbf{U}]$ is differentiable with respect to $\mathbf{U}$. Though the three conditions are common sense, the work in [29] showed that neither the MUSCL [30] scheme with flux limiters nor WENO schemes [31] satisfy these requirements. Consequently, the use of MUSCL and WENO schemes, including all WENO variants, is not recommended for linearized analysis. Local artificial diffusivity (LAD) schemes (e.g., the Jameson-Schmidt-Turkey (JST, [32]) scheme or more recent high-order versions [33]) do satisfy every condition and were found in [29] to provide convergent adjoint solutions to the non-linear inviscid Burger's equation.

2.5 Analytical Solution of Zero-Frequency Resolvent Operator for Shock-Laden Mean Flows

In this section, the work from [34] and [35] are used to investigate a convergent resolvent mode computation procedure using the Green's function operator corresponding to the forward and adjoint problems as described in [34]. This work is meant to extend that of [29] to equations more relevant to CFD and to provide a semi-analytical resolvent calculation for verifying whether numerical methods give rise to meaningful linearized operators when shocks or other discontinuities are present. To the authors' knowledge, no comparable analysis exists. Due to the complexity of the analysis, only results from the quasi-one-dimensional equations are presently available.

2.5.1 Inviscid Quasi-One-Dimensional Equations

The inviscid quasi-one-dimensional (quasi-1d) equations representing the compressible flow through a converging-diverging nozzle are given as

$$\mathbf{R}(\mathbf{U}, h) = \frac{d}{dx}(\mathbf{F}h) - \frac{dh}{dx}\mathbf{P}, \tag{2}$$

where,

$$\mathbf{U} = \begin{bmatrix} \rho \\ \rho q \\ \rho E \end{bmatrix}, \ \mathbf{F} = \begin{bmatrix} \rho q \\ \rho q^2 + p \\ \rho q H \end{bmatrix}, \ \mathbf{P} = \begin{bmatrix} 0 \\ p \\ 0 \end{bmatrix}, \tag{3}$$

and $H = E + \frac{p}{\rho} = \frac{\gamma}{\gamma - 1}\frac{p}{\rho} + \frac{1}{2}q^2$. Additionally, if the solution to the quasi-1d equations contains a discontinuity (at $x = x_s$) then the Rankine-Hugoniot jump conditions are given by

$$[\mathbf{F}]_{x_s^-}^{x_s^+} = 0. \tag{4}$$

2.5.2 Forward Green's Function

The Green's function operator $\boldsymbol{G}(x, \xi)$ of the forward linearized problem is derived, exactly as presented in [34], by first constructing three homogeneous solutions to the linear problem $\boldsymbol{L}\mathbf{u}_i = 0$, where $\boldsymbol{L}\mathbf{u}_i = d(h(x)\ \boldsymbol{A}(x)\ \mathbf{u}_i)/dx - dh/dx\ \boldsymbol{B}(x)\ \mathbf{u}_i$, $\boldsymbol{A}(x) = d\mathbf{F}/d\mathbf{U}$, $\boldsymbol{B}(x) = d\mathbf{P}/d\mathbf{U}$ and $\mathbf{P} = [0, p, 0]^T$. Next, a solution to an impulse-type-forcing of the form $\mathbf{f}_i(\xi)\ \delta(x - \xi)$ is constructed as a piece-wise solution of two homogeneous solutions $a_{1,i}\mathbf{u}_i + a_{2,i}\mathbf{u}_i$ (with summation over repeated indices and $\delta(x)$ the Dirac delta distribution), one before and the other after the location where the impulse type forcing is applied. The vector $\mathbf{f}_i$ is then computed by integrating the equation representing the linear problem across the location where the impulse-type-forcing is applied $\int_{\xi^-}^{\xi^+} \boldsymbol{L}\mathbf{u}_i(x, \xi)\ dx = \mathbf{f}_i(\xi)$. Then, the Green's function operator is constructed using the vectors $\mathbf{u}_i(x, \xi)$ and $\mathbf{f}_i(\xi)$ as shown below

$$\boldsymbol{G}(x, \xi) = (\mathbf{u}_1(x, \xi) \mid \mathbf{u}_2(x, \xi) \mid \mathbf{u}_3(x, \xi))\ \ (\mathbf{f}_1(\xi) \mid \mathbf{f}_2(\xi) \mid \mathbf{f}_3(\xi))^{-1}, \tag{5}$$

where vertical lines indicate the partitioning of the row vector and array into three columns.

2.5.3 Adjoint Green's Function

Similar to the construction of the forward Green's function, the Green's function for the adjoint problem can be constructed using three linearly independent gradient functions ($\mathbf{g}_i$) and corresponding adjoint solutions ($\mathbf{v}_i$), such that $\boldsymbol{L}^\dagger\mathbf{v}_i(x, \eta) = \mathbf{g}_i(\eta)\delta(x - \eta)$. To accomplish this, three objective functions, $J_1(x) = \int_{-1}^{1} p(x)\ dx$,

$J_2(x) = \int_{-1}^{1} T(x)\, dx$, $J_3(x) = \int_{-1}^{1} q(x)\, dx$ are defined. The corresponding gradients ($dp/d\mathbf{U}$, $dT/d\mathbf{U}$ and $dq/d\mathbf{U}$) of the objective function integrands (k_i such that $J_i = \int_{-1}^{1} k_i(x)\,\mathrm{d}x$) are then used to evaluate,

$$\begin{aligned} dJ_{i,j} = I_{i,j}(\xi,\eta) = \int_D \Big\{ & \mathbf{u}_j(x,\xi)\cdot \frac{\partial k_i(\eta)}{\partial \mathbf{U}} \delta(x-\eta) \\ & - k_i(\eta)\,\delta(x-\eta)\,\delta(x-x_s^+)\,\delta_j(\xi) \\ & + k_i(\eta)\,\delta(x-\eta)\,\delta(x-x_s^-)\,\delta_j(\xi) \Big\}\, dx \end{aligned} \tag{6}$$

$$\begin{aligned} = & \,\mathbf{u}_j(\eta,\xi)\cdot \mathbf{g}_i(\eta) \\ & - k_i(\eta)\,\delta(\eta - x_s^+)\,\delta_j(\xi) \\ & + k_i(\eta)\,\delta(\eta - x_s^-)\,\delta_j(\xi), \end{aligned} \tag{7}$$

where δ_j represents the linearized displacement in the shock location and the indices i and j represent the choice of objective function and forcing (ρuh, H or p_0) used to compute the perturbation $I_{i,j}$, respectively. Then, the operator $\big(I_{i,1}(\xi,\eta) \mid I_{i,2}(\xi,\eta) \mid I_{i,3}(\xi,\eta)\big)$ is used to evaluate

$$\mathbf{v}_i^T(\xi,\eta) = \big(I_{i,1}(\xi,\eta) \mid I_{i,2}(\xi,\eta) \mid I_{i,3}(\xi,\eta)\big)\; (\mathbf{f}_1(\xi) \mid \mathbf{f}_2(\xi) \mid \mathbf{f}_3(\xi))^{-1}, \tag{8}$$

which is used to obtain the Green's function for the adjoint problem, as shown in Eq. (5). Here, $\mathbf{u}_i(x,\xi)$ is replaced by $\mathbf{v}_i(x,\eta)$ and $\mathbf{f}_i(\xi)$ with $\mathbf{g}_i(\eta)$ to evaluate $\boldsymbol{G}_s^\dagger(x,\eta)$, as shown below

$$\boldsymbol{G}_s^\dagger(x,\eta) = (\mathbf{v}_1(x,\eta) \mid \mathbf{v}_2(x,\eta) \mid \mathbf{v}_3(x,\eta))\;(\mathbf{g}_1(\eta) \mid \mathbf{g}_2(\eta) \mid \mathbf{g}_3(\eta))^{-1}. \tag{9}$$

The adjoint solution corresponding to any given $\mathbf{g}(\eta)$ can then be computed as

$$\mathbf{v}(x) = \int_D \mathbf{G}^\dagger(x,\eta)\,\mathbf{g}(\eta)d\eta \;-\; \mathbf{\Phi}^{-T}(x) \begin{Bmatrix} \delta_1(x) \\ 0 \\ \delta_3(x) \end{Bmatrix} \big[(p \mid T \mid q)\,\mathbf{\Gamma}^{-1}\,\mathbf{g}\big]_{x_s^-}^{x_s^+}, \tag{10}$$

where $\mathbf{G}^\dagger(x,\eta) = \mathbf{\Phi}^{-T}(x) \begin{bmatrix} \mathbf{u}_1(\eta,x)^H \\ \mathbf{u}_2(\eta,x)^H \\ \mathbf{u}_3(\eta,x)^H \end{bmatrix}$, $\mathbf{\Phi}(x) = (\mathbf{f}_1(x) \mid \mathbf{f}_2(x) \mid \mathbf{f}_3(x))$ and $\mathbf{\Gamma}$ $(x) = (\mathbf{g}_1(x) \mid \mathbf{g}_2(x) \mid \mathbf{g}_3(x))$.

2.5.4 Procedure to Construct the Zero-Frequency Resolvent Operator

The zero frequency resolvent modes (forcing modes $\hat{f}$ and response modes $\hat{r}$) are constructed by solving for the forcing $\hat{f}$, that will maximize the gain σ using the

Green's function operators of the forward and adjoint problems, which were derived in the previous subsections, as follows

$$\sigma^2 = \max_{\hat{f}} \left(\frac{\langle \hat{r}, \hat{r} \rangle}{\langle \hat{f}, \hat{f} \rangle} \right) = \max_{\hat{f}} \left(\frac{\langle \boldsymbol{G} * \hat{f} \,,\, \boldsymbol{G} * \hat{f} \rangle}{\langle \hat{f} \,,\, \hat{f} \rangle} \right) = \max_{\hat{f}} \left(\frac{\langle \hat{f} \,,\, \boldsymbol{G}^{\dagger} * \boldsymbol{G} * \hat{f} \rangle}{\langle \hat{f} \,,\, \hat{f} \rangle} \right). \tag{11}$$

A normalized power iteration method is used to solve the eigenvalue problem $\boldsymbol{G}^{\dagger} * \boldsymbol{G} * \hat{f} = \sigma^2 \hat{f}$. This produces a solution to the most-amplified-forcing ($\hat{f}$) and most-receptive-response ($\hat{r}(x) = \int_D \boldsymbol{G}(x, \xi) \hat{f}(\xi) d\xi$) mode corresponding to the linear system $\boldsymbol{L}\hat{r} = \hat{f}$.

2.5.5 Zero-Frequency Resolvent Analysis of Shock-Laden Quasi-1d Flow Through a CD Nozzle

In this section, the zero-frequency resolvent modes corresponding to steady flow through a duct of cross-section $h(x)$ and governed by the quasi-one-dimensional Euler Eq. (2) are computed using the procedure described in Sect. 2.5.4 and compared against the same modes computed using a finite volume method (FVM) fitted with the JST scheme as described in [36].

Flow solution

Two steady quasi-1d isentropic flows, one supersonic and the other subsonic, through a CD nozzle are considered. The geometric definition of the CD nozzle is given by

$$h(x) = \begin{cases} 2 & (-1 \leq x \leq -1/2) \\ 1 + \sin^2(\pi x) & (-1/2 < x < 1/2) \\ 2 & (1/2 \leq x \leq 1). \end{cases} \tag{12}$$

In case of the supersonic flow, the inlet is set to a non-dimensional (as per [34]) stagnation enthalpy $H_{\text{in}} = 4$, stagnation pressure $p_{0,\text{in}} = 2$ and Mach number $M_{\text{in}} = 3$. And, in case of the subsonic flow the inlet is set to the same stagnation conditions, but a different exit condition where the static pressure $p_{ex} = 1.98$. Finally, a flow with the same inlet conditions and a different exit static pressure $p_{ex} = 1.6$ is considered. This exit boundary condition generates a steady quasi-1d flow that features a shock in the diverging section of the CD nozzle.

The isentropic flows are chosen to verify the derivation for the zero-frequency resolvent modes corresponding to the quasi-1d flow through a CD nozzle, given in Sect. 2.5.4. The shock laden case is selected to investigate a discrete numerical methodology (similar to section [36]) that can converge to the verified zero-frequency resolvent mode solutions computed as per Sect. 2.5.4.

Resolvent modes corresponding to shock free base-flows

The zero-frequency resolvent ($\boldsymbol{R}(\omega = 0)$) modes corresponding to the shock free isentropic flows are computed by solving the eigenvalue problem that originates when computing the SVD of the resolvent operator, given by

$$\boldsymbol{R}^H(\omega = 0)\ \boldsymbol{R}(\omega = 0)\ \mathbf{f} = \boldsymbol{G}^\dagger * \boldsymbol{G} * \mathbf{f} = \sigma^2 \mathbf{f}, \tag{13}$$

(here $*$ represents the convolution operation), and are in agreement, as shown in Figs. 10, 11, 12 and 13, with the results of the resolvent analysis procedure described in [36].

The first step towards a numerical procedure capable of producing convergent resolvent modes is to show that the numerical procedure detailed in Sect. 2.4 can produce converging adjoint modes corresponding to the quasi-1d shock laden base flow described in the previous section. Figures 14 and 15 show that using a linear finite volume method, with a shock capturing mechanism that is a composite differentiable function of derivatives of the forward flow solution, it is possible to build a numerical procedure that can converge to the adjoint ($\boldsymbol{v}$) and forward solutions ($\mathbf{u}$; here convergence is only in a pointwise sense except at $x = x_s$), computed as per [34], corresponding to a shock laden base-flow, a specific gradient ($\mathbf{g}$) and forcing ($\mathbf{f}$; as defined in Sect. 2.5.2) function, respectively.

Resolvent operators constructed using the forward ($\boldsymbol{L}$) and adjoint ($\boldsymbol{L}^\dagger$) operators, used in the previous section to obtain Figs. 14 and 15, do not converge to the resolvent modes computed using Eq. (11; as seen in Fig. 16). This suggests that the zero-frequency resolvent modes constructed using the equations

$$(\boldsymbol{L}^{-1})^\dagger\ \boldsymbol{L}^{-1}\ \mathbf{f} = \sigma^2 \mathbf{f} \quad \& \quad \mathbf{r} = \boldsymbol{L}^{-1}\ \mathbf{f} \tag{14}$$

are not equal to the modes computed using the equations

$$\boldsymbol{G}^\dagger * \boldsymbol{G} * \mathbf{f} = \sigma^2 \mathbf{f} \quad \& \quad \mathbf{r} = \boldsymbol{G} * \mathbf{f} \tag{15}$$

when the shock laden base-flows are considered. To better understand why the shock-laden flows have this issue we consider two questions

1. Are the resolvent modes computed as per Eq. (15) physical? That is, in the limit of vanishing viscosity do modes computed as per a viscous extension of Eq. (15) approach the inviscid modes computed as per Eq. (15); and shown in [redacted]on Fig. 16.
2. Do the modes computed using the numerical procedure, as per Eq. (14), converge to the physical analytical adjoint modes computed as per Eq. (15).

To answer these questions the resolvent modes computed as per (15) need to be extended to consider viscous quasi-1d flows. To this end, and as the first step, an operator $\boldsymbol{G}_{viscous}$ needs to be constructed using three linearly independent linear forward solutions ($\mathbf{u}_i$), source vectors ($\mathbf{f}_i$) and the linearized quasi-1d viscous

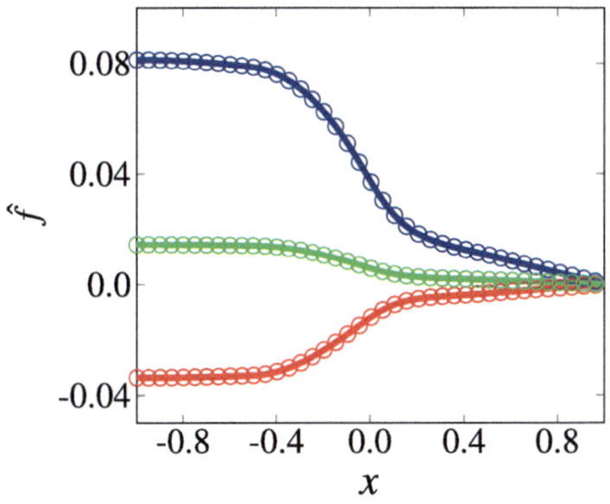

Fig. 10 Match between forcing modes computed using as per [34] (solid lines) compared with a central FVM + JST based discrete (circles) implementation for supersonic isentroipic flow through a CD nozzle

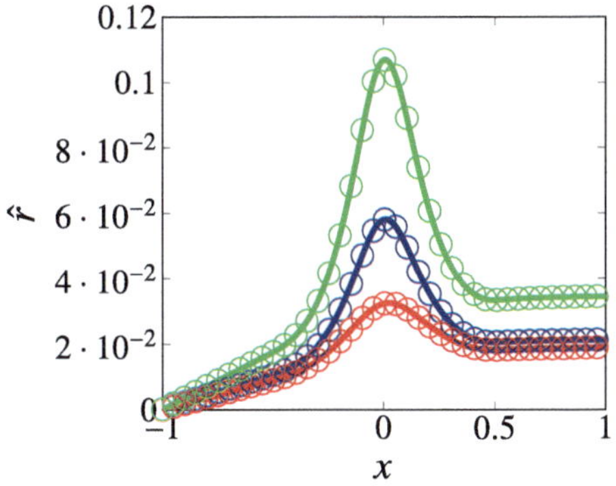

Fig. 11 Match between response modes computed using as per [34] (solid lines) compared with a central FVM + JST based discrete (circles) implementation for supersonic isentroipic flow through a CD nozzle

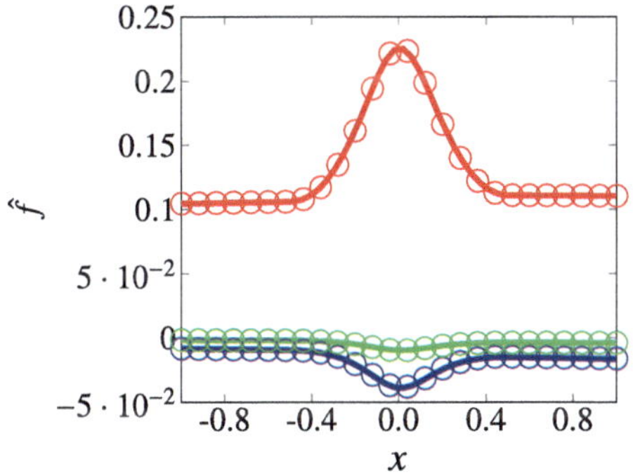

Fig. 12 Match between forcing modes computed using as per [34] (solid lines) compared with a central FVM + JST based discrete (circles) implementation for subsonic isentroipic flow through a CD nozzle

operator $\boldsymbol{L}_{viscous}$. The Figs. 17 and 18 show three such linearly independent linear forward solution and source vectors, respectively. The complete viscous extension of the procedure to analytically compute resolvent modes, similar to Eq. (15), and the development of a convergent numerical resolvent mode computation procedure, which converges to the analytical results as per Eq. (15), is currently ongoing.

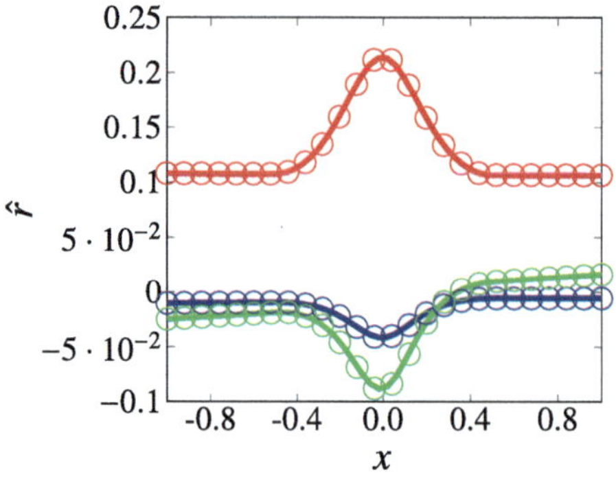

Fig. 13 Match between response modes computed using as per [34] (solid lines) compared with a central FVM + JST based discrete (circles) implementation for subsonic isentroipic flow through a CD nozzle

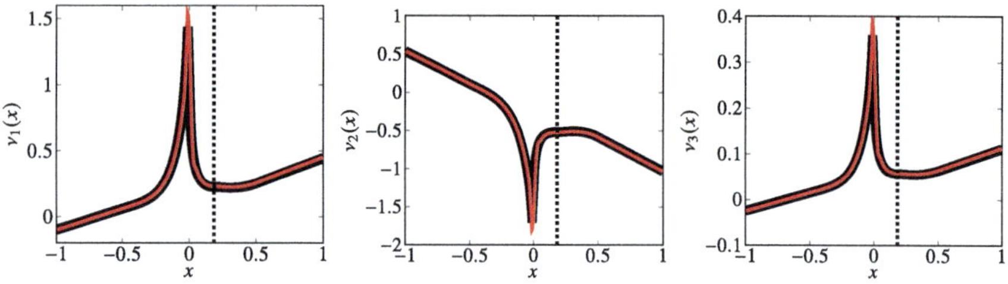

Fig. 14 Converged adjoint mode solution (FVM + JST based implementation, represented as ▬) and solution as per [34] , (represented as ▬), corresponding to the transonic shock laden flow through a quasi 1d model of flow through a CD nozzle

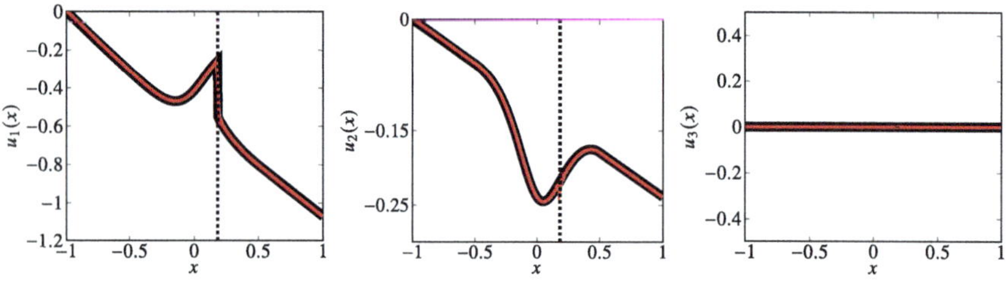

Fig. 15 Pointwise converged (except at $x = x_s$) forward mode solution (FVM + JST based implementation represented as ▬ and solution as per [34] represented as ▬), corresponding to the transonic shock laden flow through a quasi 1d model of flow through a CD nozzle

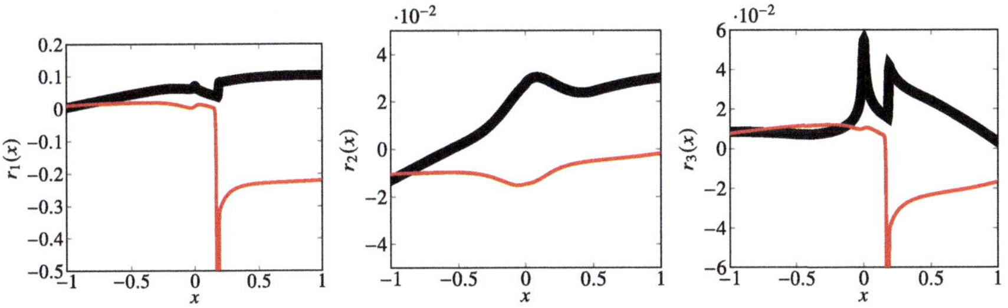

Fig. 16 Figure shows a comparison between the resolvent modes computed using Eq. (11) (represented as ▬) and using a FVM+viscous-dissipation based implementation (represented as ▬)

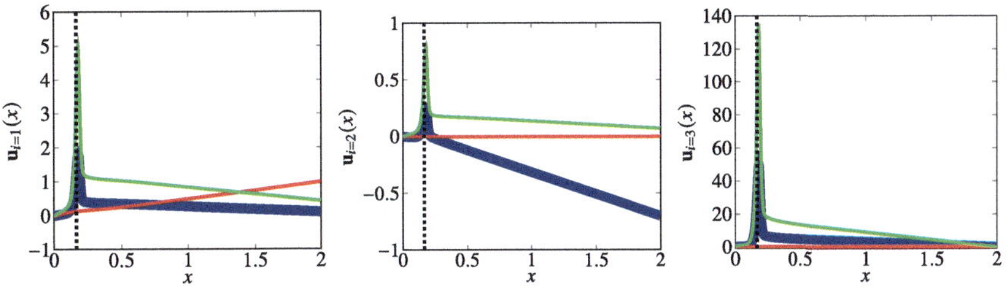

Fig. 17 Three linearly independent viscous forward mode solutions ($\mathbf{u}_{i=1,2,3}$) over the diverging section of the CD nozzle. Here the diverging section of the CD nozzle extends from $x = 0$ to $x = 2$. In each plot, the three line colors, blue, red and green represent the first, second and third component of each vector $\mathbf{u}_i$, respectively. The location of the shock is represented by the dashed black line

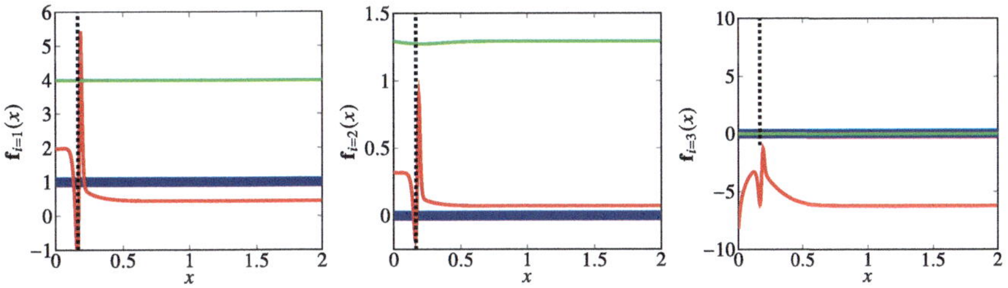

Fig. 18 Three linearly independent viscous forward mode source vectors ($\mathbf{f}_{i=1,2,3}$) over the diverging section of the CD nozzle. Here the diverging section of the CD nozzle extends from $x = 0$ to $x = 2$. In each plot, the three line colors, blue, red and green represent the first, second and third component of each vector $\mathbf{f}_i$, respectively. The location of the shock is represented by the dashed black line

3 Conclusion

The reduction of jet noise is a long sought-after goal and significant progress for subsonic jet JNR has been made through a wavepacket–resolvent approach. When applied to supersonic jets, the presence of shocks at under- or overexpanded conditions introduces discontinunities in the flow and numerical method requirements that may corrupt a linearized analysis. In this paper, we examine use the inviscid Burgers' and Euler equations as shock-laden jet surrogates to determine (a) suitable numerical methods whose linearization yields meaningful forward and adjoint sensitivities (and, consequently, meaningful resolvents) and (b) develop semi-analytical results suitable to verifying such methods. Our analysis shows that MUSCL, WENO, or any other method with stencil switching and/or flux limiting is not suitable for linearized analysis despite being appropriate for the forward solution. Localized artificial diffusivity methods are shown to work. Using the quasi-one-dimensional Euler equations, we developed a semi-analytical theory to verify these claims and preliminary results were presented. We show in particular that although forward and adjoint Green's functions provide convergent results when shocks are present, the zero-frequency resolvent modes do not yet converge when constructed as the product of the adjoint and forward operators that separately demonstrate pointwise convergence (except at the shock) to the solutions. This issue will be the topic of future work.

Acknowledgements This work was sponsored by the Office of Naval Research (ONR), under grant number N00014-19-1-2431. Dr. Steve Martens is the program officer. The simulations were supported by the NSF XSEDE program using allocation CTS20006 & TG-CTS090004 and performed on the TACC Frontera & Stampede2 computers, respectively.

References

1. Jordan, P., Colonius, T.: Wave packets and turbulent jet noise. Annu. Rev. Fluid Mech. **45**, 173–195 (2013)
2. Ponton, M.K., Seiner, J.M.: The effects of nozzle exit lip thickness on plume resonance. J. Sound Vib. **154**(3), 531–549 (1992)
3. Williams, J.F., Maidanik, G.: The mach wave field radiated by supersonic turbulent shear flows. J. Fluid Mech. **21**(4), 641–657 (1965)
4. Norum, T.D., Seiner, J.M.: Broadband shock noise from supersonic jets. AIAA J. **20**(1), 68–73 (1982)
5. Seiner, J., Ukeiley, L., Jansen, B., Kannepalli, C., Dash, S.: Noise reduction technology for F/A-18 E/F aircraft. AIAA Paper 2004-2972, Presented at the 10th AIAA/CEAS Aeroacoustics Conference, Manchester, GREAT BRITAIN (2004)
6. Bridges, J., Brown, C.: Parametric testing of chevrons on single flow hot jets. AIAA Paper 2004-2824, Presented at the 10th AIAA/CEAS aeroacoustics conference, Manchester, GREAT BRITAIN(2004)
7. Kim, J., Bodony, D.J., Freund, J.B.: Adjoint-based control of loud events in a turbulent jet. J. Fluid Mech. **741**, 28–59 (2014)
8. Michalke, A.: Instability of a compressible circular free jet with consideration of the influence of the jet boundary layer thickness. NASA TM-75190 (1977)
9. Crighton, D.G., Gaster, M.: Stability of slowly diverging jet flow. J. Fluid Mech. **77**(2), 397–413 (1976)
10. Gudmundsson, K., Colonius, T.: Instability wave models for the near-field fluctuations of turbulent jets. J. Fluid Mech. **689**, 97–128 (2011)
11. Lesshafft, L., Semeraro, O., Jaunet, V., Cavalieri, A.V., Jordan, P.: Resolvent-based modeling of coherent wave packets in a turbulent jet. Phys. Rev. Fluids **4**(6), 063901 (2019)

12. Farrell, B.F., Ioannou, P.J.: Stochastic forcing of the linearized navier-stokes equations. Phys. Fluids A **5**(11), 2600–2609 (1993)
13. Farrell, B.F., Ioannou, P.J.: Statistical state dynamics: a new perspective on turbulence in shear flow. Zonal Jets Phenomenology, Genesis, and Phys., pp. 380–400 (2019)
14. Garnaud, X., Lesshafft, L., Schmid, P.J., Huerre, P.: The preferred mode ofincompressible jets: linear frequency response analysis. J. Fluid Mech. **716**, 189–202 (2013)
15. Schmid, P.J.: Nonmodal stability theory. Annu. Rev. Fluid Mech. **39**, 129–162 (2007)
16. Sipp, D., Marquet, O.: Characterization of noise amplifiers with global singular modes: the case of the leading-edge flat-plate boundary layer. Theoret. Comput. Fluid Dyn. **27**(5), 617–635 (2013)
17. Beneddine, S., Sipp, D., Arnault, A., Dandois, J., Lesshafft, L.: Conditions for validity of mean flow stability analysis. J. Fluid Mech. **798**, 485–504 (2016)
18. Visbal, M.R., Gaitonde, D.V.: On the use of higher-order finite-difference schemes on curvilinear and deforming meshes. J. Comput. Phys. **181**(1), 155–185 (2002)
19. Bodony, D.J., Zagaris, G., Reichert, A., Zhang, Q.: Provably stable overset grid methods for computational aeroacoustics. J. Sound Vib. **330**, 4161–4179 (2011)
20. Martín, M.P., Taylor, E.M., Wu, M., Weirs, V.G.: A bandwidth-optimized weno scheme for the effective direct numerical simulation of compressible turbulence. J. Comput. Phys. **220**(1), 270–289 (2006)
21. Nonomura, T., Terakado, D., Abe, Y., Fujii, K.: A new technique for freestream preservation of finite-difference weno on curvilinear grid. Comput. Fluid **107**, 242–255 (2015)
22. Comte-Bellot, G., Corrsin, S.: Simple eulerian time correlation of full-and narrow-band velocity signals in grid-generated,'isotropic'turbulence. J. Fluid Mech. **48**(2), 273–337 (1971)
23. Li, Z., Ju, Y., Zhang, C.: Hybrid central-weno scheme for the large eddy simulation of turbulent flows with shocks. Numer. Heat Transf. Part B: Fundam. **72**(2), 170–189 (2017)
24. Moin, P., Squires, K., Cabot, W., Lee, S.: A dynamic subgrid-scale model for compressible turbulence and scalar transport. Phys. Fluids A **3**(11), 2746–2757 (1991)
25. Germano, M., Piomelli, U., Moin, P., Cabot, W.H.: A dynamic subgrid-scale eddy viscosity model. Phys. Fluids A **3**(7), 1760–1765 (1991)
26. Murthy, S.R., Bodony, D.J.: Resolvent analysis based jet-noise-reduction of a biconical tactical jet nozzle. AIAA Paper 2023-4518, Presented at the 2023 AIAA Aviation Forum (2023)
27. Liu, J., Corrigan, A.T., Kailasanath, K., Heeb, N.S., Munday, D.E., Gutmark, E.J.: Computational study of shock-associated noise characteristics using LES., AIAA Paper 2013-2199, Presented at the 19th AIAA/CEAS Aeroacoustics Conference, Berlin, Germany (2013)
28. Liu, J., Corrigan, A. T.: Large-eddy simulations of supersonic jet noise generation using wall modeling, 2018 AIAA/CEAS Aeroacoustics Conference, p. 3943 (2018)
29. Adjoint-based sensitivity of shock-laden flows (2022)
30. Leveque, R.J.: Finite volume methods for hyperbolic problems. Cambridge Texts in Applied Mathematics. Cambridge Univ. Press (2002)
31. Jian, G.-S., Shu, C.-W.: Efficient implementation of weighted ENO schemes. J. Comput. Phys. **126**, 202–228 (1996)
32. Jameson, A., Schmidt, W., Turkel, E.: Numerical solution of the euler equations by finite volume methods using runge-kutta time-stepping schemes. AIAA Paper 81-1259, Presented at the 14th AIAA Fluid and Plasma Dynamics Conference Palo Alto, CA pp. 23–25 (1981)
33. Kawai, S., Lele, S.K.: Localized artificial diffusivity scheme for discontinuity capturing on curvilinear meshes. J. Comput. Phys. **227**, 9498–9526 (2008)
34. Giles, M.B., Pierce, N.A.: Analytic adjoint solutions for the quasi-one-dimensional euler equations. J. Fluid Mech. **426**, 327–345 (2001)
35. Giles, M.B.M.B., Pierce, N., Giles, M., Pierce, N.: Adjoint equations in cfd-duality, boundary conditions and solution behaviour. In: 13th computational fluid dynamics conference, p. 1850 (1997)
36. Murthy, S.R., Bodony, D.J.: Resolvent analysis of a biconical tactical jet nozzle. In: 28th AIAA/CEAS aeroacoustics 2022 conference, p. 2969 (2022)

Wall-Modelled Large-Eddy Simulation of a Horizontal Axis Wind Turbine Blade Section

Lucy Mackie, Hao Xia, and Rob Watson

Abstract Horizontal axis wind turbines remain the dominant technology for extraction of wind power and this is expected to persist into the future as net-zero deadlines approach. Therefore, the continual improvement of their efficiency is imperative, for which a comprehensive understanding of their aerodynamics is necessary. This study presents wall-modelled large-eddy simulation of a 2.5D HAWT blade section model. At zero angle of attack, computed wall shear stress, surface pressure, and lift and drag statistics compare excellently with existing experimental data. The method is later applied to a near stall case and discussion is made on its application to future work. This study forms part of a wider project to investigate the potential for noise reduction geometry on wind turbine blades.

1 Introduction

Renewable energy generation has seen a rapid increase since the 1990s, with the percentage of final energy consumption from renewables almost doubling to 11.5% by 2019 [1]. In the United Kingdom, renewables have seen similar growth, primarily driven by an increase in wind power capacity which accounted for 24% of the UK's energy generation in 2020, making it the 4th largest generator of electricity from wind globally [2, 3]. Despite this, it is estimated that to achieve net zero targets, global wind capacity will need to increase by a further 375% by 2030 [1]; therefore, further research and development of wind power technologies will be required.

CUFS-2024-20, Cambridge, United Kingdom, 4-5 March 2024.

L. Mackie (✉) · H. Xia · R. Watson
Aeronautical and Automotive Engineering, Loughborough University, Loughborough, Leicestershire LE11 3TU, England
e-mail: l.j.mackie@lboro.ac.uk

H. Xia
e-mail: h.xia@lboro.ac.uk

R. Watson
e-mail: r.a.watson@lboro.ac.uk

J. C. Tyacke and N. R. Vadlamani (eds.), *Proceedings of the Cambridge Unsteady Flow Symposium 2024*, https://doi.org/10.1007/978-3-031-69035-8_11

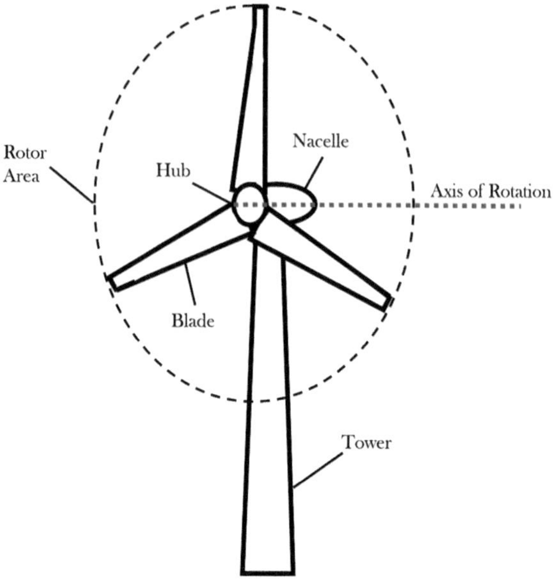

Fig. 1 Diagram of a typical HAWT structure highlighting the axis of rotation

Although there are several existing devices for extracting wind power, the most prolific is the horizontal axis wind turbine (HAWT); these feature blades driven by the wind to rotate about an axis parallel with the ground and have typically converged on a single design concept as shown in Fig. 1. While this technology is relatively mature compared to alternatives such as the vertical axis wind turbine (VAWT), improvements to their design remains ongoing as the typical efficiency of a HAWT may be as low as half their theoretical maximum of 59.3%, known as the Betz limit [4, 5].

The flow over an aerofoil has been extensively characterised in literature and is an important area to understand when considering HAWT design [6–8]. Incoming air causes a pressure distribution over the aerofoil surface, generating lift and drag forces. As angle of attack (α) increases, lift and drag will both increase to the stall point, after which the lift will decrease and drag increase rapidly [7]. The location of this point will vary between aerofoils, meaning it is important to characterise the airflow across a range of α in order to optimise performance. HAWTs typically operate with their blades close to this stall point, at which several complex flow features can be observed on the aerofoil suction side. This begins with a laminar boundary layer at the leading edge, followed by separation, transition to turbulence, and reattachment; along the aerofoil, this turbulent boundary layer then grows before separating near the TE [6, 7]. Despite the breadth of literature describing the flow over aeroplane aerofoils both experimentally and through simulation, there is a lack of study on HAWT specific aerofoils, which have different design requirements. Understanding of these features allows for their relative locations to be manipulated through aerofoil optimisation. For example, due to the changeability of conditions in which HAWTs operate, they are designed to have a relatively wide laminar bucket. To achieve this, HAWT aerofoils feature a sharp leading edge, promoting turbulent transition close to this point and producing a suction peak at higher lift coefficients; this results in a low sensitivity of drag to changes in α [7].

A key issue of HAWTs is that of noise, which has been identified as one factor limiting their implementation, as proposed wind farms are often met with local opposition to the perceived persistent disturbance they would bring [9]. This study forms part of a larger project on the modelling of TE waves or serrations for the study of their impact on aerodynamic noise generation and HAWT performance. Furthermore, the presence of wind farms has been identified as having an impact on local ecosystems, with Pearce-Higgins et al. observing decreased presence in seven out of 12 monitored species of birds, suggesting these areas are deliberately avoided by wildlife [10]. This presents further motivation to minimise HAWT noise given a key motivation for their installation is the reduction of anthropological impact on the environment. The focus of the present study is to develop a robust and efficient method of modelling a HAWT appropriate aerofoil section to gain full understanding of the aerodynamic features and validate this with existing qualitative and quantitative experimental data. This is an imperative initial step as, despite the breadth of literature relating to aerofoil aerodynamics, only limited LES calculations have been presented for HAWT appropriate aerofoils and Reynolds numbers [4]. In future work, the described methods will be applied to models with increased geometric fidelity and a range of TE geometries.

The following Section of this report details the numerical methods and computational setup of the HAWT aerofoil section. Section 3 presents a discussion of the success of the model at both $\alpha = 0°$ and $\alpha = 15°$. Finally, a summary of the presented and proposed work is given in Sect. 4.

2 2.5D WMLES of Aerofoil Section

A 2.5D model refers to a geometry which is unchanging in one dimension, allowing for the investigation of 3D flow effects without significantly increasing computational requirements [8]. This is particularly important for eddy resolving methods such as LES, as a third dimension is required to allow turbulent structures to develop. This modelling approach is common in aerofoil simulation, meaning domain, mesh and solver inputs have been extensively documented in literature, providing useful reference for the model detailed below [8, 11, 12]. The aerofoil selected for this study is the S809 developed for use on medium sized commercial HAWTs [7]. As is common with HAWT aerofoils, this is thicker than those commonly used in aerospace application and is designed to have a relatively restrained maximum lift coefficient

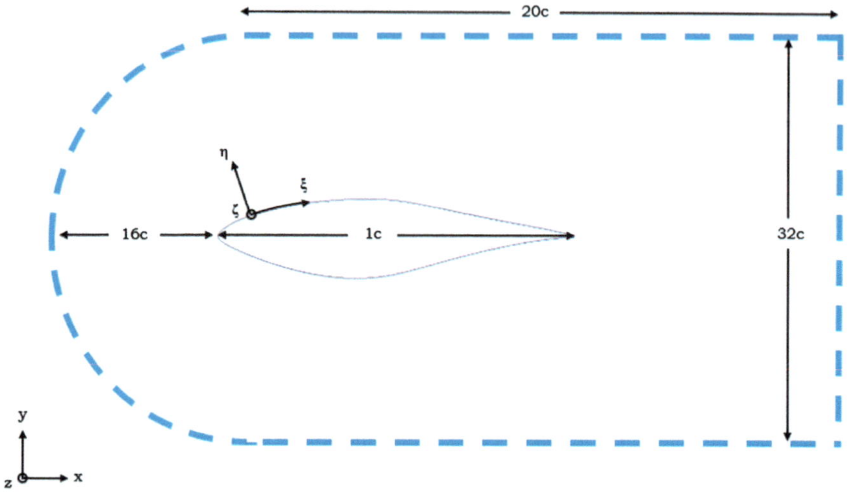

Fig. 2 2D slice of the computational domain showing the aerofoil profile, domain size (not to scale) and coordinate systems used. Spanwise (ζ) extent of the domain is 0.2c

and wide drag bucket [7]; the profile of the S809 aerofoil can be seen in Fig. 2. The aerofoil was simulated at first zero angle of attack ($\alpha = 0°$), then at a near stall case ($\alpha = 15°$).

2.1 Numerical Methods

Large-eddy simulation (LES) refers to a method of approximating solutions to the Navier-Stokes (NS) equations where the largest eddies are resolved directly on the computational grid and the eddies which are smaller than a defined filter size are modelled using a sub-filter scale model. Typically the filtering operation is done by projecting the solution onto a relatively coarse grid; the sub-filter model is then commonly referred to as a sub-grid scale (SGS) model. LES incurs higher computational cost than lower turbulence fidelity methods such as Reynolds-Averaged Navier-Stokes (RANS); this is due, in part, to the need to resolve the field temporally as well as spatially, and due to the need for higher mesh refinement to resolve sufficient turbulent kinetic energy [13]. However, this was deemed an acceptable trade-off given the significant increase in information about the instantaneous flow field, particularly high frequency turbulent fluctuations, which will be necessary for future aeroacoustic analysis later in this project. Furthermore, the accuracy of RANS methods is seen to significantly reduce with more complex flows - such as those with separated regions, three dimensional flow and streamline curvature. Some of these complexities are expected to be significant in the cases considered in this study, therefore the use of LES is also expected to produce more accurate statistics of the mean flow field. While grid requirements can be manageable in the free stream, in the near wall region even the largest eddies become small enough that the grid requirements become impractically large [13]. This requirement is reduced through the use of a wall model which, as estimated by Choi and Moin [14], causes the grid requirements to scale with approximately Re, as opposed to $Re^{13/7}$ for wall resolved LES. Furthermore, compared with hybrid RANS/LES methods, wall modelled LES (WMLES) does not necessitate the solving of additional transport equations for the near wall region, making it relatively less computationally expensive [15]. With similar motivation, an algebraic SGS model was selected: the wall-adapting local eddy-viscosity (WALE) model was employed as it has been shown to perform well in near wall and high shear regions, as well as providing good accuracy in aeroacoustic analysis [16, 17]. The turbulent eddy viscosity defined by this model can be found in [18].

As part of the WMLES approach used here, the friction velocity is computed to apply a computed wall shear stress at the boundary. The friction velocity is calculated using an all y^+ treatment which involves iteratively solving Reichardt's model of the law of the wall

$$u^+ = \frac{1}{\kappa}\ln\left(1+\kappa y^+\right) + C\left(1 - e^{-y^+/y_m^+} - \frac{y^+}{y_m^+}e^{-by^+}\right) \tag{1}$$

with an input velocity equal to the wall tangential velocity at the second cell away from the wall:

$$\tau_w = \rho u_\tau^2 \tag{2}$$

This study was completed with the Star-CCM+ CFD code, using a finite volume discretisation. The freestream Mach number of the flows was found to be on the limit of compressibility ($M = 0.3$), so a low Mach solver—Pressure-Implicit with Splitting of Operators (PISO)—was used. The convective fluxes were formulated through a second order upwind approach to balance stability and accuracy. The interface values of the primitive variables are calculated as:

$$\phi_{f,i} = \phi_i + \left(\mathbf{x}_f - \mathbf{x}_i\right)\cdot(\nabla\phi)_{f,i} \tag{3}$$

Then the transported scalar, ϕ, is upwinded in the direction of travel of the mass flux. The gradients required for the inviscid and diffusive fluxes were calculated through:

$$\nabla\phi_f = (\phi_R - \phi_L)\mathbf{A} + \overline{\nabla\phi} - (\overline{\nabla\phi}\cdot\mathbf{ds})\mathbf{A} \tag{4}$$

where ϕ_L and ϕ_R refer to the left and right cell values, $\mathbf{A} = \frac{\mathbf{a}}{\mathbf{a}\cdot\mathbf{ds}}$ (here $\mathbf{a}$ is the surface area vector), $\overline{\nabla\phi}$ is the mean of $\nabla\phi_L$ and $\nabla\phi_R$ and $\mathbf{ds}$ is the distance across the cell. These gradients were limited by the Venkatakrishnan limiter. The time discretisation was done using a dual-time stepping formulation of a backwards differencing approach; BDF2 in general, with a single BDF1 stage used for initialisation. Due to the use of the implicit BDF formulation, the timestep was chosen for accuracy and physical relevance, rather than stability. This was set at 1.02×10^{-3} in nondimensional time by the free stream velocity and chord length to ensure a reasonable range of frequencies was captured. This fixed timestep size corresponds to a convective Courant number of less than one over the majority of the domain. Before the averaging process was begun, the flow was allowed to develop through an initial transient period of $tU_\infty/c > 25$ where t, U_∞ and c are the time, freestream velocity and chord length respectively. Sampling was then conducted over a period of $tU_\infty/c = 27$.

2.2 Computational Setup

The present model consists of the S809 aerofoil in a C-shaped domain. The dimensions of this are shown in Fig. 2, where c denotes the aerofoil chord length of 1 m, and were selected from literature such that the results should not be influenced by the boundaries, particularly at higher α [19]. The outer edges (indicated by the blue dashes in Fig. 2) were defined as Free Stream, the aerofoil surface as a no-slip wall and the front and back surfaces as periodic planes to allow the flow to pass freely across them. The domain extends $0.2c$ in the spanwise direction, exceeding that employed by Asada and Kawai for their study of an aerofoil near stall, for which a domain thickness of $0.05c$ was deemed sufficient [6]. Figure 3 shows the variation in correlation coefficient of the fluctuating velocity components in the spanwise direction from the centre of the domain for both the $\alpha = 0°$ and $\alpha = 15°$ cases, taken from behind the trailing edge (TE) of the aerofoil, indicating the flow to be decorrelated by the edge of the domain and showing the spanwise extent to be sufficient.

Turbulent energy spectra for each case taken at the same location behind the aerofoil TE are shown in Fig. 4. In each Figure there is a section following the $-5/3$ law, indicating the successful capturing of the inertial subrange and energy cascade from larger to smaller eddies, building confidence in the accuracy of the physics captured. Both Figs. 3 and 4 were used to inform appropriate mesh sizing to efficiently capture the appropriate scales of eddies.

A trimmer (Cartesian) meshing strategy was adopted for this model using primarily isotropic hexahedral elements, with the highest refinement directly around the aerofoil and in the near wake region, which was angled in the direction of the free stream and sized according to the wake size indicated by initial RANS simulations. Anisotropic elements were applied in the wake region to further increase refinement in the vertical direction. A cross section of the mesh in the whole domain can be seen in Fig. 5. Also highlighted is an image of the mesh at the TE showing the aerofoil surface mesh refinement; growth off the surface through the volume; and isotropic wake elements which begin just before the TE. So as not to significantly increase mesh requirements in the boundary layer, first cell height off the aerofoil was adjusted to give a Δ_η^+ in the log law region. A summary of the mesh statistics for each simulation case can be seen in Table 1 where the wall units are calculated based on the maximum friction velocity on the aerofoil surface; the increased number of elements in the near stall case is due to the expected larger wake region.

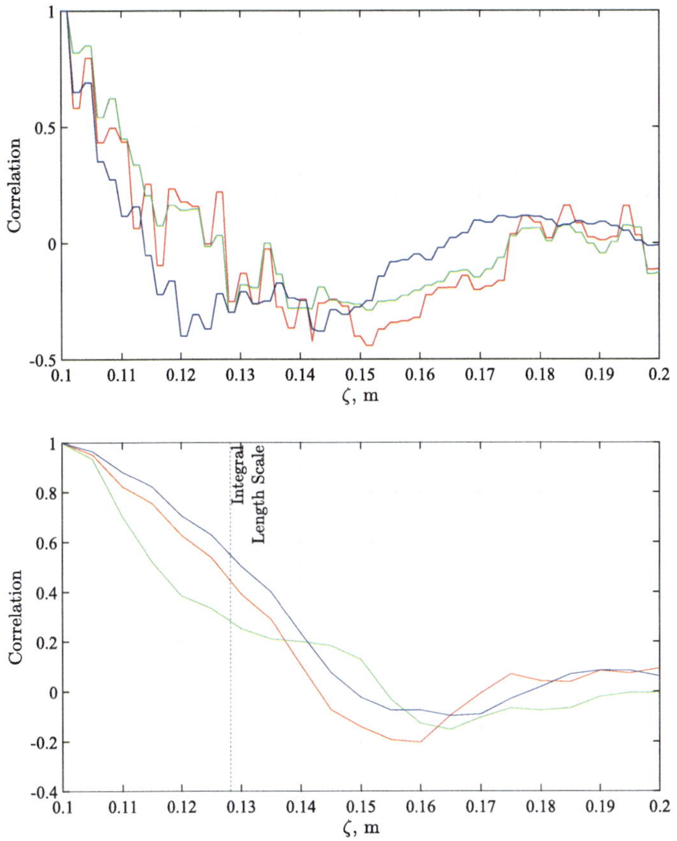

Fig. 3 Spanwise two-point correlation coefficient of the fluctuating velocity components taken from behind the aerofoil trailing edge for the $\alpha = 0°$ (top) and $\alpha = 15°$ (bottom) cases

2.3 *Validation of $\alpha = 0°$ Case*

For validation of the methods described in Sect. 2.1, key quantitative outputs were compared with experimental values obtained by Somers [7] during the development of the S809 aerofoil. The present model produces a calculated C_L value of 1.56×10^{-1} and C_D value of 6.99×10^{-3} at $\alpha = 0°$ giving a percentage error from Somers' experimental values [7] of 11.4% and 26.4% respectively. Additionally, Fig. 6 shows excellent agreement between the simulated and experimental C_p plots for this case, in particular capturing the short region of adverse pressure gradient between $0.3 < x/c < 0.5$ which was identified as a key design goal [7]. A key region of discrepancy between the simulation and experiment is between $0.45 < x/c < 0.55$ on the pressure side (darker circles) and $0.47 < x/c < 0.6$ on the suction side (lighter circles): here the experimental data shows a steep decrease in C_p corresponding with the locations of laminar separation and turbulent reattachment. On the suction side

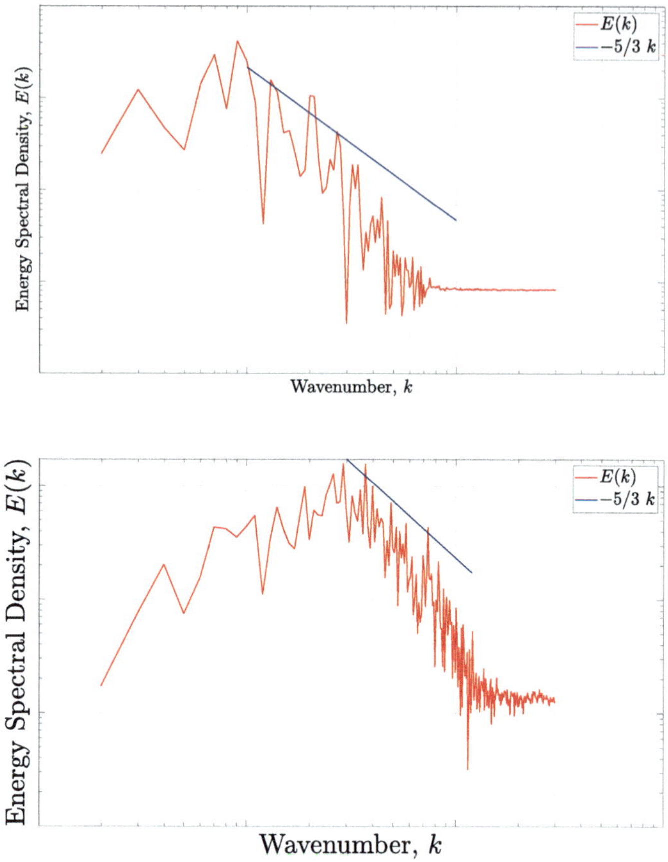

Fig. 4 Turbulent energy specra for the $\alpha = 0°$ (top) and $\alpha = 15°$ (bottom) cases taken from a location behind the aerofoil TE

this trend is matched, although slightly later along the chord length, indicating this phenomenon was more closely captured. However, on the pressure side, the simulated C_p values show a continuous, steady decrease in this region, suggesting no separation was modelled and highlighting an area where further improvement could be made. The proximity of these quantitative results to the experiment builds confidence in the validity of the model inputs and subsequent qualitative discussion.

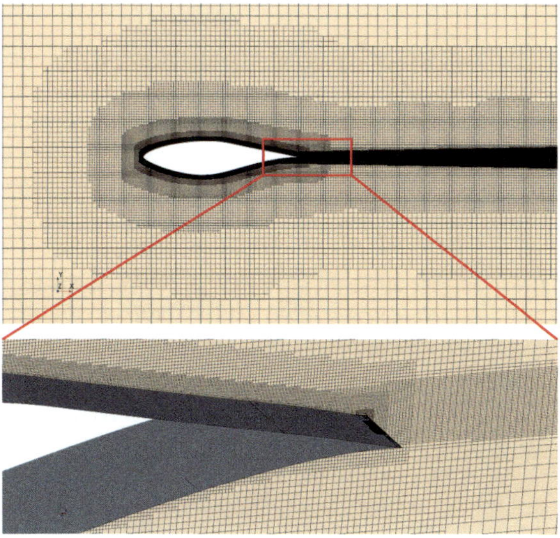

Fig. 5 Image of mesh showing the refinement around the aerofoil for $\alpha = 0°$ case and highlighting the surface mesh and wake refinement at the TE

Table 1 Summary of mesh metrics for each simulation

α	No. grid points	Δ_ξ^+	Δ_η^+	Δ_ζ^+
0	22.4M	24	8	24
15	26.4M	32	19	32

3 Results

3.1 Laminar Separation Bubble at $\alpha = 0°$

Figure 7 shows the instantaneous flow field coloured by streamwise velocity showing increased velocity over both the pressure and suction surfaces and a clear transition between the laminar and turbulent regions along the aerofoil surface. Towards the TE, the formation of hairpin vortices can also be observed off the aerofoil surface. Finally, a thin turbulent wake can be seen which extends beyond the Figure but decays before the end of the domain, indicating the domain size to be appropriate so as not to affect the airflow around the aerofoil. The small size of the wake also suggests a comparatively small contribution by form drag to the overall drag value, corroborated through the C_D monitor set up in the simulation which calculated 60% of the drag value to be from which, which may explain the lower accuracy of C_D than C_L compared to experiment.

A contour plot of streamwise wall shear stress (WSS) on the aerofoil suction side is shown in Fig. 8 with a comparison of the location of the laminar separation bubble

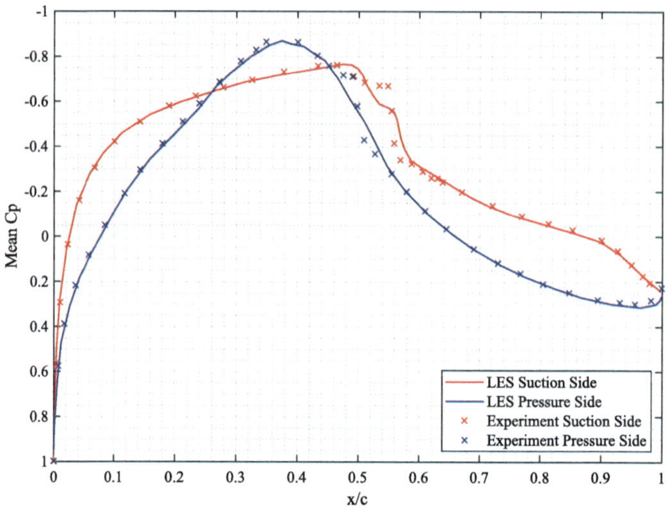

Fig. 6 C_p plot showing variation across the aerofoil surface comparing simulation output (green circles) with experimental values (grey lines) [7] at $\alpha = 0°$

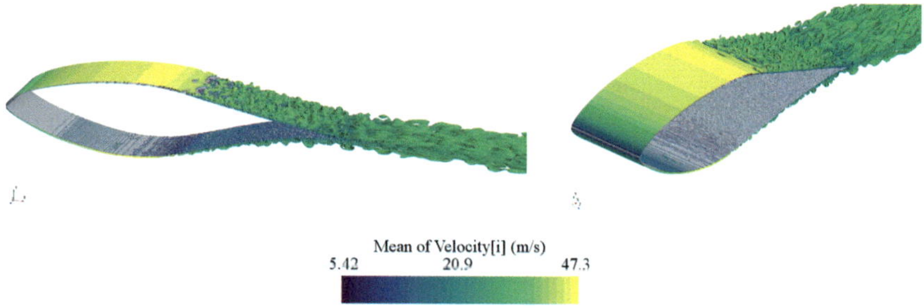

Fig. 7 Instantaneous flow field coloured by streamwise velocity at $\alpha = 0°$. Isosurface is the second invariant of the velocity gradient tensor ($\hat{Q} = 0.0173$ where $\hat{Q}$ is Q normalised by the free stream velocity and chord length)

with that identified in wind tunnel testing; streamlines constrained to the aerofoil surface are displayed in black. The location of negative WSS and absent streamlines closely matches the laminar separation and turbulent reattachment locations identified by Somers [7], reinforcing the notion that this phenomenon was accurately modelled and building confidence in overall model accuracy. By contrast, Fig. 9 shows the same outputs for the aerofoil pressure side. Here, the constant attachment of streamlines to the aerofoil surface and minimum WSS not going noticeably below zero indicate a lack of separation in the region indicated through experiment; this corroborates the observation made based on the C_p plot shown in Fig. 6. However, the location of laminar/turbulent transition is in approximately the correct region. Also shown in both Figs. 8 and 9 are outputs from comparative RANS simulations;

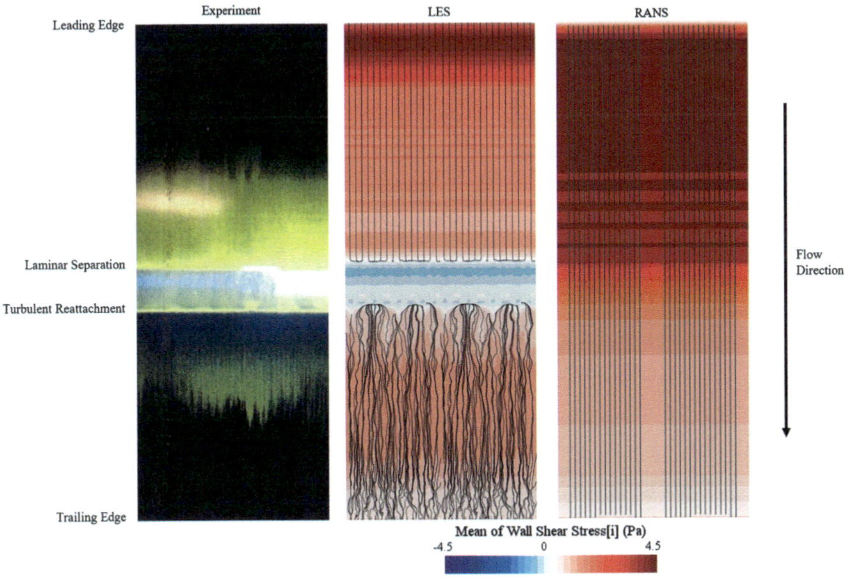

Fig. 8 Comparison of separation and reattachment locations on aerofoil suction side with with oil flow visualisation at $Re = 1 \times 10^6$ [7]

given that no transition model was used, the laminar separation bubble was not captured in the simulation. Additionally, the streamlines present on the LES surface highlight the extent of the rich physics that is captured by this simulation, compared to the lack of equivalent detail in the RANS streamlines. These observations provide a visual justification for the increased use of computational resources for the LES calculation.

A further way to corroborate the accurate capturing of the transition location is through the use of surface pressure fluctuation traces such as shown in Fig. 10; here p' denotes instantaneous pressure fluctuation from the mean value. As described by Wang and Moin [20], minimal fluctuations indicate attached laminar flow; low amplitude, high frequency fluctuations indicate attached turbulent flow; and the opposite indicates detached flow. Comparing Figs. 8 and 10 again highlights the model's accuracy with pressure fluctuations being minimal in the laminar region ($x/c \leqslant 0.4$, shown here in puple) and high frequency in the turbulent region ($x/c \geqslant 0.6$, yellow here). Shown in blue in Fig. 10 is the pressure fluctuation at the location of the separation bubbles as indicated by Somers. On the suction side these are low frequency across the time history, indicating the bubble has been accurately captured by the simulation. However, on the pressure side, the fluctuations at the separation bubble location are more indicative of turbulent flow, again highlighting the failure of the model to capture this feature.

The proximity of the above qualitative and quantitative results to experimental values have built sufficient confidence that the applied method outlined in Sect. 2.2

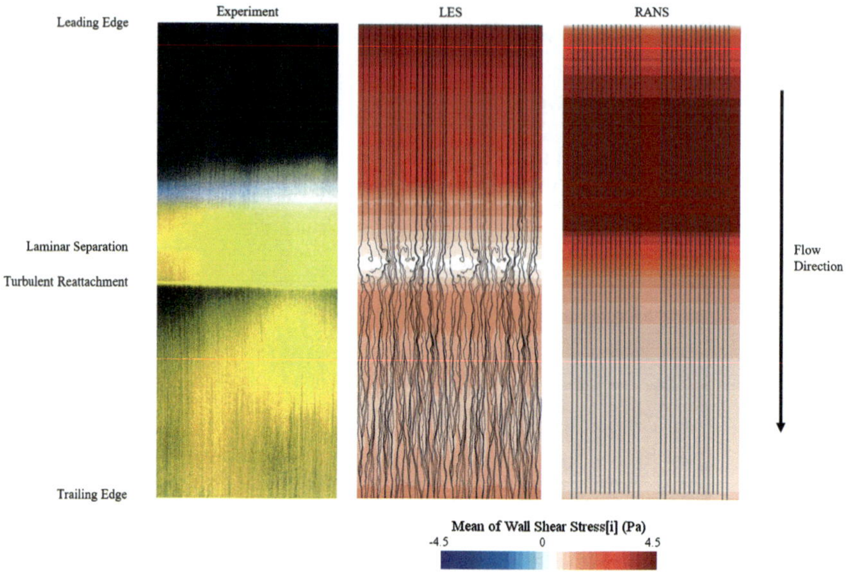

Fig. 9 Comparison of separation and reattachment locations on aerofoil pressure side with with oil flow visualisation [7]

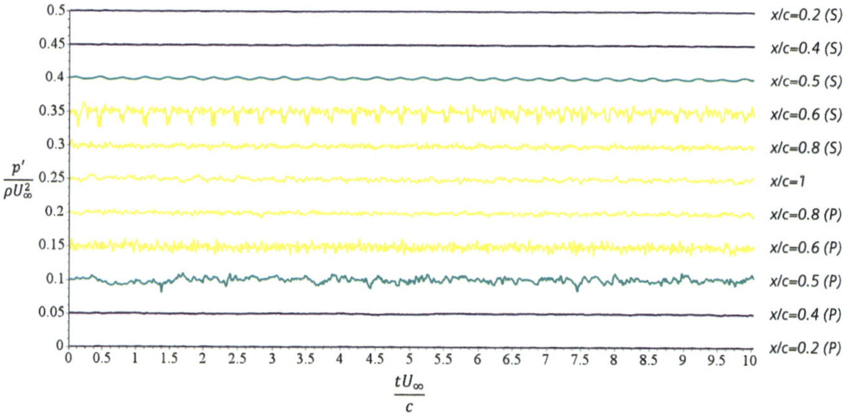

Fig. 10 Time history of surface pressure fluctuations at various points along the aerofoil surface sampled after initial transient period (individual curves are separated by a 0.05 offset)

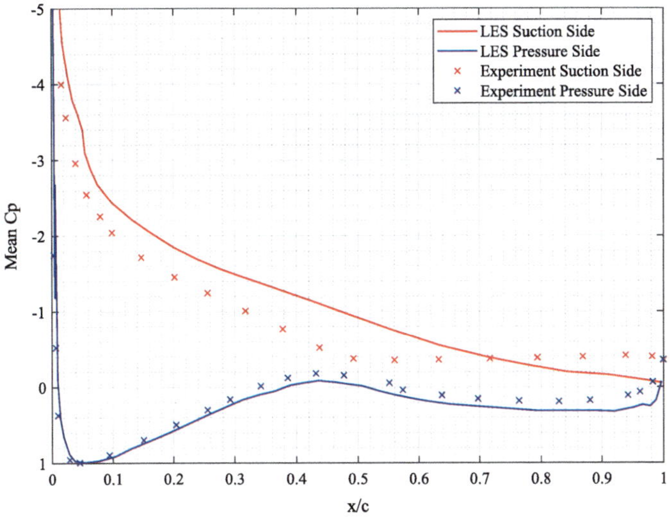

Fig. 11 C_p plot showing variation across the aerofoil surface comparing simulation output (green circles) with experimental values (grey lines) [7] at $\alpha = 15°$

is appropriate for the HAWT aerofoil at $\alpha = 0°$. Consequently, the model has been adapted to simulate the same aerofoil at $\alpha = 15°$, which is closer to HAWT operation and is likely to generate more complex aerodynamic features as described in Sect. 1.

3.2 *Near Stall Case*

In practice, HAWTs are operated at much higher angles of attack, close to their stall point; therefore, following the development and validation of the $\alpha = 0°$ model, the free stream velocity direction was changed to 15° which is the angle of maximum lift generation as identified by Somers [7].

As with the $\alpha = 0°$ case, initial assessment of model quality was done through comparison of C_L, C_D and C_p values with experiment. The initially calculated C_L and C_D values of 1.076 and 0.264 are within 0.5% and 282.6% of the experiment respectively; the corresponding C_p plot can be seen in Fig. 11. Although the simulated C_L value being almost identical to the experimental value at first builds significant confidence in the model's accuracy, examination of the corresponding C_p plot shows this to be misleading. Lift force is given by the area enclosed by a C_p plot; therefore, although the areas enclosed by the experimental and simulated lines may be near identical, there is significant discrepancy between the actual curves, particularly on the suction side across the whole surface.

Furthermore, there is significant discrepancy between the simulated and experimental C_D values, with the simulated value being much higher indicating the aero-

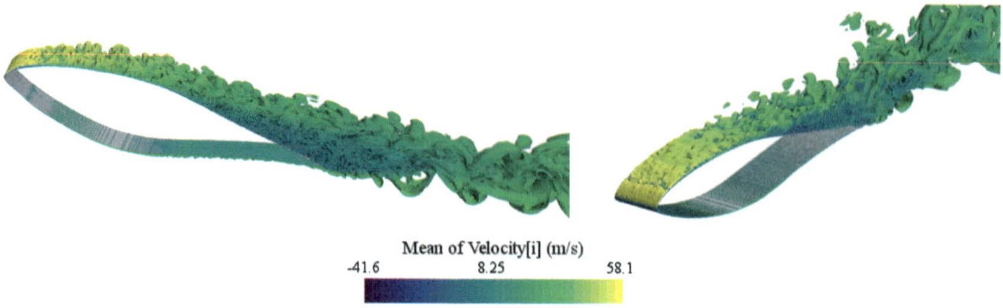

Fig. 12 Instantaneous flow field coloured by streamwise velocity at $\alpha = 15°$. Isosurface is the second invariant of the velocity gradient tensor ($\hat{Q} = 0.0173$ where $\hat{Q}$ is Q normalised by the free stream velocity and chord length)

foil has begun to stall. This is corroborated qualitatively with Fig. 12 which shows a large region of turbulence and small recirculation region behind the aerofoil. Further investigation at different values of α is planned to identify the angle at which this significant increase in drag begins to occur. It is possible that a source of inaccuracy in the model could be a consequence of limited spanwise extent, given Manolesos and Papadakis [8] recommended a 10 times increase compared to the present model, providing an additional potential avenue for investigation in future study.

On the other hand, qualitatively the model does show promise. Figure 12 shows the instantaneous flow field around the aerofoil from which several aerodynamic features mentioned by Asada and Kawaii [6] can be identified. A laminar region at the leading edge of the suction side is seen which transitions to a turbulent boundary layer which grows in size over the length of the aerofoil; a small region of recirculating flow can be seen at the TE.

4 Summary and Future Work

Overall, this study has shown that the methods described can successfully be applied to a HAWT appropriate aerofoil with reasonable accuracy at both $\alpha = 0°$ and in the near stall condition, although the latter model requires improvement. Additionally, it has highlighted the need for a higher fidelity method to be employed, as the LES model was able to successfully capture the laminar separation bubble and downstream turbulent structures not present in the comparative RANS simulations. The focus of subsequent work will be to apply this method to a full three dimensional model of a HAWT blade to further characterise the aerodynamic features and provide a baseline model for noise reduction modifications which will encourage the proliferation of HAWTs and their associated social, environmental and economic benefits.

References

1. International Energy Agency. Renewables. Online (2022)
2. Office of National Statistics. Wind energy in the uk: June 2021. Online (2021)
3. International Renewable Energy Agency. Country rankings. Online (2022)
4. Rajamohan, S., Vinod, A., Aditya, M.P., Vadivudaiyanayaki, H.G., Arıcı, M., Nižetić, S., Le, T.T., Hidayat, R., Nguyen, D.T.: Approaches in performance and structural analysis of wind turbines—a review. Sustain. Energy Technol. Assess. **53**, 102570 (2022)
5. Karthikeyan, N., Murugavel, K.K., Kumar, S.A., Rajakumar, S.: Review of aerodynamic developments on small horizontal axis wind turbine blade. Renew. Sustain. Energy Rev. **42**, 801–822 (2015)
6. Asada, K., Kawai, S.: Large-eddy simulation of airfoil flow near stall condition at reynolds number 2.1×10^6. Phys. Fluids **30**(8), 085103 (2018)
7. Somers, D.M.: Design and experimental results for the s809 airfoil. Technical Report WE711110, National Renewable Energy Laboratory and Airfoils Incorporated, State College, Pennsylvania (1997)
8. Manolesos, M., Papadakis, G.: Investigation of the three-dimensional flow past a flatback wind turbine airfoil at high angles of attack. Phys. Fluids **33**(8), 85–106 (2021)
9. Smith, E., Klick, H.: Explaining nimby opposition to wind power. In: American political science association annual meeting, pp. 1–19 (2007)
10. Pearce-Higgins, J.W., Stephen, L., Langston, R.H.W., Bainbridge, I.P., Bullman, R.: The distribution of breeding birds around upland wind farms. J. Appl. Ecol. **46**(6), 1323–1331 (2009)
11. Salman Siddiqui, M., Khalid, M.H., Badar, A.W., Saeed, M., Asim, T.: Parametric analysis using cfd to study the impact of geometric and numerical modeling on the performance of a small scale horizontal axis wind turbine. Energies **15**(2), (2022)
12. Tabib, M., Rasheed, A., Salman Siddiqui, M., Kvamsdal, T.: A full-scale 3d vs 2.5d vs 2d analysis of flow pattern and forces for an industrial-scale 5mw nrel reference wind-turbine. In: 14th deep sea offshore wind R&D conference, EERA DeepWind **137**, pp. 477–486 (2017)
13. Bose, S.T., Park, G.I.: Wall-modeled large-eddy simulation for complex turbulent flows. Annu. Rev. Fluid Mech. **50**(1), 535–561 (2018)
14. Choi, H., Moin, P.: Grid-point requirements for large eddy simulation: Chapman's estimates revisited. Phys. Fluids **24**(1), 011702 (2012)
15. Piomelli, U.: Wall-layer models for large-eddy simulations. Prog. Aerosp. Sci., **44**(6), 437–446 (2008). Large Eddy Simulation—Current Capabilities and Areas of Needed Research
16. Lasota, M., Šidlof, P., Kaltenbacher, M., Schoder, S.: Impact of the sub-grid scale turbulence model in aeroacoustic simulation of human voice. Appl. Sci. **11**(4), (2021)
17. Yang, C., Feng, H., Peng, Y.: Noise characteristic analysis and sound sources identification for rod–airfoil interaction using different subgrid-scale models. E3S Web of Conferences **233**, 04036 (2021)
18. Nicoud, F., Ducros, F.: Subgrid-scale stress modelling based on the square of the velocity gradient tensor. Flow Turbul. Combust. **62**(3), 183–200 (1999)
19. Salman Siddiqui, M., Rasheed, A., Kvamsdal, T.: Numerical assessment of rans turbulence models for the development of data driven reduced order models. Ocean Eng. **196**, 106799 (2020)
20. Wang, M., Moin, P.: Computation of trailing-edge flow and noise using large-eddy simulation. AIAA J. **38**(12), 2201–2209 (2000)

Large-Eddy Simulations of Thermal Ribbed Duct Flows

H. Garg, L. Wang, and C. Fureby

Abstract Ribbed channel flows are encountered in numerous engineering applications to promote turbulence and enhance heat transfer in various cooling passages, e.g., in turbine blades and at combustor walls. Numerical simulations have become an increasingly useful tool for providing insights into complex turbulence and flow separation, which are critical to the design and optimization of such channels. In this numerical study we use the experimental case of Wang, [1, 2], as a reference set-up for thermal ribbed duct flows in the fully developed turbulent flow regime of $4000 \leq Re \leq 24000$, where Re is the Reynolds number. Wang, [1, 2], performed detailed, non-intrusive, experimental studies to characterize the turbulent flow and its impact on the thermal performance, thus providing an experimental data base against which numerical simulations based on Reynolds Averaged Navier Stokes (RANS) and Large Eddy Simulations (LES) can be validated. In this paper we specifically report on RANS and LES, using different turbulence models, subgrid models, and grids, at Re=4000, 8000, 12000, 16000, 22000, and 24000, using a code developed within openFoam, [3]. The turbulence models include the RANS k-ϵ and k-ω SST models, and the LES subgrid models include Smagorinsky (SMG), One Equation Eddy Viscosity (OEEVM), Wall-Adapting Local Eddy viscosity (WALE), and Localized Dynamic K Equation (LDKM) models, whereas the grid sizes range from 1.0 to 25.0 million cells, resolving between 86% and 94% of the turbulent kinetic energy, suggesting that all grids are acceptable for LES. The LES results are generally in better agreement with the experimental data than the RANS results. The LES using the WALE and LDKM models are also in closer agreement with the experimental data than the LES using the SMG and OEEVM models. The LES also reveal a highly unsteady and very complex flow between and above the ribs.

CUFS-2024-22, Cambridge, United Kingdom, 4-5 March 2024.

H. Garg (✉) · L. Wang · C. Fureby
Department of Energy Sciences, Lund University, Ole Römers väg 1, 221 00 Lund, Sweden
e-mail: himani.garg@energy.lth.se

L. Wang
e-mail: lei.wang@energy.lth.se

C. Fureby
e-mail: christer.fureby@energy.lth.se

J. C. Tyacke and N. R. Vadlamani (eds.), *Proceedings of the Cambridge Unsteady Flow Symposium 2024*, https://doi.org/10.1007/978-3-031-69035-8_12

1 Introduction

Ribbed duct flows are commonly encountered in engineering applications to enhance turbulence and heat transfer in cooling passages. These channels with raised ribs are found in turbine blades, guide vanes, electronic equipment and microprocessors, heat exchangers, and combustor walls. More specifically, turbine blade and guide vanes cooling technology will play a critical role in the next generation of propulsion and power generation gas turbines. Accurate prediction of blade metal temperature can avoid the use of excessive compressed bypass air and allow higher turbine inlet temperature, increasing fuel efficiency and decreasing emissions. Electrical equipment and microprocessors distribute power and generate and/or dissipate heat due to the resistance and impedance of the various electrical components in any given circuit. As an alternative to transitional heat sinks, indirect channel cooling with structured channels is of interest for a growing number of applications as weight restrictions become more important to improve sustainability. By increasing the heat transfer area of heat exchangers using dimples, pins, ribs, and other surface protrusion elements, the heat transfer can be significantly improved. The internal, liner walls, in combustors must be cooled by the surrounding air otherwise their structural integrity will break down due to the very high combustion temperatures. On the outside of the liner ribs are often used to enhance the heat transfer rates between the cold surrounding air and the liner wall itself, which is often equipped with film cooling holes to facilitate the formation of an internal cooling film directly protecting against the hot combustion product temperatures. In short, ribbed duct flows constitute an important technical building block whose function needs to be fully understood, and the design and optimization need to be facilitated by well validated computational fluid dynamic methods.

Walls rib-roughened cooling elements typically consist of straight or curved walls on which raised ridges or ribs, which disturb the flow and induce turbulence, are attached. The flow disturbance and increased turbulence levels increase the heat transfer rates between the fluid and the wall. To investigate the impact of roughness on heat transfer and friction factor, turbulent flow is often simulated in channels equipped with continuous, evenly spaced, transverse ribs, serving as a simplified model. The flow characteristics downstream of a rib exhibit significant complexity, characterized by downstream flow separation, subsequent reattachment, and relaxation within the mainstream, along with the emergence of strong secondary flow caused by geometric discontinuity. The presence of reverse flow and a high level of turbulence intensity further complicate the flow within the separated shear layer.

Ensuring optimal geometry in ribbed channels is critical for gas turbine engines to avoid overheating and subsequent damage to engine components. Hence, investigating ribbed channels plays a significant role in enhancing their thermal performance.

Many experimental investigations have been conducted to improve heat transfer rates for turbine blades, considering various parameters such as passage aspect ratio [4], pitch ratio [5], blockage ratio [6], rib angle of attack [7], and number of ribbed walls [8]. In general, these experimental studies showed that repeated ribs with pitch ratio of $p/e = 10$, where p is the distance between two consecutive ribs and e is the rib height, provides the optimum thermal performance in terms of both heat transfer and friction factor. Regarding the local heat transfer distribution, different methods using holographic interferometry [9], Thermochromic Liquid Crystal (TLC) [10], and naphthalene sublimation [11] displayed that the first peak in heat transfer occurred at $3 < x/e < 4$, which corresponds to the reattachment point between consecutive ribs. Usually, the second peak of the Nusselt number just upstream the second rib has also been observed [9–11]. With large rib height and the aspect ratio of channel close to unity, the flow becomes more three-dimensional, which makes it difficult to predict the heat transfer coefficient using the one-dimensional similarity analysis. PIV measurements [12] revealed persistent velocity dips in the velocity profiles within the outer boundary layer, indicating that flow redevelopment after the reattachment point at ($x/e = 3.6$) deviates significantly from the canonical equilibrium boundary layer. Using Galilean decomposition, different vortical structures which are embedded in the separated shear layer were identified in the instantaneous velocity fields.

Numerical and analytical studies using Direct Numerical Simulations (DNS), Reynolds-Averaged Navier-Stokes (RANS) [13, 14], or Large-Eddy Simulations (LES) techniques [15] have been carried out to investigate flow separation characteristics in ribbed channels. The numerical study conducted by Ashrafian *et al.* [16] examined turbulent flow in a channel roughened with transverse square rods at a Reynolds number of $Re_\tau = 400$ using DNS. These rods had a height, e, of $e^+ = eU_b/\nu = 13.6$ wall units, where U_b is the bulk velocity and ν is the viscosity and were symmetrically arranged with a constant pitch about the mid-plane of the channel. The study revealed the presence of two co-rotating vortices in the mean streamline pattern between consecutive rods. Additionally, streaky structures along the streamwise direction were observed in the buffer region, indicating a transitionally rough nature of the flow. Earlier research [17–19] extensively examined turbulent flow in channels where the bottom wall was equipped with square bars as roughness elements. These studies explored a broad range of cavity length to rib-height ratios, referred to as p/e. Notably, they observed that the maximum form drag, and minimum skin frictional drag occurred at $p/e = 7$ [17]. As p/e increased, streamwise coherence decreased, reaching its minimum value at $p/e = 7$. Additionally, by analyzing invariants, researchers discovered that flow over a rough wall exhibited greater isotropy compared to a smooth wall [19]. Moreover, the performance of LES in this flow configuration was also evaluated [19]. The numerical study conducted by Vijiapurapu & Cui [20] examined time-averaged flow patterns, which were reasonably accurate when compared to experimental data, using different turbulence models.

Numerical simulations play a crucial role in unravelling complex flow, including the interaction between ribs and the main flow, and their impact on overall heat trans-

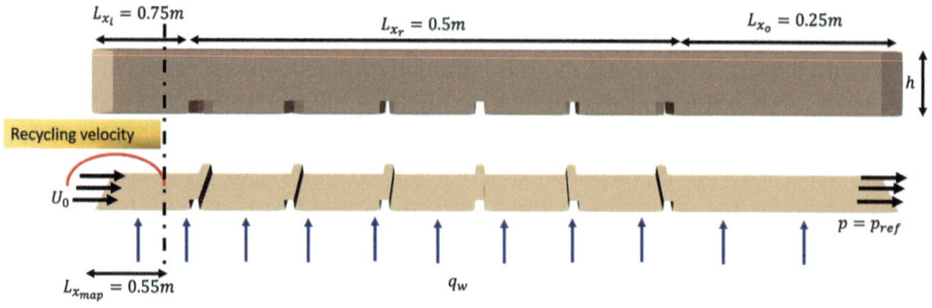

Fig. 1 Computational domain: schematic of ribbed channels [9]

fer performance. Despite extensive research on ribbed channel flows, many aspects of the flow physics remains not fully understood, such as various instantaneous and statistical features, and dynamics. In the present study, to gain insight into the accuracy of LES in predicting the physical process of separation, wall-modelled LES [34] is employed. This simulation method enables the measurement of resolved velocity, $\overline{u}_i$, skin-friction factor, c_f, and Nusselt number, Nu.

2 Computational Methodology

This investigation focuses on ribbed channel flows in a square test section with a length of 1.50 m according to Fig. 1, being experimentally studied using different non-intrusive techniques in [1, 2, 11, 12, 21, 41]. The bottom wall of the channel is equipped with ribs of square cross-section, each having a height equal to 20% of the channel height, h=0.05 m, placed at regular intervals of pitch ratio of $p/e = 10$. The flow is assumed to be incompressible, Newtonian, and heated with a uniform wall heat flux $q_w = 1000 W/m^2$ from below.

The Reynolds number, Re, based on the channel height, h, is varied within the range of 4000 to 24000 by changing the air mass flow through the channel. The coordinates for the flow are x (streamwise), y (wall-normal), and z(spanwise), with corresponding velocity components u, v, and w. At the inlet, at $x = 0$, a uniform velocity boundary condition, U_b, with a constant temperature, T_0, is imposed. A fixed pressure boundary condition, and a zero-gradient condition for velocity and temperature are applied at the outlet.

In the experiments, [21], the inlet flow is known to be turbulent due to the characteristics of the blower used to generate the mass flow of air through the channel. However, when simulations are performed using velocity inlet and pressure outlet, the flow naturally remains laminar in the channel before approaching the ribs. To address this issue and generate fully developed turbulent flow, matching the flow in the experiment, before the ribs, the inlet velocity field is recycled from an arbitrary plane located at a distance $L_{x_{map}} = 0.55$ m downstream of the inlet. Cyclic boundary conditions are imposed for the velocity at the inlet and the arbitrary plane downstream. The recycled velocity then enters the rest of the channel continuously and flows toward the outlet using mapping boundary conditions in openFoam, [3].

The advantage of this approach over that of having a separate channel flow simulation, [22] or using synthetic turbulence, [23, 24], is that it combines everything within one simulation and geometry set-up, eliminating the need for any precursor simulations. It is however important to note that while a constant temperature is fixed at the inlet and the bottom wall remains heated throughout the simulations, only the velocity field is recycled.

To simulate fully developed turbulent flow through rib-roughened cooling channels, two RANS models, the $k - \epsilon$, [29], and $k - \omega$ SST, [30], models, and four LES subgrid models, One Equation Eddy Viscosity (OEEVM) [25], Smagorinsky (SMG) [26], Wall-Adapting Local Eddy viscosity (WALE) [27], and localized Dynamic K Equation (LDKM) [28] models, have been used. The RANS equations govern the transport of the averaged flow quantities, with the whole range of scale modelled. The RANS-based modelling approach therefore greatly reduces the required computational effort and resources and is widely adopted for practical engineering applications. The standard $k - \epsilon$ model is adopted, where the turbulent kinetic energy, k, and its dissipation rate, ϵ, are obtained by solving their transport equations. The $k - \omega$ SST model used is an empirical model based on transport equations for the turbulent kinetic energy, k, and the specific dissipation rate, ω. In LES, the governing equations are the low pass filtered conservation equations of mass, momentum, and energy as represented by,

$$\frac{\partial \overline{u}_i}{\partial x_i} = 0, \tag{1}$$

$$\frac{\partial \overline{u}_i}{\partial t} + \frac{\partial}{\partial x_j}(\overline{u}_i \overline{u}_j) = -\frac{1}{\rho}\frac{\partial \overline{p}^*}{\partial x_i} + \frac{\partial}{\partial x_j}\left(\nu \frac{\partial \overline{u}_i}{\partial x_j} - \left(\overline{u_i u_j} - \overline{u}_i \overline{u}_j\right)\right) - \beta(T - T_0)g_i, \tag{2}$$

$$\frac{\partial \overline{e}}{\partial t} + \frac{\partial}{\partial x_j}(\overline{u}_j \overline{e}) = -\nu \frac{\partial \overline{u}_i}{\partial x_j}\left(\frac{\partial \overline{u}_i}{\partial x_j} + \frac{\partial \overline{u}_j}{\partial x_i}\right) - \frac{\partial}{\partial x_i}\left(\frac{\lambda}{C_p}\frac{\partial \overline{e}}{\partial x_j} - \left(\overline{u_i e} - \overline{u}_i \overline{e}\right)\right), \tag{3}$$

where filtered variables are denoted with an overbar. $\overline{u}_i$ is the velocity, $\overline{p^*}$, the hydrodynamic pressure, ν, the kinematic viscosity, $\tau_{ij} = \overline{u_i u_j} - \overline{u}_i \overline{u}_j$, the subgrid stress tensor, e the internal energy, $q_j = \overline{u_j e} - \overline{u}_j \overline{e}$, the subgrid heat flux vector, modelled with a gradient transport model [31], λ the thermal conductivity, and C_p the specific heat capacity. The hydrodynamic pressure is formulated as $p^* = p - \rho g_i x_i$, where p is the total pressure and $\rho g_i x_i$ the hydrostatic pressure, to account for induced buoyancy forces, whose effects are modelled using the Boussinesq approximation [32]. In the present case, the Rayleigh number is small and Re is high, buoyancy forces can be neglected, and the Navier-Stokes Equations simplify to their incompressible form. The subgrid stress tensor is modelled using the Boussinesq hypothesis, $\tau_{ij} = -\nu_k(\partial \overline{u_i}/\partial x_j + \partial \overline{u_i}/\partial x_j)$, where ν_k is the subgrid eddy viscosity, and, similarly, the subgrid heat flux vector is modelled as $q_j = -(\nu_k/Pr_t)\partial \overline{e}/\partial x_j$, where

$Pr_t \sim 0.85$ is the turbulent Prandtl number [33]. An algebraic subgrid wall model, based on Spalding's law-of-the-wall, is here used to compute ν_k at the wall to provide a turbulent flow profile adjacent to the wall, following [34, 35].

2.1 Numerical Methods

The numerical solution of Equations (1) to (3) is carried out using the finite volume method as implemented in openFoam. For all LES models the discretization uses linear reconstruction of the convective fluxes and central differencing of the viscous fluxes terms with second-order stencils. Temporal discretization is performed using a second-order accurate semi-implicit Adams-Bashforth method [36], ensuring numerical stability with a maximum CFL number of 0.5.

The Pressure Implicit Splitting of Operators (PISO) algorithm [36] is utilized to achieve pressure-velocity coupling. The analysis presented in this study focuses on the flow behaviour between the third and fourth consecutive ribs, where the flow is considered fully established. Time-averaged mean flow profiles are considered over a statistically converged period of 20 flow-through times, excluding the initial 10 flow-through times to account for transient effects. The stability of the mean values over consecutive time intervals verifies the statistical convergence of the time-averaged results.

3 Results and Discussions

3.1 Grid Sensitivity Study

To perform a mesh sensitivity study, we have constructed four grids (M_1 to M_4) with increasing resolution, varying cell counts from 1 to 25 million cells. The grids used are primarily based on resolving the geometrical features of the rib-roughened cooling channels. This approach however does not take into account the scales of the flow, and to evaluate how well the different grids resolve the integral, Taylor, and Kolmogorov scales, we estimate these length scales from conventional approximations. The integral length scale is controlled by the geometry, whereas the Taylor and Kolmogorov scale are determined by the dissipation rate, ϵ, of the turbulent kinetic energy, k, and the kinematic viscosity of the fluid, ν. The integral scale, l_I, is estimated by a characteristic length of the domain, here taken as the channel height, $l_I = h$. The Taylor scale is an intermediate length scale in the part of the inertial subrange where viscous effects are evident but not entirely dominating, and can be estimated as, $\lambda = (15\nu u'^2/\epsilon)^{1/2}$. The Kolmogorov scale is in the dissipation range, where viscosity dominates, and can be estimated by $\eta = (\nu^3/\epsilon)^{1/4}$. By combining the expression for the Kolmogorov scale with that of dissipation, $\epsilon \approx k^{3/2}/l_I$,

Table 1 Estimated turbulent length scales

Turbulent length scales	Value [mm]
Integral length scale, l_I	10
Taylor microscale, λ	1.37
Kolmogorov scale, η	0.09

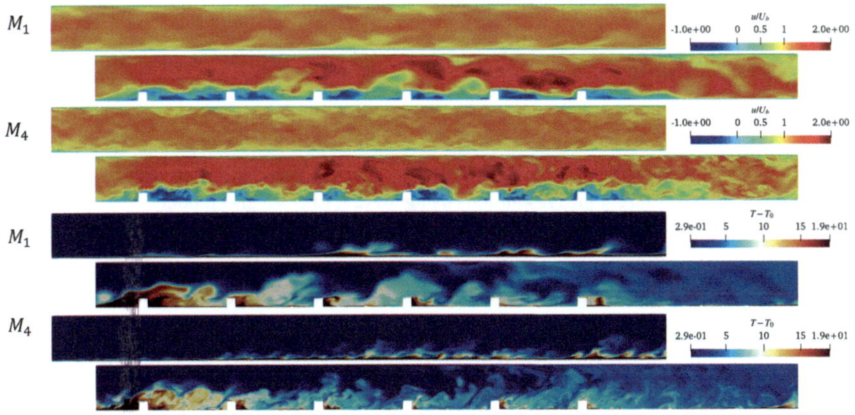

Fig. 2 Normalized instantaneous streamwise velocity and temperature variation in a slice through the rib-roughened cooling channel at $Re = 22000$, for mesh M_1 and M_4 using WALE LES subgrid model

using $k = 3/2u'^2$, we find that $\eta/l_I = C_1 Re_l^{-3/4}$, where $C_1 \approx 0.86$ is a constant and Re_l is the integral scale Re number. The fluctuating velocity component, u', can be approximated using the turbulent intensity, I as $u' = IU_b$ where $I \approx 10\%$ for high-turbulence intensity applications such as heat exchangers and gas turbines. A similar relation can be obtained for the Taylor microscale such that $\lambda/l_I = C_2\sqrt{15}Re_l^{-1/2}$, where $C_2 \approx 0.74$ is a constant. In Table 1, the estimated integral, Taylor, and Kolmogorov scales are presented for the rib-roughened cooling channel based on the relations above. The bulk mesh size of our four meshes lies between (0.96–0.38 mm) and the surface mesh size is between (0.35–0.02 mm).

These estimates clearly show that the Taylor scale is resolved on all four grids, just barely on the 1 million cell mesh, and with margin on the 25 million cell mesh. In LES it is anticipated that the Taylor scale should be resolved, and hence we can initially assume that the meshes are acceptable for LES and should thus also be acceptable for RANS.

To visually analyse the impact of the grid resolution based on the resolved normalized mean velocity, $\langle u \rangle / U_b$, and mean temperature increase, $\langle T \rangle - T_0$, comprehensive visualizations are used for the instantaneous streamwise velocity and temperature distributions across the whole channel for grids M_1 and M_4, Fig. 2. As expected, instantaneous visualization of the normalized streamwise velocity, u/U_b shows the

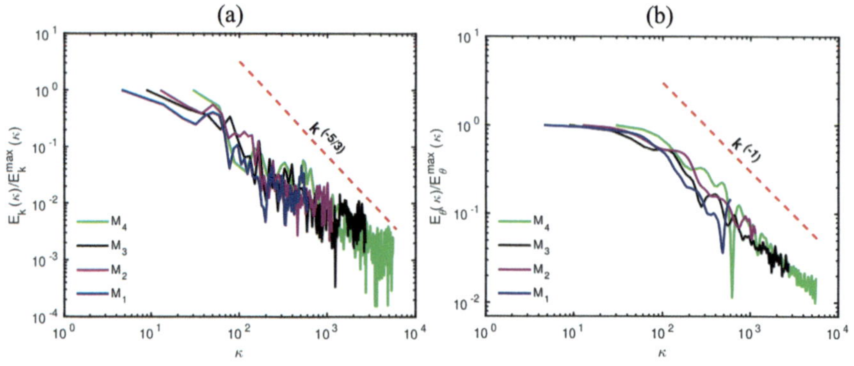

Fig. 3 **a** Kinetic energy spectrum and **b** temperature variance spectrum of flow over rib-roughened cooling channel, for different grid resolutions using WALE LES subgrid model

presence of recirculation zones downstream of the ribs, where the flow separation and reverse flow occur. With increasing grid resolution, these recirculation regions become more and more complex, with large and small vortical structures interacting. Near the heated bottom wall and before the ribs a turbulent thermal boundary layer is observed to develop. With increasing grid resolution finer structures are observed in the thermal boundary layer. Between the ribs, a more concentrated region of higher temperature is observed upstream of the first rib and in the recirculation regions between the subsequent ribs strong thermal mixing, discharging warm fluid into the bulk flow, appears. This indicates the influence of the heat source on the surrounding flow. Upon comparing the instantaneous visualizations, it is clear that M_4 resolves more flow structures and details than M_1. However, the coarse grid, M_1, also captures the main flow features.

Figure 3 presents instantaneous turbulent kinetic energy spectra, $E_k(k)$, and temperature variance spectra, $E_\theta(k)$, normalized by its maximum value, $E_k^{max}(k)$ and $E_\theta^{max}(k)$, respectively, for all four LES predictions along a line between third and fourth consecutive ribs using Fast Fourier Transform (FFT) analysis. For rib-roughened cooling channels, it could be observed that these spectra generally follow the well-known inertial subrange behaviour, with $E_k = C_k \epsilon^{2/3} k^{-5/3}$, where $C_k \approx 1.5$ is the Kolmogorov constant. We also note that the length of the inertial subrange increases with finer grids, thus facilitating resolving a wider range of scales. For E_θ, we found a k^{-1} subrange as expected from the theory [38, 39] for $Pr > 0.7$. This means that heat conduction has no noticeable influence on small scale temperature fluctuations compared to the momentum diffusion, while velocity scales are damped by viscous dissipation. Another observation that stands out is that a rather high grid resolution is required to resolve the thermal boundary layer compared to the flow boundary layer to obtain accurate statistics for the temperature and the temperature variance spectra.

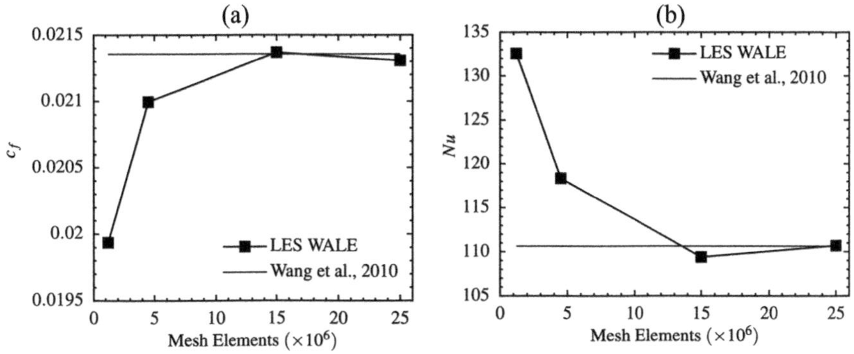

Fig. 4 **a** Friction factor and **b** Nusselt number estimation for rib-roughened cooling channel, for different grids using WALE LES subgrid model. Experimental data from wang *et al.* [21] is also included for comparison

To further evaluate the grid sensitivity, the global thermal and flow resistance performance, evaluated using the Nusselt number, $Nu = \alpha h/\lambda$ and the skin friction coefficient, $c_f = 2\tau_w/\rho U_b^2$ are also examined. Here, h represents the channel height, ρ denotes density, τ_w indicates wall shear stress, U_b stands for bulk velocity, α signifies the convective heat transfer coefficient, and λ represents thermal conductivity. Assuming flow periodicity in the streamwise direction, the investigated domain lies between the third and fourth rib, in which the fully developed turbulence is deemed to be reached. Hereafter, the results for both c_f and Nu are integrated over the surface within the investigated domain. The thermal and hydraulic performance for rib-roughened cooling channels is presented in Fig. 4. Experimental data is provided from Wang *et al.* [21]. It can be inferred from the results that both Nu and c_f converge as the grid is refined. On average, for mesh M_4, the predicted values of c_f and Nu deviate by only 2% to 6% from their corresponding experimental data. For most purposes, an adequate grid resolution for wall-modeled LES would typically fall within the range of M_3 to M_4 (15M to 25M). Based on these findings, we will only consider meshes beyond M_3 for further analysis.

3.2 *Turbulence and Subgrid Model Sensitivity*

The influence of turbulence and subgrid closure models in RANS and LES was investigated using the RANS and LES results available here. For this purpose, numerical simulations were performed using two different RANS turbulence models ($k - \epsilon$, $k - \omega$ SST) and four different LES subgrid models (SMG, OEEVM, WALE, and LDKM) at fully turbulent conditions at $Re = 22000$. Based on the grid resolution study performed, grid M_3 is chosen for this study. Whenever available, experimental data is also presented for comparison with the numerical results.

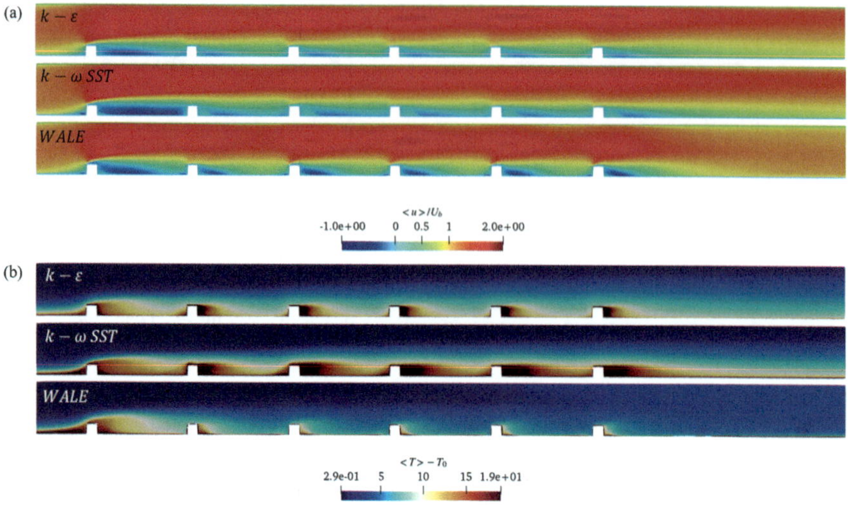

Fig. 5 **a** Normalized mean streamwise velocity and **b** mean temperature variation in a slice through the rib-roughened cooling channel at a fixed $Re = 22000$ and for two turbulence models ($k - \epsilon$, $k - \omega$SST) and one LES subgrid model (WALE)

Figure 5 shows the normalized mean velocity, $\langle u \rangle / U_b$, and mean temperature increase, $\langle T \rangle - T_0$, across the full channel at fixed $Re = 22000$ for two turbulence models ($k - \epsilon$, $k - \omega$ SST) and one LES subgrid model (WALE). In analyzing the mean flow profiles, a noticeable discrepancy arises regarding the length of the recirculation region. The considered RANS models exhibit an overestimation of the recirculation zone's length, potentially leading to inaccuracies in predicting the extent and behaviour of recirculation zones, compared to LES. In comparing the mean temperature and velocity distributions along the channel, distinct disparities emerge between the turbulence models and LES subgrid models. Both $k - \epsilon$ and $k - \omega$ SST models reveal a concentrated temperature region adjacent to the ribs, implying limitations in capturing the fine-scale flow dynamics. In contrast, the WALE subgrid LES model showcases a more even temperature distribution near the ribs, suggesting its ability to qualitatively simulate intricate turbulent fluctuations.

Figure 6 shows the normalized mean velocity, $\langle u \rangle / u_c$ at different downstream positions, plotted against the height above a ribbed wall at fixed $Re = 22000$. Here onwards, the symbol $\langle ... \rangle$ will represent spatially and temporally averaged quantities and u_c represent the centerline velocity along the channel. The numerical results are validated against local mean velocity profiles obtained from experiments by Wang *et al.* [21] representing specific flow regions. Near the rib's leeward corner ($x/e = 0.1$), the fluid is nearly stationary. Above the rib's tip ($1.0\langle y/e \rangle 1.5$), there is a rapid increase in the normal velocity gradient. A backward flow occurs at $x/e = 2$, followed by reattachment at $x/e = 3.6$, where the velocity gradient on the wall is zero. Downstream of the reattachment point, in the outer part of the boundary layer at $x/e = 6, 8$, there are persistent dips in the "velocity profiles, indicating that the

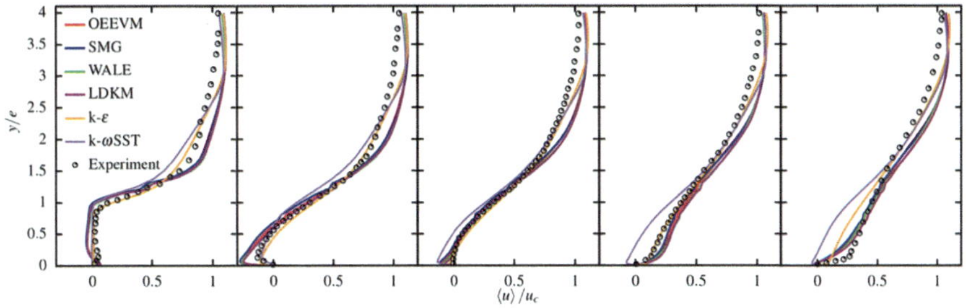

Fig. 6 Local estimate of the normalized velocity profile, $\langle u \rangle / u_c$, along the centerline for $k - \epsilon$ and $k - \omega$SST turbulence models and four different subgrid models between two ribs at $Re = 22000$. Here u_c is the centerline velocity. Experimental data from Wang *et al.* [21] is also included for comparison

redeveloping boundary layer is still subject to a moderately adverse pressure gradient. Figure 6 shows that, for WALE and LDKM, the $\langle u \rangle / u_c$ profiles are closest to experimental data, particularly in capturing the reattaching bubble position with increasing grid resolution. However, a deviation from experimental results [21] is observed in the mean stream flow region, far above the boundary layer, which is due to uncertainties in the experimental setup and limited PIV measurement resolution. The $\langle u \rangle / u_c$ profiles obtained from the $k - \epsilon$ model successfully predict the reattachment point and redeveloping region accurately. However, in the case of the $k - \omega$ SST model, there is a surprising deviation from experimental results. It is worth noting that even with a fine mesh, some discrepancies compared to the experimental results persist, primarily attributed to uncertainties in the experimental measurements (within 5–10% far from the wall) and the limited resolution of the PIV technique employed in the experiments, [12, 21]. Therefore, relying solely on the comparison of local flow properties with experimental data may not yield conclusive findings.

The Darcy friction factor and Nu number estimations are further provided in Fig. 7 using four different SGS models (SMG, OEEVM, WALE, and LDKM) and two different turbulence models. As expected, all LES models allow us to capture more detailed information about the flow, including the effects of turbulence on heat transfer. In Fig. 7a, the friction factor predictions are very close to the experimental ones. Overall, the deviations of predicted c_f from their corresponding experimental data range between 2% and 6%. As seen from Fig. 7b absolute deviations of $Nu/Pr^{1/3}$ do not exceed 6%. In contrast, predictions from both turbulence model for c_f and Nu are scattered and severely either over-predict or under-predict the values by roughly 25%. The statement therefore can be made that SMG, OEEVM, WALE, and LDKM predict better c_f and Nu for flow and heat transfer in a range of complicated practical engineering applications. Finally, it is worth remarking that WALE and LDKM give by far the best accuracy, with deviations less than 2% for both cases. This is interesting and non-trivial at first sight, as the origin of these models is based on different assumptions.

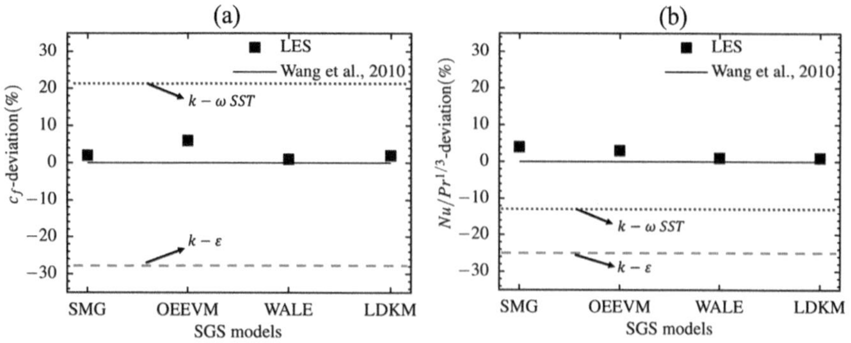

Fig. 7 **a** Darcy friction factor and **b** Nu deviation for rib-roughened cooling channel compared to experimental data using different SGS models and for $k-\epsilon$ and $k-\omega$ SST turbulence models

3.3 Reynolds Number Dependence

To investigate the correlation between Re number and flow behaviour in rib-roughened channels, Fig. 8a displays the mean surface streamlines, which show the overall flow characteristics based on Re. The coloration of the background represents the normalized value of the instantaneous streamwise velocity, u_i/U_b. Hereafter, all results are obtained from WALE LES subgrid model. The introduction of ribs causes a reduction in the channel's cross-sectional area, leading to acceleration of the main-stream flow around the obstacles. Regardless of Re, flow separation occurs at the sharp edge of the upstream rib. This results in the formation of a separation bubble, enclosed by the free shear layer and the ribbed wall. Following reattachment down-stream, a new boundary layer gradually develops until it reattaches to the vertical surface of the subsequent rib.

It is important to note that the streamlines just above the crest of the upstream rib remain straight and parallel to the bottom wall. However, above the top surface of the downstream rib, the streamlines become more convex. Figure 8a illustrates the presence of separation bubbles within the cavity between the ribs, with their sizes varying in accordance with Re number. For small Re (e.g., $Re = 1000$), the separation bubble enlarges, and some streamlines reattach very close to the vertical wall of the following rib. As Re increases, the separation bubble diminishes in size, and the flow reattaches to the floor, leading to the redevelopment of streamlines into a new boundary layer. However, for high turbulent Re numbers ($Re > 1000$), the influence of Re number on separation bubble size appears negligible.

Figure 8b displays the distributions of normalized local Nu numbers between consecutive continuous ribs at various Re numbers. As mentioned earlier, the flow initiates separation at the upstream rib's edge, resulting in the formation of an expand-ing free shear layer that reunites downstream of the floor. At the concave corner just behind the upstream rib, localized areas of reduced fluid velocity and domi-nant conduction create hot spots (labeled A-B). The jet-like shear layer bifurcates

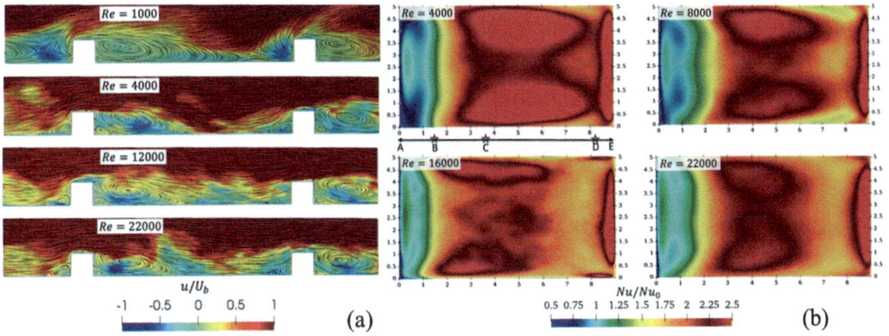

Fig. 8 **a** Surface line integral convolution for different Re between two neighbouring ribs, coloured with instantaneous normalized streamwise velocity. **b** Map of normalized local Nu number between two successive continuous ribs for different Re. Results are shown for WALE LES subgrid model

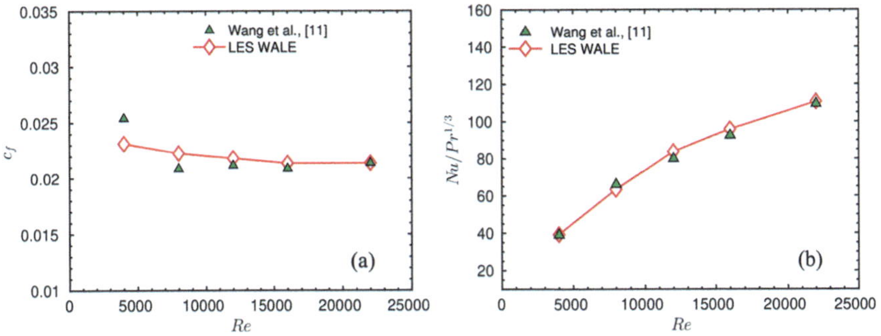

Fig. 9 **a** friction factor estimation and **b** $Nu/Pr^{1/3}$ for different Re numbers between two ribs in a rib-roughened cooling channel. Experimental data is provided for validation by Wang *et al.* [11]

at the reattachment point (C), with one portion redirecting upstream into a primary recirculation zone (B-C) adjacent to the upstream rib. This redirection induces swift entrainment of cooler fluid from the shear layer, significantly augmenting the Nu number. Simultaneously, the remaining portion of the flow re-establishes downstream (C-D), forming a fresh boundary layer that contributes to the reduction in Nu number. Remarkably, the heat transfer coefficient peaks at the reattachment point, reaching its highest point. Progressing downstream, the flow experiences separation due to an adverse pressure gradient arising from the subsequent rib. This separation triggers reattachment on the vertical wall, generating a secondary recirculation zone (D-E) preceding the rib. This secondary recirculation zone leads to a further escalation in heat transfer coefficients. Notably, in Fig. 8b, two symmetrical humps emerge about the centreline. These humps are potentially induced by secondary flows at higher Re numbers.

The global friction factor and Nu number within the rib-roughened cooling channel were determined over a range of $4000 \leq Re \leq 24000$, encompassing the fully developed turbulent regime. The Darcy friction factor, denoted as c_f, was graphed against Re, as shown in Fig. 9a. The outcomes from the LES with WALE closely align with those from Wang *et al.* [11]. Across the examined Re number range, as Re increases, c_f tends towards an asymptote, consistent with expectations for high Re number flows. A minor inconsistency between experimental and LES WALE predictive data is observed, which is particularly noticeable at $Re = 4000$. This discrepancy could be attributed to uncertainties inherent in both numerical and experimental data. Moving on to Fig. 9b, it displays the graph of the averaged normalized Nu number as a function of Re number. Evidently, the presence of rib turbulators significantly enhances the average Nu number compared to a smooth duct. In the light of limited available information regarding Nu number between two ribs, a comparison with experimental data [11] is conducted. In accordance with Wang *et al.*'s experimental findings, the ratio $Nu/Pr^{1/3}$ demonstrates an increasing trend with rising Re numbers. Both c_f and $Nu/Pr^{1/3}$ exhibit strong agreement with the experimental data, thus affirming the validity of the numerical simulation. However, within the tested Re number range, a slight deviation in the slope of the experimental dataset becomes apparent at $Re \leq 8000$, possibly stemming from uncertainties associated with the experimental set-up (within 4–6%).

4 Conclusions

The present investigation utilizes high-fidelity Large Eddy Simulations to explore flow structures, turbulent characteristics, and heat transfer efficiency within the flow separation zone of a cooling channel that features square ribs for surface roughening. A direct comparison is drawn with the prior study conducted by Wang et al. [1, 2, 11, 12, 21]. The Re number, which is determined based on the bulk mean velocity and the hydraulic diameter of the channel, is intentionally varied within the range from 4000 to 24000. The ribs inside the channel obstruct 20% of its height and are symmetrically positioned with a spacing equivalent to 10 times the rib height between them. By capitalizing on the periodic nature of the flow, the domain under scrutiny extends from the third to the fourth ribs.

Examination of the instantaneous velocity distribution unveils that the separated shear layer is primarily influenced by coherent vortices, and the precise point of reattachment is notably reliant on the mesh resolution. When comparing the outcomes obtained from meshes M3 and M4, it becomes evident that the friction coefficient and Nu number are in good agreement with experimental data, acknowledging the inherent uncertainties associated with both approaches. A contrast between RANS (Reynolds-Averaged Navier-Stokes) models and SGS (Sub-Grid Scale) models suggests that WALE (Wall-Adapting Local Eddy-Viscosity) and LDKM (Localized Dynamic K Equation Model) exhibit superior performance and align well with experimental findings, deviating by no more than 2%. Nevertheless, models like SMG (Smagorinsky) and OEEVM (One-Equation Explicit Algebraic Reynolds Stress Model), despite their inherent dissipative nature, predict thermal and hydraulic performance with commendable accuracy, within a deviation of 10% from experimental results. Numerical assessment of thermal and hydraulic performance is based on the Nu number and the Darcy friction factor, which are then compared with previously published experimental data across a range of *Re* numbers. These trends are effectively captured, encompassing the turbulent range for the rib-roughened cooling channel.

Acknowledgements The authors greatly appreciate the Swedish National Infrastructure for Computing (SNIC) for providing the computer time, partially funded by the Swedish Research Council through grant agreement no. 2018-05973.

References

1. Sunden, B., Wang, L.: Experimental investigation of the effect of rib shape on local heat transfer in a square duct by means of liquid crystal thermography. Advan. Heat Transf. Eng. 226, (2005)
2. Wang, L., Sunden, B.: Experimental investigation of local heat transfer in a square duct with various-shaped ribs. Heat Mass Transf, **43**, 759 (2007)
3. Weller, H.G., Tabor, G., Jasak, H., Fureby, C.: A tensorial approach to computational continuum mechanics using object-oriented techniques. Comput. Phys **12**, 620 (1998)
4. Han, J. C.: Heat transfer and friction characteristics in rectangular channels with rib turbulators. J. Heat Transf. **110**, 321 (1998)
5. Rau, G., Cakan, M., Moeller, D., Arts, T.: The Effect of Periodic Ribs on the Local Aerodynamic and Heat Transfer Performance of a Straight Cooling Channel. J. Turbomachinery, **120**, 368 (1998)
6. Taslim, M.E., Li, T., Kercher, D.M.: Experimental Heat Transfer and Friction in Channels Roughened With Angled, V-Shaped, and Discrete Ribs on Two Opposite Walls. J. Turbomachinery (1996)
7. Astarita, T., Cardone, G.: Convective heat transfer in a square channel with angled ribs on two opposite walls. Exp. Fluids **34**, 625 (2003)
8. Chandra, P.R., Alexander, C.R., Han, J.C.: Heat transfer and friction behaviors in rectangular channels with varying number of ribbed walls. Int. J. Heat Mass Transf. **46**, 481 (2003)
9. Liou, T.M., Hwang, J.J.: Effect of ridge shapes on turbulent heat transfer and friction in a rectangular channel. Int. J. Heat Mass Transf. **36**, 931 (1993)
10. Cho, H.H., Wu, S.J., Kwon, H.J.: Local heat/mass transfer measurements in a rectangular duct with discrete ribs. J. Turbomach. **122**, 579 (2000)
11. Wang, L., Sundén, B.: An experimental investigation of heat transfer and fluid flow in a rectangular duct with broken V-shaped ribs. Exp. Heat Transf. **18**, 179 (2005)
12. Wang, L., Hejcik, J., Sundén, B.: PIV measurement of separated flow in a square channel with streamwise periodic ribs on one wall. ASME. J. Fluid Eng. **129**, 834 (2007)
13. Khalid, A., Xie, G., Sunden, B., Enhanc, J.: Numerical simulations of flow structure and turbulent heat transfer in a square ribbed channel with varying rib pitch ratio. Heat Mass Transf. **23**, 155 (2003)
14. Baruskov, A.V., Terekhov, V.V., Terekov, V.I.: Numerical simulation of a separated flow in a ribbed channel by the RANS and LES methods. J. Phys. **2039**, 012003 (2021)
15. Kügele, S., Mathlouthi, G.O., Renze, P., Grützner, T.: Numerical simulation of flow and heat transfer of a discontinuous single started helically ribbed pipe. Energies **15**, 7096 (2022)
16. Ashrafian, A., Andersson, I.H., Manhart, M.: DNS of turbulent flow in a rod-roughened channel. Int. J. Heat Flow **25**, 373 (2004)
17. Leonardi, S., Orlandi, P., Smalley, R.J., Djenidi, L., Antonia, R.A.: Direct numerical simulations of turbulent channel flow with transverse square bars on one wall. J. Fluid Mech. **491**, 229 (2003)

18. Leonardi, S., Orlandi, P., Djenidi, L., Antonia, R.A.: Structure of turbulent channel flow with square bars on one wall. Int. J. Heat Fluid Flow **25**, 384 (2004)
19. Leonardi, S., Tessicini, F., Orlandi, P., Antonia, R.A.: Direct numerical and large-eddy simulations of turbulent flows over rough surface. AIAA J. **44**, 2482 (2006)
20. Vijiapurapu, S., Cui, J.: Performance of turbulence models for flows through rough pipes. Appl. Mathematic. Modelling **34**, 1458 (2010)
21. Wang, L., Salewski, M., Sunden, B.: Turbulent flow in a ribbed channel: Flow structures in the vicinity of a rib. Exp. Therm. Fluid Sci. **34**, 165 (2010)
22. Mukha, T., Liefvendahl, M.: The generation of turbulent inflow boundary conditions using precursor channel flow simulations. Comp. Fluids **156**, 21 (2017)
23. Shur, M.L., Spalart, P.R., Strelets, M.K., Travin, A.K.: A hybrid rans-les approach with delayed-des and wall-modelled les capabilities. Flow Turb. Comb. **93**, 63 (2014)
24. Schmidt, S., Breuer, M.: Source term based synthetic turbulence inflow generator for eddy-resolving predictions of an airfoil flow including a laminar separation bubble. Comp. Fluids **146**, 1 (2017)
25. Schumann, U.: Subgrid scale model for finite difference simulations of turbulent flows in plane channels and annuli. J. Comput. Phys. **18**, 376 (1975)
26. Smagorinsky, J.: General circulation experiments with the primitive equations: I. the basic experiment. Mon. Weather Rev. **91**, 99 (1963)
27. Nicoud, F., Ducros, F.: Subgrid-scale stress modelling based on the square of the velocity gradient tensor. Flow Turbul. Combust. **62**, 183 (1999)
28. Germano, M., Piomelli, U., Moin, P., Cabot, W.H.: Subgrid-Scale Stress Modelling Based on the Square of the Velocity Gradient Tensor. Phys. Fluids A. **3**, 1760 (1991)
29. Launder, B.E., Spalding, D.B.: A dynamic subgrid-scale eddy viscosity model. In: Lectures in mathematical models of turbulence. Accadem. Press, London, England (1972)
30. Menter, F.R.: Two-equation eddy-viscosity turbulence models for engineering applications. AIAA J. **32**, 1598 (1994)
31. Moin, P., Squires, K., Cabot, W., Lee, S.: A dynamic subgrid-scale model for compressible turbulence and scalar transport. Phys. Fluids **3**, 2746 (1991)
32. Boussinesq, J.: Théorie de l'écoulement tourbillonnant et tumultueux des liquides dan les lits rectilignes a grande section. (1897)
33. Yakhot, V., Orszag, S.A., Yakhot, A.: Heat transfer in turbulent fluids—i. pipe flow. Int. J. Heat Mass Transf. **30**, 15 (1987)
34. Larsson, J., Kawai, S., Bodart, J., Moreno, I.B.: Large eddy simulation with modeled wall-stress: recent progress and future directions. Mech. Eng. Rev. **3**, 15 (2016)
35. Fureby, C.: Large eddy simulation of high-Reynolds-number wall bounded flows. IAA J. **42**, 457 (2004)
36. Peinado, J., Ibáñez, J., Arias, E., Hernández, V.: Adams–bashforth and adams–moulton methods for solving differential riccati equations. Comput. & Math. Applicat. **60**, 3032 (2010)
37. Caretto, L.S., Gosman, A.D., Patankar, S.V., Spalding, D.B.: Two calculation procedures for steady, three-dimensional flows with recirculation. Proc. Third Int. Conf. Num. Meth. Fluid Mech.: II Problems of Fluid Mechanics. Springer Berlin Heidelberg (1973)
38. Corrsin, S.: On the spectrum of isotropic temperature fluctuations in an isotropic turbulence. J. Appl. Phys. **22**, 469 (1951)
39. Obukhov, A.M.: Structure of the temperature field in a turbulent flow. Izv. Akad. Nauk SSSR, Ser. Geogr. Geofiz. **13**, 58 (1949)
40. Lumley, L.J., Newman, R.G.: The return to isotropy of homogeneous turbulence. J. Fluid Mech. **82**, 161 (1997)
41. Choi, S.K., Lumley, L.J.: The return to isotropy of homogeneous turbulence. J. Fluid Mech. **436**, 59 (2001)

Near-Wall Flow Characteristics of Flapping Foils at Re = 1,000,000

Andhini N. Zurman-Nasution, Gabriel D. Weymouth, and Bharathram Ganapathisubramani

Abstract Studies of swimming animals contribute to recent and future developments of efficient propulsion for autonomous underwater vehicles but mostly focus on flapping foils at low Reynolds numbers (Re). Flapping foils for large-size underwater vehicles can experience from intermediate to high Re at which the ratio of inertia to viscous effect is important. In this work, we study a NACA0016 foil undergoing propulsive heave and pitch motions using 3-D simulations of wall-resolved LES and an Immersed Boundary Method. We characterize the instantaneous vorticities and phase-averaged flow fields, especially the near-wall parameters for both velocity and pressure. Similar to $Re = 10,000$, we found that flapping foils at $Re = 1,000,000$ and a Strouhal number of $St = 0.3$ experience cyclic behaviours. Flapping foils at higher Re can still experience laminar flow mostly at the pressure side, depending on the phases and locations, but turbulent on the suction side. Their near-wall flow characteristics show dynamic pressure gradients and advection speeds dictated by the kinematics and Re. We found that vortex breakdown coincides with the advection path. The inner layer advection at high Re can be faster than the outer edge of the boundary layer due to the advection of smaller vortices. At lower Re, the advection speed is related to a large LEV and is similar for inner and outer-edge boundary layers.

CUFS-2024-28, Cambridge, United Kingdom, 4-5 March 2024

A. N. Zurman-Nasution (✉) · B. Ganapathisubramani
University of Southampton, Southampton, UK
e-mail: A.N.Zurman-Nasution@soton.ac.uk

B. Ganapathisubramani
e-mail: G.Bharath@soton.ac.uk

G. D. Weymouth
TU Delft, Delft, The Netherlands
e-mail: G.D.Weymouth@tudelft.nl

J. C. Tyacke and N. R. Vadlamani (eds.), *Proceedings of the Cambridge Unsteady Flow Symposium 2024*, https://doi.org/10.1007/978-3-031-69035-8_13

1 Introduction

Flapping foil studies are recently recognised along with the development of Autonomous Underwater Vehicles (AUVs) to seek the possibilities of more efficient performances. However, no study focused on the flow details around flapping foils and especially at intermediate to higher Reynolds numbers ($Re \gg 500{,}000$).

We already know small fish swim at the range of body-length-based Reynolds numbers $Re_B \leq 10{,}000$. However, fast-swimmers such as scombridae and sharks operate at a minimum $Re_B = 1{,}000{,}000$, while large cetaceans such as whales swim at even higher at $Re_B > 10{,}000{,}000$ [4]. The tail or fluke of a whale can experience a chord-based Reynolds number up to $Re \approx 1{,}000{,}000$. The bio-inspired flapping foil designs for large-size underwater vehicles are expected to operate at the same Re or even higher and reach the same hydrodynamic performance as the large cetaceans. The fact that propulsion made by flapping foils is more efficient, highly maneuverable, and less noisy compared to propellers are attractive features to implement by industries. For this reason, it is essential to study not only the general hydrodynamic performances of a flapping foil at $Re \approx 1{,}000{,}000$ but also the flow details for future acoustics and drag-reduction studies.

In several studies, we know that most animals prefer to operate at a certain range of Strouhal number $0.2 \leq St \leq 0.4$ [11]. Our previous studies proved that in this St range, the flows are relatively two-dimensional and the foil generates less secondary or spanwise perturbation in spite of the large leading-edge vortex generated [14]. The flow around the foil looks laminar and two-dimensional at $St \approx 0.3$ up to $Re = 50,000$. Currently, there are no detailed studies of flapping foil reaching $Re = 1,000,000$ both in an experiment and three-dimensional (3D) simulations because of complex experimental setups and high numerical cost.

In flapping foil flow at lower Re, the kinematics are major parameters that drive the hydrodynamic performance, whereas geometry parameters such as aspect ratio or sweep angle are secondary [16]. This phenomenon contrasts with the flow characteristics around a stationary foil. Several studies have proven that the flow around a stationary foil at $Re = 1{,}000{,}000$ is inertia-dominated and stays turbulent once perturbed as shown by Vinuesa et al. [13]. They showed that the near-wall flow exhibits turbulent boundary layer characteristics and is less affected by adverse pressure gradients when the outer region reaches a steady flow condition. In this study, we are interested in investigating the dominating characteristics of flapping foils at $Re = 1{,}000{,}000$, either the inertia or the kinematics and compared to the lower Re case. Triantafyllou et al. [12] published a graph about the increase of near-wall velocity of swimming RoboTuna that indicates a relaminarisation phenomenon. However, we found no further investigation that intensively studied the near-wall flow of flapping foils especially at high Re.

2 Methodology

We use an immersed-boundary implicit-LES solver called the Boundary Data Immersion Method (BDIM) to generate simulations of flapping foils with prescribed sinusoidal motions. The solver has been validated in several publications for flapping foil studies [9, 16].

In this study, we use a NACA0016 foil with chord length C, thickness $D = 0.16C$, and pivot point at $0.25C$ from the leading edge. The three-dimensional (3D) simulations are extruded from two-dimensional (2D) simulations after 15 cycles to promote fast perturbation for 3-D flow. Periodic conditions are applied at the side boundaries. We present simulation results at two Reynolds numbers $Re = 10{,}000$ and $Re = 1{,}000{,}000$ for comparison. The foil span at $Re = 1{,}000{,}000$ is 1% of chord length to fulfill the minimum of $100\nu/u_\tau$ as suggested by Jiménez & Moin [7], and 20% for $Re = 10{,}000$. The foil is placed in a uniform freestream flow with a constant speed of U_∞, a fluid density ρ and the chord-based Reynolds number $Re = U_\infty C/\nu$, where ν is kinematic viscosity.

The foil is placed within a uniform Cartesian grid area and surrounded by stretched domain (see Fig. 1b). The stretched domain extends from $X = -3C$ to $X = 11C$ and $Y = \pm 3C$, where X, Y, Z are global coordinates to streamwise, lateral-wise and spanwise directions. For simulations at $Re = 10{,}000$, we use a resolution of $C/\Delta y = 700$ to reach $y^+ = yu_\tau/\nu \approx 1$, $C/\Delta x = 2C/\Delta y$ and $C/\Delta z = 2C/\Delta y$. u_τ is friction velocity and x, y, z are local coordinates for tangential, normal and spanwise distances respectively to the solid boundary. For $Re = 1{,}000{,}000$, the resolutions are $C/\Delta y = 8555$ to reach the laminar sublayer or linear part of the boundary layer or within $3 \leq y^+ \leq 10$, $C/\Delta x = 4C/\Delta y$ and $C/\Delta z = 2C/\Delta y$.

2.1 Flapping Kinematics

For both Re cases, flapping foils experience the same kinematics. They consists of a combination of simple heave $H(t)$ and pitch $\theta(t)$ motions, with functional form

$$H(t) = A \sin(2\pi f t) \tag{1}$$

$$\theta(t) = \theta_0 \sin(2\pi f t + \psi) \tag{2}$$

$$\theta_0 = \sin^{-1}\left(A/(0.75C)\right) \tag{3}$$

where f is the frequency of flapping, $A = 0.25C$ is trailing-edge amplitude and θ_0 is pitch amplitude, and heave-pitch phase difference $\psi = 90°$. To isolate the amplitude from the frequency, the kinematics is characterized by Strouhal number $St = 2Af/U_\infty$. The Strouhal number chosen is $St = 0.3$ based on Triantafyllou et al. [11] at which animals mostly swim for optimum performance and Zurman-Nasution et al. [15] for its two-dimensionality of the flow structures. The reduced frequency is kept constant at $fC/U_\infty = 0.6$, and the foil experiences maximum angle of attacks

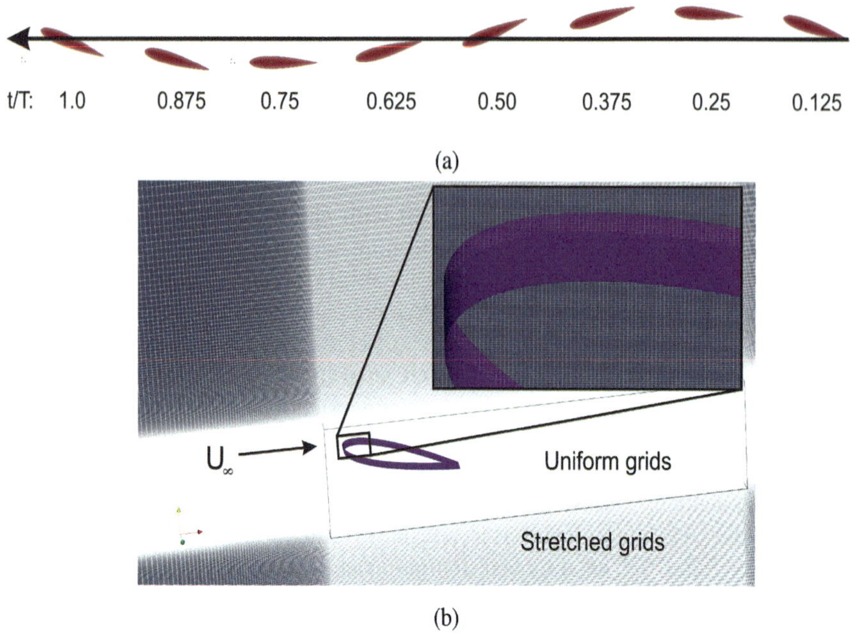

Fig. 1 **a** The flapping phases for one period (T) cycle, and **b** grid arrangement of a BDIM simulation

along the cycle of $\max(|\alpha(t)|) = 23.8°$ where $\alpha(t) = -(\tan^{-1}\frac{\mathrm{d}H(t)}{\mathrm{d}t} - \theta(t))$. The flapping foil cycles are depicted in Fig. 1a where we define the evolution of cycle time t over one period T.

The presented results in this study are based on span- and phase-averaged. Forces and power are presented in thrust-force (C_T), lift-force (C_L), and power (C_{pow}) coefficients as

$$C_T = -\frac{F_X}{0.5\rho S_p U_\infty^2}, \quad C_L = \frac{F_Y}{0.5\rho S_p U_\infty^2},$$
$$C_{pow} = \frac{P_w}{0.5\rho S_p U_\infty^3}, \quad \eta = \frac{\overline{C_T}}{\overline{C_{pow}}} \tag{4}$$

where measured thrust force ($-F_X$), lift force (F_Y), power P_w, and efficiency η are calculated from the integration of pressure and viscous forces over the foil and S_p is the foil planform area.

2.2 *Validation with Experiments*

We compare the general hydrodynamic performance with available experimental data presented by [6] for harmonic pitch-heave motion. Since flapping foils at $Re > 1000$

Table 1 Hydrodynamic performance comparison with experimental data interpolated in [6]

Case	Re	St	α_{max}	fC/U_∞	ψ	θ_0	$\overline{C_T}$	$\overline{C_L}$	η
Experiment, NACA0012	$\approx 3x10^4$	0.3	23.8°	0.15	90°	67.13°	≈ 0.45	≈ 0	≈ 0.4
Simulation, NACA0016[a]	$1x10^4$	0.3	23.8°	0.6	90°	19.47°	0.6	0	0.4
Simulation, NACA0016[a]	$1x10^6$	0.3	23.8°	0.6	90°	19.47°	0.61	0	0.4

[a] Current study
The overline represents the cycle average

are dominated more by inertia than viscous effect, their performance variations are independent of the Reynolds numbers [10], especially drag, thrust coefficient C_T, and most importantly the efficiency (η) prediction. In Table 1, our efficiencies are similar for the same kinematics (St and α_{max}) presented in the experimental results, explaining the similarity of thrust to power ratio. Zero C_L shows that no side force is produced because the chosen kinematics produce no deflected wakes that normally happen at higher St or with additional pitch angles. The discrepancies in thrust coefficient C_T are marginally contributed from NACA profiles ($\approx 5\%$ based on La Mantia & Dabnichki [8]) and mostly because our study uses higher reduced frequency with lower pitch and heave amplitudes. The wake structures for our study show an inverted Bénard-von Kaŕmań street, known as a thrust-producing case similar to other experimental results mentioned in [3, 5] for similar St.

3 Pressure Evolution of a Flapping Foil

Each side of the flapping foil experiences alternating suction and pressure. In Fig. 2, we see that the suction side is approximately located at phase $0.27 \leq t/T \leq 0.84$ when the pressure coefficients are negative. The pressure side starts after $t/T > 0.84$ and continues to the next cycle. Small areas near the leading and trailing edges might experience suction at moderate magnitudes starting from $t/T = 0.84$ to $t/T = 0.18$. The alternating C_p is driven by sinusoidal kinematics of pitch and heave. The C_p distribution will guide the discussions in the following sections.

4 Vortex Structures and Local Velocity Evolutions

In this section, we compare the evolution of vortex structures for $Re =$ 10,000 and $Re =$ 1,000,000. We will present the local velocity profiles and their evolution through the phases per cycle (t/T). The phases are presented only for the upper part of the foil, whereas the lower part experiences the opposite phase at half cycle

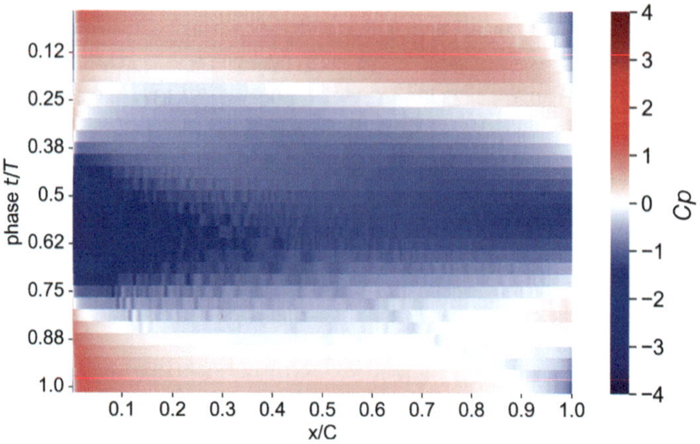

Fig. 2 A typical pressure coefficient C_p distribution of one cycle phase T along the chord length C of one side of a flapping foil. The pressure side is in red, and the suction side is in blue. The example is for $Re = 1,000,000$ and $St = 0.3$, but $Re = 10,000$ show similar contours. Phases (t/T) refer to the upper side of the foil in Fig. 1a

or $t/T + 0.5$. All velocity and pressure data are span- and phase-averaged across cycles, whereas the vorticities and λ_2 contours are instantaneous data of the last cycle.

4.1 Flapping Foil at *Re* = 10,000

In Fig. 3, we can see a pronounced leading edge vortex (LEV) on the suction side of the foil at phase $t/T = 0.75$ while the remaining areas remain laminar. At the same time, the other side of the foil experiences pressure. No LEV is generated at the pressure side except for the previous LEV advected close to the trailing edge.

The separation bubble or the LEV is generated near the leading edge and is advected from the leading towards the trailing edge. The velocity profiles retain similar shapes when no LEV is passing by. However, the flow is disturbed during the generation of LEV or at any x/C locations of the foils passed by the LEV. The local tangential velocities on the foil can experience inverse flow caused by the laminar separation bubble.

We notice that the flow in some areas is faster than the freestream, while others are slower. We will discuss it in more detail later in Sect. 5. At phases $t/T = 0.5$ and 0.75, the spanwise vorticity magnitude on the upper part decreases near the trailing edge similar to their local velocities. On the other hand, the opposite side of the foil experience increased velocities where the trailing edge shoots a jet that produce a roll-up vortex forming the next inversed Von Karḿán street.

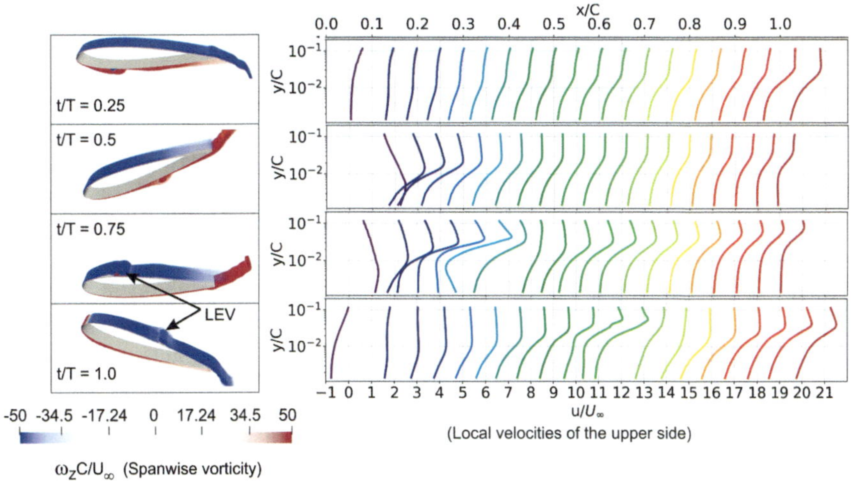

Fig. 3 The four phases of spanwise vorticity are presented in volume render (left figures), and local streamwise velocities u scaled with freestream velocity U_∞ (right figures) for $Re = 10{,}000$. The first velocity profiles are probed at $x = 0.05C$ and then every $0.05C$, while the velocity magnitudes are shifted by $+1u/U_\infty$ for presentation. Each color represents a single x/C for comparison among phases. Vertical axes are local normal directions presented in logarithmic. Vorticity is presented as instantaneous, whereas the velocity profile is span- and phase-averaged

4.2 *Flapping Foil at Re = 1,000,000*

Similar to the results at a lower Re, the flapping foil at $Re = 1{,}000{,}000$ experiences alternating suction and pressure on both sides of the foil per cycle. However, we now notice multiple small-size LEVs. They start to appear earlier at phase $t/T = 0.5$ in Fig. 4.

In the beginning, the LEVs start as laminar separation bubbles instead of turbulent and 2-D or without spanwise perturbation (see inset figure *1). They start to grow and break down into 3-D vortices when advected towards the trailing edge (inset figure *2). They generate more turbulence downstream (at $t/T = 0.75$) in the form of multiple turbulent separation bubbles. Similar to lower Re, local velocities increase as the LEVs appear or pass by, especially near the freestream flow. The locations impacted by the separation bubbles are larger at high Re than at low Re. The trailing edge areas and their velocity profiles are permanently dominated by turbulent flow throughout the cycle.

The upper side of the foil at $t/T = 0.25$ surprisingly generates no vortices at such a high Re despite the high-inertia influence from the freestream. The velocities at this phase also confirm that the profiles are undisturbed and are similar to the laminar states of the same phase at a lower Re. The flow is relaminarised at the same phase every cycle, except at a small region near the trailing edge where the flow is always turbulent. The smoothed or laminar velocity regions start as early as the $t/T = 1.0$ when this foil side experiences pressure.

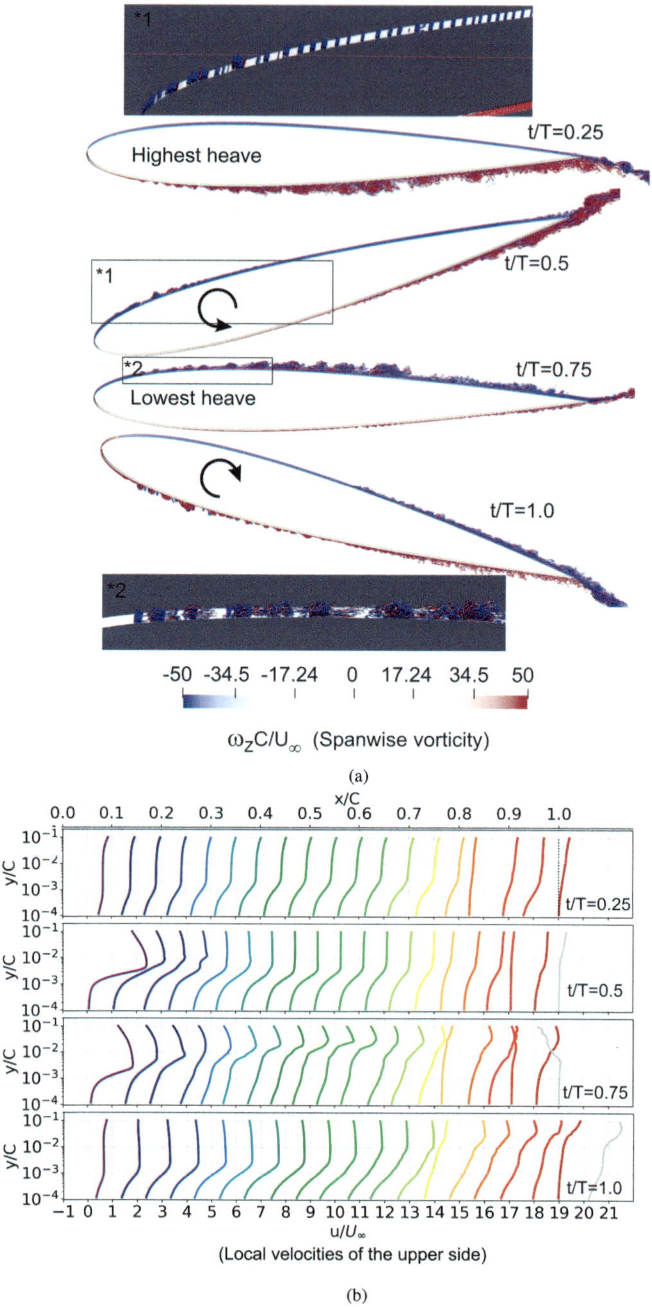

Fig. 4 **a** Spanwise vorticities are presented in volume render with insets presenting 1% of $min(\lambda_2)$ contour, and **b** u/U_∞ for Re = 1,000,000. The insets with λ_2 contour are instantaneous figures. Vorticity and velocity descriptions are similar to Fig. 3

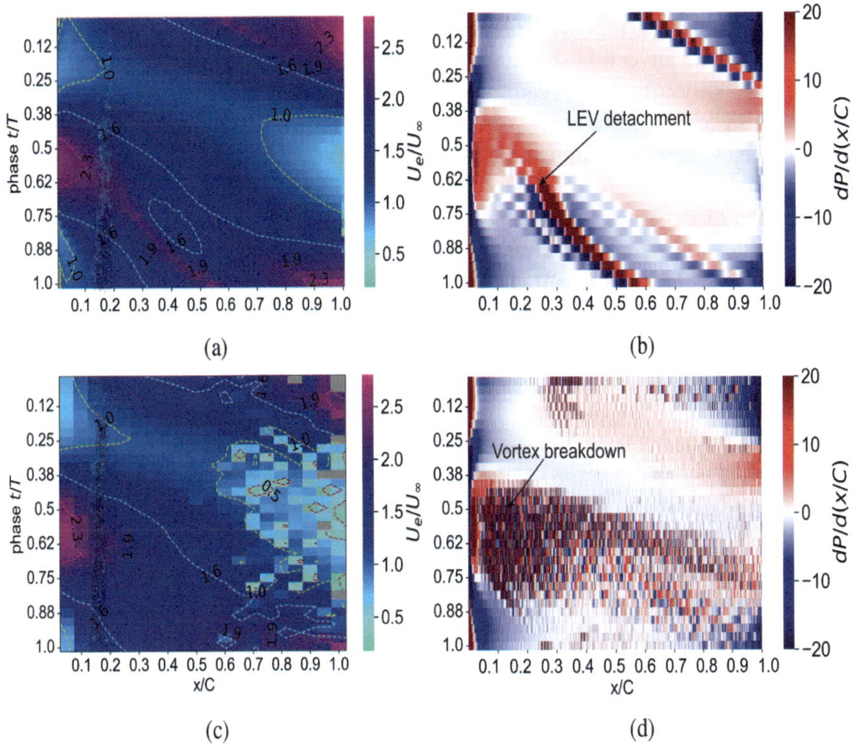

Fig. 5 Heatmaps of **a, c** boundary-layer outer edge velocities are scaled to the freestream velocity or U_e/U_∞ and **b, d** near-wall pressure gradient or $dP/d(x/C)$. No available data is in grey. Cases at **a, b** are $Re = 10{,}000$ and **c, d** $Re = 1{,}000{,}000$. Horizontal axes are locations on the foil x/C, and vertical axes are phases t/T. Red-white-blue regions in $dP/d(x/C)$ map represent adverse, zero and favourable pressure gradients. Heatmaps are based on span-averaged results and 10 cycles of phase averaging

5 Velocities at the Edge of the Boundary Layer and the Evolutions of Near-Wall Pressure Gradients

In this section, we are looking at the freestream or the outer flow influence using the heatmap graph of the local outer-edge boundary layer velocity U_e scaled to the freestream velocity U_∞. U_e is measured as the maximum velocities from near wall location to approximately $y = 0.1C$ height where the flow has indicated to be irrotational. At the same time, we compare them with a parameter of the inner boundary layer, i.e., the near-wall pressure gradient. For pressure gradients, pressure P data are probed at the third point closest to the solid boundary where they are no longer affected by the immersed boundary region [9, p.112].

The heatmap graphs of U_e/U_∞ in Fig. 5a, c confirm that some regions experience higher velocities than the freestream, especially near the leading edge at around $t/T = 0.5$ and trailing edge at around $t/T = 1$. The high-velocity regions coincide with the timing of LEV generation and the path where the vortices advected downstream. The main advection speed is about 0.5 cycle, which can be easily detected by the same velocity contours spread from the leading to the trailing edges with increasing phases. The advection process is repeated every cycle. The half-cycle advection speed is the evidence that sinusoidal kinematics drive the velocities at the outer edge of the boundary layer.

The regions experiencing lower velocities than the freestream coincide with the pressure side at the front half-chord where their velocity profiles show laminar characteristics. We can see clearly at the lower Re heatmap, that the lowest velocity affected by the kinematics starts at the leading edge region around $t/T = 1$. The low-velocity region near the trailing edge at $t/T = 0.5$ explains the phenomenon of ceased vorticities in Fig. 3 for the lower Re case. This low velocity region balances out the jet coming from the opposite or pressure side.

The difference between U_e/U_∞ at low and high Reynolds numbers is the area of the separation or high-velocity regions, especially around location $0.4 \leq x/C \leq 0.7$. The higher Re case maintains a constant advection speed from the leading edge at location $t/T = 0.5$ towards the trailing edges at $t/T = 1$. The separation region is relatively larger for high Re at mid-chord locations. At lower Re, the highest U_e/U_∞ at mid-chord is narrower and concentrated on the LEV advection path. The LEV speed is slower than the main advection speed after detachment at lower Re, due to its size. The detachment is followed by a small zone of favourable pressure gradient marked by a circle of $U_e/U_\infty \geq 1.6$ at around $x/C = 0.4$ and $t/T = 0.75$ in Fig. 5a.

The flapping foil experiences high-low velocity magnitudes once per cycle, similar to one peak per cycle for lift coefficients. However, it experiences twice adverse (APG) and favourable pressure gradients (FPG) at the majority of locations (x/Cs), similar to the peak of thrust coefficients (see lift and drag curve comparison in Andersen et al. [2]). The APG locations coincide with locations where the $dU_e/d(x/C)$ switch signs from positive to negative and vice versa. The APG also promotes vortex breakdown, especially when it occurs at high magnitudes near the leading edge.

The pressure gradient $dP/d(x/C)$ maps for lower and higher Res are completely driven by vortex breakdown as seen in Fig. 5b, d. The red-blue fluctuations locate the advection of vortices. A massive vortex breakdown near the leading edge happens earlier in the main APG region at higher Re. The speed of these small vortices advected downstream is faster than the half-cycle advection speed seen at the U_e/U_∞ in Fig. 5c. At the lower Re, the advection speed of $dP/d(x/C)$ does not seem to differ from its U_e/U_∞. At lower Re, the LEV detachment is advected at the same path as the locations of high U_e/U_∞ magnitudes.

We conclude that a large-size LEV at lower Re is advected at the same speed as the freestream. Meanwhile, multiple smaller LEVs at higher Re appear and break down earlier than the lower Re case and are advected faster and spread in wider areas than the velocities at the outer edge of the boundary layer.

6 Concluding Remarks

The flow characteristics of a flapping foil are highly driven by kinematics, affecting not only its hydrodynamic performance but also the velocities and pressure gradients located between the outer edge and the near-wall regions (inner boundary layer). As the Reynolds number increases from $Re = 10{,}000$ to $Re = 1{,}000{,}000$, the inertia increase speeds up the breakdown of vortices and promotes more turbulence. However, we still found laminar regions for flapping foils at optimum Strouhal number $St = 0.3$ and $Re = 1{,}000{,}000$, especially on the foil side experiencing pressure.

The leading-edge vortices (LEVs) are generated on the suction side of the foil and are advected downstream every half cycle. Their advection speed is influenced by sinusoidal kinematics and Re. The advection of the outer-edge boundary layer for $Re = 1{,}000{,}000$ is well spread about a half cycle, whereas it is affected by a single LEV detachment at $Re = 10{,}000$ in the mid-chord region. There is an additional advection at $Re = 1{,}000{,}000$ due to the earlier vortex breakdown. Ambrogi et al. [1] studied the oscillating velocity profiles with suction-blowing flow over a flat plate and found a hysteresis phenomenon during the transition from APG to FPG and vice versa that indicates phase lags at the inner layer from the freestream forcing. The local velocity profiles in our studies at higher Re indicate similar behaviours because of relatively the same reduced frequency.

For future studies, we will look in detail and quantify how much the vortex breakdown affects differently the advection speeds of the inner part (influenced by friction velocity) and the outer part of the boundary layer (influenced by freestream flow). Despite its proven existence at $Re = 1{,}000{,}000$, we still have no details of the exact cause of the relaminarisation phenomenon found on the pressure side. We will investigate which parts of the kinematics (flapping frequency, amplitude, or both) mainly drive this phenomenon. Does the convex shape of the foil also play a role in flow relaminarisation? And, how does this phenomenon change with increasing Strouhal numbers?

Acknowledgements The Author would like to thank IRIDIS High-Performance Computing Facility with its associated support services at the University of Southampton and the Defence Science and Technology Laboratory DSTL UK for project funding.

References

1. Ambrogi, F., Piomelli, U., Rival, D.E.: Characterization of unsteady separation in a turbulent boundary layer: mean and phase-averaged flow. J. Fluid Mech. **945**, (2022)
2. Andersen, A., Bohr, T., Schnipper, T., Walther, J.H.: Wake structure and thrust generation of a flapping foil in two-dimensional flow. J. Fluid Mech. **812**, (2016)
3. Bratt, J.B.: Flow pattern in the wake of an oscillating foil. Tech. rep, Aeronautical Research Council (1950)
4. Eloy, C.: Optimal strouhal number for swimming animals. J. Fluids Struct. **30**, 205–218 (2012)

5. Godoy-Diana, R., Aider, J.L., Wesfreid, J.E.: Transitions in the wake of a flapping foil. Phys. Rev. E: Stat., Nonlin, Soft Matter Phys. **77**(1 Pt 2), 016308 (2008)
6. Hover, F.S., Haugsdal, O., Triantafyllou, M.S.: Effect of angle of attack profiles in flapping foil propulsion. J. Fluids Struct. **19**(1), 37–47 (2004)
7. Jiménez, J., Moin, P.: The minimal flow unit in near-wall turbulence. J. Fluid Mech. **225**, 213–240 (2006)
8. La Mantia, M., Dabnichki, P.: Effect of the wing shape on the thrust of flapping wing. Appl. Math. Model. **35**(10), 4979–4990 (2011)
9. Maertens, A.P., Weymouth, G.D.: Accurate cartesian-grid simulations of near-body flows at intermediate reynolds numbers. Comput. Method. Appl. Mech. Eng. **283**, 106–129 (2015)
10. Senturk, U., Smits, A.J.: Reynolds number scaling of the propulsive performance of a pitching airfoil. AIAA J. **57**(7), 2663–2669 (2019)
11. Triantafyllou, G.S., Triantafyllou, M.S., Grosenbaugh, M.A.: Optimal thrust development in oscillating foils with application to fish propulsion. J. Fluid. Struct. **7**(2), 205–224 (1993)
12. Triantafyllou, M.S., Techet, A.H., Zhu, Q., Beal, D.N., Hover, F.S., Yue, D.K.: Vorticity control in fish-like propulsion and maneuvering. Integr. Comp. Biol. **42**(5), 1026–31 (2002)
13. Vinuesa, R., Negi, P.S., Atzori, M., Hanifi, A., Henningson, D.S., Schlatter, P.: Turbulent boundary layers around wing sections up to rec=1,000,000. Int. J. Heat Fluid Flow **72**, 86–99 (2018)
14. Zurman-Nasution, A.N., Ganapathisubramani, B., Weymouth, G.D.: Influence of three-dimensionality on propulsive flapping. J. Fluid Mech. **886**, 1–14 (2020)
15. Zurman-Nasution, A.N., Ganapathisubramani, B., Weymouth, G.D.: Effects of aspect ratio on rolling and twisting foils. Phys. Rev. Fluids *6*(1), (2021a)
16. Zurman-Nasution, A.N., Ganapathisubramani, B., Weymouth, G.D.: Fin sweep angle does not determine flapping propulsive performance. J. R. Soc. Interface **18**(178), 20210174 (2021)

The Evolution of Backflow with Vortex Clusters in Wall-Bounded Flows

Xue Chen, Yongmann M. Chung, and Minping Wan

Abstract This study investigates the evolution of the backflow regions in wall-bounded flows which is a rare phenomenon induced by the large-scale motions in the viscous wall region following Chen et al. (J. Fluid Mech., vol. 966, 2023, A38) which revealed the unclear process and cause of the splitting, merging, and wall-detaching behaviours of the backflow regions. The backflow region can go through multiple life cycles of growth and decay which was linked to the regeneration in spanwise vortices in vortex clusters, also known as hairpin packets. Observations in this study show extensive evidence of the backflow regions being induced underneath strong spanwise rotations as the heads of hairpin vortices including cane- and arch-shaped asymmetric vortices. The low-speed structure is bounded between the quasi-streamwise vortices as the legs of the hairpins, similar to the structure organisation depicted in Adrian et al. (*J. Fluid Mech.*, vol. 422, 2000, pp. 1–54). The growth and decay of the backflow regions in their life cycles follow the development of the associated hairpin vortices. It is observed that multiple backflow regions can occur in streamwise alignment under clustered vortices where the new backflow region develops downstream of the existing one. The observations match the conceptual model for the large-scale motions and hairpin packets wrapped around the low-speed structure attached to the wall where its tail can be strong enough to cause flow reversal underneath the heads of the trailing vortices in the cluster.

CUFS-2024-29, Cambridge, United Kingdom, 4-5 March 2024

X. Chen (✉) · M. Wan
Department of Mechanics and Aerospace Engineering, Southern University of Science and Technology, Shenzhen 518055, China
e-mail: chenx3@sustech.edu.cn

M. Wan
e-mail: wanmp@sustech.edu.cn

Y. M. Chung
School of Engineering, University of Warwick, Coventry CV4 7AL, UK
e-mail: y.m.chung@warwick.ac.uk

J. C. Tyacke and N. R. Vadlamani (eds.), *Proceedings of the Cambridge Unsteady Flow Symposium 2024*, https://doi.org/10.1007/978-3-031-69035-8_14

1 Introduction

The rare backflow of near-wall negative instantaneous streamwise velocity and the associated strong local fluctuations of wall friction are of great importance for the control of flow separation, drag reduction and the understanding of the interaction between the large-scale motions (LSM) and the near-wall cycle. Early observations of the existence of backflow ([13, 22, 25]) were found controversial to Eckelmann [14], "with certainty, there are no negative velocities near the wall" and in Colella & Keith [13], "The probability density of wall-shear stress fluctuations...exhibits no flow reversals at the wall." The vast number of reports on backflow in the high-resolution direct numerical simulations (DNS) data and the recent experimental observations [4, 5, 27] have shown that the backflow is a genuine phenomenon in all kinds of wall-bounded flows, from canonical wall-bounded flows (channel [6, 7, 9, 10, 20, 23], pipe [9, 11, 12, 17, 18, 21], duct [30, 31] and boundary layer flows [15, 26, 29]) including flows over wing sections and in the torus [10, 12], and they were observed in both fully turbulent flows and the late stage of transition [28].

The spatial extents of the backflow regions have been measured in the channel, pipe and duct flows [5, 7, 9, 21, 23, 26, 30]. The average spatial extents of the backflow were suggested to be around 20 wall units in the wall-parallel direction by Lenaers et al. [23] and were found to be slightly stretched in the spanwise direction on average by Cardesa et al. [7]. As shown in Fig. 1 ([9]), the wall-parallel span of the backflow increases subtly with the Reynolds number whereas their wall-normal span of approximately 1 wall unit is independent of the Reynolds number. The average lifespan of the backflow regions in canonical wall-bounded flows was reported to be around 7 to 9 wall units. The stronger and larger backflow regions can live much longer with a lifespan $T^+_{life} > 40$ [9, 10] (T_{life} denotes the lifespan of the backflow regions). The larger backflow regions were also found to be splitting owing to the near-wall streaks which separate the backflow regions spanning over the near-wall high- and low-speed streaks. The near-wall streaks also affect the backflow regions as part of the large-scale low-speed structure where the high-speed streaks can very occasionally, obstruct the trailing end of the low-speed structure away from the wall during the formation of backflow. The backflow regions forming wall-detached were found to be only associated with the critical wall-shear stress point after they had become attached to the wall.

The backflow is formed as part of the large-scale low-speed structures adjacent to the wall during their interaction with a large-scale high-speed structure originating upstream beyond the buffer layer [4, 9, 17, 18, 23, 30]. The backflow regions in semi-ellipsoidal shapes are located at the tail of the low-speed structure, under a strong spanwise vortex that is responsible for the continuous penetration of large-scale high-speed structures towards the wall, as shown in Fig. 2, The formation mechanism of backflow during the interaction between the large-scale high- and low-speed structures in the near-wall region closely resembles the vortex clusters of hairpin vortices in wall turbulence (see Fig. 25 in Adrian et al. [2]). The hairpin vortices wrap around the low-speed structures and induce the streamwise-elongated low-speed regions

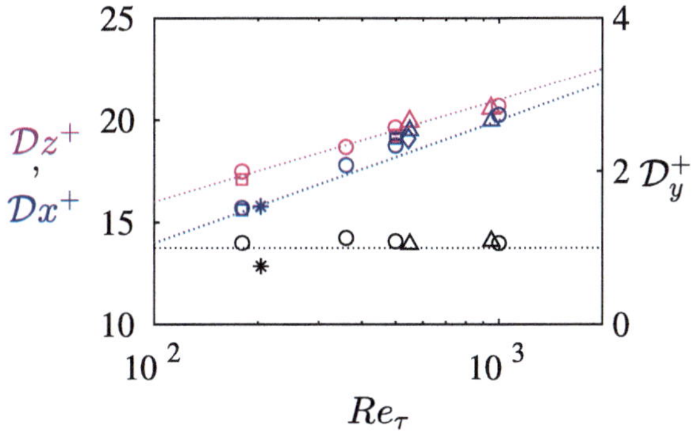

Fig. 1 The average spatial extents of the backflow regions in the streamwise direction, Dx (blue), spanwise direction, Dz (magenta) and wall-normal direction, Dy (black) in the pipe: [21] ('▽'), [9] ('○'); channel: [23] ('◇'), [7] ('△'), [9] ('□'); and duct: [30] ('$*$') (figure from Ref. [9])

with $Q2$ ejection events whereas the heads of the hairpin, i.e., strong spanwise vortices are responsible for the high-speed region of $Q4$ sweep events. Recent Studies have observed the backflow in the vicinity of the quasi-streamwise leg of the hairpin vortices (see figures 18–20 in Wu et al. [28] and Fig. 1 in Zaripov et al. [30]). In Chen et al. [9], the backflow regions were found to be able to live outstandingly, going through multiple life cycles of growth and decay, and reappearing near a decayed backflow region, which was linked to the regeneration of hairpin vortices. However, clear evidence of coincident life cycles of the backflow regions, large-scale structures and hairpin vortices is still lacking.

In this study, we extend the investigations of Chen et al. [9] on the evolutions of the large and long-lived backflow regions going through multiple life cycles of growth and decay. The spatial and temporal evolutions of the backflow regions organised by nearby strong vortical structures are presented. The observations show the formation process of multiple backflow regions under clustered vortices, possibly linked to the regeneration of streamwise-aligned hairpin vortices (Fig. 3).

2 Methodology

This study used the same direct numerical simulations (DNS) database of fully-developed pipe flows exploited in [8] and [9] which was conducted in the solver, `Nek5000` [16] using the high-order spectral element method. Here, the streamwise, wall-normal and spanwise (azimuthal) directions are denoted as x, y and θ, respectively with the corresponding instantaneous velocity components denoted as U, V and W where the fluctuation terms are given in lower cases. In the results, the space is non-dimensionalised by the characteristic length, which is the pipe radius, R; and the normalisations of time in wall units using ν/u_τ^2 is denoted by the superscript '+'. The

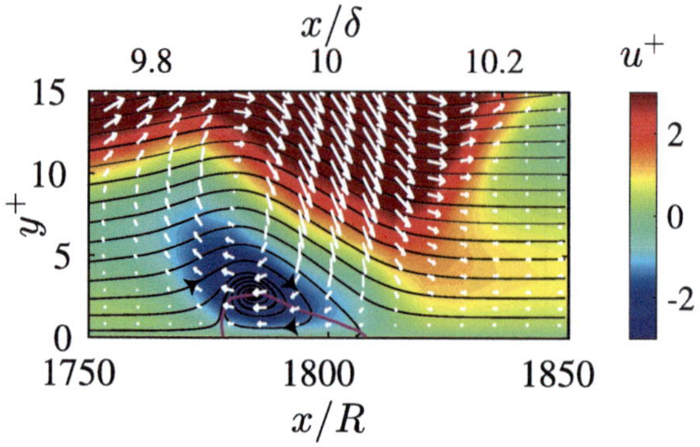

Fig. 2 The streamwise-wall-normal contours of streamwise velocity fluctuation, u^+ around an instantaneous backflow region outlined by the magenta isocontour of $U = 0$. The black streamlines indicate the in-plane instantaneous velocities, U and V, and the white vectors indicate the in-plane velocity fluctuations, u and v. The swirl centre is indicated by '⊗'

Fig. 3 The average lifespan of the backflow regions in the pipe [9], channel [7, 9], duct [31] and boundary layer [5] (from Ref. [9])

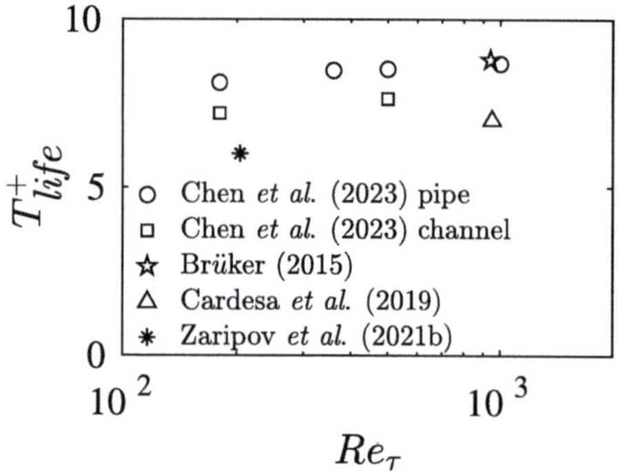

computational setups can be found in [8] and [9]. The near-wall spatial resolution of $\Delta y^+_{min} \approx 0.15$ is sufficient to revolve the backflow regions with an average wall-normal extent of 1 wall unit, and the temporal resolution of the data is kept below $\Delta t^+ < 2$ for the tracking of backflow region evolutions through their life cycles.

3 Results and Discussion

Inspections on the backflow regions resolved in the pipe flow at Reynolds numbers up to 1000 suggested that in fully-developed flows, the backflow regions are more commonly formed beneath the strong spanwise rotations belonging to arch vortices. In the results, we show examples from the pipe flow at a low Reynolds number of $Re_\tau = 360$ for clarity. This dataset gave more backflow regions compared to $Re_\tau = 180$ [9] while still being low so that the associated vortices are more organised. Figure 4 shows a cropped window of the instantaneous contour of the streamwise velocity fluctuation, u at a low Reynolds number $Re_\tau = 360$ with four backflow regions as labelled, and the three-dimensional ($3D$) vortices around the backflow regions are shown by the isosurfaces from $Q-$criterion using a very high threshold of $Q = 15$, i.e., intense vorticity. The backflow regions are located at the upstream part of the stronger and wider near-wall low-speed streaks owing to the large-scale low-speed structures carrying the backflow at their tail-tips [9, 17, 18, 23, 30]. The backflow regions BF-I, II and IV are formed under inversely arched vortices similar to the observations in Zaripov et al. [30]. Downstream of these backflow regions and the associated arch vortices, the quasi-streamwise vortices on the sides of the low-speed structures are possibly the legs of downstream vortices belonging to the same cluster (Figs. 5, 6 and 7).

The observations of backflow regions forming beneath vortices with a closer resemblance to 'horseshoe' or arch rather than typical hairpin-shaped vortices are expected since besides the possible effect of low Reynolds number [19, 24], the backflow regions were found at the upstream tail of the low-speed structures where it is wrapped by the upstream smaller vortices in a cluster, in arch shapes, with their head of the strong spanwise rotation within the buffer layer [17, 23, 30]. Furthermore, in fully-developed flows, the more distinctly hairpin-shaped vortices usually appear away from the wall while vortices in the near-wall region are more chaotic with disorganised small-scale clusters [1, 3]. As shown by the sketch in Fig. 8, in a hairpin packet, the larger vortices would be wrapped around the more downstream and the thicker part of a low-speed structure [2] and the backflow regions are unlikely to be found under a typical hairpin vortex with very long quasi-streamwise legs in which their heads have elevated further away from the wall. Therefore, only the vortices with their main parts residing in the viscous wall region are displayed in the later part of the results.

A particularly long-lived backflow region in the dataset with $T^+_{life} \approx 40$, around 5 times longer than the average lifespan of the backflow regions as shown in Fig. 3 [9] is shown in Fig. 5. This is also a particularly large backflow region compared to the average wall-normal span of a backflow region of 20 wall units [7, 9, 23, 30]. The complete life cycle of the backflow region is shown in Fig. 5b with its temporal projections superpositioned on the wall-parallel and spanwise-temporal planes. This backflow region shows the typical evolution of a backflow region where they grow and decay in size with spanwise elongation [7, 9]. Figure 5a, c show the instantaneous backflow region after it has just formed where the vortices are removed in (c) for a clearer view of the vortices associated with the backflow. In Fig. 5c, the backflow forms at a cane-shaped, i.e., asymmetric hairpin vortex which is connected to the long quasi-streamwise vortex potentially as the other leg. The backflow forms just under the middle of the head of the hairpin vortex. The backflow region grows and decays with the hairpin vortex. In the middle of its lifespan, the backflow region has grown in size and elongated in the spanwise direction together with the head of the hairpin vortex which has developed into a shape closer to a hairpin with its quasi-streamwise legs being closer within the spanwise extent of its large head. The backflow region soon diminishes with the associated vortices. In Fig. 5e, the spanwise rotations diminished and the backflow region shrunk and soon disappeared with the

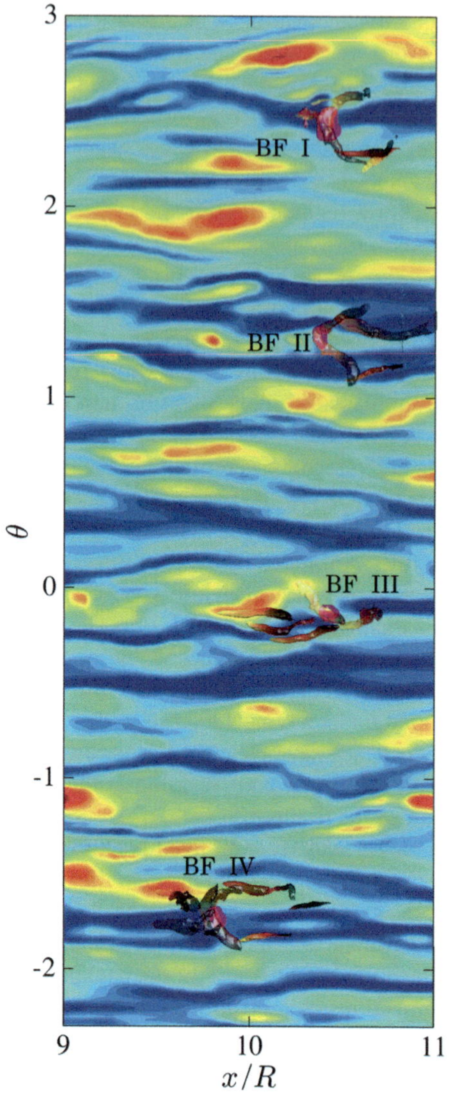

Fig. 4 Contour of the streamwise velocity fluctuation, u on a cropped wall-parallel plane at $y^+ = 5$ with the colour scale of $-0.3 < u/U_b < 0.3$ and the backflow region coloured in magenta in a cropped snapshot of the pipe at $Re_\tau = 360$. The overlaid vortices associated with the backflow regions are defined at $Q = 15$. The backflow regions are slightly enlarged for visibility

vortex while younger quasi-streamwise vortices and cane-shaped hairpin vortices are developing nearby and are about to induce new backflow regions.

A similar evolution process of the backflow with its associated hairpin vortex is shown by a backflow region with an even longer lifespan of $T^+_{life} \approx 50$ in Fig. 6. The life cycle of this backflow region in Fig. 6b shows that the backflow region goes through multiple growth and decay in size which is related to the size and strength of the spanwise vorticity above the backflow, i.e., the head of the hairpin vortex shown in Fig. 6c, d. The backflow is located just at the trailing end of a low-speed structure joined by an upstream high-speed region. The hairpin vortex above the backflow is

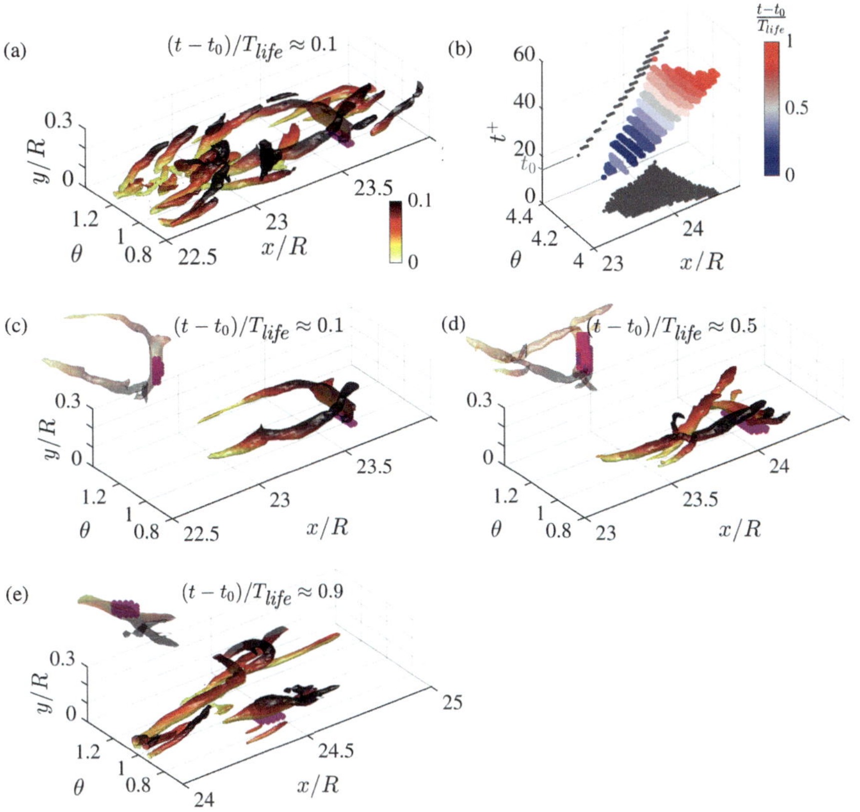

Fig. 5 **a, c, d, e** Isosurfaces of $Q = 15$ in the pipe at $Re_\tau = 360$ at **a, c** the beginning, **d** the middle and **e** the end of the lifespan of a backflow region. The isosurfaces are coloured by their wall-normal elevation. **b** The complete life cycle of this backflow region with spatial projections. t_0 denotes the time of the first detected appearance of a backflow region. In **c, d, e** the vortices are trimmed around the vortex associated with the backflow region, and the top view is shown faded

in arch shape and the downstream quasi-streamwise legs of the upstream vortices in the cluster are flanked at the low-speed structure, being responsible for the strong $Q2$ ejections in the low-speed region induced between them. This backflow region remains relatively small and less spanwise elongated compared to the one in Fig. 5 possibly because the associated arch-shaped hairpin vortex remains such shape rather than developing into a classical hairpin shape.

Last, we show the first direct observation of multiple streamwise-aligned backflow regions induced by clustered vortices. As labelled in Fig. 7, backflow region BF-a appears earlier upstream of BF-b which is induced beneath strong spanwise vortices clustered and joined by quasi-streamwise vortices on the sides of the induced low-speed structure. In Fig. 7, BF-b appears when BF-a is roughly at its matured stage [9] in the middle of its lifespan, and grows under the developing spanwise vortex above BF-b. In Fig. 7d, the vortex associated with BF-a which was mostly a purely spanwise

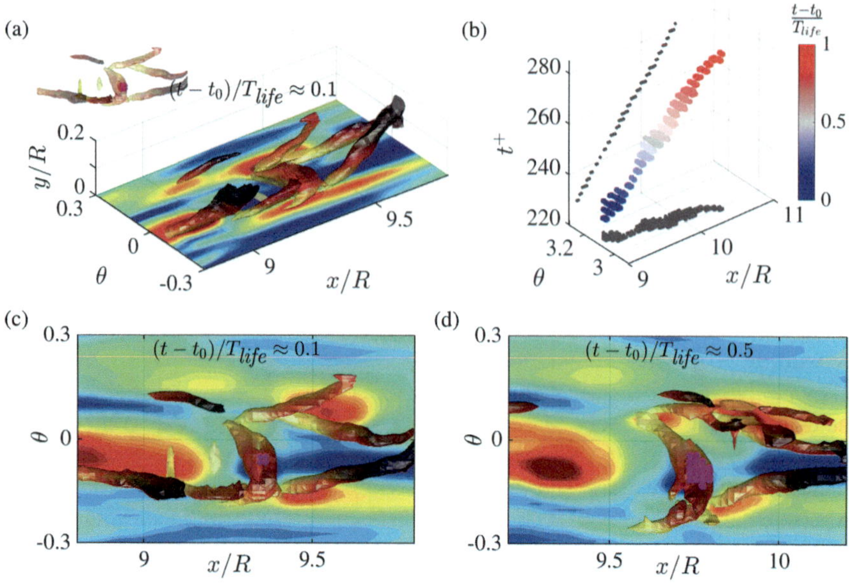

Fig. 6 **a, c, d** Isosurfaces of $Q = 15$ (coloured by their wall-normal elevation of $0 < y/R < 0.1$) in the pipe at $Re_\tau = 360$ at **a, c** the beginning and **d** the middle of the life cycle of a backflow region. The wall-parallel contours of $-0.05 < u/U_b < 0.05$ at $y^+ = 1$ are overlaid. The vortices are trimmed around the vortex associated with the backflow region, and the top view is shown faded. **b** The complete life cycle of the backflow region with spatial projections. In **c, d** the backflow regions attached to the wall under the vortices are plotted on top of the vortices for visibility

vortex similar to BF-b in Fig. 7 has become an arch-shaped vortex, and BF-b and its associated spanwise vortex will follow with a similar evolution. Although the hairpin vortices in the near-wall region of fully-developed flows are not in ideal shapes, it is clear that a vortex cluster in the viscous wall region around a low-speed structure can cause streamwise-aligned regions of backflow with the associated critical points of wall-shear stress in the downstream direction under the regenerated new vortices in the cluster as illustrated in Fig. 8. In fact, in our ongoing works, when the near-wall vortex clusters become more organised in turbulence transitions, the regeneration of both the hairpin vortices and the associated backflow region beneath the hairpin heads will be much more evident.

4 Conclusion

The DNS data of fully turbulent pipe flows with high spatial and temporal resolutions were used to resolve backflow regions in the near-wall region and to capture vortices and vortex clusters inducing the backflow and the associated large-scale low-speed structures. Instantaneous visualisations were presented to show the dynamics of the

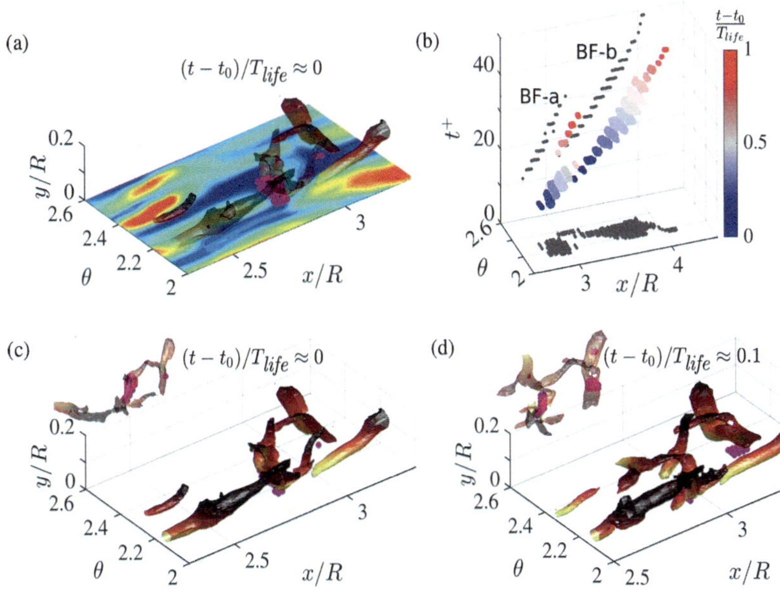

Fig. 7 **a, c, d** Isosurfaces of $Q = 15$ (coloured by their wall-normal elevation of $0 < y/R < 0.1$) in the pipe at $Re_\tau = 360$ at different life stages of the backflow regions, BF-a and BF-b with the time labelled for BF-b. The vortices are trimmed around the vortex associated with the backflow region, and the top view is shown faded. The wall-parallel contours of $-0.05 < u/U_b < 0.05$ at $y^+ = 1$ are overlaid in (**a**). **b** The complete life cycle of the two backflow regions with spatial projections

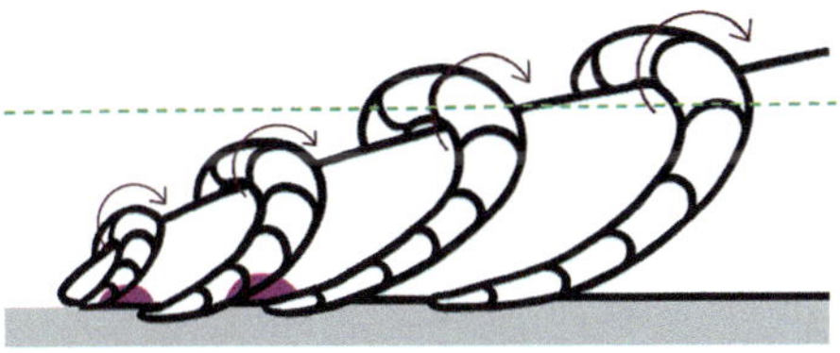

Fig. 8 Conceptual model of the a hairpin packet wrapped around the resultant low-speed structure similar to [2] with the backflow regions (magenta) induced by the strong spanwise rotations at the heads of the vortices in the near-wall region. The dashed line represents the upper boundary of the buffer region

vortex clusters and hairpin vortices in the viscous wall region when backflow occurs. The evolution of the backflow with the associated vortex clusters was shown in detail. It was observed that the backflow is induced by strong spanwise rotations by arch-shaped vortices including hairpin vortices including cane- and arch-shaped forms, and also inversely arched spanwise vortices. The low-speed regions upstream of the backflow are flanked by quasi-streamwise vortices as the legs of the hairpin vortices in a cluster. As the hairpin vortex grows, its legs become closer and the underneath backflow regions grow with the enlarged head. A 'tongue' feature develops downstream from the existing hairpin and elongates into a new quasi-streamwise vortex.

It is found to be uncommon for backflow regions to occur under developed hairpin vortices in ideal hairpin shapes because only the ones with their heads that have not elevated far from the wall, residing within the buffer layer cause the backflow, and this matches the conceptual model of hairpin packets wrapped around the low-speed structures [2] where the smaller upstream arch-shaped hairpins located at the trailing end of the low-speed structure are responsible for the backflow. Visualisations showed that multiple streamwise-aligned backflow regions can occur under the aligned heads of hairpins in a vortex cluster so that backflow may follow the regeneration of hairpin vortices in the packet.

4.1 Future Work

The formation and evolution of backflow regions are also investigated in turbulent pipe flows subjected to shear-dependent rheology. In this upcoming work, the low-Reynolds-number shear-thinning flow sustains in a weakly turbulent state similar to transition owing to the shear-thinning effect, i.e., decreasing local Reynolds numbers away from the wall which reduces the turbulent intensity and thus the population and complexity of vortical structures away from the near-wall region. Such flow is found to be ideal for observing the synchronized evolution process of the hairpin vortices and the induced backflow regions underneath their heads. In the weakly turbulent flow, a much more vivid dynamics of the concurrent backflow generation following the auto-regeneration of organised hairpin vortices in a packet as depicted in Fig. 8 is presented.

References

1. Adrian, R., Liu, Z.-C.: Observation of vortex packets in direct numerical simulation of fully turbulent channel flow. J. Vision **5**, 9 (2002)
2. Adrian, R., Meinhart, C., Tompkins, C.D.: Vortex organisation in the outer region of the turbulent boundary layer. J. Fluid Mech. **422**, 1–54 (2000)
3. Adrian, R.J.: Hairpin vortex organization in wall turbulence. Phys. Fluids **19**(4), 041301 (2007)
4. Bross, M., Fuchs, T., Kähler, C.J.: Interaction of coherent flow structures in adverse pressure gradient turbulent boundary layers. J. Fluid Mech. **873**, 287–321 (2019)
5. Brücker, C.: Evidence of rare backflow and skin-friction critical points in near-wall turbulence using micropillar imaging. Phys. Fluids **27**(3), 031705 (2015)
6. Cardesa, J.I., Monty, J.P., Soria, J., Chong, M.S.: Skin-friction critical points in wall-bounded flows. J. Phys: Conf. Ser. **506**, 012009 (2014)
7. Cardesa, J.I., Monty, J.P., Soria, J., Chong, M.S.: The structure and dynamics of backflow in turbulent channels. J. Fluid Mech. **880**, R3 (2019)
8. Chen, X., Chung, Y.M., Wan, M.: The uniform-momentum zones and internal shear layers in turbulent pipe flows at Reynolds numbers up to $Re_\tau = 1000$. Int. J. Heat Fluid Flow **90**, 108817 (2021)
9. Chen, X., Chung, Y.M., Wan, M.: Backflow structures in turbulent pipe flows at low to moderate Reynolds numbers. J. Fluid Mech. **966**, a38 (2023)

10. Chin, C., Vinuessa, R., Örlü, R., Cardesa, J.I., Noorani, A., Schlatter, P., Chong, M.S.: Flow topology of rare backflow events and critical points in turbulent channels and toroidal pipes. J. Phys: Conf. Ser. **1001**, 012002 (2018)
11. Chin, R.C., Monty, J.P., Chong, M.S., Marusic, I.: Conditionally averaged flow topology about a critical point pair in the skin friction field of pipe flows using direct numerical simulations. Phys. Rev. Fluid. **3**, 114607 (2018)
12. Chin, R.C., Vinuesa, R., Örlü, R., Cardesa, J.I., Noorani, A., Chong, M.S., Schlatter, P.: Backflow events under the effect of secondary flow of Prandtl's first kind. Phys. Rev. Fluids **5**, 074606 (2020)
13. Colella, K.J., Keith, W.L.: Measurements and scaling of wall shear stress fluctuations. Exp. Fluids **34**(2), 253–260 (2003)
14. Eckelmann, H.: The structure of the viscous sublayer and the adjacent wall region in a turbulent channel flow. J. Fluid Mech. **65**(3), 439–459 (1974)
15. El Khoury, G.K., Schlatter, P., Brethouwer, G., Johansson, A.V.: Turbulent pipe flow: statistics, Reynolds-dependence, structures and similarities with channel and boundary layer flows. J. Phys: Conf. Ser. **506**, 012010 (2014)
16. Fischer, P.F., Lottes, J.W., Kerkemeier, S.G.: nek5000 Web page (2008). http://nek5000.mcs.anl.gov
17. Guerrero, B., Lambert, M.F., Chin, R.C.: Extreme wall shear stress events in turbulent pipe flows: spatial characteristics of coherent motions. J. Fluid Mech. **904**, A18 (2020)
18. Guerrero, B., Lambert, M.F., Chin, R.C.: Precursors of backflow events and their relationship with the near-wall self-sustaining process. J. Fluid Mech. **933**, A33 (2022)
19. Head, M.R., Bandyopadhyay, P.: New aspects of turbulent boundary-layer structure. J. Fluid Mech. **107**, 297–337 (1981)
20. Hu, Z.W., Morfey, C.L., Sandham, N.D.: Wall pressure and shear stress spectra from direct simulations of channel flow. AIAA J. **44**(7), 1541–1549 (2006)
21. Jalalabadi, R., Sung, H.J.: Influence of backflow on skin friction in turbulent pipe flow. Phys. Fluids **30**, 065104 (2018)
22. Johansson, G.: An experimental study of the structure of a flat plate turbulent boundary layer, using laser-Doppler velocimetry. PhD thesis, Chalmers University of Technology, Sweden (1988)
23. Lenaers, P., Li, Q., Brethouwer, G., Schlatter, P., Örlü, R.: Rare backflow and extreme wall-normal velocity fluctuations in near-wall turbulence. Phys. Fluids **24**(3), 035110 (2012)
24. Robinson, S.K.: Coherent motions in the turbulent boundary layer. Annu. Rev. Fluid Mech. **23**, 601–639 (1991)
25. Spalart, P.R., Coleman, G.N.: Numerical study of a separation bubble with heat transfer. Euro. J. Mech. B/Fluids **16**(2), 169–189 (1997)
26. Vinuesa, R., Örlü, R., Schlatter, P.: Characterisation of backflow events over a wing section. J. Turbul. **18**(2), 170–185 (2017)
27. Willert, C.E., Cuvier, C., Foucaut, J.M., Klinner, J., Stanislas, M., Laval, J.P., Srinath, S., Soria, J., Amili, O., Atkinson, C., Kähler, C.J., Scharnowski, S., Hain, R., Schröder, A., Geisler, R., Agocs, J., Röse, A.: Experimental evidence of near-wall reverse flow events in a zero pressure gradient turbulent boundary layer. Exp. Thermal Fluid Sci. **91**, 320–328 (2018)
28. Wu, X., Cruickshank, M., Ghaemi, S.: Negative skin friction during transition in a zero-pressure-gradient flat-plate boundary layer and in pipe flows with slip and no-slip boundary conditions. J. Fluid Mech. **887**, A26 (2020)
29. Zaki, A., Abdelrahman, M.A., Ayad, S.S., Abdellatif, O.E.: Effects of leading edge slat on the aerodynamic performance of low Reynolds number horizontal axis wind turbine. Energy **239**, 122338 (2022)
30. Zaripov, D.I., Ivashchenko, V., Mullyadzhanov, R., Li, R., Markovich, D., Kähler, C.: On a mechanism of near-wall reverse flow formation in a turbulent duct flow. J. Fluid Mech. **923**, A20 (2021)
31. Zaripov, D.I., Ivashchenko, V., Mullyadzhanov, R., Li, R., Markovich, D., Kähler, C.: Reverse flow phenomenon in duct corners at a low Reynolds number. Phys. Fluids **33**, 085130 (2021)

Unsteady Flow and Mass Transfer Induced by Rayleigh-Bénard-Marangoni Convection

Jan Wissink, Fatima Ali, and Herlina Herlina

Abstract Evaporative cooling at the water surface is usually modelled by imposing a constant heat flux at the surface. This boundary condition allows for variations in the water surface temperature T, which then induce variations in surface tension. The resulting Marangoni forces tend to move surface water from low surface tension (high T) regions to high surface tension (low T) regions. To study the combined effect of buoyancy and Marangoni forces on interfacial mass transfer, (fully resolved) direct numerical simulations have been performed. The simulations were carried out for a fixed macro Rayleigh number of $Ra_L = 21200$ and a variety of Marangoni (Ma) numbers to assess the relative importance of buoyancy and Marangoni forces on the air-water mass transfer. It is known that both forces tend to reinforce one another even though the underlying physical mechanisms are different. This is highlighted in the present results showing that Marangoni forces, acting at the water surface, induce very efficient mixing of dissolved gases (as well as heat) in a well-defined layer adjacent to the surface. Buoyant convection, on the other hand, tends to cause deep penetration of plumes of cold, saturated water into the bulk.

1 Introduction

The motivation behind this study is to gain a better understanding of the heat and mass transfer process across the air-water interface in natural water bodies. A specific example of this phenomenon is the heat and atmospheric gas transfer across the

CUFS-2024-PAPER ID, Cambridge, United Kingdom, 4-5 March 2024.

J. Wissink (✉) · F. Ali
Brunel University London, Kinston Lane, Uxbridge UB8 3PH, UK
e-mail: jan.wissink@brunel.ac.uk

F. Ali
e-mail: fatima.ali2@brunel.ac.uk

H. Herlina
Karlsruhe Institute of Technology, Karlsruhe D76131, Germany
e-mail: herlina@kit.edu

J. C. Tyacke and N. R. Vadlamani (eds.), *Proceedings of the Cambridge Unsteady Flow Symposium 2024*, https://doi.org/10.1007/978-3-031-69035-8_15

interface of lakes and reservoirs. It is vital to accurately predict the transfer rate in order to generate a definitive global heat and greenhouse gas budget. In this context, wind shear is commonly recognized as the primary factor responsible for generating turbulence, which supports the interfacial heat and mass flux [1, 2]. Typically, other sources of turbulence, such as buoyancy, are often overlooked. However, there has been a growing interest in buoyancy-driven heat and gas transfer resulting from surface cooling, especially in lakes and ponds with low wind speeds [3–5]. During this process, surface cooling, often caused by evaporation, leads to the formation of plumes and sheets of relatively dense, cold water. These cold-water (gas-saturated) plumes plunge down while being replaced by warmer bulk water that is not fully saturated, thus facilitating the heat and gas transfer across the water surface.

Surface cooling is known to create an unstable density gradient and can also cause variations in surface tension due to localized changes in temperature. These variations in surface tension give rise to Marangoni forces, which induce flows from regions of low surface tension (high temperature) to regions of high surface tension (low temperature). Consequently, buoyancy- and/or Marangoni-induced convective instabilities lead to the formation of convection cells at the surface [6, 7]. The characteristic pattern of these convection cells reveals the existence of one or more regions with elevated temperature within their structure [8]. Additionally, it demonstrates that individual cells are separated by narrow regions characterized by lower temperatures. Numerous theoretical and experimental studies have been conducted to explore the onset of the underlying instability and the resulting formation of convection cell patterns [9–12]. Many of these investigations were motivated by applications in the field of chemical engineering and involved the utilization of a relatively thin fluid layer that was confined by a solid wall with a no-slip boundary condition. These studies demonstrated that within such thin fluid layers, horizontal temperature gradient induced Marangoni forces, in conjunction with buoyancy forces, promote the mixing of cooler surface water with warmer water from the upper bulk region [13, 14].

In previous numerical simulations of buoyancy-driven heat transfer across the air-water interface two approaches were commonly used: prescribing a constant temperature [8] or a constant heat flux [15, 16]. Unlike simulations where the surface temperature remains constant, constant-heat-flux simulations generate variations in surface temperature that results in the generation of Marangoni forces. It should be noted that in previous numerical investigations on the effect of these forces on interfacial mass transfer, the Marangoni forces were surfactant-concentration-induced [17, 18]. Even tiny amounts of surfactants were found to significantly inhibit interfacial mass transfer. Despite the potentially significant impact of Marangoni forces on heat and mass transfer coefficients, the surface-temperature-induced ones were ignored in previous numerical studies.

Recently, a numerical study of the combined effect of buoyancy and Marangoni forces on developing interfacial heat transfer, was performed by Wissink and Herlina [19]. It was found that at a fixed Rayleigh number the Marangoni-force-induced-mixing of water from the surface and from the upper bulk resulted in a progressively earlier onset of the so-called Rayleigh-Bénard Marangoni (RBM) instability with increasing Marangoni number. Hence, it is to be expected that also the impact of Marangoni forces on interfacial mass transfer will be non-negligible.

In the paper we will present results of a series of fully resolved three-dimensional direct numerical simulations (DNS) that study the initial development of the RBM instability and its effect on interfacial mass transfer at a Rayleigh number of $Ra_L = 21200$ and a variety of Marangoni (Ma) and Schmidt (Sc) numbers. The results presented will show the development of convection cells from initially random disturbances added to the temperature field at $t = 10$ s, thereby highlighting the influence of Marangoni-forces on the instantaneous surface mass transfer velocity. Also, the effect of Marangoni forces on the scaling of the horizontally-averaged mass transfer K_L as a power of the Schmidt number will be investigated.

2 Numerical Aspects

The in-house numerical code employed was especially developed to produce fully resolved simulations of the subsurface flow and scalar fields. It allows for the application of a dual mesh, where the evolution of the velocity and temperature fields are calculated on the base mesh, while the evolution of the low-diffusivity scalars are calculated on a refined mesh. To normalise the components of the Navier-Stokes equations, the length scale $L = 0.01$ m is used together with the velocity scale $U = \kappa/L$ m/s, where κ is the thermal diffusivity. As a result, using the Einstein summation convention, the continuity equation reads

$$\frac{\partial u_k}{\partial x_k} = 0 \tag{1}$$

and the momentum equations read

$$\frac{\partial u_i}{\partial t} + \frac{\partial u_i u_j}{\partial x_j} = -\frac{\partial p}{\partial x_i} + Pr \frac{\partial^2 u_i}{\partial x_j \partial x_j} + Ra_L\, Pr\, T^\star \delta_{i3} \quad i = 1, \ldots, 3, \tag{2}$$

where u_1, u_2, u_3 are the normalised velocity components in the x_1, x_2, x_3 direction, respectively, with $x_1 = x, x_2 = y, x_3 = z$, p is the universal pressure, δ_{ij} is the Kronecker delta,

$$T^\star = \frac{T_{b,0} - T}{L(\partial T/\partial z)|_S} \tag{3}$$

is the nondimensional temperature, where $T_{b,0}$ is the intital temperature in the bulk, which is normalised using the heat flux at the surface and the length scale L (note that the subscript "S" denotes the water surface),

$$Ra_L = \frac{\alpha\ (\partial T/\partial z)|_S\, g L^4}{\kappa \nu} \tag{4}$$

is the Rayleigh number, where α is the thermal expansion rate, $g = -9.81$ m/s^2 is the gravitational acceleration, ν is the kinematic viscosity, and

$$Pr = \frac{\nu}{\kappa} = 7 \tag{5}$$

is the Prandtl number corresponding to a temperature of $T = 20°$C. The Marangoni number, finally, is defined by

$$Ma_L = \frac{(\partial \sigma/\partial T)\ (\partial T/\partial z)|_S\, L^2}{\mu \kappa}, \tag{6}$$

where μ is the dynamic viscosity.

The scalar transport equations are solved in parallel with the Navier-Stokes equations. They consist of the convection-diffusion equations for the temperature

$$\frac{\partial T^\star}{\partial t} + \frac{\partial u_j T^\star}{\partial x_j} = \frac{\partial^2 T^\star}{\partial x_j \partial x_j}, \tag{7}$$

and the normalised scalar concentrations $C^\star = C/C_S$ (where C_S is the saturation concentration), given by

$$\frac{\partial C^\star}{\partial t} + \frac{\partial u_j C^\star}{\partial x_j} = \frac{Pr}{Sc} \frac{\partial^2 C^\star}{\partial x_j \partial x_j}, \tag{8}$$

where Sc is the Schmidt number defined by the ratio of the momentum diffusivity of the fluid and the mass diffusivity of the dissolved substance D.

2.1 *Discretisations and Boundary Conditions*

The discretisation of the momentum equations is carried out using fourth-order central discretisations of the convection and diffusion, while the Poison equation for the pressure is solved using the conjugate gradient method with diagonal preconditioning.

The convection-diffusion equations for the temperature and the normalised scalar concentrations $C^\star = C/C_S$ (where C_S is the saturation concentration) are solved using a fifth-order WENO scheme for the convection and a fourth-order central discretisation for the diffusion. The code is parallelised by dividing the computational mesh in several blocks of identical size. Each block is allocated to a unique processing core, while communication between cores is done using the MPI protocol. The cubic computational domain, of size $5L \times 5L \times 5L$, is periodic in the horizontal directions. The surface is assumed to be flat with zero vertical velocity, fully saturated scalar concentrations and a fixed temperature flux to model evaporative cooling. The effect of Marangoni forces on the horizontal velocity components is modelled by

$$\begin{cases} \frac{\partial u}{\partial z} = -Ma_L \left.\frac{\partial T^\star}{\partial x}\right|_S \\ \frac{\partial v}{\partial z} = -Ma_L \left.\frac{\partial T^\star}{\partial y}\right|_S \end{cases}, \tag{9}$$

where the temperature derivatives are taken at the water surface. At the bottom, zero flux conditions are employed for all scalars in combination with a free-slip condition for the velocity.

2.2 Grid Resolution

In all simulations a $200 \times 200 \times 252$ mesh was used to discretise the $5L \times 5L \times 5L$ computational domain. In the horizontal directions the mesh was chosen to be uniform, while in the vertical direction the mesh was gradually refined towards the surface in order to resolve the thin thermal and concentration boundary layers.

In Table 1 an overview is presented of the simulations discussed in this paper. All simulations employ the same Rayleigh number, $Ra_L = 21{,}200$, combined with a variety of Marangoni numbers. The Prandtl number was set to $Pr = 7$ and the Schmidt numbers employed in the simultaneously solved scalar transport equations were $Sc = 50, 100, 200$. Note that all passive scalars were solved by refining the base mesh with a factor of 2, while the temperature was solved on the $200 \times 200 \times 252$ base mesh.

Figure 1 shows the maximum ratio over time of the geometric mean of the grid cell size

$$\overline{\Delta} = \sqrt[3]{\delta x \times \delta y \times \delta z}, \tag{10}$$

and the scaled horizontally-averaged Batchelor length scale at $Sc = s$ (or $Pr = s$)

Table 1 Overview of macro Marangoni numbers

Run	D1	D2	D3	D4	D5	D6	D7	D8	D9
Ma_L	−1050	−250	0	250	550	800	1050	2700	5250

In all simulations, $Ra_L = 21200$ and the refinement factor for the scalar mesh with $Sc = 50, 100, 200$ was $R = 2$

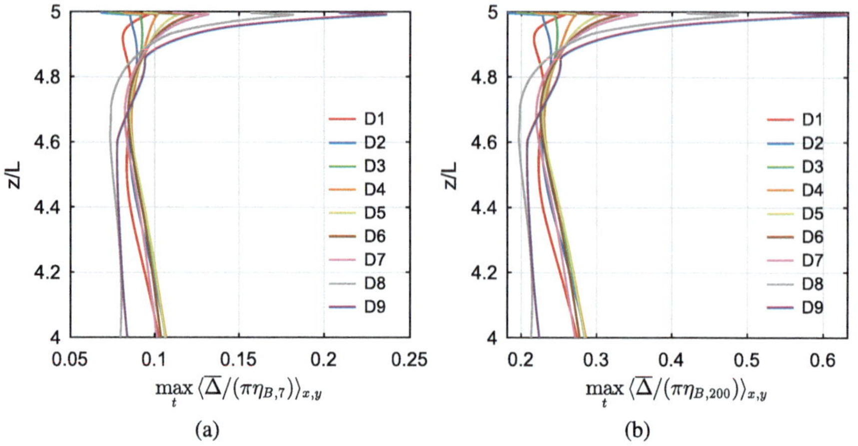

Fig. 1 Ratio of geometric mean of grid cells to **a** Batchelor scale at $Sc = 16$ (for base mesh) and **b** Batchelor scale at $Sc = 100$ (for two times refined mesh)

$$\langle \pi \ \eta_{B,s} \rangle_{x,y}, \tag{11}$$

where δx, δy, δz are the mesh sizes in the x, y, z-directions. It can be seen that in all simulations, the ratio $\overline{\Delta}/\langle \pi \ \eta_{B,s} \rangle_{x,y}$ is less than one for both $s = 7$ (resolved on the base mesh) and $s = 200$ (resolved on the two times refined mesh). Additionally (not shown here), in all simulations the linear (diffusive) part of the concentration boundary layer, located adjacent to the surface, is fully resolved by at least 4 grid points. Hence, according to the Grötzbach criterion, the scalar fields adjacent to the surface are sufficiently well resolved. Please note that for the convection part of the transport equations, a WENO scheme is used that is capable of resolving very steep gradients.

3 Results

The paper focuses on studying the mass transfer during the initial development of the RBM instability and the simulations were stopped some time after the first plumes started to fall down.

3.1 Instantaneous Mass Transfer

The RBM instability combines the effects of buoyancy forces and Marangoni forces. The former are responsible for the formation of thermal plumes that results from the accumulation of cold (relatively heavy) water near the surface. This process continues until sufficient potential energy is obtained to overcome diffusive forces resulting in cold water plumes penetrating relatively deep into the bulk. Marangoni forces, on the other hand, result from local differences in surface tension, where a force is generated moving fluid from low surface tension (high $T|_S$) regions to high surface tension (low $T|_S$) regions. For a more detailed description see [19]. Both instabilities tend to reinforce one another even though the Marangoni forces only act at the surface and, hence, do not result in deep penetrative convection.

The surface heat flux continuously cools the instantaneous surface temperature $T|_S$, which results in an initial thickening of the thermal boundary layer. Simultaneously, also the concentration boundary layer thickens in time, which, in turn, results in a reduction of the instantaneous mass transfer velocity

$$k_L = \left| \frac{-1}{C - C_b} \frac{1}{Re\,Sc} \left. \frac{\partial C}{\partial z} \right|_S \right| . \tag{12}$$

Some time after the RBM instability starts to develop from the initially random disturbances, eventually $T|_S$ becomes increasingly non-uniform. In areas where the thermal boundary layer grows, generally also the concentration boundary layer becomes thicker so that k_L reduces. On the other hand, in areas where warm, unsaturated fluid from the (upper) bulk moves upwards both $T|_S$ and k_L become larger. Figure 2 illustrates the effect of the RBM instability in simulation D9 by comparing snapshots of the instantaneous surface temperature T and the instantaneous mass transfer velocity k_L at $Sc = 100$ for $t = 20, 28, 35, 58$ s, where the first two time-instances $t = 20, 28$ s correspond to the times where the horizontally-averaged k_L is minimum and maximum, respectively (see also Fig. 4). In this time-interval, increasingly well-defined convection cells can be seen to develop. Each convection cell consists of one or more upflow areas where relatively warm water from the lower thermal boundary layer flows to the surface, where it continues to flow in the radial direction as it cools down. At locations where two convection cells meet, the now relatively cold water is transported back down into the upper bulk. In time, the size of the convection cells can be seen to increase significantly. At $t = 20$ s, areas of low $T|_S$ can be seen to correlate well with areas of low k_L, while areas of high $T|_S$ correlate well with areas of high k_L. At later times, this correlation very gradually reduces. At $t = 58$ this deterioration can be seen most clearly. It is (partially) linked to an increased variation of k_L as reflected in the absence of k_L maxima in the middle of large convection cells (where peaks in T can clearly be observed) and the presence of local peaks of high instantaneous mass transfer in most of the smaller convection cells (which do correspond to peaks in $T|_S$). The peaks in the small convection cells are associated with small vortical structures immediately underneath the water surface that actively promote the local upflow of unsaturated, warm water from the lower thermal boundary layer resulting in the emergence of relatively small (often elongated) convection cells with a more intense mixing than in the larger cells). The emergence of these small surface-parallel) vortices are typical for the Marangoni instability and result in a locally increased vertical mixing.

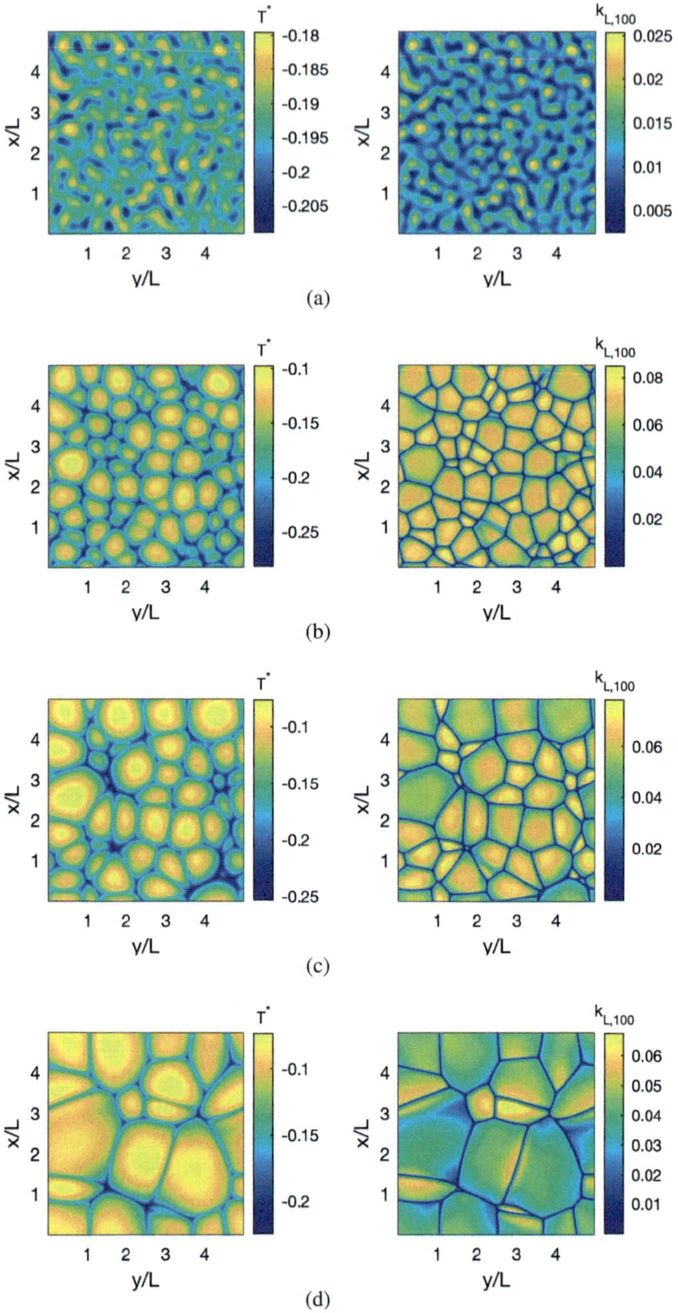

Fig. 2 Snapshots from simulation D9 of the instantaneous surface temperature and mass transfer velocity at $Sc = 100$ and **a** $t = 20$ s., **b** $t = 28$ s., **c** $t = 35$ s. and **d** $t = 58$ s

Figure 3 shows snapshots from simulation D9 of the temperature T and the concentration C_{100} at $Sc = 100$ in the plane $y/L = 2.5$, taken at the same time instances as the snapshots of the instantaneous surface temperature and mass transfer in Fig. 2. It can be seen that at $t = 20$ s the thermal boundary layer shows almost no disturbances (due to the relatively large thermal diffusivity with $Pr = 7$), while the concentration boundary layer at $Sc = 100$ shows very small disturbances. In time these disturbances grow and at $t = 28$ s these disturbances also become clearly visible in the thermal boundary layer in the form of diffused thermal plumes. Because of the relatively high Schmidt number of $Sc = 100$, the concentration contours are much less diffused and clearly show the typical mushroom-shape that is often associated with the initial formation of thermal plumes driven by buoyant convection. In time, these plumes gradually merge such that their number reduces while their size increases. As a result, at $t = 58$ s only two plumes remain visible in the plane $y/L = 2.5$, while due to significant differences in diffusion, the correlation between T and C_{100} gradually reduces.

3.2 Horizontally-Averaged Statistics

Figure 4 shows the development in time of the horizontally-averaged mass transfer velocity

$$K_L = \langle k_L \rangle_{x,y} \tag{13}$$

for all simulations D1-D9 listed in table 1. It can be seen that at a fixed Rayleigh number Ra_L, an increase in the Marangoni number results in an earlier onset of the RBM instability, as identified by K_L starting to deviate from its laminar value that is identified by the lower envelope of the curves (up to $t \approx 95$ s). While the positive Marangoni numbers used in simulations D4-D9 result in a promotion of the RBM instability, negative Marangoni numbers (simulations D1,D2) can be seen to progressively inhibit the RBM instability as the onset of growth in K_L is delayed compared to the purely buoyant simulation D3, where $Ma_L = 0$.

Figure 5 investigates (at a fixed $Ra_L = 21200$) the dependence of K_L/U on (a) Ma_L for $Sc = 50, 100, 200$ and (b) Sc for a variety of Marangoni numbers Ma_L at the time where K_L/U is maximum (see Fig. 4). In Fig. 5a, it can be seen that with increasing Schmidt number, not only K_L/U reduces for fixed Ma_L, but also that at each fixed Sc, the rate of increase in K_L/U with increasing Ma_L is reduced. Figure 5b indicates the existence of a power-law relation

$$\frac{K_L}{U} \propto Sc^n \tag{14}$$

for each fixed Marangoni number (with n dependent on Ma_L). For the purely buoyant simulation D3, with $Ma_L = 0$, the slope of $n = -0.49$ is very close to the theoretical slope of $n = -0.5$ obtained in [8], which is valid for simulations with a free-slip

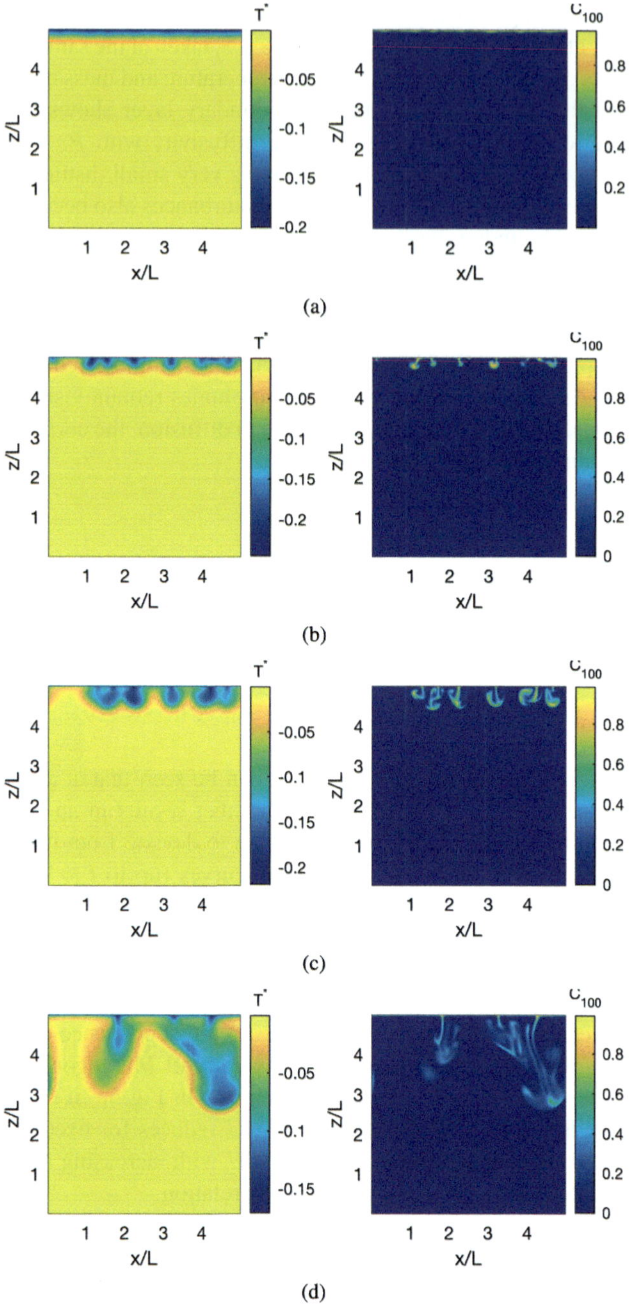

Fig. 3 Snapshots from simulation D9 of the instantaneous temperature T and concentration C_{100} at $Sc = 100$ and **a** $t = 20$ s., **b** $t = 28$ s., **c** $t = 35$ s. and **d** $t = 58$ s

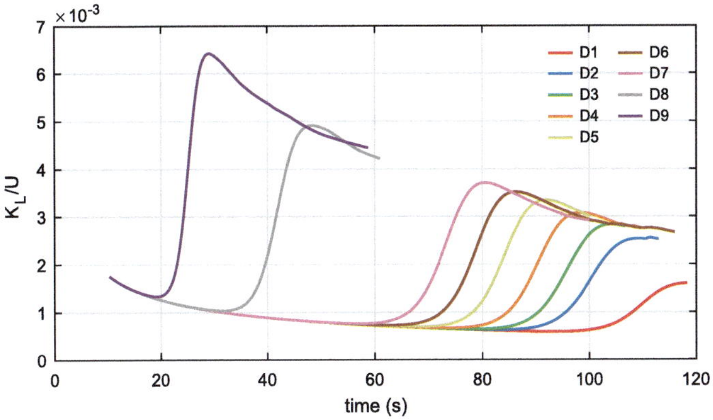

Fig. 4 Temporal evolution of K_L for various Marangoni numbers at fixed $Ra_L = 21200$ and $Sc = 100$

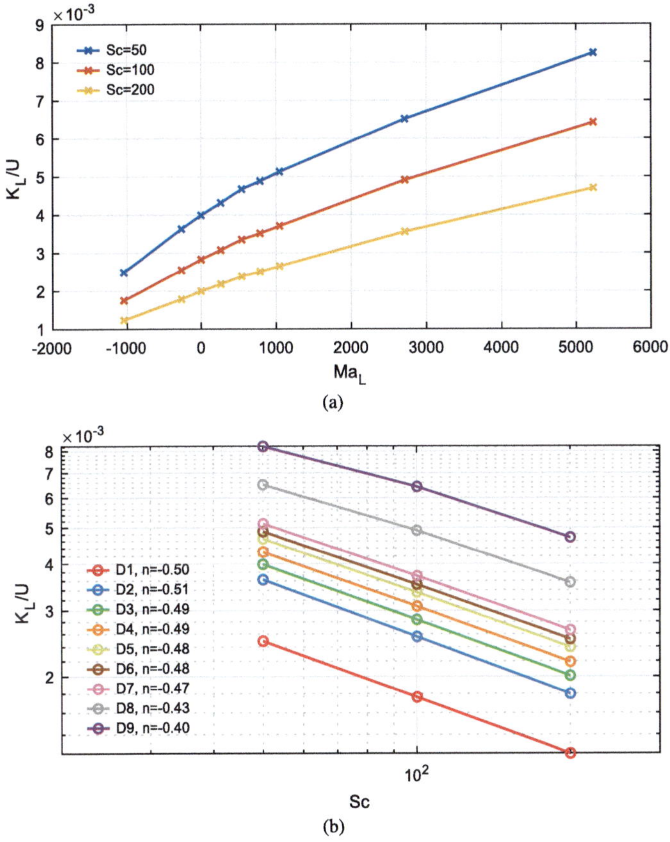

Fig. 5 Variation of K_L (at the time where K_L is maximum) a) with Ma_L for different Sc. b) with Sc for different Ma_L. The fitted slope n ($K_L \propto Sc^n$) is obtained using the data points of $Sc = 50, 100, 200$. In all simulations $Ra_L = 21200$

surface boundary condition for the velocity (such as obtained when $Ma_L = 0$). For $Ma_L \neq 0$, the slope n can be seen to noticeably deviate from -0.50. This is especially clear for larger Ma_L, where the slope gradually changes from $n \approx -0.50$ to $n = -0.40$ for simulation D9 with $Ma_L = 5250$.

In our previous work [18] it was shown that progressive damping of near surface turbulence leads to a gradual change of slope from $n = -0.5$ to $n = -2/3$ (corresponding to the value obtained for a no-slip surface boundary condition). Similarly, here it is also expected that a change in Ma_L from $Ma_L = 0$ to increasingly negative values of Ma_L will result in a similar change in slope n.

4 Conclusions

Fully resolved numerical simulations were performed to study interfacial mass transfer driven by an emerging Rayleigh-Bénard-Marangoni instability induced by employing a fixed surface heat flux to model evaporative cooling. The study focusses on the effect Schmidt and Marangoni numbers and uses a fixed Rayleigh number of $Ra_L = 21200$.

In our previous paper [19] it was shown that with increasing Ma_L, the speed at which the RBM instability develops significantly increases. It was also shown that, due to the fact that Marangoni forces act only at the surface, the (cold, saturated) fluid pushed down into the lower boundary layer by the Marangoni forces tends to linger close to the surface until sufficient cold fluid has accumulated for buoyancy forces to take charge and transport the saturated, cold fluid further down into the bulk. In this paper it was shown that the generation of shear by the presence of Marangoni forces caused a change in the scaling of the mass transfer velocity with Schmidt number, where the slope in the power law $K_L \propto Sc^n$ changes from $n \approx -0.50$ at $Ma_L = 0$ (free-slip condition) to $n = -0.40$ at $Ma_L = 5250$.

The combined effect of Rayleigh and Marangoni number on the above scaling is yet to be determined and will be the focus of a follow-up study.

Acknowledgements The simulations were performed on the supercomputer bwUniCluster 2.0 supported by the state of Baden-Württemberg, Germany through bwHPC.

References

1. Wanninkhof, R., Asher, W.E., Ho, D.T., Sweeney, C., McGillis, W.R.: Advances in quantifying air-sea gas exchange and environmental forcing. Ann. Rev. Marine Sc. **1**, 213–244 (2009)
2. Garbe, C.S., Rutgersson, A., Boutin, J., de Leeuw, G., Delille, B., Fairall, C.W., Gruber, N., Hare, J., Ho, D.T., Johnson, M.T., et al.: Transfer across the air-sea interface. Ocean Atmos. Interact. Gases Part pp. 55–112 (2014)
3. Rutgersson, A., Smedman, A.: Enhanced air-sea CO2 transfer due to water-side convection. J. Marine Syst. **80**(1–2), 125–134 (2010)

4. Podgrajsek, E., Sahlée, E., Rutgersson, A.: Diurnal cycle of lake methane flux. J. Geophys. Research: Biogeosci. **119**(3), 236–248 (2014)
5. MacIntyre, S., Crowe, A.T., Cortés, A., Arneborg, L.: Turbulence in a small arctic pond. Limnol. Oceanogr. **63**(6), 2337–2358 (2018)
6. Spangenberg, W.G., Rowland, W.R.: Convective circulation in water induced by evaporative cooling. Phys. Fluids **4**(6), 743–750 (1961)
7. Maroto, J.A., Pérez-Munuzuri, V., Romero-Cano, M.S.: Introductory analysis of Bénard-Marangoni convection. Eur. J. Phys. **28**(2), 311 (2007)
8. Wissink, J.G., Herlina, H.: Direct numerical simulation of gas transfer across the air-water interface driven by buoyant convection. J. Fluid Mech. **787**, 508–540 (2016)
9. Pearson, J.R.A.: On convection cells induced by surface tension. J. Fluid Mech. **4**(5), 489–500 (1958)
10. Chandrasekhar, S.: Hydrodynamic and hydromagnetic stability. Courier Corporation (2013)
11. Bodenschatz, E., Pesch, W., Ahlers, G.: Recent developments in rayleigh-bénard convection. Ann. Rev. Fluid Mech. **32**(1), 709–778 (2000)
12. Nepomnyashchy, A., Legros, J.C., Simanovskii, I.: Interfacial convection in multilayer systems. Springer (2006)
13. Nield, D.A.: Surface tension and buoyancy effects in cellular convection. J. Fluid Mech. **19**(3), 341–352 (1964)
14. Toussaint, G., Bodiguel, H., Doumenc, F., Guerrier, B., Allain, C.: Experimental characterization of buoyancy-and surface tension-driven convection during the drying of a polymer solution. Int. J. Heat Mass Transf. **51**(17–18), 4228–4237 (2008)
15. Handler, R.A., Saylor, J.R., Leighton, R.I., Rovelstad, A.L.: Transport of a passive scalar at a shear-free boundary in fully developed turbulent open channel flow. Phys. Fluids **11**(9), 2607–2625 (1999)
16. Fredriksson, S.T., Arneborg, L., Nilsson, H., Zhang, Q., Handler, R.A.: An evaluation of gas transfer velocity parameterizations during natural convection using DNS. J. Geophys. Res. Oceans **121**(2), 1400–1423 (2016)
17. Khakpour, H.R., Shen, L., Yue, D.K.P.: Transport of passive scalar in turbulent shear flow under a clean or surfactant-contaminated free surface. J. Fluid Mech. **670**, 527–557 (2011)
18. Wissink, J.G., Herlina, H., Akar, Y., Uhlmann, M.: Effect of surface contamination on interfacial mass transfer rate. J. Fluid Mech. **830**, 5–34 (2017)
19. Wissink, J.G., Herlina, H.: Surface-temperature-induced Marangoni effects on developing buoyancy-driven flow. J. Fluid Mech. **962**, A23 (2023)

On the Role of the Spanwise Vortices Aligned Over an Airfoil Surface for the Flow Control

Kozo Fujii, Yoshiaki Abe, and Takuto Ogawa

Abstract The mechanisms of both dielectric barrier discharge plasma actuators (DBD-PA) under burst mode and synthetic jet (SJ) actuators for flow separation control over an airfoil have been clarified through a series of studies with high-fidelity flow simulations. For flows with relatively low Reynolds numbers, the presence of small coherent spanwise vortices induced by intermittent actuations was found to be a crucial factor for effective flow control with both devices. It should be noted that during the transition from separated to attached flows, small coherent spanwise vortices appeared aligned over the airfoil surface for a certain period of time. This article summarises the results in Phys. Fluids 35, 065114 (2023) and Actuators 12(8), 322 (2023).

1 Introduction

An improvement in the capability to suppress flow separation brings a great benefit to systems that have fluid flow in or around them. The typical and conventional devices for separation control are mechanical "flap" and "slat" implemented on aircraft wings, which, however, necessitate complicated systems and result in increasing the structural weight. Recently, a "microdevice" was developed with a new aircraft design concept, which is a simple and small device with low energy consumption.

This study focuses on the separation control at low Reynolds number ($\mathrm{Re} = \mathcal{O}(10^4)$) flows around the airfoil and the use of two active-control devices: a

CUFS-2024-3, Cambridge, United Kingdom, 4-5 March 2024

K. Fujii (✉)
Tokyo University of Science, Tokyo 1258585, Japan
e-mail: fujii@rs.tus.ac.jp

Y. Abe · T. Ogawa
Tohoku University, Miyagi 9808577, Japan
e-mail: yoshiaki.abe@tohoku.ac.jp

T. Ogawa
e-mail: takuto.ogawa@tohoku.ac.jp

J. C. Tyacke and N. R. Vadlamani (eds.), *Proceedings of the Cambridge Unsteady Flow Symposium 2024*, https://doi.org/10.1007/978-3-031-69035-8_16

dielectric barrier discharge plasma actuator (DBD-PA or PA) under burst mode and a synthetic jet (SJ). The PA generates plasma to put kinetic energy inside the boundary layer by means of direct current, alternative current, radio frequency, microwave, arc, corona, and spark electric discharge. The SJ, also known as a zero-net-mass-flux jet, consists of a cavity and an orifice connected to the cavity, of which the bottom oscillates with a small amplitude and produces weak and periodic flow from the orifice exit. Both SJ [3, 5, 10, 23, 29, 30, 36] and PA [8, 11, 12, 19, 26, 33] are categorised into microdevices, which are small ($\mathcal{O}(10)$ mm), light-weight, more simple, and have lower energy consumption than conventional devices. Note that DBD-PA under bust mode demonstrated better flow control authority than under continuous mode for the flows over an airfoil at angles of attack near the stall [21, 25, 28]. Both computational and experimental studies identified important flow mechanisms behind it, and the guidelines for parameter selection were proposed [14, 15, 27].

Thus far, the separation control capabilities and mechanisms of these microdevices have been exhaustively studied numerically and experimentally. Both SJ and PA generate spatially and temporally localised flows, although their mechanisms for generating disturbances are different. Consequently, it is anticipated that a similarity exists between the separation-control mechanisms utilising the PA and the SJ, which has not been thoroughly investigated. This study aims to elucidate a mechanism for separation control around the airfoil using the PA in burst mode and the SJ. The flow field around the NACA0015 airfoil was assumed to be in the low Reynolds number regime (Re $=$ 63,000), and a large-eddy simulation (LES) was used to see how the PA in burst mode and the SJ controlled separation. By comparing in depth the mechanisms that maintain the suppression of the separated flow, the role of spanwise vortices over an airfoil was clarified.

The rest of this article is organised as follows. Section 2 describes a formulation of the problem, including the computational models of the PA and the SJ, and Sec. 3 represents the computational cases, flow solver, and grids. Section 4 discusses the performance of the PA and the SJ for separation control and flow fields, including the role of the coherent vortices over the airfoil surface. Section 5 concludes the article.

2 Problem Specifications

2.1 Separated Flow

This study focuses on the separation control around the NACA0015 airfoil at the chord Reynolds number of 63,000. The angle of attack (AoA) was set to be 12°, wherein the flow is separated near the leading edge, i.e., approximately 2.0% of the chord length from the leading edge [7]. Such a flow condition helps us clarify the mechanism of amplification of extremely small disturbances from the SJ or the PA and the resultant modification of the entire flow.

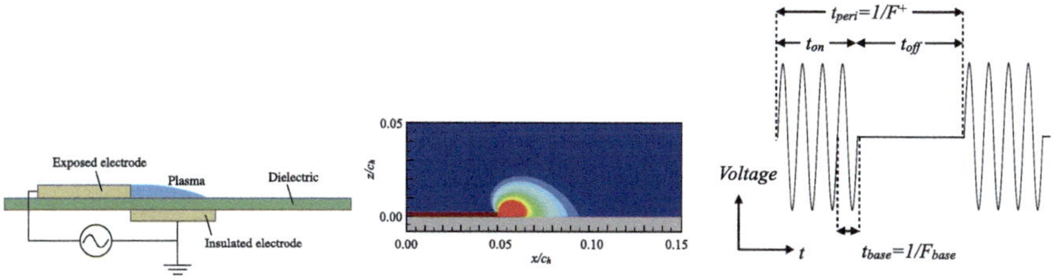

(a) Schematic of the PA (b) Body force (S_x) (c) Input voltage

Fig. 1 Computational model of the PA. Figures are taken from Abe et al. [4]

2.2 *Configuration of the PA*

The PA consists of two electrodes and a dielectric, of which schematics are illustrated in Fig. 1a. A high AC voltage generates plasma in between two electrodes, which induces a flow over the PA. The body force distribution S_x is illustrated in Fig. 1b as an example. The base frequency of AC voltage is denoted as f_{base} hereinafter, which is approximately 60 Hz in the experiment. The burst mode is activated by multiplying a higher-frequency sinusoidal wave by a lower-frequency square wave (i.e., a step function) in the time direction. The on and off phases are periodically forced by the input voltage in addition to the base-frequency oscillation (see Fig. 1c) [9]. The lower frequency of the square wave is called a "burst frequency" in the previous studies, i.e., f. The burst frequency f is normalised to F^+ as follows:

$$F^+ = \frac{f c_{\mathrm{h}}}{u_\infty}, \tag{1}$$

where c_{h} and u_∞ are the chord length and uniform velocity, respectively [18, 23]. F^+ is the same normalisation as the Strouhal number, St, based on the same reference length and velocity. The more details, including the definition of the momentum coefficient C_μ, can be referred to in Abe et al. [4], Sato et al. [26], Aono et al. [6] and Fujii [15]. The remainder of this paper focuses on the PA in burst mode, which will be often referred to as the PA for brevity.

2.3 *Configuration of the SJ*

Two-dimensional shapes along the span were adopted for the orifice and cavity of the SJ for simplicity, as shown in Fig. 2. The SJ was installed at the leading edge of the airfoil, and its orifice was normal to the airfoil surface. The bottom of the cavity oscillates periodically, and the flow inside the cavity and orifice was numerically

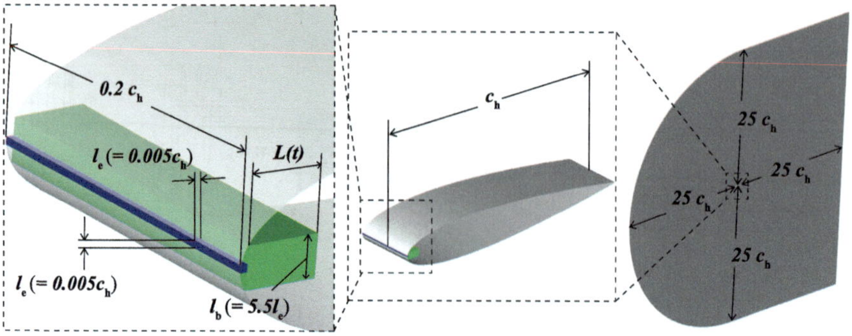

Fig. 2 Schematic illustration of the computational domain and the SJ installed to the leading edge of a NACA0015 airfoil. Figures are taken from Abe et al. [3]

solved. The orifice width l_e was set to be 0.5% of the chord length, which was often used in previous studies [23, 35]. The orifice height was set to be l_e; the cavity depth is denoted by $L(t)$ (t is the nondimensional time), which was initially set as $L(0) = L_0 = 10l_e$; and the width of the cavity bottom was set as $l_b = 5.5l_e$. According to Eq. (1), F^+ stands for the nondimensional value of actuation frequency f. The bottom of the cavity oscillates in a translational motion. The more details, including C_μ, can be referred to in Abe et al. [3].

2.4 Case Description

The computational cases are summarized in Table 1. In order to compare the capabilities of the PA in burst mode and the SJ, both of the strong and weak input momentum cases were considered. The momentum coefficient, C_μ, was similar to or smaller than that in the previous studies, e.g., $C_\mu = 3.5 \times 10^{-3}$ in [5]. Note that the weak C_μ values are different between the SJ and the PA: $C_\mu = 2.00 \times 10^{-5}$ and 5.15×10^{-5} are set for the SJ and the PA, respectively. This is because the PA under burst mode with $C_\mu = 2.00 \times 10^{-5}$ was not able to control the separation while the SJ with the same C_μ successfully suppressed the separation. Since the present study aims to clarify a common mechanism for suppressing the separated flow, it was preferred that the similar controlled flows be compared between the SJ and the PA. The PA under burst mode with $C_\mu = 5.15 \times 10^{-5}$ was able to suppress the separation, and the input disturbance of $C_\mu = 5.15 \times 10^{-5}$ is close to the minimum input controllable case using the PA.

Table 1 Computational cases. Both of the PA and the SJ are installed to the leading edge

Case name	Input momentum (C_μ)	F^+
Strong input (PA)	2.00×10^{-3}	1.0, 6.0, 10, 15, 20, 30
Strong input (SJ)	2.00×10^{-3}	1.0, 6.0, 10, 15, 20, 30
Weak input (PA)	5.15×10^{-5}	1.0, 6.0, 10, 15, 20, 30
Weak input (SJ)	2.00×10^{-5}	1.0, 6.0, 10, 15, 20, 30

The table is taken from Abe et al. [4]

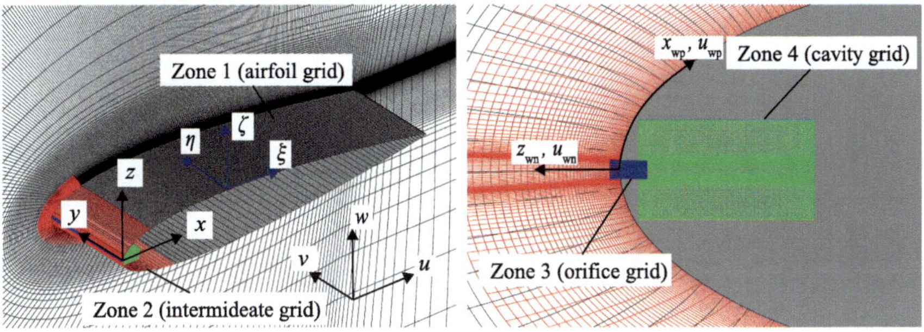

Fig. 3 Computational grids (every 5 points are visualised) and coordinates. Figures are taken from Abe et al. [3]

3 Methodology

3.1 Flow Solver

In the present study, LANS3D [16] was employed for the series of computations. The governing equations were the three-dimensional compressible Navier-Stokes equations in body-fitted coordinates. A sixth-order compact finite-difference scheme evaluated the spatial derivatives of the convective and viscous terms [20]. The metrics and Jacobian for the coordinate transformation were evaluated by the symmetric conservative forms [1, 2]. The metrics form can avoid a freestream preservation error on moving and deforming grids even with the high-order compact scheme, which was utilized for simulating the cavity deformation of the SJ. The body force and energy terms for the simulations of the PA were added to the governing equation. A tenth-order filtering [17] was used with a filtering coefficient of 0.495, and a numerical oscillation was suppressed. Time integration is conducted by a backward second-order difference formula converged by the five subiterations [31] of the lower-upper symmetric alternating direction implicit and symmetric Gauss-Seidel (ADI-SGS), and the second-order accuracy in time is ensured. An implicit LES approach (ILES [31, 32]) was adopted for turbulent modelling. The grids were constructed based on the zonal grid approach [13]. The grids for the SJ are visualised in Fig. 3 and summarised in Table 2, while the PA simulations utilised Zones 1 and 2 only. The validation and verification of the solver and grids can be found in [3, 4].

Table 2 Number of grid points. (N_ξ, N_η, N_ζ) represents the number of grid points in each direction

Zone name	Description	N_ξ	N_η	N_ζ	Number of grid points
Zone 1	Airfoil grid	795	134	179	19,068,870
Zone 2	Intermediate grid	253	134	91	3,085,082
Zone 3	Orifice grid	45	134	75	452,250
Zone 4	Cavity grid	157	134	214	4,502,132

The table is taken from Abe et al. [3]

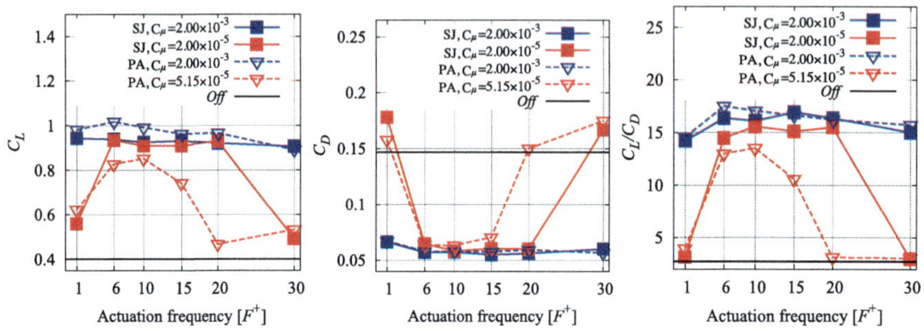

(a) C_L (b) C_D (c) C_L/C_D

Fig. 4 Time-averaged C_L, C_D, and C_L/C_D are summarized. Solid and dashed lines represent the results of the SJ and the PA, respectively. The blue and red lines are strong and weak C_μ cases, respectively. Figures are taken from Abe et al. [4]

4 Results and Discussion

4.1 Time-Averaged Flows of Controlled Cases

The horizontal axis in Fig. 4 shows the actuation frequency F^+, and the vertical axis shows the time-averaged aerodynamic coefficients. The red, blue, and black lines, respectively, show the strong, weak, and noncontrolled cases. Solid and dashed lines, respectively, show the results of the SJ and the PA. There is a localised range of frequencies where the lift-to-drag ratio (C_L/C_D) is recovered, which is not significantly different between two devices. For either strong or weak input momentum, C_L/C_D was largely recovered in both PA and SJ at approximately $6 \leq F^+ \leq 20$. To investigate the difference in the actuation frequencies, the rest of this article mostly focuses on $F^+ = 1.0$ and 6.0.

Next, the time- and spanwise-averaged chordwise velocity, u/u_∞, in the cases of $F^+ = 1.0$ and 6.0 are shown in Fig. 5. As shown in Fig. 5a–d, attached flows were obtained by both PA and SJ in the cases with the strong C_μ. A separation bubble was formed on the leading edge, with the size of the bubble being smaller in $F^+ = 6.0$ than 1.0. With the weak C_μ, the separation cannot be suppressed with $F^+ = 1.0$ as in Fig. 5e and f, while the flows are attached with the separation bubble in Fig. 5g and h.

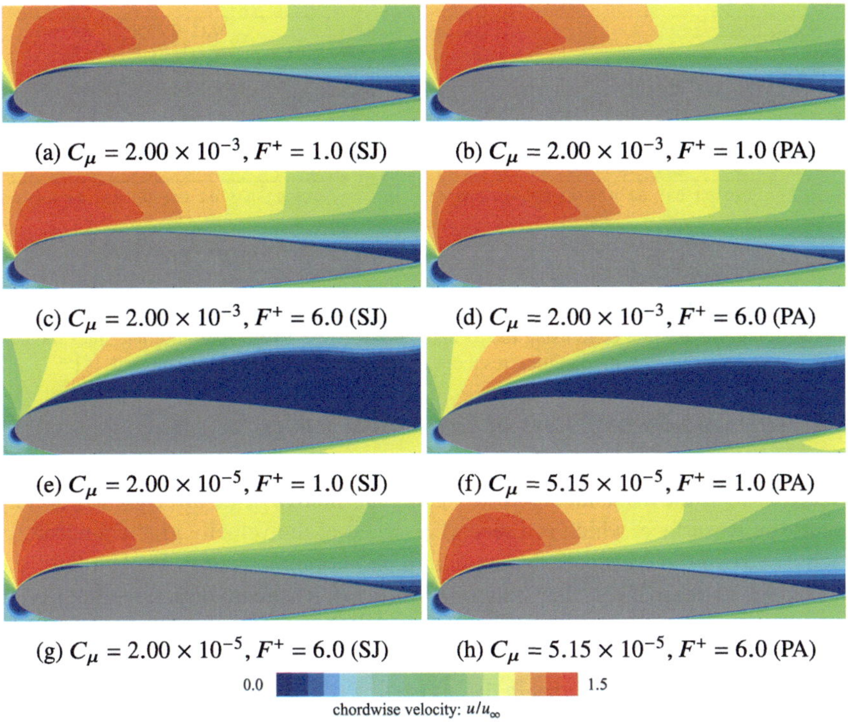

Fig. 5 Time- and spanwise-averaged fields of u/u_∞ colored between 0.0 and 1.5. Figures are taken from Abe et al. [4]

A similar trend was seen in the reversed flow area represented in Fig. 6. In Fig. 6, the distance from the wall at which the x-component of the velocity, u, becomes zero is plotted against the chordwise location. Figure 6 shows that the size of the separation bubble decreases when the SJ replaces the PA under most of the actuation conditions except for the weak input case with $F^+ = 1.0$. This difference can be explained by the position of the turbulent transition in the next subsection. Meanwhile, when the input case is weak with $F^+ = 1.0$, the separation is not suppressed by either the PA or SJ actuation. In contrast to controllable cases involving separation bubbles, the size of the reversed flow region decreases when the PA is utilised as opposed to the SJ. This paper does not include an investigation of the reversed flow region in uncontrollable cases. However, the future study might focus on investigating the present difference from the viewpoint of input fluctuation by the SJ and PA.

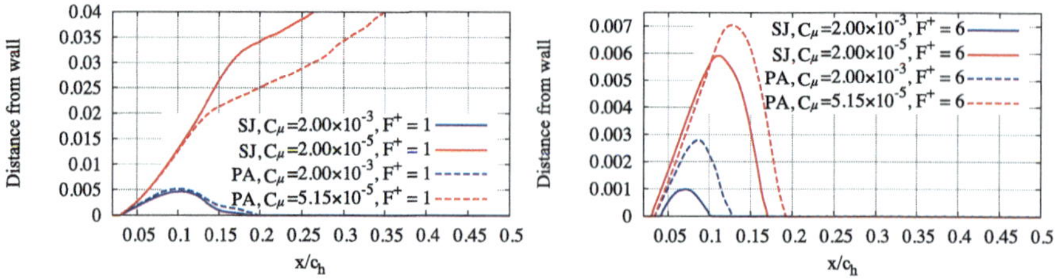

(a) $F^+ = 1.0$ (b) $F^+ = 6.0$

Fig. 6 Reversed flow region near the leading edge. Solid and dashed lines show the reversed flow region around the leading edge using the SJ and the PA, respectively. Red and blue lines indicate the weak and strong input momentum, respectively. Figures are taken from Abe et al. [4]

4.2 Instantaneous Flows of Controlled Cases

Figure 7 shows the instantaneous flow fields. The chordwise (x-directional) vorticity colors the isosurface, which is a second invariant of the velocity gradient tensor. In each case, two-dimensional (spanwise uniform) vortex structures are emitted from the separated shear layer, then convected in the downstream direction, amplifying three-dimensional fluctuations. The separated shear layer is mostly laminar when separation is suppressed. After transition, it reattaches to the airfoil surface, and the turbulent boundary layer forms in the aft portion of the airfoil. Thus, the separation bubble is a laminar separation bubble (LSB) in those cases. Indeed, the surface pressure in the LSB shows a plateau distribution [3, 4]. Note that in the cases of strong C_μ, the SJ introduces three-dimensional fluctuation near the leading edge, whereas the PA introduces only two-dimensional fluctuation, effectively enhancing turbulent transition.

It is noteworthy that in the $F^+ = 1.0$ cases of Fig. 7a and c, a large spanwise vortex structure can be slightly observed and seems to convect in the turbulent boundary layer with smaller three-dimensional vortices. This is more clearly captured in the PA case of Fig. 7c than the SJ case of Fig. 7a. Meanwhile, those spanwise structures cannot be identified in the $F^+ = 6.0$ cases from the instantaneous vortical fields. Therefore, the next subsection attempts to extract and identify those coherent vortex structures from the turbulent boundary layer by phase averaging. To assess the effect of exchanging momentum near the airfoil surface, the Reynolds shear stress is also decomposed into periodic and turbulent components according to the actuation frequency F^+.

4.3 Coherent Vortex Structures and Chordwise Momentum Exchange in Phase-Averaged Flow Fields

The instantaneous quantity $f(t)$ is decomposed into overall average $\overline{f}$; phase fluctuation $\tilde{f}$; and turbulent fluctuation f'' as follows:

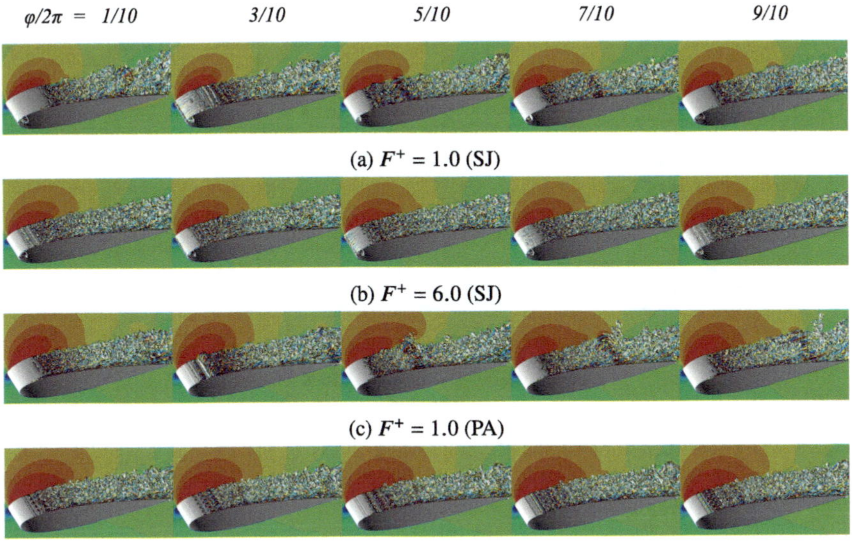

(a) $F^+ = 1.0$ (SJ)

(b) $F^+ = 6.0$ (SJ)

(c) $F^+ = 1.0$ (PA)

(d) $F^+ = 6.0$ (PA)

Fig. 7 Representative instantaneous flows in the controlled cases at different phases $\varphi/2\pi$. Isosurfaces show the second invariant of the velocity gradient tensor (colored by chordwise velocity u/u_∞ for 0.0 to 1.5.)

$$f(t) = \overline{f} + f'(t) = \overline{f} + \tilde{f}_\varphi + f''(t) = \langle f \rangle_\varphi + f''(t). \quad (2)$$

where the total and phase averaging operators are defined as $\overline{\bullet}$ and $\langle \bullet \rangle_\varphi$, respectively. Therefore, the following decomposition holds:

$$\langle f'g' \rangle_\varphi = \langle \tilde{f}_\varphi \tilde{g}_\varphi \rangle_\varphi + \langle f''g'' \rangle_\varphi, \quad (3)$$

where $\langle \tilde{f}_\varphi \tilde{g}_\varphi \rangle_\varphi$ and $\langle f''g'' \rangle_\varphi$ are called periodic and turbulent component, respectively. Replacing f and g by u and w in Eq. (3) defines the phase decomposition of the Reynolds stress $\langle u'w' \rangle_\varphi$. The spanwise average was also conducted in addition to the phase averaging procedure.

Figures 8 and 9 show $\langle \tilde{u}\tilde{w} \rangle_\varphi$ and $\langle u''w'' \rangle_\varphi$ in the phases from $\varphi/2\pi = 1/10$ to $9/10$. The black contour lines show the second invariant of the velocity gradient tensor in each phase. Figures 8 and 9 successfully extract the coherent spanwise vortex structures in the turbulent boundary layer. The number of those vortex structures depends on the actuation frequency F^+. The coherent vortices are generated downstream of the LSB near the leading edge. They formed a large single vortex in the $F^+ = 1.0$ cases of Fig. 8; six vortices in the $F^+ = 6.0$ cases of Fig. 9. Therefore, both SJ and PA generate the coherent vortex structures in the separation-controlled (attached) flows, which are concealed in the turbulent boundary layer but can be successfully identified by the phase-averaging procedure.

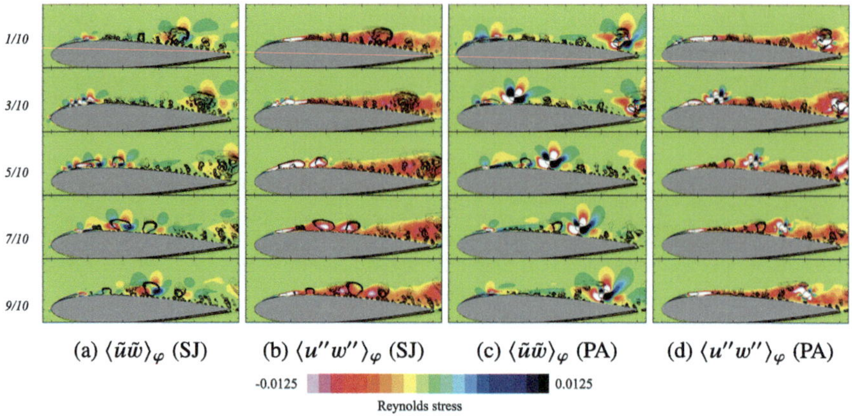

Fig. 8 Phase decomposition of the Reynolds shear stress in the case of $C_\mu = 2.0 \times 10^{-3}$, $F^+ = 1.0$. Contour color shows periodic $\langle\tilde{u}\tilde{w}\rangle_\varphi$ and turbulent $\langle u''w''\rangle_\varphi$ components. Black contours show the second invariant of the velocity gradient tensor in the phase- and spanwise-averaged flow field. Figures are taken from Abe et al. [4]

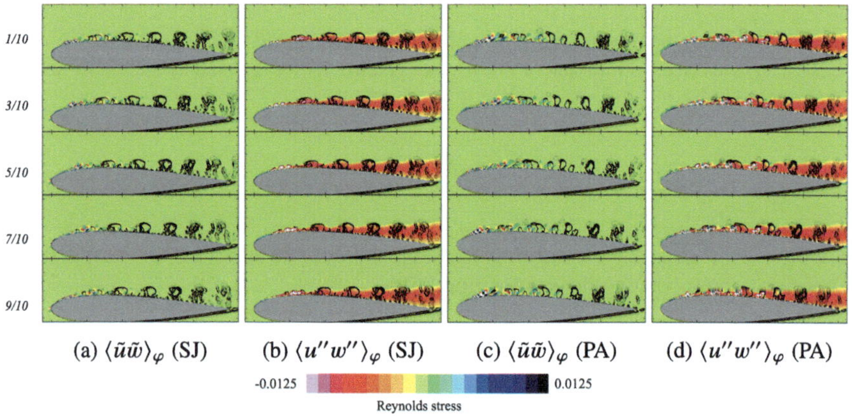

Fig. 9 Phase decomposition of the Reynolds stress in the case of $C_\mu = 2.0 \times 10^{-3}$, $F^+ = 6.0$. The description follows Fig. 8. Figures are taken from Abe et al. [4]

The contour color illustrates that the turbulent component $\langle u''w''\rangle_\varphi$ is dominant over most of the airfoil surface for both PA and SJ. The turbulent component is also locally amplified around the coherent vortices, which are entrained by the coherent vortices in the downstream direction. On the other hand, the periodic component $\langle\tilde{u}\tilde{w}\rangle_\varphi$ is stronger in the PA cases than in the SJ cases (see Fig. 8a and c for examples). This is caused by the spanwise-uniform fluctuation introduced by the PA, wherein the PA does not induce three-dimensional fluctuation and thus the spanwise vortex remains strongly. The strength of such a coherent vortex can affect the entrainment of the turbulent component of the Reynolds shear stress. Indeed, the turbulent components around the coherent vortices in Figs. 8d and 9d are stronger than they are in Figs. 8b and 9b, respectively. Such a difference leads to a more effective exchange of chordwise momentum using the PA, and thus the aerodynamic performances are better in the PA than in the SJ, as shown in Fig. 4.

In summary, the phase decomposition of the Reynolds shear stress in the separation-controlled flows revealed that the turbulent component mainly contributes to the effective exchange of the chordwise momentum in both SJ and PA cases. Meanwhile, in the strong C_μ cases, the coherent vortex structures and periodic component of the Reynolds shear stress are stronger in the PA than in the SJ due to the spanwise-uniform fluctuation induced by the PA. Since such coherent vortex structures strongly remain in the downstream direction, the flow is better suppressed by PA than SJ in the strong C_μ cases.

4.4 Formation of the Coherent Vortex

In this subsection, the formation of the coherent vortex is discussed for the strong input case with $F^+ = 6.0$ with the SJ. Figure 10 shows a space-time combined visualization of the coherent vortex motion and the turbulent component of the Reynolds stress. The φ-axis indicates the phase angle based on $F^+ = 6.0$ for $0 \leq \varphi \leq 8\pi$ (four periods), and the x/c_h-axis indicates the spatial chordwise direction. At each phase, the phase- and span-averaged flow fields are visualised, where the isosurface shows a second invariant of the velocity gradient tensor. Therefore, the isosurface in Fig. 10 shows the convection of the coherent vortex structures, which are illustrated by the black contours in Fig. 9a and b. The isosurface is colored by the turbulent component of the Reynolds stress at each phase: $-\langle u''w''\rangle_\varphi$.

In Fig. 10a, the LSB near the leading edge periodically expands and contracts, from which the small vortex structures are emitted. Those small vortex structures convect in the downstream direction, and two of them are merging at $x/c_\mathrm{h} \simeq 0.25$ and $\varphi \simeq 4\pi$ while the other structures are gradually dissipated, which eventually form a single coherent vortex (see the black-dotted circle). The resultant coherent vortex remains strong further downstream and is then periodically released from the trailing edge. As such, there are clear merging and diffusion processes of small vortex structures that are originally generated from the separated shear layer, which form a single coherent vortex in the period of F^+. Note that similar coherent vortex structures were identified in the recent experimental studies [24, 34].

4.5 Role of the Coherent Vortex

A coherent vortex within the turbulent boundary layer has been identified so far in controlled cases at AoA=12° through the application of a phase and spanwise averaging procedure. Subsequently, the significance of the coherent vortex was primarily

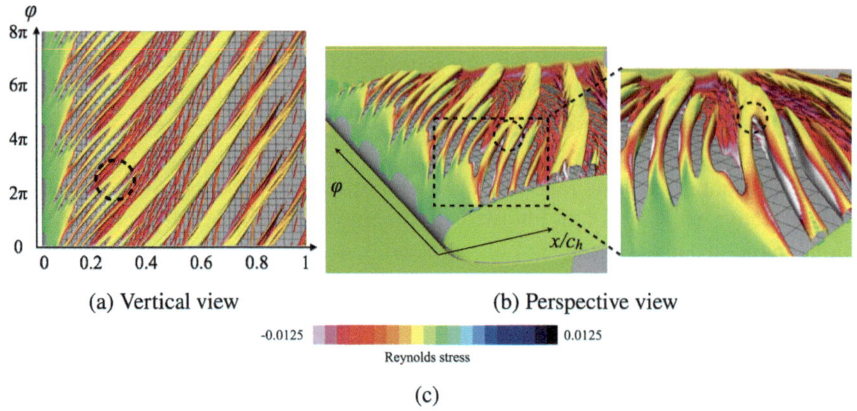

Fig. 10 Time-space visualisation of the convecting coherent vortex structures in the strong input case ($C_\mu = 2.0 \times 10^{-3}$) with $F^+ = 6.0$. φ indicates the phase between 0 and 8π (four periods), and x/c_h indicates the chordwise coordinate. An isosurface represents a second invariant of the velocity gradient tensor of the phase- and spanwise-averaged flows (of which value is 1.0 and normalized by the freestream velocity), i.e., $\langle u\rangle_\varphi$, $\langle v\rangle_\varphi$, and $\langle w\rangle_\varphi$, which is colored by the turbulent component of the Reynolds stress at each phase, $-0.0125 \le -\langle u''w''\rangle_\varphi/u_\infty^2 \le 0.0125$. Figures are taken from Abe et al. [3]

examined in terms of the decomposition of the Reynolds stress and the entrainment of momentum from the freestream.

In contrast, certain lower AoA conditions utilising the PA control, such as AoA=10°, demonstrated the maintenance of laminar coherent vortices in the aft portion of the suction surface. The angle of attack of 10 degrees is slightly below the stall angle. At this AoA, the flow remains attached to the surface with a small separation near the leading edge, without any control. As shown in Fig. 11a, the laminar coherent vortices can be identified even in instantaneous flows when it is under control. Similar coherent structures can also be observed in high-performance airfoils without control in a low Reynolds number regime.

Meanwhile, the controlled flow becomes turbulent at higher AoA values, such as 14 degrees in Fig. 11b. The AoA= 14° is greater than the stall angle, at which point the flow separates in the vicinity of the leading edge without control. Comparable to the AoA=12° cases that were examined in previous sections, coherent vortices are generated in the turbulent boundary layer when the PA control is applied.

As such, the form of the coherent vortices might change depending on the flow conditions, with both laminar and turbulent structures seen in distinct AoA cases. It should be noted, however, that a more comprehensive analysis of the control strategy could be achieved through the careful observation of the coherent vortices and controlled flow in addition to systematic or statistical turbulent analyses, which are not elaborated upon in our discussion. This paper primarily examined the coherent vortices in the turbulent boundary layer and highlighted their significance in terms of turbulent mixing. Hence, the role of those vortices should be carefully addressed in the different regimes, such as the cases with laminar coherent vortices.

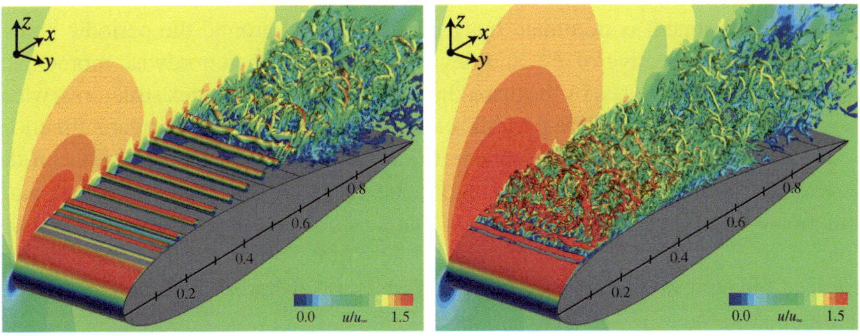

(a) 10°, F^+ = 12 (PA) (b) 14°, F^+ = 18 (PA)

Fig. 11 Representative instantaneous flows in the controlled cases at AoA=10° and 14°. Isosurfaces show the second invariant of the velocity gradient tensor (colored by the chordwise velocity u/u_∞ for 0.0 to 1.5.) Figures are taken from Ogawa et al. [22]

5 Conclusions

This study investigated numerically the separation control mechanisms of the DBD-PA (PA) under burst mode and the SJ around a NACA0015 airfoil with a chord-Reynolds number of 63, 000. Both PA and SJ were installed on the leading edge with the same order of input momentum ($C_\mu = \mathcal{O}(10^{-3})$ and $\mathcal{O}(10^{-5})$) and the same actuation frequencies between $F^+ = 1.0$ and 30.

The optimum range for F^+ is between 6.0 and 20 for the current flow and actuation conditions. This range achieves a high lift-to-drag ratio, C_L/C_D, in both the time- and spanwise-averaged flow fields. Typically, the controlled (attached) flows have an LSB near the leading edge, and a turbulent boundary layer develops downstream of the LSB. Promoting a turbulent transition to minimise the size of the LSB near the leading edge is therefore one of the most effective methods for mitigating separation and enhancing aerodynamic performance. Furthermore, in these separation-controlled flow fields, the LSB expands and contracts with a frequency of F^+, releasing small spanwise-uniform vortex structures periodically. These structures are eventually diffused and merged to form a large periodic coherent vortex, which is clearly visualised in the phase- and spanwise-averaged flow fields.

Regarding the mechanism to maintain the attached flows, it is important to continuously inject momentum in the chordwise direction near the airfoil surface. This is achieved by entraining the chordwise momentum from the freestream that exists above the turbulent boundary layer (chordwise momentum exchange) because the present input momentum C_μ is too small to directly compensate for the lack of chordwise momentum in the separated flow, i.e., $\mathcal{O}(10^{-5}\text{-}10^{-3})$ of the freestream momentum, and thus, it cannot directly alter the direction of the freestream. The chordwise momentum exchange was evaluated by the phase decomposition (based on F^+) of the Reynolds stress into periodic and turbulent components, where the turbulent component is dominant but locally enhanced around the periodic coherent vortices, irrespective of F^+ and devices. Therefore, the chordwise momentum exchange is mainly caused by the three-dimensional turbulent vortex structures, while they are also convected by the periodic coherent vortex in the downstream direction.

Meanwhile, the form of the coherent vortices might change depending on the flow conditions, with both laminar and turbulent structures seen in distinct AoA situations. Therefore, it should be noted that a more comprehensive analysis of the control strategy can be achieved through the careful observation of the coherent vortices and controlled flow in addition to systematic or statistical turbulent analyses introduced so far.

We believe that these mechanisms for separation control are applicable to more general devices as long as they are operated with a small input momentum, such as $C_\mu = \mathcal{O}(10^{-3} - 10^{-5})$, and the separated flow has a Reynolds number of Re $= \mathcal{O}(10^4)$ and a low AoA.

Acknowledgements This research was supported by a JSPS Grant-in-Aid for Scientific Research No. 18H03816 as well as No. 24K01074 in Japan. This work was also supported by MEXT as "Program for Promoting Researches on the Supercomputer Fugaku" (JPMXP1020230320) and by JST FOREST program (JPMJFR2124).

References

1. Abe, Y., Iizuka, N., Nonomura, T., Fujii, K.: Conservative metric evaluation for high-order finite difference schemes with the GCL identities on moving and deforming grids. J. Comput. Phys. **232**, 14–21 (2013)
2. Abe, Y., Iizuka, N., Nonomura, T., Fujii, K.: Geometric interpretations and spatial symmetry property of metrics in the conservative form for high-order finite-difference schemes on moving and deforming grids. J. Comput. Phys. **260**, 163–203 (2014)
3. Abe, Y., Nonomura, T., Fujii, K.: Flow instability and momentum exchange in separation control by a synthetic jet. Phys. Fluids **35**, 065114 (2023)
4. Abe, Y., Nonomura, T., Sato, M., Aono, H., Fujii, K.: Comparison of separation control mechanisms for synthetic jet and plasma actuator. Actuators **12**(8), (2023)
5. Amitay, M., Glezer, A.: Role of actuation frequency in controlled flow reattachment over a stalled airfoil. AIAA J. **40**(2), 209–216 (2002). February
6. Aono, H., Sekimoto, S., Sato, M., Yakeno, A., Nonomura, T., Fujii, K.: Computational and experimental analysis of flow structures induced by a plasma actuator with burst modulations in quiescent air. Mech. Eng. J. **2**(4), (2015)
7. Asada, K., Ninomiya, Y., Oyama, A., Fujii, K.: Airfoil flow experiment on the duty cycle of dbd plasma actuator. In: *AIAA-2009-531* (2009)
8. Benard, N., Moreau, E., Griffin, J., L. N. C.: III. Plasma flow control - autonomous lift improvement by slope-seeking. In: *AIAA 2009-4182* (2009)
9. Benton, S.I., Visbal, M.R.: Extending the Reynolds number range of high-frequency control of dynamic stall. AIAA J. **57**(7), 2676–2681 (2019)
10. Chen, F., Yao, C., Beeler, G., Bryant, R., Fox, R.: Development of synthetic jet actuators for active flow control at nasa langley. In: *AIAA-2000-2405* (2000)
11. Corke, T., Post, M., Orlov, D.: Single dielectric barrier discharge plasma enhanced aerodynamics: physics, modeling and applications. Exp. Fluids **46**, 1–26 (2009). https://doi.org/10.1007/s00348-008-0582-5

12. Corke, T.C., Enloe, C.L., Wilkinson, S.P.: Dielectric barrier discharge plasma actuators for flow control, Annual Review of Fluid Mechanics (2010)
13. Fujii, K.: Unified zonal method based on the fortified solution algorithm. J. Comput. Phys. **118**, 92–108 (1995)
14. Fujii, K.: High-performance computing-based exploration of flow control with micro devices. Phylosophical Transactions of The Royal Society A **372**, 20130326 (2014)
15. Fujii, K.: Three flow features behind the flow control authority of DBD plasma actuator: Result of high-fidelity simulations and the related experiments. Appl. Sci. **8**(4), 546 (2018)
16. Fujii, K., Endo, H., Yasuhara, M.: Activities of Computational Fluid Dynamics in Japan: Compressible Flow Simulations, High Performance Computing Research and Practice in Japan. JOHN WILEY & SONS, (1990)
17. Gaitonde, D.V., Visbal, M.R.: Padé-type higher-order boundary filters for the navier-stokes equations. AIAA J. **38**(11), 2103–2112 (2000)
18. Greenblatt, D., Wygnanski, I.J.: The control of flow separation by periodic excitation. Prog. Aerosp. Sci. **36**, 487–545 (2000)
19. Kotsonis, M., Ghaemi, S., Veldhuis, L., Scarano, F.: Measurement of the body force field of plasma actuators. J. Phys. D Appl. Phys. **44**, 045204 (2011)
20. Lele, S.K.: Compact finite difference schemes with spectral-like resolution. J. Comput. Phys. **103**(1), 16–42 (1992)
21. Ogawa, T., Asada, K., Sato, M., Tatsukawa, T., Fujii, K.: Computational study of the plasma actuator flow control for an airfoil at pre-stall angles of attack. Appl. Sci. **12**(18), 9073 (2022)
22. Ogawa, T., Asada, K., Yakeno, A., Fujii, K.: Revisiting burst drive conditions of dbd plasma actuator for airfoil flow control. In *AIAA Scitech 2024*, Orlando, Florida, 8–12 January (2024)
23. Okada, K., Oyama, A., Fujii, K., Miyaji, K.: Computational study of effects of nondimensional parameters on synthetic jets. Trans. Japan Soc. Aeronaut. Space Sci. **55**(1), 1–11 (2012)
24. Rathay, N., Amitay, M.: Interaction of synthetic jets with a massively separated three-dimensional flow field. Phys. Rev. Fluids **7**, 034702 (2022). Mar
25. Sato, M., Aono, H., Yakeno, A., Nonomura, T., Fujii, K.: Multifactorial effects of operating conditions of dielectric-barrier-discharge plasma actuator on laminar-separated-flow control. AIAA J. **53**(9), 2544–2559 (2015)
26. Sato, M., Nonomura, T., Okada, K., Asada, K., Aono, H., Yakeno, A., Abe, Y., Fujii, K.: Mechanisms for laminar separated-flow control using DBD plasma actuator at low reynolds number. Phys. Fluids **27**(11), 117101 (2015)
27. Sato, M., Okada, K., Asada, K., Aono, H., Nonomura, T., Fujii, K.: Unified mechanisms for separation control around airfoil using plasma actuator with burst actuation over Reynolds number range of 10^3-10^6. Physics of Fluids, **32**(2), (2020)
28. Sekimoto, S., Nonomura, T., Fujii, K.: Burst-mode frequency effects of dielectric barrier discharge plasma actuator for separation control. AIAA J. **55**(4), 1385–1392 (2017)
29. Smith, B.L., Glezer, A.: The formation and evolution of synthetic jets. Phys. Fluids **10**, 2281 (1998)
30. Smith, B.L., Glezer, A.: Jet vectoring using synthetic jets. J. Fluid Mech. **458**, 1–34 (2002)
31. Visbal, M.R., Rizzetta, D.P.: Large-eddy simulation on general geometries using compact differencing and filtering schemes. In: AIAA-2002-288, (2002)
32. Visbal, M., Morgan, P., Rizzetta, D.: An implicit les approach based on high-order compact differencing and filtering schemes. In: AIAA 2003-4098, June (2003)
33. Wang, J.J., Choi, K.S., Feng, L.H., Jukes, T.N., Whalley, R.D.: Recent developments in DBD plasma flow control. Prog. Aerosp. Sci. **62**, 52–78 (2013)
34. Yang, E., Ekmekci, A., Sullivan, P.: Phase evolution of flow controlled by synthetic jets over NACA0025 airfoil. J. Visualization **25**, 751–765 (2022)
35. You, D., Moin, P.: Active control of flow separation over an airfoil using synthetic jets. J. Fluids Struct. **24**, 1349–1357 (2008)
36. Zhang, W., Samtaney, R.: A direct numerical simulation investigation of the synthetic jet frequency effects on separation control of low-re flow past an airfoil. Phys. Fluids (2015)

Effects of LES/RANS Hybrid Interface Location on Oscillating Shockwave in Transonic Buffeting Flow

K. Kitamura, Y. Tsukamoto, Z.-N. Wang, F. M. Möller, and P. G. Tucker

Abstract A comparative study varying both the hybrid interface location and grid spacing was conducted on a transonic buffeting flow involving an unsteady shock-wave/turbulent-boundary-layer interaction by Zonal DES, a hybrid RANS/LES method. It was demonstrated that the hybrid interface as high as 1.5 times the boundary-layer thickness was a good choice for reproducing the unsteady nature of the buffet, on the contrary to the typically recommended value of y^+ around 100 (corresponding to 3.4% the boundary-layer thickness). With this interface location of y^+ being 60, the grid spacing recommended for flat plate boundary-layer or even the reduced cell size led to notorious, early grid-induced separation upstream the shock.

K. Kitamura (✉)
Faculty of Engineering, Yokohama National University, 79-5 Tokiwadai, Hodogaya, Yokohama 240-8501, Kanagawa, Japan
e-mail: kitamura@ynu.ac.jp

Department of Engineering, University of Cambridge, Cambridge CB2 1QA, UK

Y. Tsukamoto
Graduate School of Engineering Science, Yokohama National University, 79-5 Tokiwadai, Hodogaya, Yokohama 240-8501, Kanagawa, Japan
e-mail: tsukamoto-yuta-tz@ynu.jp

Z.-N. Wang
College of Engineering and Physical Sciences, University of Birmingham, Birmingham B15 2TT, UK
e-mail: z.n.wang@bham.ac.uk

F. M. Möller
Institute of Test and Simulation for Gas Turbines, German Aerospace Center (DLR), Linder Höhe, 51147 Cologne, Germany
e-mail: felix.moeller@dlr.de

P. G. Tucker
Cambridge University Engineering Department, University of Cambridge, Cambridge CB2 1PZ, UK
e-mail: pgt23@cam.ac.uk

J. C. Tyacke and N. R. Vadlamani (eds.), *Proceedings of the Cambridge Unsteady Flow Symposium 2024*, https://doi.org/10.1007/978-3-031-69035-8_17

1 Introduction

Turbulent flow simulations are of high importance both in understanding fundamental fluid physics and in industrial applications [1]. Since turbulence simulations are typically very expensive, there are many modelling methods according to the level of fidelity [2] so that users can pick (they believe) the best suitable one. Among them, hybrid LES (Large Eddy Simulation)/RANS (Reynolds-Averaged Navier–Stokes) simulations are now gaining popularity owing to their economical treatment of attached boundary-layers by RANS and accurate modelling of flows away from the wall by LES. There are variety of such hybridized methods in literature (as reviewed extensively by [3]). Among them, we will focus on a variant of a widely-used DES (Detached Eddy Simulation), namely, ZDES (Zonal DES) [4]. Wall-Modelled LES (WMLES) also falls into a hybrid method [5] and attracting many practitioners. While it provides us with very accurate solutions, it is reportedly 10,000 times more expensive than DES-type methods [6]. Tucker [7] stated that WMLES is "not suitable when complex surface topologies of small scale are being considered," for our potential applications in the future. These are why we will employ ZDES rather than WMLES.

An important aspect of a hybrid RANS/LES is where and how to set a hybrid interface location, which can be manually selected in ZDES whereas automatically determined (and hence, cannot be controlled) in DES or DDES (Delayed DES). Research in this direction has been actively conducted [5]. Deck et al. [8, 9] compared flat plate turbulent boundary-layer cases with the interface located at "$y^+ = 100$," "$y^+ = 200$" and "0.1δ" (they are close to each other, as observed in Fig. 1) but did not mention which was better (where the superscript " $+$ " refers to a wall unit; δ is the 99% velocity boundary-layer thickness). Tucker and Davidson [10] examined the effects of the interface location and concluded that around $y^+ \sim 100$ be the best in most cases, which is later employed in their Zonal LES by [11, 12]. Those values correspond well with WMLES guideline in which inner-layer of the boundary-layer should be treated by a wall-model while the outer-layer is supposed to be resolved by LES [5, 13]. Nevertheless, these guidelines are established for shock-less, incompressible flows. In fact, for transonic buffet problems (involving shock-wave/turbulent-boundary-layer interactions), Deck et al. [4, 9] and Ishida et al. [14] recommended much higher value of $O(\delta)$ (e.g., 1.5δ, corresponding with $y^+ \sim$ 2,000, see Fig. 1). This choice appears reasonable to significantly save the otherwise prohibitively huge computational cost. It also helps a straightforward extension to more complex geometries, such as installation of aerodynamics devices over the wing (e.g., vortex generators) whose height being $O(\delta)$ [15]. We will come back later to this topic.

As for selecting grid spacings, Tucker provided a useful guideline for RANS, hybrid RANS/LES, LES, and DNS (Direct Numerical Simulation) in his book [16], based on an incompressible turbulent boundary-layer simulation over a flat plate (Table 1): This table tells us how small the grid spacing should be for accurate turbulent simulations. On the other hand, as reviewed by Deck and Renard [17], too

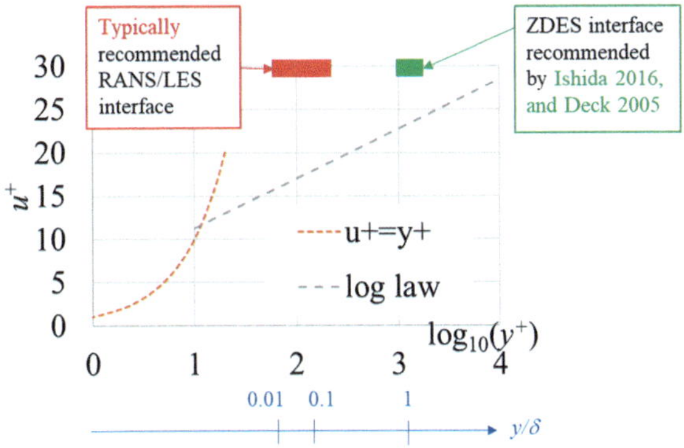

Fig. 1 Typical velocity profile versus wall-distance in turbulent-boundary-layer

Table 1 Grid spacing requirements by Tucker (adapted from [16])

Method	Δx^+ (streamwise)	$\min(\Delta y^+)$ (wall-normal)	Δz^+ (spanwise)
DNS	10–15	1	5
Wall-Resolved LES	50–130	1	15–30
Wall-Modelled LES	100–600	30–150	100–300
Hybrid RANS/LES	100–600	1	100–300
RANS	1000	1	

small spacing (say, $\Delta x < 0.3\delta$) can lead to undesirable switching from RANS to LES modes for DDES (and even ZDES [18]), and eventually "severely underprediction" of skin friction and "unphysically early flow separation". In addition, their applicability to shock-present flows is questionable. Furthermore, as will be demonstrated later in this paper, these two criteria contradict with each other at least for a particular flow condition selected here for a transonic buffet over an airfoil. Therefore, such criteria appear to be problem-dependent, and there may not be a universally acceptable one.

As reviewed above, specifically for hybrid RANS/LES methods, the hybrid interface location appears to have played a major role in the performance of the method, as well as the grid spacing. Their interrelations, however, have not been clearly investigated. This motivates our present work, with the expectation of eventually offering (i) a more specific, reliable, and accurate guideline for both the hybrid interface location and grid spacing when using a hybrid method, and then (ii) a clue for developing a new RANS/LES hybrid turbulence modelling, in the future.

We will pay a particular attention to a challenging, transonic buffeting flow which involves unsteady shock-motion and its interaction with a separating turbulent boundary-layer (see [19] for a more comprehensive review). However, we will not go on to the detailed inspection of the flow physics, as conducted in [17]; instead, we will try to capture the key physics economically by employing ZDES with a carefully selected set of hybrid interface position and grid spacing.

2 Numerical Setup

2.1 Governing Equations and Discretization

The governing equations are the three-dimensional compressible Navier–Stokes equations, discretized by a cell-centered finite volume method. More specifically, we utilized the FaSTAR (FaST Aerodynamic Routines) code developed by Japan Aerospace Exploration Agency (JAXA) [20]. FaSTAR is equipped with a variety of numerical methods available, but we selected the Green–Gauss-based weighted least squares (GLSQ) method [21] for gradient reconstruction; monotone upstream-centered schemes for conservation laws (MUSCL) interpolation for second-order spatial accuracy [22] along with Hishida (van Leer type) slope limiter [23]; the simple low-dissipation advection upstream splitting method (SLAU) as a numerical flux [24]; the lower–upper symmetric Gauss–Seidel (LU-SGS)-type implicit method [25] coupled with a second-order accurate dual time-stepping [26] for the time integration. The number of inner iterations was set to four, with the local Courant–Friedrichs–Lewy (CFL) numbers being 130 near the wall, and the time step was fixed at $dt = 0.035$ in dimensionless time.

The turbulence is modelled by Zonal DES, a hybrid RANS/LES method proposed by Deck [4]. In the Zonal DES method, the RANS region (and the LES model outside) can be manually defined by the user. Specifically, the wall distance d in SA (Spalart–Allmaras noft2 [27])-RANS is replaced with the length scale of Zonal DES: l_{ZDES}, where the mode switching of RANS and LES length scale is defined by

$$l_{ZDES} = \begin{cases} \mathrm{l_{RANS}} & \text{if } \mathrm{d} \le \mathrm{d_{input}} \\ \mathrm{l_{LES}} & \text{if } \mathrm{d} \ge \mathrm{d_{input}} \end{cases} \tag{1}$$

$\mathrm{l_{RANS}}$ is the length scale of RANS (i.e., $\mathrm{l_{RANS}} = d$: wall distance), $\mathrm{l_{LES}}$ is the length scale of LES ($\mathrm{l_{LES}} = \mathrm{C_{DES}}\Delta$), and d_{input} is set by the user. C_{DES} is the model constant, which is set to 0.65 as calibrated by Shur et al. [28]. The grid scale Δ is given by

$$\Delta = \min[\max\{\mathrm{c_w d}, 2\mathrm{c_w}\Delta_{max}, 2\Delta_{wn}\}, 2\Delta_{max}], \tag{2}$$

where Δ_{max} is the largest distance from cell-face to cell-center, and Δ_{wn} is one half of the grid scales in normal direction to the wall surface. $\mathrm{c_w}$ is the model constant, which is 0.15 in FaSTAR[28].

Figure 2 presents the model switching region with the user-defined d_{input}. As discussed earlier, the location of the RANS/LES hybrid interface remains a subject of discussion (as in the current work), with no definitive guidelines for shock-induced, separated boundary layers, such as those encountered in transonic buffeting flows. Hence, the RANS region thickness can be determined based on the methodology from previous research by Ishida et al. [14]. For our purposes, this corresponds to a thickness of 3.0 mm (1.5δ; $y^+ \sim 2000$).

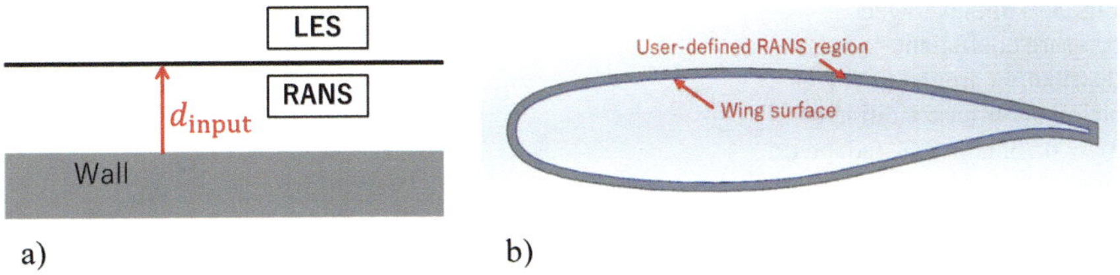

Fig. 2 Zonal DES switching schematic [15]: **a** RANS/LES switching distance as defined by the user and **b** depiction of the RANS region around the airfoil

2.2 OAT15A Supercritical Airfoil Geometry, Characteristics, and Grid

In its experimental model by Jacquin et al. [29], the chord length is $c = 230$ mm, with a maximum thickness-to-chord ratio 12.5% and a thick trailing edge having 0.50% of the chord length. The buffet onset occurred when the angle of attack exceeded $\alpha = 3.25°$, with a freestream Mach number $M = 0.73$ and Reynolds number $Re = 3.0 \times 10^6$. The main buffet frequency was observed to be around 70 Hz. As a buffet condition in this computational study, a combination of $M = 0.73$, $Re = 3.0 \times 10^6$, and $\alpha = 3.5°$ is selected.

In this study, we first created an O-topology, two-dimensional grid (112,590 cells), and then extruded it in the spanwise direction (122 cells). The near-wall grid spacing is set at $4.3 \times 10^{-6}\ c$, ensuring $y^+ < 1$. They also satisfy the grid spacing requirements of hybrid RANS/LES in other directions, i.e., $\Delta x^+ < 600$ (streamwise) and $\Delta z^+ < 300$ (spanwise) [16]. The grid convergence was already confirmed, along with its validation test by comparison with experimental results in [15].

3 Transonic Buffet Simulation Results

3.1 Baseline Case

A key flow features with a brief explanation is duplicated from [15], as a baseline case in the present work. Periodic oscillations of a shock were observed. The pressure coefficient distributions (time-averaged over three oscillation periods) are displayed in Fig. 3, along with the measured values [29]. The plateau within $0.05 < x/c < 0.35$ and the shock position in the range of $0.35 < x/c < 0.55$ are accurately represented. A source of the slight discrepancy can be explained as a much shorter duration for time-averaging in the computation (as in [17]), compared with 2300 periods in the experiment. More details are explained in [15].

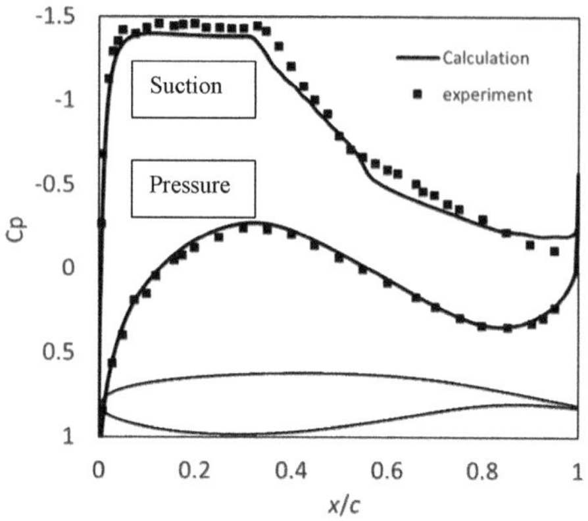

Fig. 3 Time-averaged pressure coefficient distribution around OAT15A airfoil at buffet condition of $M = 0.73$, $\alpha = 3.5°$ (adapted from [15])

3.2 Effects of Hybrid Interface Location

Here let us try to lower the hybrid interface location from 3.0 mm (1.5δ, corresponding with $y^+ \sim 2000$) (in the baseline) to 0.088 mm (0.034δ, corresponding with $y^+ = 60$ as recommended for shock-less flows by many researchers, e.g., [11]) on the same mesh. Note that these values can be obtained from the baseline simulation: For example, the (99% velocity) boundary-layer thickness, 2.0 mm, is probed from the baseline solution [14] at 0.56*c* over the upper surface at the foot of the shock (we confirmed that this thickness was not far away from 2.6 mm, which was obtained from an empirical fit curve, $0.37x/Re_x^{1/5}$ [30], for an incompressible, flat plate turbulent-boundary-layer). After some manipulation, this thickness can be converted to the wall-unit values as above. These values are summarized in Table 2, including the cases of full Unsteady RANS (URANS), and also "0.1δ" [8] which is *not* actually employed here (but shown only for the comparison purpose), falling between the Baseline and "$y^+ = 60$" cases. Note also that once the interface height is determined at a representing point (the shock location in this study), it is applied to all the locations surrounding the airfoil (Fig. 2).

Figure 4 shows the time histories of lift coefficient for the three cases of Baseline, "$y^+ = 60$", and URANS. The Baseline case exhibits clear periodicity, whereas "$y^+ = 60$" does not. The URANS case showed almost no change in time, meaning that it failed to reproduce the buffet phenomenon.

Figure 5 shows Mach number contours at four instants within one cycle of the Baseline case, demonstrating successful simulation of the back-and-forth motion of the shock ($0.35 < x/c < 0.55$). Figure 6 is the Mach contours at the same set of timesteps for "$y^+ = 60$" case. As observed, the flowfields are significantly different from Fig. 5, specifically at the separation location upstream the shock (at around $x/c = 0.3$). This is an indication of notorious "grid-induced separation (GIS)" for DES-type methods. In fact, a very similar solution was also observed in our preliminary DDES case (not shown) [15]. URANS case in Fig. 7 shows almost no sign of flow unsteadiness.

Table 2 Hybrid interface height

Case	dimensional value [mm]	Ratio to boundary-layer thickness (δ)	wall-unit (y^+)	Brief Description
URANS	∞	∞	∞	Full RANS
Baseline	3.0	1.5	2,000	Above the boundary-layer edge
"0.1δ" (*Not conducted, but shown only for reference purpose*)	(0.26)	(0.1) [8]	(177)	(Near the bottom of the outer layer (log-layer))
"$y^+ = 60$"	0.088	0.034	60	Bottom of the outer layer

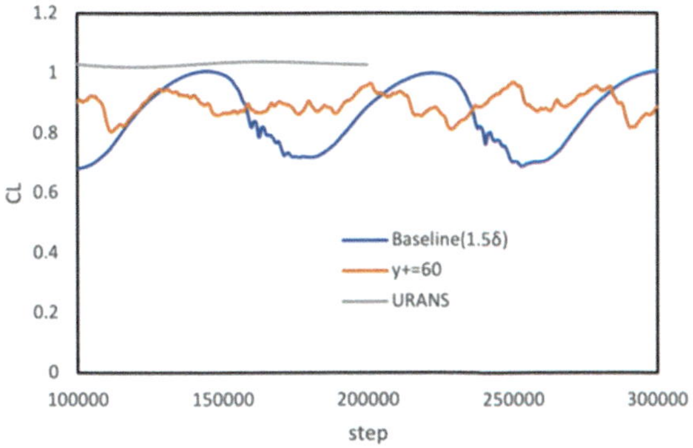

Fig. 4 Time histories of lift coefficients

The GIS is known to be caused by too small (but not small enough for LES) grid spacing in the attached boundary-layer, in which RANS is supposed to work but actually (unintentionally) switched to LES mode [17, 18]. As a result, the wall skin-friction is underestimated, and the separation line moves far upstream, as evidenced in time-averaged values (over three periods) in Fig. 8. Figure 9 shows streamwise velocity distributions upstream the shock, illustrating that Baseline and URANS solutions are almost identical at this place. This indicates that in the Baseline case the boundary-layer upstream the shock is entirely treated by RANS as desired, whereas not in the "$y^+ = 60$" case. It is also worth noting that the turbulence tripping was introduced at $x/c = 7\%$ in [17], whereas not in the present work (but the natural transition was simulated by the SA-noft2 RANS model instead). Since such a tripping will effectively feed the turbulence energy into the boundary-layer, leading it to remain attached potentially, this will be examined in the future work.

It is argued that the hybrid interface, or also called "grey area" [31, 32], is a transition zone and hence, needs a special care to make the transition smooth [11, 12, 33] because otherwise it can contaminate the flowfield. From this point of view,

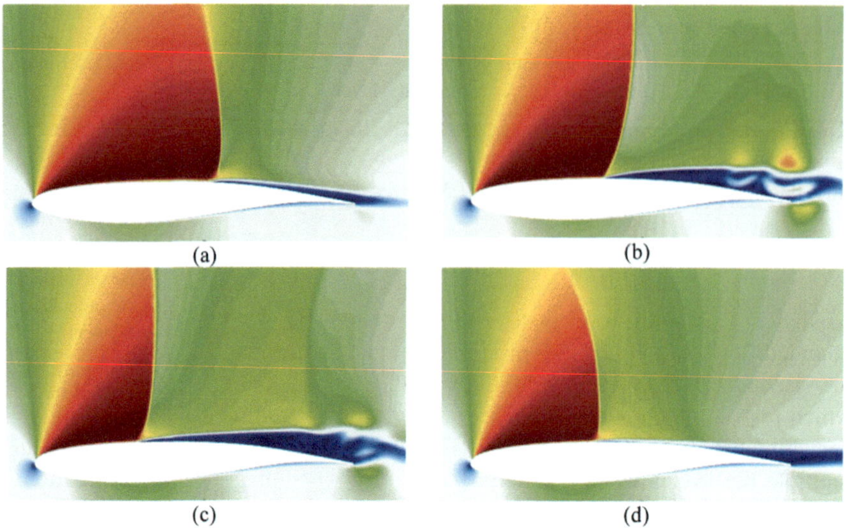

Fig. 5 Mach number contours at **a** 150,000 steps, **b** 166,000 steps, **c** 180,000 steps, and **d** 196,000 steps (Baseline)

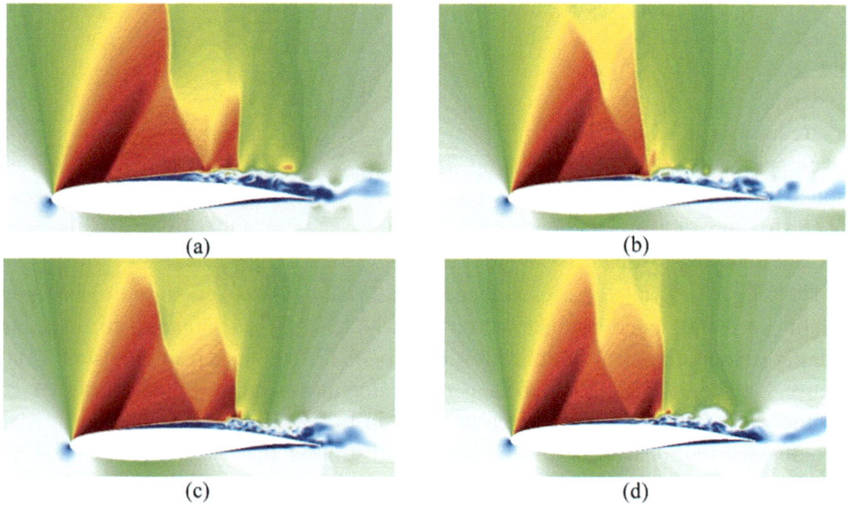

Fig. 6 Mach number contours at **a** 150,000 steps, **b** 166,000 steps, **c** 180,000 steps, and **d** 196,000 steps ($y^+ = 60$)

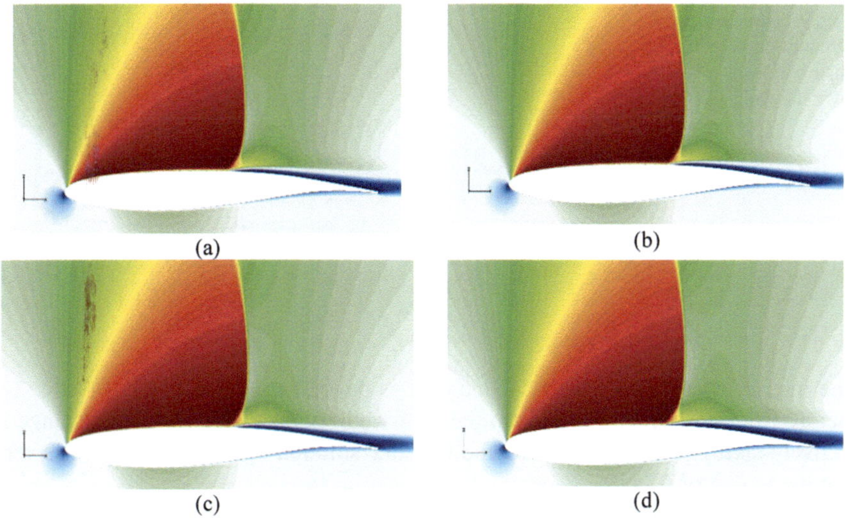

Fig. 7 Mach number contours at **a** 150,000 steps, **b** 166,000 steps, **c** 180,000 steps, and **d** 196,000 steps (URANS)

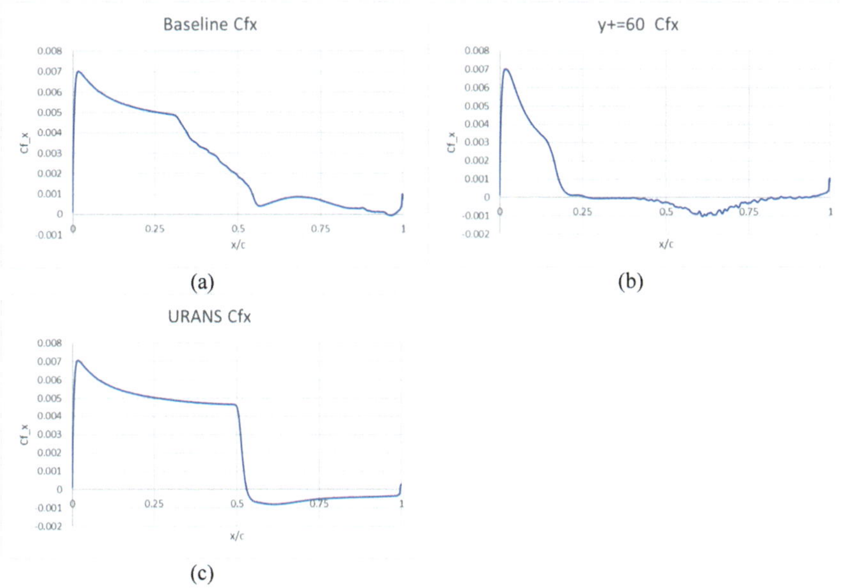

Fig. 8 Skin-friction coefficient distributions at the centerplane, **a** Baseline, **b** $y^+ = 60$, and **c** URANS

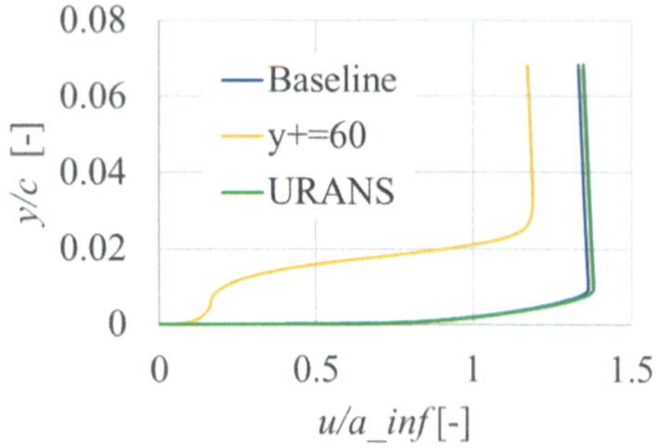

Fig. 9 Time-averaged streamwise velocity distributions at $x/c = 0.46$ over the wing

the interface location in the Baseline case is beyond the boundary-layer edge (= nearly isolated), which is far from the near-wall region where energy exchange is actively taking place. This is why the Baseline solution upstream the shock is very similar to that of RANS.

It should be noted, however, that the grid used for these three cases were originally prepared for the Baseline case, which may not be suitable for "$y^+ = 60$" case in which the LES region is extended, and the RANS region is narrowed. Thus, we will try to further reduce the grid spacing, especially in the upstream portion of the airfoil next.

3.3 *Effects of Grid Spacing*

Let us try the reduced grid spacing case, by increasing the number of cells on the upper surface from 500 to 750 in the streamwise direction, and clustered in the upstream portion. As illustrated in Fig. 10, the new grid fulfils the Tucker's guideline, i.e., $\Delta x^+ < 100$–600 [16] (Table 1), not only at the shock but also near the leading edge. However, this new grid spacing violates another criterion to prevent early separation, $\Delta x < 0.3\delta$, as reviewed in [17].

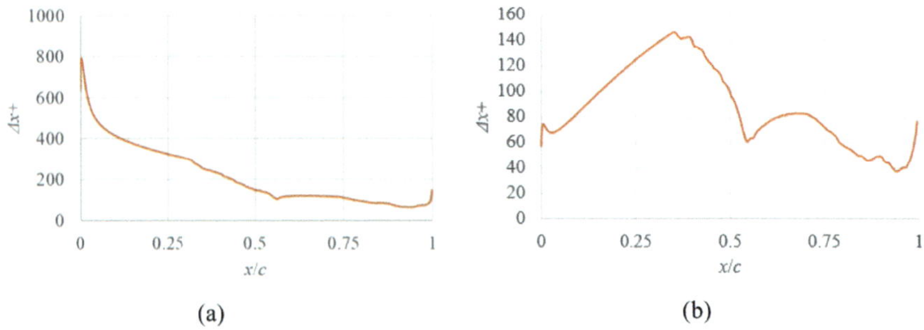

Fig. 10 Grid spacing Δx^+, **a** original, and **b** reduced

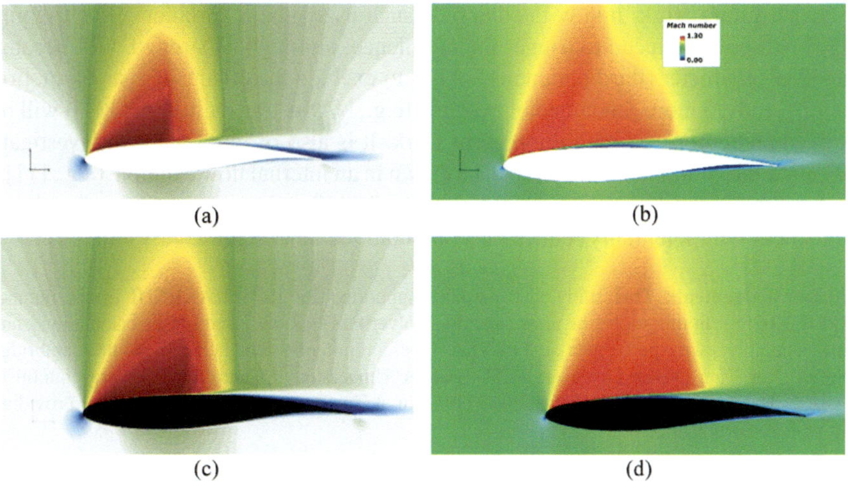

Fig. 11 Time-averaged Mach number contours, **a** Baseline (original), **b** $y^+ = 60$ (original), **c** Baseline (reduced spacing), and **d** "$y^+ = 60$" (reduced spacing)

The numerical solutions are compared with the original ones in Fig. 11. As seen, the solutions are almost unchanged from the original ones for both cases of Baseline and "$y^+ = 60$" hybrid interface height from the wall, presumably due to the violation of the second criterion ($\Delta x < 0.3\delta$). All in all, among many potential factors (interface location, transition model/trip, grid resolution, etc.) that may influence the numerical solution for the present buffet problem, we have observed the strong sensitivity to the selection of the hybrid interface location: Its effects have been found much clearer than those of the streamwise grid spacing. Influences of the other (wall-normal and spanwise) directions are to be investigated in the future work.

4 Conclusions

A comparative study varying both the hybrid interface location and grid spacing was conducted by using Zonal DES, a hybrid RANS/LES method, on a transonic buffeting flow involving an unsteady shock-wave/turbulent-boundary-layer interaction over OAT15A supercritical airfoil.

It was demonstrated that the hybrid interface as high as 1.5 times the boundary-layer thickness was a good choice for reproducing the unsteady nature of the buffet, on the contrary to the typically recommended value of y^+ around 100 (corresponding to 3.4% the boundary-layer thickness). With this interface location of y^+ being 60, the grid spacing recommended for flat plate boundary-layer or even the reduced cell size led to notorious, early grid-induced separation upstream the shock.

As a future work, inflow fluctuations (such as synthetic turbulence) or tripping [17] will be introduced to trigger the turbulence, although its strength will become another parameter (third parameter). Also, to exclude the effects of a shock motion in this study, a more fundamental example (e.g., NASA transonic bump [34]) will be solved and compared with the present work. It is also of importance to investigate whether the similar conclusions can be drawn in an internal flow example (e.g., [11]), since if this is the case the Zonal DES with the RANS region up to the boundary-layer thickness will potentially reduce the computational cost significantly.

Acknowledgements This study was partially supported by JSPS KAKENHI (grant number JP21KK0258). In this study, the numerical fluid solver FaSTAR was used for the calculations, and the calculations were performed on the JAXA Supercomputer System 3 (JSS3) and the Cambridge Service for Data Driven Discovery (CSD3) at the University of Cambridge. We are grateful to Atsushi Hashimoto at JAXA, Japan, and Yabin Liu at University of Edinburgh, UK for providing us with constructive comments.

References

1. Rolls-Royce: The Jet Engine, 5th edn. Wiley (2015)
2. Tucker, P.G., Wang, Z.-N.: J. Turbomach. **143**, 010801–010802 (2021)
3. Fröhlich, J., von Terzi, D.: Prog. Aerosp. Sci. **44**, 349 (2008). https://doi.org/10.1016/j.paerosci.2008.05.001
4. Deck, S.: S. AIAA J. **43**(7), 1556 (2005). https://doi.org/10.2514/1.9885
5. Larsson, J., Kawai, S., Bodart, J., Bermejo-Moreno, I.: Mech. Eng. Rev. **3**(1) (2016). Paper No 15-00418. https://doi.org/10.1299/mer.15-00418
6. Fukushima, Y., Kawai, S.: AIAA J. **56**(6), 2372 (2018). https://doi.org/10.2514/6.2017-0495
7. Tucker, P.G.: Advanced Computational Fluid and Aerodynamics. Cambridge Aerospace Series. Cambridge University Press, Cambridge, England, UK (2016). https://doi.org/10.1017/CBO9781139872010
8. Deck, S., Renard, N., Laraufie, R., Sagaut, P.: Phys. Fluids **26**, 025116 (2014). https://doi.org/10.1063/1.4866180
9. Deck, S., Weiss, P.-É., Pamiès, M., Garnie, E.: Comput. Fluids **48**, 1 (2011). https://doi.org/10.1016/j.compfluid.2011.03.009
10. Tucker, P.G., Davidson, L.: Comput. Fluids **33**, 267 (2004). https://doi.org/10.1016/S0045-7930(03)00039-2
11. Wang, Z.N., Tucker, P.G.: AIAA J. **61**(7), 2824 (2023). https://doi.org/10.2514/1.J062495
12. Tyacke, J.C., Wang, Z.N., Tucker, P.G.: AIAA J. **57**(3), 1215 (2019). https://doi.org/10.2514/1.J057057
13. Yasuda, H., Kawai, S.: AIAA J. **61**(2), 712 (2023). https://doi.org/10.2514/1.J062087
14. Ishida, T., Ishiko, K., Hashimoto, A., Aoyama, T., Takekawa, K.: AIAA Paper 2016-1310 (2016). https://doi.org/10.2514/6.2016-1310
15. Tsukamoto, Y., Kitamura, K.: AIAA J. **62** (8), 3052 (2024). https://doi.org/10.2514/1.J063886
16. Tucker, P.G.: Unsteady Computational Fluid Dynamics in Aeronautics. Springer, New York (2013)
17. Deck, S., Renard, N.: J. Comput. Phys. **400**, 108970 (2020). https://doi.org/10.1016/j.jcp.2019.108970
18. Ehrle, M., Waldmann, A., Lutz, T., Krämer, E.: CEAS Aeronaut. J. **11**, 1025 (2020). https://doi.org/10.1007/s13272-020-00466-7

19. Giannelis, N.F., Vio, G.A., Levinski, O.: Prog. Aerosp. Sci. **92**, 39 (2017). https://doi.org/10.1016/j.paerosci.2017.05.004
20. Hashimoto, A., Murakami, K., Aoyama, T., Ishiko, K., Hishida, M., Sakashita, M., Lahur, P.: AIAA Paper 2012-1075 (2012). https://doi.org/10.2514/6.2012-1075
21. Shima, E., Kitamura, K., Haga, T.: AIAA J. **51**(11), 2740 (2013). https://doi.org/10.2514/1.J052095
22. Van Leer, B.: J. Comput. Phys. **32**(1), 101 (1979). https://doi.org/10.1016/0021-9991(79)90145-1
23. Hishida, M., Hashimoto, A., Murakami, K., Aoyama, T.: Proceedings of 42nd Fluid Dynamics Conference/Aerospace Numerical Simulation Symposium pp. 1–10 (2010). (in Japanese)
24. Shima, E., Kitamura, K.: AIAA J. **49**(8), 1693 (2011). https://doi.org/10.2514/1.J050905
25. Sharov, D., Nakahashi, K.: AIAA J. **36**(3), 484 (1998). https://doi.org/10.2514/2.392
26. Visbal, M.R., Gordnier, R.E.: AIAA Paper 2000-2619 (2000). https://doi.org/10.2514/6.2000-2619
27. Spalart, P., Allmaras, S.: 30th Aerospace Sciences Meeting and Exhibit, pp. 439 (1992). https://doi.org/10.2514/6.1992-439
28. Shur, M.L., Spalart, P.R., Strelets, M.K., Travin, A.K.: Int. J. Heat Fluid Flow **29**(6), 1638 (2008). https://doi.org/10.1016/j.ijheatfluidflow.2008.07.001
29. Jacquin, L., Molton, P., Deck, S., Maury, B., Soulevant, D.: AIAA J. **47**(9), 1985 (2009). https://doi.org/10.2514/1.30190
30. White, F.M.: Viscous Fluid Flow, 3rd edn. McGraw–Hill, New York (2006)
31. Husson, J., Terracol, M., Deck, S., Le Garrec, T.: J. Comput. Phys. 112814. https://doi.org/10.1016/j.jcp.2024.112814
32. Spalart, P.R.: Annu. Rev. Fluid Mech. **41**, 181 (2009). https://doi.org/10.1146/annurev.fluid.010908.165130
33. Renard, N., Deck, S.: AIAA J. **53**(11), 3499 (2015). https://doi.org/10.2514/1.J054143
34. Bachalo, W.D., Johnson, D.A.: AIAA J. **24**(3), 437 (1986). https://doi.org/10.2514/3.9286

Turbomachinery

Controlling Tip Leakage Flow with Permeable Tip Treatment

Yabin Liu, Chandan Bose, and Zhong-Nan Wang

Abstract The present study focuses on the efficacy of controlling tip leakage flow through permeable tip treatment. A zonal large eddy simulation approach is employed to accurately capture the tip flow features around a linear cascade. The flow separation and shear layer interactions around the tip lead to predominant unsteadiness, significantly influencing the turbulence characteristics. Moreover, the tip separation vortex and the primary tip leakage vortex result in pressure drop, which increases the cavitation risk for water-based machines. Inspired by the exploitation of porous structures in controlling vortical structures, we propose the novel concept of porous treatment, locally distributed near the tip, to suppress unsteady flow separation and advance vortex breakdown. To that end, we investigated the effect of permeability and porosity on the turbulence characteristics and pressure drop in tip leakage flow. The influence of permeability is found to be significant, whereas the direct effect of porosity is seen to be minimal. The permeable treatment demonstrates great potential to mitigate tip flow unsteadiness and cavitation. The present work will be directly beneficial to the passive control of tip flow, a longstanding problem in turbomachinery and aerial/underwater vehicles, with a primary target of suppressing flow unsteadiness, cavitation, and noise generation.

CUFS-2024-PAPER ID, Cambridge, United Kingdom, 4-5 March 2024

Y. Liu (✉)
Institute for Energy Systems, School of Engineering, University of Edinburgh, Edinburgh, Scotland
e-mail: yabin.liu@ed.ac.uk

C. Bose · Z.-N. Wang
College of Engineering and Physical Sciences, University of Birmingham, Birmingham, England
e-mail: c.bose@bham.ac.uk

Z.-N. Wang
e-mail: z.n.wang@bham.ac.uk

J. C. Tyacke and N. R. Vadlamani (eds.), *Proceedings of the Cambridge Unsteady Flow Symposium 2024*, https://doi.org/10.1007/978-3-031-69035-8_18

1 Introduction

The narrow gap between the blade tip and casing wall in turbomachinery or aerial/underwater vehicles results in tip leakage jet (TLJ) from the blade pressure side (PS) to the blade suction side (SS), as shown in Fig. 1. The tip leakage flow results in multiple vortical structures, including primary tip leakage vortex (PTLV), tip separation vortex (TSV) and secondary induced vortex (SIV), and significantly contributes to flow unsteadiness, cavitation and noises [1–4].

Tip leakage flow inherently presents a generic unsteady mode that is found to be a potential precursor of rotating instability and dominates the measured sound spectrum in a fan [5]. Moreover, under specific circumstances such as large tip gap size and high rotating speed, an intense impingement interaction may occur between the PTLV and blades [6]. This results in significant vortex oscillations regarded to induce stall, noises and intensified pressure fluctuation [7–11]. The Particle Image Velocimetry (PIV) measurements in an axial pump [12] demonstrated the multiple-scale vortex structures induced by tip leakage flow, and the large-scale PTLV will break down and collapse into small-scale structures when developing downstream [12, 13]. Before the PTLV breakdown, there is an occurrence of a severe pressure drop around the vortex core, resulting in enlarged cavitation risks [14].

As the vortex breakdown, especially that of the large-scale PTLV, will fundamentally change the vortex scale and intensity, advancing the vortex breakdown by introducing local micro-jets near the tip has been explored. It shows promising potential to suppress the PTLV and associated cavitation [3, 15]. Bio-inspired research in natural systems, such as the flight of dandelion seeds, has demonstrated the critical role of permeability and porosity in stabilising the separated vortex structure in the wake [16, 17]. Inspired by the above, we propose a novel control method of tip leak-

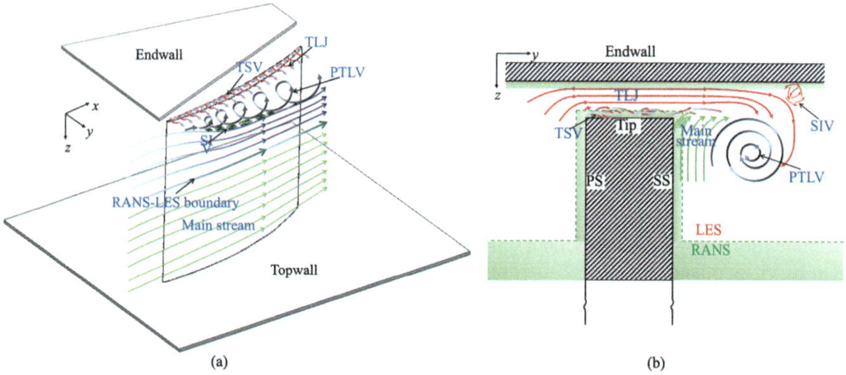

Fig. 1 Investigated cascade and schematic of ZLES configuration: **a** 3D view **b** wall-modelling and porous tip treatment, $y - z$ view. 'PTLV' denotes the primary tip leakage vortex; 'TSV' denotes the tip separation vortex; 'SIV' denotes the secondary induced vortex; and 'TLJ' denotes the tip leakage jet. 'PS' and 'SS' denote pressure and suction sides, respectively

age flow with permeable tip treatment, aiming at simultaneously advancing vortex breakdown and surprising vortex separation.

Though there are several existing studies for controlling tip leakage flow by using porosity, they focused on the influence of porosity on the tip leakage noises, and the role of permeability has not been discussed [18, 19]. Moreover, the steady-state Reynolds-averaged Navier-Stokes (RANS) simulation was used in existing numerical work [18], but this is found to be incapable of capturing crucial unsteady flow features compared to Large Eddy Simulation (LES) [5, 10]. In the present work, we applied a Zonalised LES (ZLES) approach, with wall-modelled LES applied for the tip region and RANS for the rest, to accurately predict unsteady tip flow features at a reduced computational cost. The permeability and the porosity of the tip treatment are modelled by changing the momentum equation in terms of adding the flow resistance in porous cell zone. Subsequently, the instantaneous vortex structures, time-averaged pressure, and turbulence kinetic energy are compared and discussed for tip treatments with different porosity and permeability.

2 Computational Methodology

2.1 Problem Definition

The present research is carried out on the same GE rotor B section blade in the experimental work performed by Muthanna and Wang [20, 21]. The case configuration consists of a linear cascade with a fixed stationary topwall and a moving endwall as shown in Fig. 1a.

The basic geometrical parameters are listed in Table 1. The blade chord length c is used as the reference length. The blade pitch L_y and the span L_z are $0.929c$ and $1.0c$, respectively. The tip gap size δ is set to $0.0165c$. Based on the free-stream velocity and chord length, the Reynolds number (Re) is around 4×10^5. The moving endwall speed V_{endwall} is set to model the relative movement between the end wall and the blade, and it was achieved by a moving belt in the experiment [20, 21]. In this

Table 1 Basic parameters of the investigated cascade

Parameter	Value
Blade chord length c	0.254 m
Axial chord length c_a	$0.546\,c$
Blade pitch L_y	$0.929\,c$
Blade span L_z	$1.0\,c$
Freestream velocity U_∞	26 m/s
Moving endwall speed V_{endwall}	23.6 m/s
Tip gap size δ	$0.0165\,c$

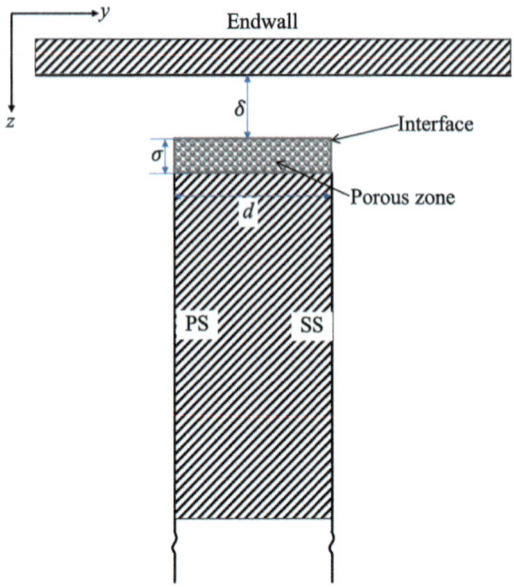

Fig. 2 Schematic of permeable tip treatment, where the span-wise scope of permeable treatment is $\sigma = 0.25\delta$. 'PS' and 'SS' denote pressure and suction sides, respectively

study, we introduce a porous zone with the thickness of $\sigma = 0.25\delta$ at the tip of the blade to investigate the effect of permeability and porosity on the tip leakage flow characteristics; see Fig. 2.

2.2 Numerical Setup

In the present work, a ZLES method is employed to resolve the energetic turbulence of tip leakage flow (Fig. 1). To make the simulation affordable, the expensive LES is zonalised in the tip region to achieve accurate prediction of tip leakage flow. In contrast, the rest of the region is modelled by RANS. As the Mach number is about 0.076, the incompressible solver is used.

The SST $k - \omega$ turbulence model is used in the RANS region where the flow field is rarely influenced by the tip leakage flow. The LES mode is activated in the tip region, which is the focus of this research. The dynamic Smagorinsky-Lilly model is used to calculate sub-grid stresses.

The ZLES is achieved by a blending function f_b. As in the Stress-Blending Eddy Simulation model [22], the turbulence stresses between RANS and LES are blended in the following way:

$$\tau_{ij}^{Turb} = f_b \tau_{ij}^{RANS} + (1 - f_b)\, \tau_{ij}^{LES}\,, \tag{1}$$

where τ_{ij}^{RANS} is the RANS portion of the turbulence stresses tensor and τ_{ij}^{LES} is the LES portion of the modelled turbulence stresses tensor.

Taking advantage of f_b, we can generically combine RANS and LES portions, and precisely define the RANS-LES boundary. The RANS and LES configuration is adjusted to obtain the optimal balance between prediction performance and calculation cost. This method has been successfully used to predict fan tip and wake flows in a bypass engine configuration [23]. This zonal approach has been implemented in ANSYS Fluent via User Defined Function by explicitly defining f_b. The region of interest near the tip is run in LES mode (f_b=0), while the rest is in the RANS mode ($f_b = 1$).

The RANS-LES boundary is determined by a streamline marked in Fig. 1a separating the tip leakage flow (run in LES by defining f_b=0) from the rest flow region (run in RANS by defining f_b=1). Figure 1b shows the wall-modelling strategy in LES region. RANS is used to model the inner parts of the boundary layer and reduce the required grid resolution of streamwise Δx^+ and spanwise Δz^+ comparing to the wall-resolved LES. The RANS layer is applied by defining f_b=1 to where a wall distance $\Delta y^+ < 100$, which approximately covers the viscous sublayer and buffer layer.

The Central Differencing scheme is employed for spatial discretization, and the Second Order Implicit backwards difference method for time integration. The `SIMPLEC` scheme is used to decouple pressure and velocity in solving the incompressible flow governing equations. In the transient simulation, the time step size is set as $\Delta t = 8 \times 10^{-6}$ s $\approx \frac{1}{1250} \times c/U_\infty$, which guarantees that the Courant number is below 1 in the region of primary interest.

The velocity inlet condition is applied at the inlet boundary, and the outflow condition, where a zero diffusion flux for all flow variables, is applied at the outlet boundary. A periodic boundary condition is adopted on the sidewalls to account for the adjacent passages. The no-slip wall condition is applied on the blade surface and the moving end wall, while a free-slip boundary condition is set on the top wall.

The porous zone at the tip of the blade is modelled by solving the Darcy-Brinkman equation [17, 24] in the porous cell zone in which the pressure loss in the flow is determined according to the porosity ϵ and the additional resistance term, which is nothing more than an added momentum sink in the governing momentum equations. A uniform permeability (κ) is considered throughout the porous zone, the effect of which is characterized by the non-dimensional parameter Darcy number ($Da = \kappa/\bar{d}^2$), where $\bar{d} = 0.0326c$ is the average of the tip thickness d as shown in Fig. 2. This study is carried out for various ϵ and Da values in the range of $0.1 - 0.9$ and $10^{-3} - 10^{-6}$, respectively.

2.3 Solver Validation

To validate the present flow solver, the mean pressure coefficient and the Reynolds normal stresses obtained from the present simulations are compared against the reference experimental results. Figure 3 shows a good agreement of the mean pressure

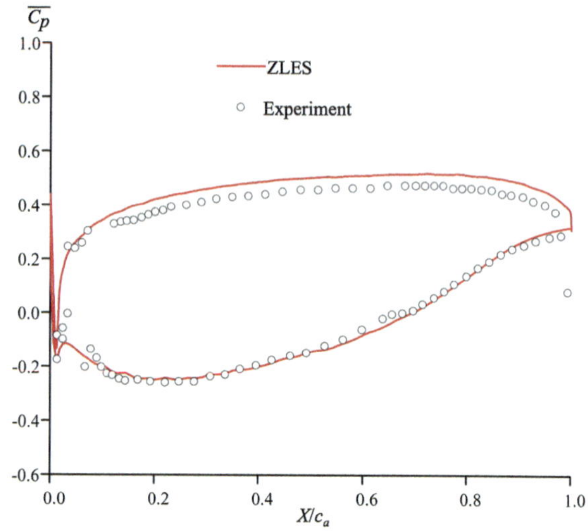

Fig. 3 Mean pressure coefficient $\bar{C}_p$ distribution on middle span. Pressure Coefficient C_p is defined as $C_p = \frac{p-p_\infty}{\frac{1}{2}\rho U_\infty^2}$, where p_∞ is the inlet pressure and U_∞ is the freestream velocity

coefficient $\bar{C}_p$ at the mid-span between the present simulation and the experimental results reported by Muthanna [20].

In the experimental study carried out by Wang [21], Reynolds normal stresses were measured on a line crossing the PTLV core on $X = 1.51c_a$, located in the wake region (see Fig. 11 in the Appendix). To validate the accuracy of the ZLES method, a comparison of Reynolds normal stresses has been made between the simulation and the experiment in Fig. 4. Moreover, the LES results reported by You et al. [2, 25] are also plotted as a reference for the comparison. Generally, the ZLES method shows an equivalent accuracy to the LES but at a reduced cost due to about 6 times less grid size (see the Appendix). Figure 5 shows that the predicted turbulence power spectral density agrees well with the experimental measurements of Wang [21]. Both the experiment and simulation show the -5/3 slope characteristic in the inertial subrange of turbulence.

3 Results and Discussions

3.1 *Vortex Structures*

We first discuss the instantaneous vortex pattern obtained from the present simulations to investigate the controlling effects of permeable tip treatment on the tip leakage flow. As demonstrated in Fig. 6, we classified the vortical structures into three groups: (1) PTLV, which occupies the largest proportion among these structures; (2) TSV inside the tip gap and between the tip gap and PTLV; (3) SIV induced by the PTLV and primarily occurs in a limited region around $x/c = 0.4$. The case

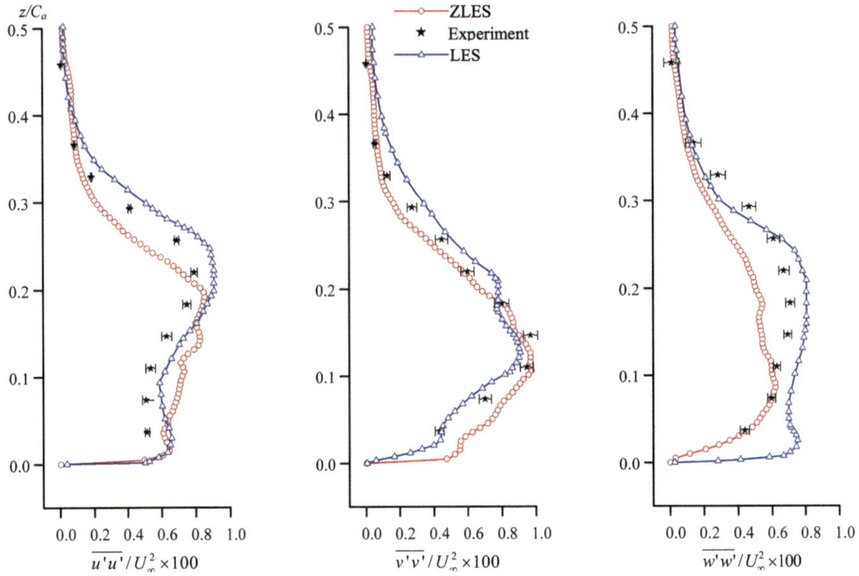

Fig. 4 Comparison of reynolds normal stress along a line crossing the PTLV core

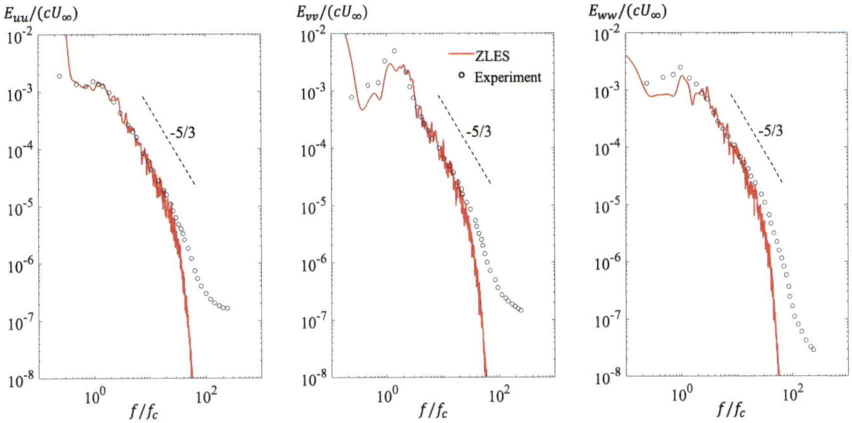

Fig. 5 Velocity power spectral density at the PTLV core point on the section of $X = 1.51c_a$

without tip treatment is denoted as baseline, and the other cases with tip treatments are characterised by permeability and porosity. Here, permeability indicates the ability of a structure to allow fluids to pass through it, while porosity is a fraction of the fluids volume over the total volume.

The permeable tip treatment significantly influences the PTLV structures in the flow passage, the TSV structures near the tip, and the pressure distribution on these structures. With the decreased permeability from $Da = 1 \times 10^{-3}$ to $Da = 1 \times 10^{-5}$,

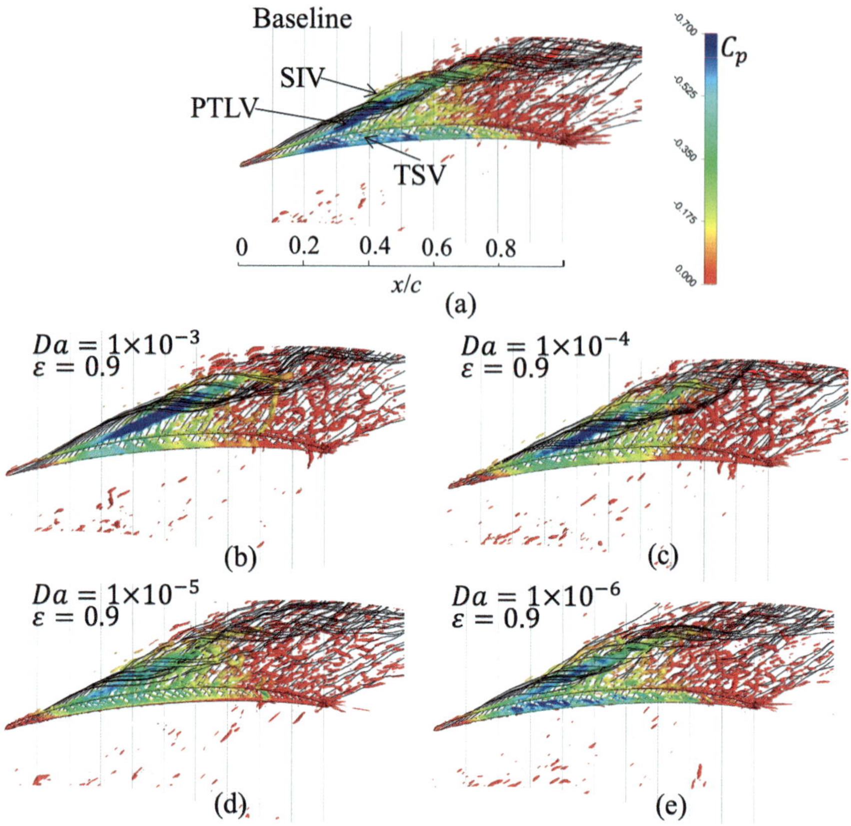

Fig. 6 Comparison of instantaneous vortex structures, defined by the iso-surface of $Qc^2/U_\infty^2 = 5 \times 10^2$ and contoured by pressure coefficient. 'PTLV' denotes the primary tip leakage vortex; 'TSV' denotes the tip separation vortex; 'SIV' denotes the secondary induced vortex

the large-scale and concentrated PTLV structure gradually shrinks and breaks down into multiple small-scale vortical structures at an advanced stage. This is demonstrated by the streamlines, released from the middle of the tip gap, as they are more condensed before the PTLV breakdown due to the PTLV's stronger capability to entrain the tip leakage jet. This advanced vortex breakdown results in less pressure drop around the PTLV. However, when the Darcy number further decreases to $Da = 1 \times 10^{-6}$, the PTLV breakdown position moves downstream, which causes more pressure drop compared to that of $Da = 1 \times 10^{-5}$.

These findings indicate that there is an optimal range of permeability to advance the PTLV breakdown and reduce the associated pressure drop. Increasing permeability results in a significantly higher flow rate through the gap, intensifying the PTLV. However, when the permeability is too low, it gets close to a solid wall, reducing the controlling effects and recovering to the original condition without porous treatment.

Similar trends are found for the TSV structures. The vortex separation near the pressure side inside the tip gap is significantly reduced between $Da = 1 \times 10^{-3}$ and $Da = 1 \times 10^{-5}$, and the pressure drop inside the tip gap is also suppressed. At $Da = 1 \times 10^{-6}$, the vortex separation near the pressure side and the low-pressure structures inside the gap are observed again, which means the influence of porous treatment is substantially limited at low permeability.

3.2 *Time-Averaged Pressure Coefficient*

The above findings suggest that the instantaneous pressure drop corresponding to the tip leakage flow structures can be suppressed in a certain range of permeability. In this section, to better understand the effects of porous tip treatment in mitigating cavitation, we further discuss the time-averaged pressure coefficient at different streamwise sections, especially along the PTLV core trajectory. The statistical results are obtained based on the simulation results between $16c/U_\infty$ and $31c/U_\infty$; the independence to the statistical period has been validated.

Figure 7 shows that the permeable tip treatment has a significant effect in suppressing the mean pressure drop inside the tip gap between $Da = 1 \times 10^{-3}$ and $Da = 1 \times 10^{-5}$. However, the mitigation effect on pressure drop along the PTLV core trajectory is less obvious except for $Da = 1 \times 10^{-5}$. This again suggests a careful consideration of the permeability range for underwater applications when considering cavitation. The mitigation effects are quantitatively presented in Fig. 8. It reaches the largest mitigation of pressure increase at $Da = 1 \times 10^{-5}$, where the maximum increase of the mean pressure coefficient is 0.303 at $x/c = 0.5$. When the Da is equal to or lower than 1×10^{-4}, the pressure drop is aggravated compared to the original condition.

The influence of porosity on the tip leakage flow characteristics is found to be minimal, as demonstrated in Fig. 9. Similar results are found for the vortex pattern and turbulence characteristics, which are not presented to avoid redundancy. This suggests that permeability is the crucial parameter that determines the flow control effects.

3.3 *Turbulence Kinetic Energy*

The permeable tip treatment also alters the spatial distribution of turbulent kinetic energy (TKE). This is illustrated by the TKE distribution on the axial sections between $x/c = 0.3$ and $x/c = 0.6$ with and without tip treatment in Fig. 10.

Higher permeability values, including $Da = 1 \times 10^{-3}$ and $Da = 1 \times 10^{-4}$, lead to increased tip leakage flow rate. As a consequence, the entraining effect from the tip leakage jet to the PTLV is enhanced and the PTLV breakdown shifts downstream (Fig. 6), hence the turbulence level in the PTLV and SIV regions is intensified. When

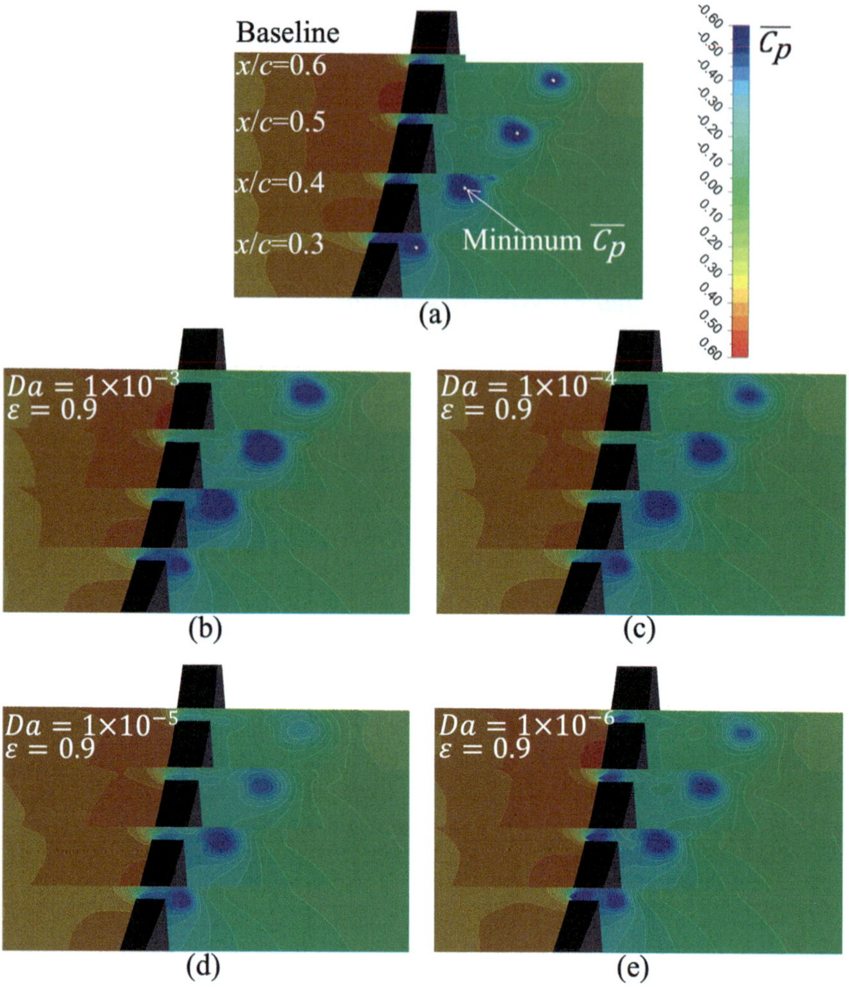

Fig. 7 Comparison of mean pressure coefficient at different Darcy numbers

Da further decreases to 1×10^{-5}, the tip permeability results in advanced vortex breakdown, so the PTLV's entraining effect is weakened, and the TKE distribution corresponding to the SIV diminishes. However, the advanced vortex breakdown causes more intense shear layer interaction between the tip leakage jet and the main stream (Fig. 6d), and the corresponding flow features become more turbulent (Fig. 10d). As the permeability further decreases, the TKE distribution becomes similar to that without tip treatment.

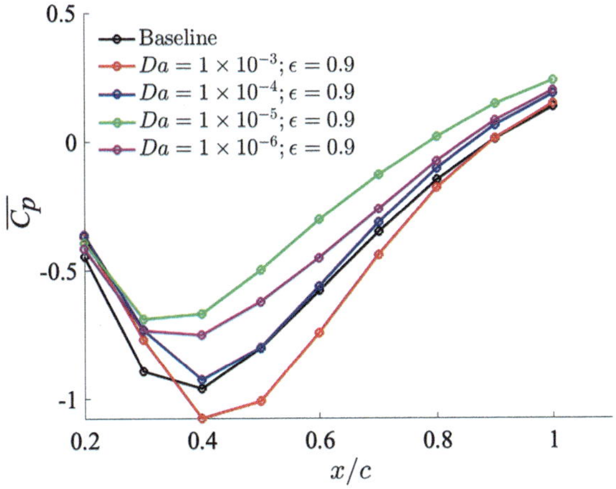

Fig. 8 Effects of permeability on suppressing pressure drop along the PTLV core trajectory

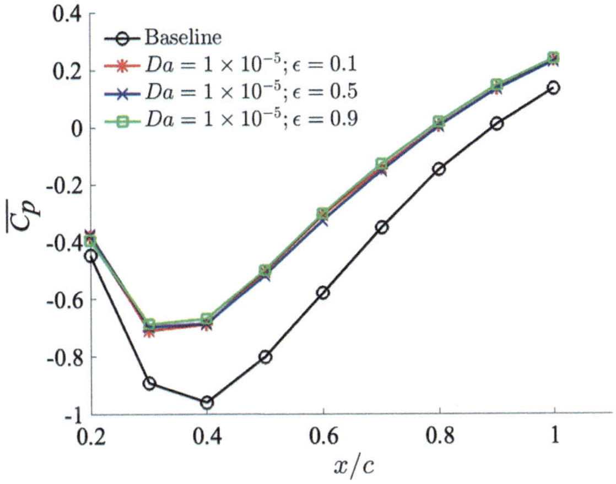

Fig. 9 Influence of porosity to the mean pressure coefficient along the PTLV core trajectory

4 Conclusion

The present research explores the new concept of controlling the tip leakage flow through permeable tip treatment. The ZLES approach employed in this research shows an equivalent accuracy at a reduced computational cost compared to the conventional LES method, as the grid number used in ZLES simulation is about one-sixth of that used in the LES method [2, 25]. The salient outcomes of this study can be summarised as follows.

Permeable tip treatment, confined in the spanwise scope of a quarter of the tip gap size, remarkably influences the tip leakage vortex features. At $Da = 1 \times 10^{-5}$, the PTLV breakdown is advanced, which weakens the entraining effect from the tip leakage jet to the PTLV. Therefore, the pressure drop and turbulence level around

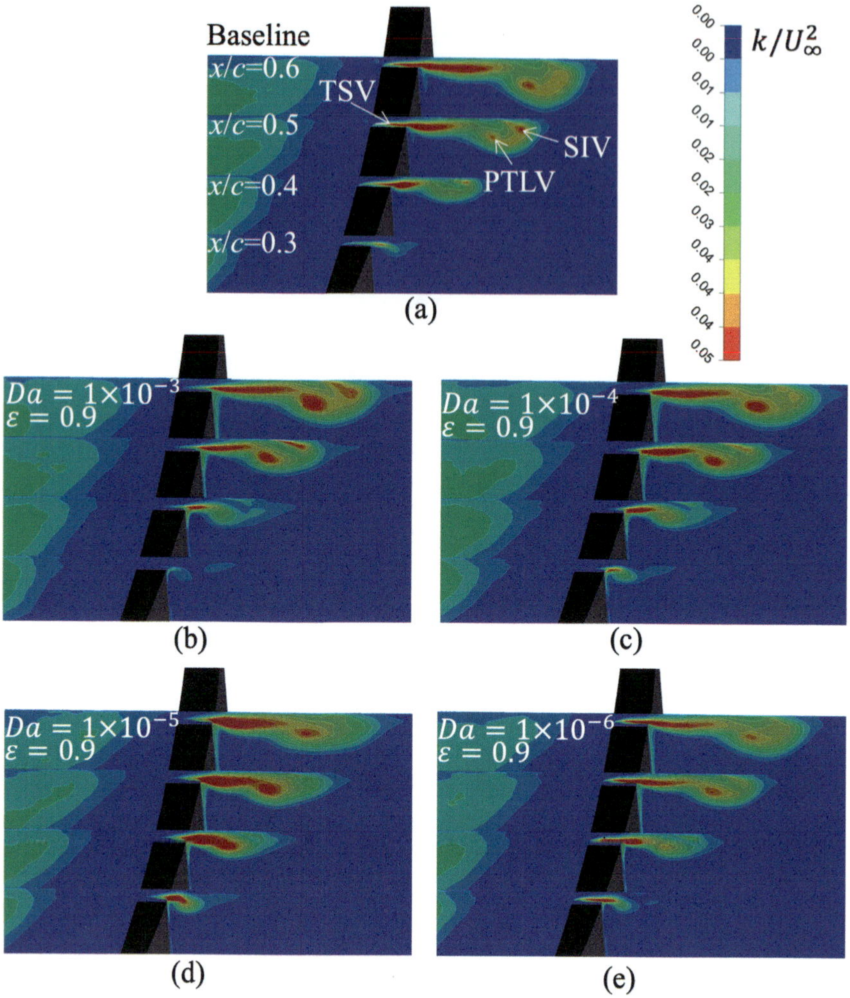

Fig. 10 Comparison of turbulence kinetic energy at different Darcy numbers. 'PTLV' denotes the primary tip leakage vortex; 'TSV' denotes the tip separation vortex; 'SIV' denotes the secondary induced vortex

the PTLV is suppressed. Meanwhile, it causes more intense shear layer interaction between the tip leakage jet and the main stream, and the flows between the blade tip and the PTLV become more turbulent.

For permeability higher than $Da = 1 \times 10^{-6}$, the permeable tip treatment always effectively mitigates the TSV-induced pressure drop and turbulence inside the tip gap, especially near the blade pressure side. Therefore, this flow control methodology has great potential to be applied in water-based systems, including hydro turbines, pumps and underwater vehicles concerned with tip cavitation; it can also be used to control noises generated from the tip gap in turbomachinery, such as compressors and fans.

The influence of porosity on the flow control effect is found to be minimal. On the contrary, it is seen to be dominated by permeability in terms of the increasing Darcy number. The present findings suggest that the permeable treatment with a specific structural design may be used in practice to address the concern of system reliability.

Acknowledgements This work has been kindly supported by the Royal Commission for the Exhibition of 1851 Brunel Fellowship and the State Key Laboratory of Hydroscience and Engineering [sklhse-2023-E-03]. We are grateful for the useful advice from Prof. Paul Gary Tucker at the University of Cambridge and Mr. Felix M. Möller at DLR.

Appendix

The original X-Y-Z coordinate system is different from the velocity coordinate composed of u (streamwise velocity), v (pitchwise velocity), and w (spanwise velocity) in the experiment [20, 21]. Therefore, the X-Y-Z coordinate has been accordingly rotated by 52.5° to be the x-y-z coordinate, which is aligned with the u-v-w coordinate, as shown in Fig. 11.

According to the requirement of the applied ZLES method, the grid resolution around the tip gap was locally refined in particular, as shown in Fig. 12. The grid dimension in the wall-modelled LES region is set as $\Delta x^+ < 60$, $\Delta y^+ < 1$ on the blade surface and $\Delta z^+ < 90$. The Δz^+ near the blade tip is refined to 1–2 to accurately capture the flow details. The pitchwise and spanwise grid resolution in the RANS region is much coarser, especially for the region far from the RANS-LES boundary, with Δx^+ and Δz^+ increased to 50–200. Therefore, the total grid elements number is only 3,531,621 for the original case without tip treatment, with the whole span considered. For the same case, more than 20 million grid elements were used in the LES studies [2, 25–27], with only half span simulated.

Independence studies of the grid resolution and the RANS-LES blending boundary location have been conducted but not demonstrated in the present paper for the sake of brevity.

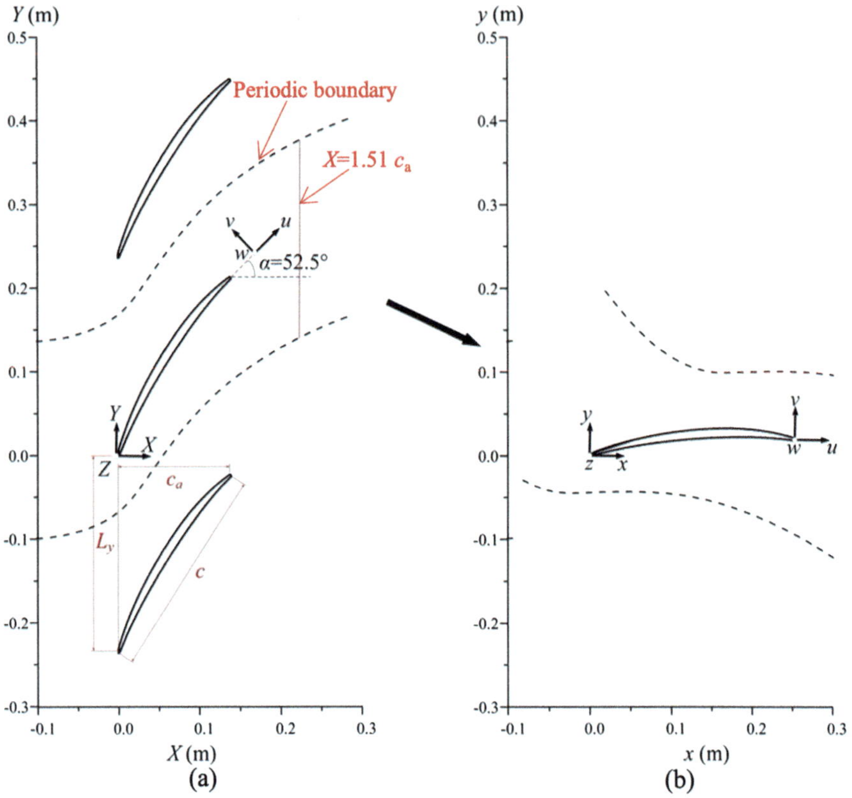

Fig. 11 Adjustment of the coordinate system (top view): **a** 5Original $X - Y - Z$ coordinate system **b** $x - y - z$ coordinate system aligned with $u - v - w$

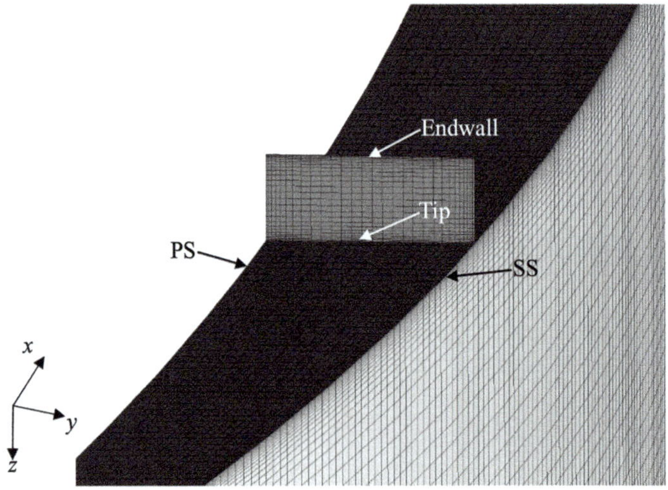

Fig. 12 Mesh on the blade and in the tip gap

References

1. Moghadam, S.M.A., Meinke, M., Schröder, W.: Analysis of tip-leakage flow in an axial fan at varying tip-gap sizes and operating conditions. Comput. Fluids **183**, 107 (2019)
2. You, D., Wang, M., Moin, P., Mittal, R.: Vortex dynamics and low-pressure fluctuations in the tip-clearance flow. J. Fluids Eng. **129**(8), 1002 (2007)
3. Liu, Y., Tan, L.: Influence of C groove on suppressing vortex and cavitation for a NACA0009 hydrofoil with tip clearance in tidal energy. Renew. Energy **148**, 907 (2020)
4. Camussi, R., Grilliat, J., Caputi-Gennaro, G., Jacob, M.C.: Experimental study of a tip leakage flow: wavelet analysis of pressure fluctuations. J. Fluid Mech. **660**, 87 (2010)
5. Boudet, J., Cahuzac, A., Kausche, P., Jacob, M.C.: Zonal large-eddy simulation of a fan tip-clearance flow, with evidence of vortex wandering. J. Turbomach. **137**(6), 061001 (2015)
6. Liu, Y., Tan, L.: Theoretical prediction model of tip leakage vortex in a mixed flow pump with tip clearance. J. Fluids Eng. **142**(2), 021203 (2020)
7. Du, J., Lin, F., Chen, J., Nie, C., Biela, C.: Flow structures in the tip region for a transonic compressor rotor. J. Turbomach. **135**(3), 031012 (2013)
8. Cameron, J.D., Bennington, M.A., Ross, M.H., Morris, S.C., Du, J., Lin, F., Chen, J.: The influence of tip clearance momentum flux on stall inception in a high-speed axial compressor. J. Turbomach. **135**(5), (2013)
9. Hewkin-Smith, M., Pullan, G., Grimshaw, S., Greitzer, E., Spakovszky, Z.: The role of tip leakage flow in spike-type rotating stall inception. J. Turbomach. **141**(6), (2019)
10. He, X., Fang, Z., Rigas, G., Vahdati, M.: Spectral proper orthogonal decomposition of compressor tip leakage flow. Phys. Fluids **33**(10), 105105 (2021)
11. Liu, Y., Tan, L.: Spatial–temporal evolution of tip leakage vortex in a mixed-flow pump with tip clearance. J. Fluids Eng. **141**(8), (2019)
12. Miorini, R.L., Wu, H., Katz, J.: The internal structure of the tip leakage vortex within the rotor of an axial waterjet Pump. J. Turbomach. **134**(3), 031018 (2011). https://doi.org/10.1115/1.4003065
13. Wu, H., Miorini, R.L., Tan, D., Katz, J.: Turbulence within the tip-leakage vortex of an axial waterjet pump. AIAA J. **50**(11), 2574 (2012)
14. Boulon, O., Callenaere, M., Franc, J.P., Michel, J.M.: An experimental insight into the effect of confinement on tip vortex cavitation of an elliptical hydrofoil. J. Fluid Mech. **390**, 1 (1999)
15. Liu, Y., Tan, L.: Method of C groove on vortex suppression and energy performance improvement for a NACA0009 hydrofoil with tip clearance in tidal energy. Energy **155**, 448 (2018)
16. Cummins, C., Seale, M., Macente, A., Certini, D., Mastropaolo, E., Viola, I.M., Nakayama, N.: A separated vortex ring underlies the flight of the dandelion. Nature **562**(7727), 414 (2018)
17. Bose, C., Bruce, C., Viola, I.M.: Porous plates at incidence. arXiv preprint arXiv:2303.13296 (2023)
18. Khorrami, M.R., Li, F., Choudhari, M.: Novel approach for reducing rotor tip-clearance-induced noise in turbofan engines. AIAA J. **40**(8), 1518 (2002)
19. Palleja-Cabre, S., Saraceno, I., Paruchuri, C.C., Joseph, P.: Reduction of tip-leakage noise by using porosity. In: 28th AIAA/CEAS Aeroacoustics 2022 Conference, p. 3061 (2022)
20. Muthanna, C.: The effects of free stream turbulence on the flow field through a compressor cascade. Ph.D. thesis, Virginia Tech (2002)
21. Wang, Y.: Tip leakage flow downstream a compressor cascade with moving end wall. Ph.D. thesis, Virginia Tech (2000)
22. Frank, T., Menter, F.: Validation of URANS SST and SBES in ANSYS CFD for the turbulent mixing of two parallel planar water jets impinging on a stationary Pool. jets, pp. 1–9 (2015)
23. Tucker, P.G., Wang, Z.N.: Eddy resolving strategies in turbomachinery and peripheral components. J. Turbomach. **143**(1), (2021)
24. Neale, G., Nader, W.: Practical significance of Brinkman's extension of Darcy's law: coupled parallel flows within a channel and a bounding porous medium. Can. J. Chem. Eng. **52**(4), 475 (1974)

25. You, D., Wang, M., Moin, P., Mittal, R.: Large-eddy simulation analysis of mechanisms for viscous losses in a turbomachinery tip-clearance flow. J. Fluid Mech. **586**, 177 (2007)
26. You, D., Mittal, R., Wang, M., Moin, P.: Computational methodology for large-eddy simulation of tip-clearance flows. AIAA J. **42**(2), 271 (2004)
27. You, D., Wang, M., Mittal, R., Moin, P.: Study of flow in tip-clearance turbomachines using large-eddy simulation. Comput. Sci. Eng. **6**(6), 38 (2004)

Study on Tip Endwall Contouring for a Steam Turbine Regulating Stage

Hao Sun, Yan Liu, Bin Wang, Jinguang Yang, Rong Xie, and Wei Wang

Abstract The regulating stage of high-pressure steam turbines typically features a relatively small blade height, which leads to high secondary flow losses. Various methods have been proposed in the literature to mitigate these losses and improve stage performance, including the use of endwall contouring. This study investigates endwall contouring using Widosinski curve for a regulating stage. Subsequently the Widosinski curve is parameterized to optimize the endwall contouring. By employing a multi-objective genetic algorithm, an optimized endwall contouring curve is obtained. Numerical results show a significant reduction in secondary flow losses for both the stator and rotor. Compared to the original flat endwall contouring, for the stator, the secondary flow loss is reduced by 20.1%, 39.6%, and 68.3% for the 100%, 33%, and 20% working loads respectively. For the rotor, it is correspondingly reduced by 57.3%, 38.2%, and 29.2%. Moreover, the total-total efficiency increases by 2.145%, 4.096%, and 5.421% for the aforementioned three load conditions. Through an analysis of the internal flow field, it is observed that the endwall contouring effectively suppresses vortex intensity in the channel and minimizes expansion along the span of the regulating stage. As a result, secondary flow losses are reduced. The findings of this study hold practical implications for the optimization of real steam turbines.

1 Introduction

For high-pressure steam turbines with big capacity, the blade height of the regulating stage is relatively small, resulting in significant secondary flow losses at blade ends. These losses have a substantial impact on the efficiency of the regulating stage.

CUFS-2024-PAPER ID, Cambridge, United Kingdom, 4-5 March 2024.

H. Sun · Y. Liu (✉) · B. Wang · J. Yang · R. Xie · W. Wang
Dalian University of Technology, No.2 Lingong Road, Ganjingzi District, Dalian, Liaoning Province, China
e-mail: yanliu@dlut.edu.cn

J. C. Tyacke and N. R. Vadlamani (eds.), *Proceedings of the Cambridge Unsteady Flow Symposium 2024*, https://doi.org/10.1007/978-3-031-69035-8_19

Additionally, steam turbines used in ships should operate safely and efficiently across a wide range of operating conditions. This requirement is particularly important for ensuring efficiency during cruising conditions and maintaining safety and efficiency at full speed. Therefore, the performance of the regulating stage in marine steam turbines holds critical significance for the economical and reliable operation of the turbine unit. Addressing the issue of reducing secondary flow losses at blade ends and improving the efficiency of marine steam turbines across a wide range of operating conditions, both in military and civilian applications, has become an urgent problem that requires attention [1].

There are a few ways to reduce endwall secondary flow losses for regulating stages of steam turbines including the meridional endwall contouring and increase in height of blades with the use of partial admission. The meridional endwall contouring technique was initially proposed by Deich [2] in the early 1960s with the aim of improving the flow characteristics near the endwall. The specific approach involves shaping the outer endwall, located after the blade throat, into a contracting form to converge the flow and to reduce the expansion pressure beyond the throat. The contraction reduces flow separation in the expansion section. Moreover, the passage contraction accelerates the flow, preventing the thickening of the boundary layer on the suction side, pressure side, and endwall of the blade. It effectively alters the streamline curvature and reduces the transverse pressure gradient, thereby controlling and eliminating secondary flow. Subsequent experiments conducted by Kopper and Milano [3] on low aspect ratio cascades showed that compared to a flat-endwall cascade configuration, the total mass-averaged loss for a contraction endwall cascade was decreased by approximately 17%. These findings were further used to validate the secondary flow loss model proposed by Morris and Hoare [4]. Additionally, Arts [5]and Boletis [6] independently conducted experimental and numerical studies on the same annular cascade. Compared to the prototype configuration, the total flow losses of the contraction endwall cascade were reduced by 15% and 8.7% at two exit measurement planes, respectively, compared to the prototype configuration. Petrovic [7] conducted optimization of the endwalls in a turbine stage to feature meridional contraction surfaces, resulting in a 2% increase in stage efficiency.

An et al. [8] conducted numerical simulations to explore the impact of seven different meridional profile redesigns on a low-pressure turbine meridional passage. Their findings revealed that the modifications to the profile had a substantial and appropriate effect on outlet energy loss. By making suitable adjustments, it was possible to reduce these losses. In 2021, Song et al. [9] utilized sine curves to recontour the meridional profile of a large meridional expansion turbine. The results demonstrated that the meridional profile exerted a significant influence on the separation location of the end-wall boundary layer, as well as the position and intensity of internal vortices. An appropriate profile adjustment had the potential to substantially decrease losses and alleviate blade thermal loads.

Although meridional contraction offers a solution to reduce secondary flow losses in high-pressure steam turbines, existing research in this field often overlooks the alterations in flow patterns induced by meridional contraction. This oversight can lead to the turbine stage deviating from its intended design operating point, which is an undesirable outcome in engineering practice. Furthermore, there is a notable scarcity of studies focused on optimizing the meridional profile for marine steam turbine stages that operate under variable speeds and multiple operating conditions.

In this study, a parametric approach is employed to optimize a meridional profile for a steam turbine stage. The objective is to achieve an optimal meridional profile that exhibits exceptional performance across multiple operating conditions for a marine steam turbine without deviating its design condition. To accomplish this, a multi-objective optimization process is conducted using an optimization algorithm.

2 Geometry Model and Numerical Simulation Methods

2.1 Geometry Model

A steam turbine stage studied in the paper operates under three commonly encountered conditions, as outlined in Table 1. Among these conditions, Condition 1 corresponds to the rated operating condition with full-speed. Condition 2 represents an operating condition at 70% of the rated speed, while Condition 3 reflects an operating condition at 57.5% of the rated speed. It is worth noting that both Condition 2 and Condition 3 are characterized as low-load operating conditions.

Figure 1 depicts the schematic diagram of the axial flow passage in the governing stage. The design of the governing stage features an equal-diameter configuration. The essential geometric parameters of the governing stage are presented in Table 2. Notably, the stator's relative blade height is 0.4, indicating its suitability for endwall contraction to enhance the stage's efficiency.

Table 1 Steam boundary conditions for three different operating conditions

Parameters	Inlet total pressure/KPa	Inlet total temperature/K	Outlet pressure/KPa	Mass flow rate/kg/s
Condition 1	2280	493	1520	34.6
Condition 2	1008	460	701.2	15.6
Condition 3	660	455	455	10.3

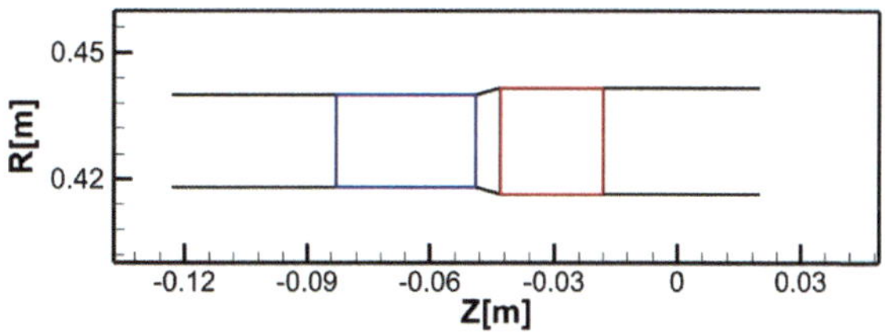

Fig. 1 Schematic diagram of the prototype meridian channel

Table 2 Key geometric parameters of regulating stage

Parameters	Unit	Stator	Rotor
Blade number	\	66	143
Blade height	mm	22	25
Axial chord length	mm	34	25
Relative blade height	\	0.4	0.974
Inlet geometry angle	deg	10	57.5

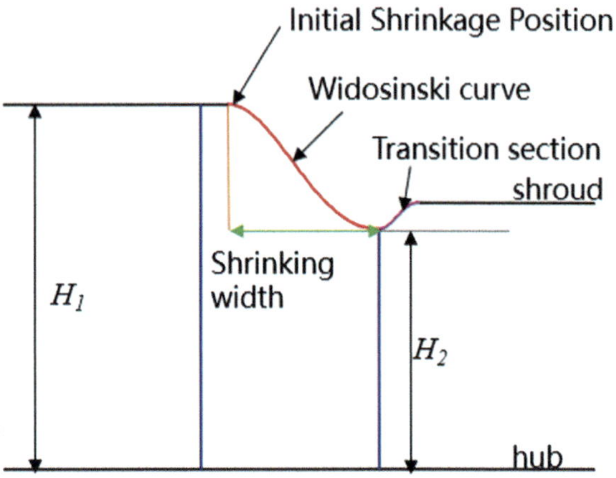

Fig. 2 Schematic diagram of the meridian channel

2.2 *Contraction Treatment of Stator*

The method of meridional profile modification involves shrinking the upper end wall of the stationary blade, optimizing the profile of the transition section to achieve G1 continuity. The meridional profile adopts the Widosinski curve [10], which is expressed as shown in Eq. (1):

$$R = \frac{R_0}{\sqrt{1 - \left[1 - \left(\frac{R_0}{R_1}\right)^2\right]\left[\frac{1-\left(\frac{X}{X_0}\right)^2}{1+\frac{1}{3}\left(\frac{X}{X_0}\right)^2}\right]}} \tag{1}$$

R represents the radius at the axial position X of the curve, R_0 is the inlet radius of the profile, R_1 is the outlet radius of the profile, X_0 is the axial length of the Widosinski curve.

The key variables for the stationary blade meridional profile modification are primarily two: starting contraction position of the profile (X_{start}), contraction ratio of the end wall of the stationary blade profile DH $\left(DH = \frac{H_1}{H_2}\right)$, Fig. 2 provides an illustrative diagram of the meridional profile modification for the stationary blade.

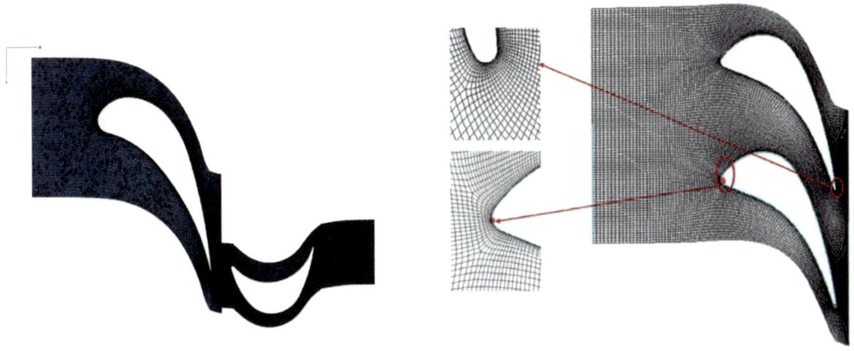

(a) Single-passage computational domain (b) Structured mesh for blades

Fig. 3 Computational domain

Static blade meridional profile modification typically involves the contraction of the upper end wall. The usual approach is to maintain the static blade's outlet blade height unchanged, to raise the inlet blade height of the static blade, and to adopt a contraction profile in the middle. However, this method leads to an increase in the throat area of the static blade, resulting in a higher flow rate for the regulating stage. While this approach may improve the overall performance at a certain operating point, it can cause a deviation in the operating point of the regulating stage, which is not suitable for practical engineering applications.

Therefore, in this study, when increasing the inlet blade height, the static blade's outlet blade height is appropriately reduced to maintain a consistent throat area of the static blade. This approach ensures that the flow rate of the regulating stage remains nearly unchanged, avoiding deviations from the design point and ensuring the accuracy of the mechanism research.

2.3 Numerical Simulation Methods

The commercial software Numeca is used to conduct numerical simulations of the flow field in the governing stage in this study. The Fine/turbo module is employed to solve the three-dimensional Reynolds-averaged Navier-Stokes (RANS) equations. The turbulence model used is the Spalart-Allmaras one-equation model. The inlet conditions are specified in terms of total temperature and total pressure, while the outlet boundary is set with an average static pressure. The working fluid considered is water vapor, which is treated as a real gas. The IGG/autogrid5 module is used to generate a hexahedral structured mesh, as shown in Fig. 3. The grid quality satisfies the following criteria: $orthogonality > 36°$ and $expansionratio < 2.0$. The first layer of the mesh around the hub and blade tip is refined for better accuracy.

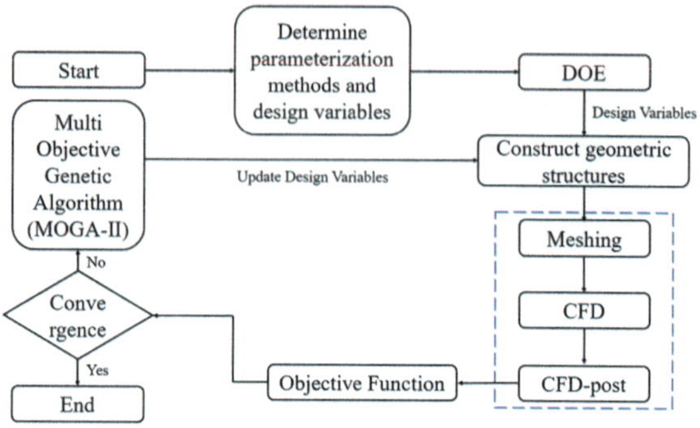

Fig. 4 Design optimization platform for meridional endwall contouring

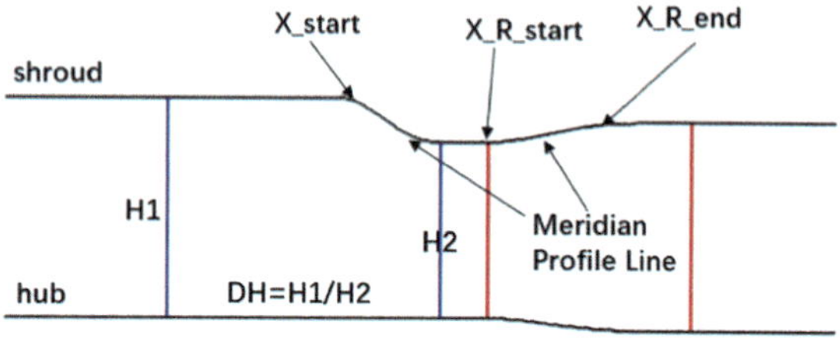

Fig. 5 Meridian flow channel optimization variables definition diagram

3 Optimization Method

3.1 Construction of the Optimization Platform

Figure 4 presents the optimization platform used for axial shaping design. It is based on thc principlcs of multi objective design optimization and combines endwall shaping parameterization, three-dimensional automatic simulation calculation methods for axial turbines, and intelligent optimization methods. The Multi-Objective Genetic Algorithm-II (MOGA-II) [11] is adopted as the optimization method for axial shaping. MOGA-II utilizes intelligent and efficient multi-search elitism, allowing it to preserve excellent (Pareto or non-dominated) solution sets without prematurely converging to local optima. The elitism mechanism improves the convergence of the algorithm, ensuring that the fitness of each new generation is greater than that of the previous generation. It is widely used by researchers.

The entire optimization process is illustrated in Fig. 4. First, parameterizing the modeling method for a given endwall contouring and establishing design variables. Then, applying the Design of Experiments (DOE) method to randomly generate initial values for the design variables within their specified ranges. Utilizing these samples to create geometric structures and performing Computational Fluid Dynamics (CFD) evaluations, obtaining aerodynamic performance results.Finally, evaluating whether the optimization objective function meets constraint conditions; if not, updating design variables using the MOGA-II algorithm. After 100 iterations of this process, an optimal result can be achieved.

Table 3 Variables optimized and its ranges

Parameters	Unit	Range of values
Position of onset of contraction of stator(X_start)	m	−0.083–−0.056
Shrinkage ratio of stator endwall(DH)	\	1.0–1.7
Starting position of expansion of rotor(X_R_start)	m	−0.049–−0.043
Rotor cutoff expansion position(X_R_end)	m	−0.043–−0.018

3.2 Optimization Variables, Objectives, and Constraints

Figure 5 illustrates the definition and location of each optimization variable in the meridional endwall contouring design optimization process. The optimization variables are listed in Table 3. Both the contraction and expansion lines are represented by Widosinski curve, and their expressions can be found in Eq. (1). Table 3 provides the range of values for these four optimization variables.

The optimization objective is to maximize the overall efficiency of the governing stage at three operating points. The objective function and the total overall efficiency η_{tt} of the governing stage, as defined in Eq. (2).

$$max F(X) = [\eta_{tt}(x)] \qquad \eta_{tt} = \frac{h^*_{in} - h^*_{out}}{h^*_{in} - h^*_{s\_out}} \tag{2}$$

In the equations, "h^*" represents the total specific enthalpy, "h^*_{in}" represents the specific enthalpy at the inlet, and "h^*_{out}" represents the specific enthalpy at the outlet. "$h^*_{s\_out}$" represents the isentropic total enthalpy at the outlet.

In the optimization process, constraints are used to limit the governing stage's inlet and outlet flow rates and the total pressure ratio to ensure that the overall design optimization does not deviate from the design point conditions and to achieve the goal of improving the efficiency of the governing stage. Mass flow and total pressure ratio constraints are given below.

$$\frac{|m_{out} - m_{in}|}{m_{in}} \le 0.005 \qquad 0.95PR_{ori} \le PR \le 1.05PR_{ori} \tag{3}$$

4 Results and Analysis

4.1 Optimal Meridional Contour and Performance Comparison

After over 1700 iterations of three-dimensional design optimization, the shaping of the meridian has converged. Figure 6a and b illustrate the convergence history of

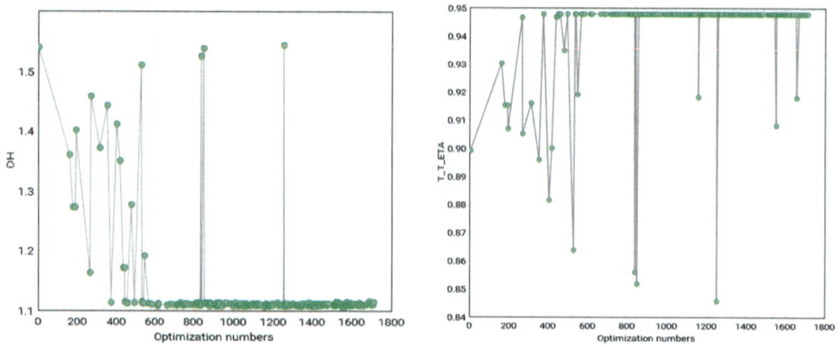

(a) The convergence history of the optimization variable DH

(b) The convergence history of the total-total efficiency in operating condition 1

Fig. 6 Convergence history of optimization variables

the design optimization variable DH and the optimization objective, which is the overall efficiency at operating point 1. From the graphs, it can be observed that after multiple generations of optimization, the optimization results have converged. Finally, the optimized starting contraction position of the stationary blade is at 0.8 times the axial chord length of the stationary blade row. The starting expansion position of the moving blade is at the inlet of the stationary blade row, and the ending expansion position of the moving blade is at the outlet of the moving blade row. The shrinking ratio is 1.11. Figure 7 provides a schematic comparison of the geometric changes in the meridional flow passage before and after optimization.

Figure 8 illustrates the variation of performance parameters before and after the meridional shaping optimization. Table 4 provides the numerical changes in performance before and after the meridional shaping optimization. It can be seen that the flow rate remains relatively unchanged for each operating point, with a relative error of less than 4%, which fall within the range of CFD calculation accuracy. It can be observed that compared to the original design, the flow rate increases. This is attributed to the implementation of a contraction-expansion channel, which enlarges the flow area, consequently increasing the mass flow rate. The overall efficiency of the three operating points is improved after optimization. The overall efficiency at operating point 1 is increased by 2.145%, at operating point 2 by 4.096%, and at operating point 3 by 5.421%. The performance at all three operating points has been improved significantly, especially at low-load conditions where the performance improvement is more pronounced.

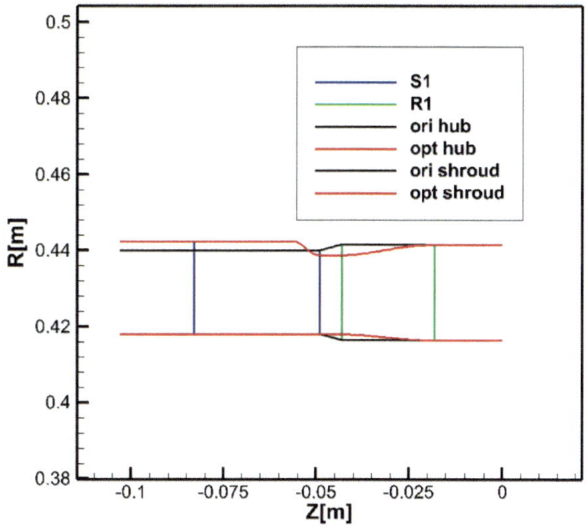

Fig. 7 Geometric comparison diagram of meridional flow channels

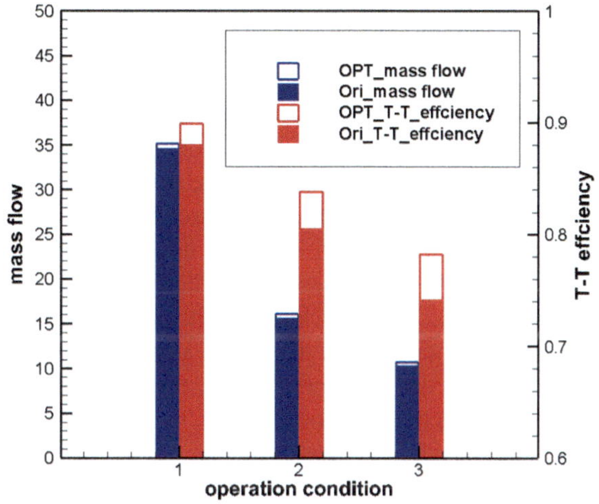

Fig. 8 Performance comparison before and after meridional modifications for three conditions

Table 4 Aerodynamic performance changes before and after optimizations

Parameters	Condition 1	Condition 2	Condition 3
Prototype CFD mass flow rate(kg/s)	34.59	15.627	10.333
Optimized CFD mass flow rate(kg/s)	35.17	16.1	10.744
Prototype CFD total-total efficiency	0.8806	0.80515	0.74185
Optimized CFD total-total efficiency	0.89949	0.83813	0.78207
Prototype CFD output power (kW)	2563.8	848.56	502.83
Optimized CFD output power (kW)	2640.6	904.41	547.45

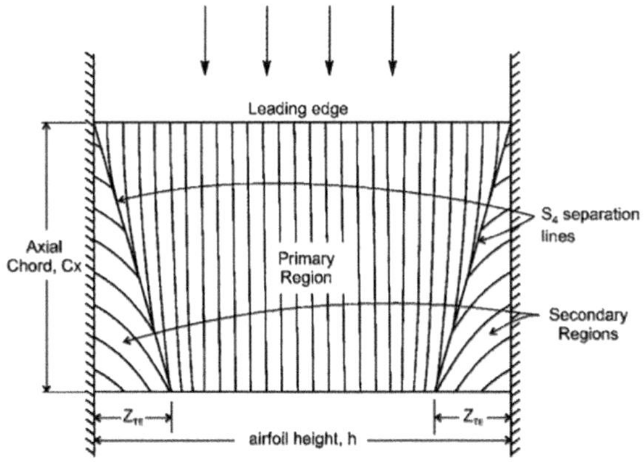

Fig. 9 Suction surface definition for the Benner loss breakdown scheme

4.2 Secondary Flow Analysis

In the study of secondary flow losses, Benner [12, 13] proposed a method to determine the spanwise location Z_{TE} (shown in Fig. 9) of the separation line of trailing edge vortex through the trailing edge vortex passage. Furthermore, a novel approach was proposed to reevaluate the decomposition of profile losses and secondary flow losses. In this approach, profile losses were defined as the losses generated within the surface boundary layer of the main region, specifically in areas where the blade surface boundary layer is essentially unaffected by secondary flows. The calculation methods for profile losses Y_P and secondary flow losses Y_S are presented in Eq. (4).

Benner believed that secondary flow losses are influenced by flow acceleration, endwall surface area, thickness of the inlet endwall boundary layer, and load distribution. The total load on the profile and the surface friction coefficient has minimal impact on secondary flows. Hence, they are not accounted for in the equation for calculating secondary flow losses. In the equation, $cos\gamma$ represents the load distribution on the blade. $\frac{cos\beta_1}{cos\beta_2}$ represents the gas flow acceleration, c_{ax}/h, $cos\beta_2$ represents the influence of endwall surface area, δ^*/h represents the ratio of mass flow rate in the endwall region to the mass flow rate in the channel, δ^*represents the displacement thickness of the inlet endwall boundary layer, the expression is given by Eq. (6),where x is the reference length parameter, typically taken as 0.5 times the axial chord length, ρ_1 represents the inlet density of the blade row, ω_1 represents the reference velocity at the inlet of the blade row, μ_1 represents the inlet viscosity coefficient of the blade row.Z_{TE} represents the spanwise location of the trailing edge vortex separation line, h/c represents the aspect ratio of the blade and the tangential load coefficient F_t represents the total blade load, as shown in Eq. (5).

$$Y_P = Y_{mid}\left(1 - \frac{Z_{TE}}{h}\right) \qquad Y_s = \begin{cases} \dfrac{0.038+0.41tanh(1.2\delta^*/h)}{\sqrt{cos\gamma}\left(\frac{cos\beta_1}{cos\beta_2}\right)(h/c)^{0.55}\left(\frac{c\cdot cos\beta_2}{Cax}\right)^{0.55}} & h/c \le 2 \\ \dfrac{0.052+0.56tanh(1.2\delta^*/h)}{\sqrt{cos\gamma}\left(\frac{cos\beta_1}{cos\beta_2}\right)(h/c)\left(\frac{c\cdot cos\beta_2}{Cax}\right)^{0.55}} & h/c > 2 \end{cases} \tag{4}$$

$$\frac{Z_{TE}}{h} = \frac{0.1\,(F_t)^{0.79}}{\sqrt{\frac{cos\beta_1}{cos\beta_2}}\left(\frac{h}{c}\right)^{0.55}} + 32.7\left(\frac{\delta^*}{h}\right)^2 \qquad F_t = 2\frac{s}{C_{ax}}cos^2\beta_m\,(tan\beta_1 + tan\beta_2) \tag{5}$$

$$\delta^* = \frac{\delta}{8} = \frac{0.0463x}{(\rho_1\omega_1 x/\mu_1)^{0.2}} \tag{6}$$

As shown in Eq.(7), building upon the Benner secondary flow loss model described above, the effect of the contraction of the meridional passage is taken into account by introducing the upper wall meridional contraction angle 'θ' to enhance the predictive accuracy of secondary flow losses beneath the meridional shaping. For the case of $h/c \le 2$, the revised model for secondary flow losses is also presented as Eq. (7).

$$Y_s = \begin{cases} cos^b\,(\theta - a)\,\dfrac{c+dtanh(1.2\delta^*/h)}{\sqrt{cos\gamma}\left(\frac{cos\beta_1}{cos\beta_2}\right)(h/c)^{0.55}\left(\frac{ccos\beta_2}{Cax}\right)^{0.55}} & h/c \le 2 \\ \dfrac{0.052+0.56tanh(1.2\delta^*/h)}{\sqrt{cos\gamma}\left(\frac{cos\beta_1}{cos\beta_2}\right)(h/c)\left(\frac{ccos\beta_2}{Cax}\right)^{0.55}} & h/c > 2 \end{cases} \qquad \theta = atan\left(\frac{h_1 - h_2}{C_{ax}}\right) \tag{7}$$

In the equation, 'a', 'b', 'c', and 'd' are all underdetermined constants. Among these, the constant 'a' represents the optimal upper wall meridional contraction angle. For the three cases, by fitting and employing a linear programming approach with constants 'a', 'b', 'c', and 'd' as variables to minimize variance, the values are determined as follows: $a = 7.909$, $b = 9.028$, $c = 0.03989$, $d = 0.4304$. In order to further evaluate the changes in internal losses before and after meridional shaping, the individual losses within the stages were calculated for each operating condition. Since the tip clearance of the moving blades was not considered, only the losses due to blade profile (Y_p) and secondary flow (Y_s) were involved.

Figure 10 presents a schematic representation of the changes in internal losses within the stages.while Table 5 provides the specific numerical variations.

From Fig. 10, it can be observed that the internal losses are highest for operating condition 3 and lowest for operating condition 1. The internal losses of the stationary and moving blades within the stage have both decreased, with a significant reduction in secondary flow losses. This indicates that an appropriate meridional configuration can effectively reduce secondary flow losses. From the graph, it can be observed that the internal losses are highest for operating condition 3 and lowest for operating condition 1. The internal losses of the stationary and moving blades within the stage

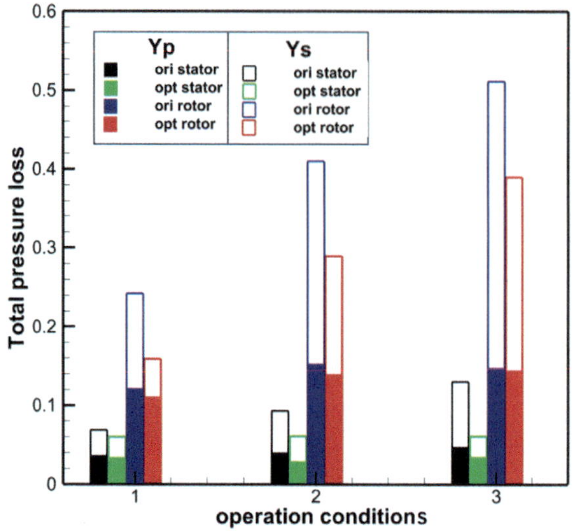

Fig. 10 Comparison of internal losses before and after meridional modifications for three conditions

Table 5 Comparison of internal losses before and after meridian modifications

Parameters	Total loss_Ori	Total loss_Opt	Blade loss_Ori	Blade loss_Opt	Secondary flow loss_Ori	Secondary flow loss_Opt
Condition 1 stator	0.06862	0.06028	0.03646	0.03460	0.03216	0.02568
Condition 1 rotor	0.24234	0.15863	0.12107	0.10688	0.12127	0.05175
Condition 2 stator	0.09314	0.06118	0.04045	0.02937	0.05269	0.05269
Condition 2 rotor	0.40961	0.28911	0.15237	0.13017	0.25724	0.15894
Condition 3 stator	0.12952	0.06101	0.04770	0.03507	0.08182	0.02593
Condition 3 rotor	0.51172	0.38966	0.14782	0.13200	0.36391	0.25766

have both decreased, with a significant reduction in secondary flow losses. This indicates that an appropriate meridional configuration can effectively reduce secondary flow losses.

From the data given in Table 5, it can be seen that the internal losses are reduced after adopting the meridional shaping optimization. Among them, operating condition 3 shows the most significant reduction in internal losses, with a 68% decrease in stationary blade secondary flow losses and a 28.2% decrease in moving blade secondary flow losses. This results in a significant improvement in performance.

4.3 Flow Field Analysis

In order to further understand the mechanism of secondary flow reduction through meridional contouring, the analysis of internal flow field adjustments is conducted for three operating conditions.

Figure 11 presents entropy distribution contour maps at the plane of 10% axial chord downstream of the original and optimized stator blade outlets for the three conditions. It can be observed from the figure that Condition 1 has the lowest loss while Condition 3 exhibits the highest loss. High entropy augmentation is primarily distributed near the trailing edge and at the boundary layers of the upper and lower endwalls. For the former part, this is attributed to the significant loss generated by the trailing edge vortex and passage vortex in the flow field.

With the implementation of meridional endwall contouring, the entropy increase for all three conditions has been reduced, resulting in diminished losses within the internal stator passages. This reduction is mainly evident in the lower endwall region. Therefore, it can be concluded that the strength of the optimized trailing edge vortex and passage vortex has been suppressed, leading to a decrease in the losses associated with internal secondary flows within the stator.

Figure 12 shows the entropy distribution contours at the plane of 20% axial chord downstream of the rotor blade outlet before and after the optimization for Condition 2. The contours are similar for the other two conditions. Comparing to the entropy distributions for the stator, it can be found that the locations of high entropy augmentation regions are shifted. This phenomenon is attributed to the centrifugal force caused by the rotor rotation, which accelerates the development of passage and trailing edge vortices from the lower endwall towards the blade tip.

After the implementation of meridional contour optimization, the scale of the high entropy augmentation regions for all three operating conditions has been reduced. The radial height of the high entropy augmentation region has also decreased. This indicates that the meridional contour optimization further suppresses the development of passage vortices within the rotor blades, subsequently reducing the secondary flow losses within the rotor passages.

Based on the comprehensive flow field analysis presented above, it is evident that appropriate meridional contouring can effectively suppress the generation and development of endwall passage vortices, thereby reducing their strength and scale. Consequently, this optimization leads to a decrease in the secondary flow losses within the blade passages and contributes to an enhancement in the efficiency of the stage.

Figure 13 presents the static pressure distribution diagrams at the 95% blade height cross-section of the rotating and stationary blades before and after optimization under various operating conditions. The horizontal axis represents the axial relative position, while the vertical axis represents the dimensionless static pressure. Regarding the stationary blade section $P = P_s/P_{in}^*$, P_S is the local static pressure. P_{in}^* the total pressure at the stage inlet. Regarding the rotor section $P = P_S/P_{int}^*$, P_{int}^* is the total pressure at the inlet of the moving blade.

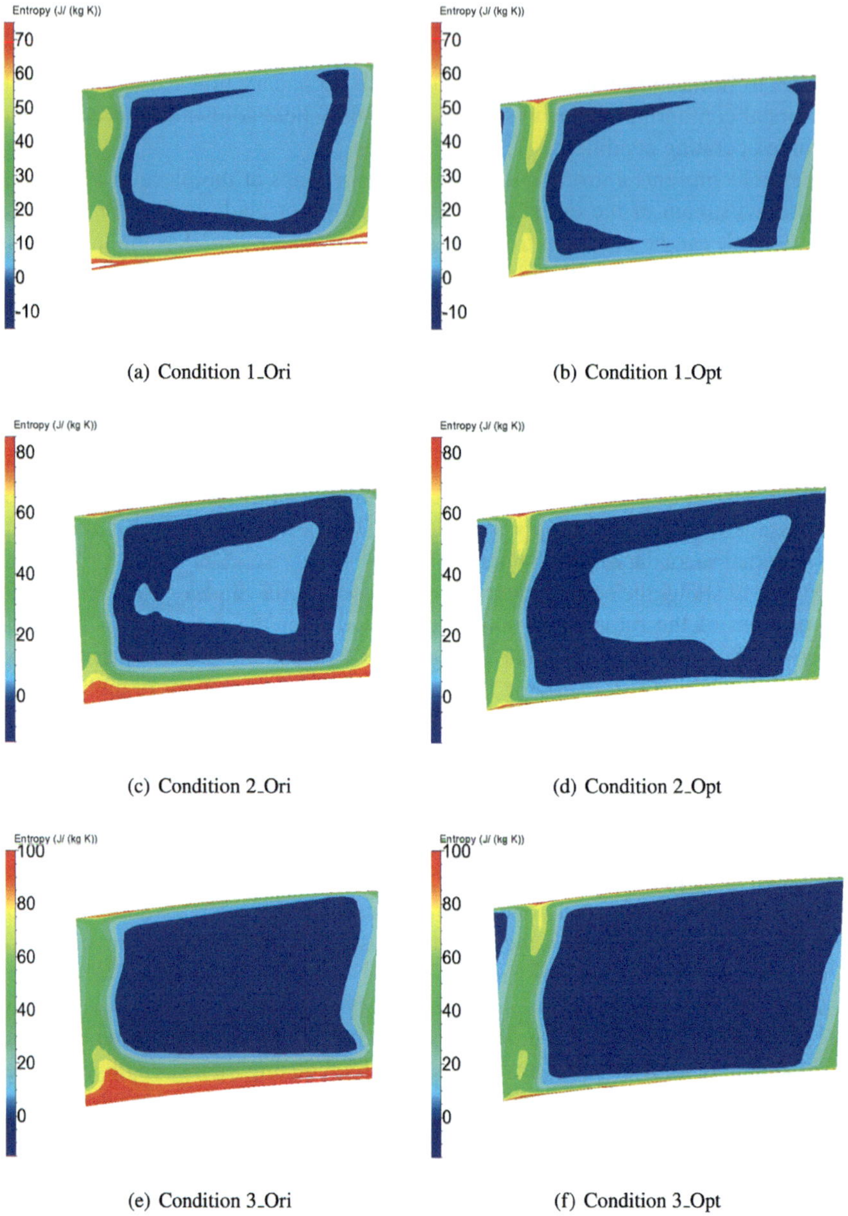

Fig. 11 Entropy distributions at the plane of 10% axial chord downstream of the original and optimized stator outlets

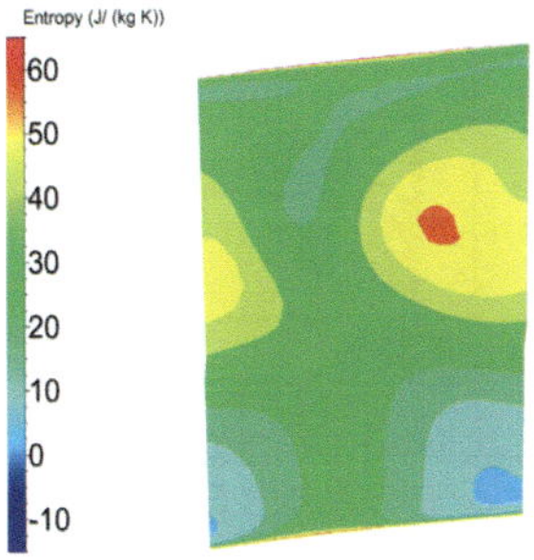

(a) Condition 2_Ori

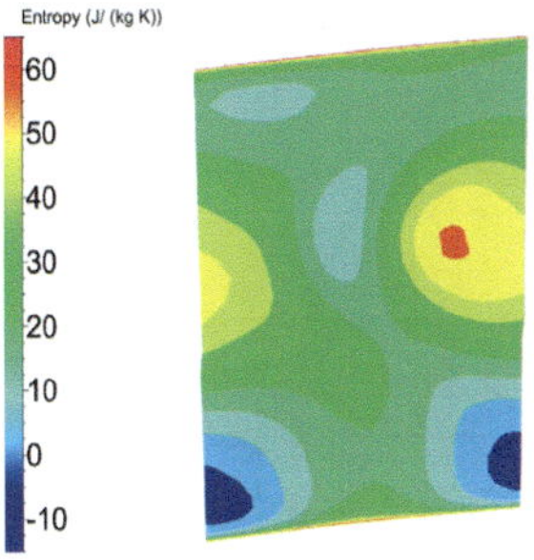

(b) Condition 2_Opt

Fig. 12 Entropy distribution nephograms at the plane of 20% axial chord downstream of the rotor blade outlet for Condition 2

From Fig. 13, considering the 95% blade height cross-section of the stationary blade it can be seen that, the load on the stationary blade is primarily concentrated in the mid-to-aft section of the blade. The static pressure on the pressure side remains relatively constant before 60% of the axial chord length and then rapidly decreases in the latter half. Before optimization, the static pressure on the suction side gradually decreases before 50% of the axial chord length, followed by a short region of adverse pressure gradient. The boundary layer on the suction side grows rapidly at this point. Beyond 95% of the axial chord length, both the suction and pressure sides exhibit significant fluctuations in static pressure due to the complex vortex system near the trailing edge.

After employing the meridional contraction profile, the static pressure on the suction side of the stator blade is elevated before 50% of the axial chord length, resulting in reduced transverse pressure gradients and decreased transverse secondary flow. This alteration is critical in mitigating losses associated with secondary flows. In comparison to the pre-optimized state, the post-optimized pressure distribution on the pressure side of the blade is more uniform and exhibits notable reduction. Additionally, beyond 65% of the axial chord length, the static pressure on the suction side increases, and transverse secondary flow at the trailing edge diminishes.

Simultaneously, it can be observed from the graph that before 50% of the axial chord length for the stationary blade and after 65% of the axial chord length for the moving blade, meridional contour optimization results in a noticeable reduction in pressure difference between the pressure and suction sides. This effectively minimizes channel vortices and transverse secondary flow, contributing to the reduction of entropy losses as depicted in Fig. 12.

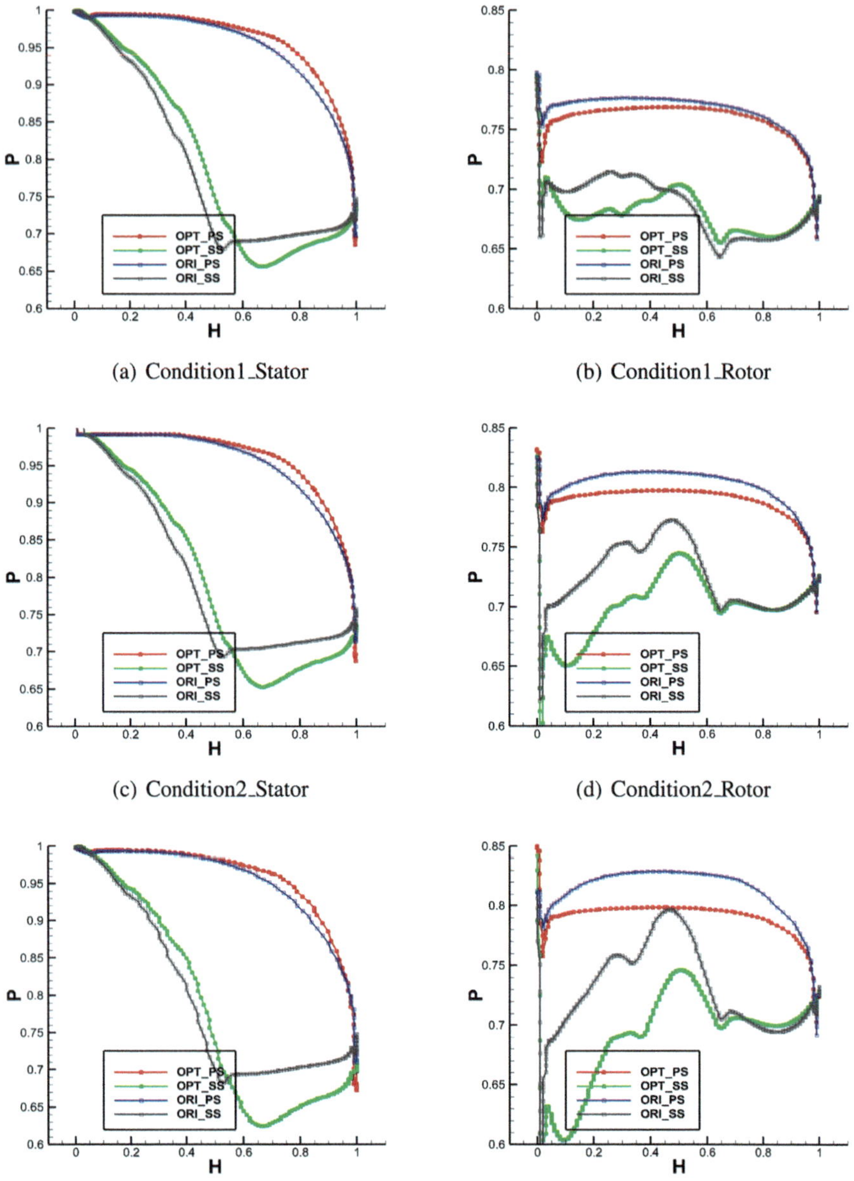

(a) Condition1_Stator (b) Condition1_Rotor

(c) Condition2_Stator (d) Condition2_Rotor

(e) Condition3_Stator (f) Condition3_Rotor

Fig. 13 Static pressure distribution diagram at 95% blade height cross-section

5 Conclusions

Meridional endwall contouring using Widosinski curve for a steam turbine regulating stage was studied in this study. In addition, a multi-objective optimization method was applied using the MOGA-II optimization algorithm for multiple operating conditions. Consequently an optimal meridional configuration was obtained. Based on the results obtained, the following conclusions can be drawn:

(1) There exists an optimal meridional contraction ratio for a steam turbine regulating stage. For the current studied stator blade of the regulating stage, the optimal meridional contraction ratio is 1.11, in which the regulating stage performs best with less secondary losses.
(2) For a contracted meridional endwall, there also exits an optimal starting contraction position of the stationary blade being at 0.8 times the axial chord length of the stationary blade rows, while the starting expansion position of the moving blade is at the inlet of the stationary blade rows, and the ending expansion position of the moving blade is at the outlet of the moving blade rows. Under this meridional configuration, the overall efficiency is increased by 2.145%, 4.096%, and 5.421% respectively for Condition 1, 2 and 3. The performance of the stage significantly is improved under all three operating conditions, particularly at low load conditions.
(3) With regard to secondary flow losses, the internal losses of both the stationary and moving blade rows are reduced for the contoured top endwall configuration. Specifically for Condition 3 with the lowest loading, there exhibits the highest reduction in internal losses, with a 68% decrease in stationary blade internal secondary flow losses and a 32.4% decrease in moving blade internal secondary flow losses. This explains why the contoured top endwall configuration can improve performance of the steam turbine stage.

References

1. Ni, Z., Li, C., Liu, F., et al.: Experimental study on reducing secondary flow loss in turbine cascade. J. Eng. Thermophys. **10**(4), 383–386 (1989)
2. Deich, M.E., Zaryankin, A.E., Fillipov, G.A., et al.: Method of increasing the efficiency of turbine stages with short blades. Teploenergetika **2**, 240–254 (1960)
3. Kopper, F.C., Milano, R., Vanco, M.: Experimental investigation of endwall profiling in a turbine vane cascade. AIAA J. **19**(8), 1033–1040 (1981)
4. Morris, A.W.H., Hoare, R.G.: Secondary loss measurements in a cascade of turbine blades with meridional wall profiling. In: American society of mechanical engineers, winter annual meeting, Houston, Tex, p. 1975 (1975)
5. Arts, T.: Effects of tip endwall contouring on the three dimensional flow field in an annular turbine nozzle guide vane: Part 2 - numerical investigation. volume Volume 1: Aircraft Engine; Marine; Turbomachinery; Microturbines and Small Turbomachinery of Turbo Expo: Power for Land, Sea, and Air, page V001T03A031 (1985)

6. Boletis, E.: Effects of tip endwall contouring on the three-dimensional flow field in an annular turbine nozzle guide vane: Part 1-experimental investigation. J. Eng. Gas Turbines Power **107**(4), 983–990 (1985)
7. Petrovic, M., Dulikravich, G.S., Martin, T.J.: Maximizing multistage turbine efficiency by optimizing hub and shroud shapes and inlet and exit conditions of each blade row. Int. J. Turbo Jet Eng. **17**(4), 267–278 (2000)
8. An, B., Han, W., Wang, S. et al.: Effect of large expansion angle meridional flow path profile on losses. J. Propul. Technol. pp. 211-214 (2001)
9. Song, Y., Meng, F., Cao, F., et al.: Study on the influence of meridional profile modification on the flow and heat transfer performance of 1.5-stage large meridional expansion turbine end region. J. Propul. Technol. **42**(1786-1797), (2021)
10. Zhang, W., Ren, D., Zhou, F.: Optimization design of 600mw steam turbine regulating stage. Turbine Technol. (27-32), (1990)
11. Ennil, A.B., Al-Dadah, R., Mahmoud, S., et al.: Minimization of loss in small scale axial air turbine using cfd modeling and evolutionary algorithm optimization. Appl. Therm. Eng. **102**, 841–848 (2016)
12. Benner, M.W., Sjolander, S.A., Moustapha, S.H.: An empirical prediction method for secondary losses in turbines-part 1: a new loss breakdown scheme and penetration depth correlation (2006)
13. Benner, M.W., Sjolander, S.A., Moustapha, S.H.: An empirical prediction method for secondary losses in turbines-part 2: a new secondary loss correlation (2006)

Compressor Disc Cavity Simulations

S. M. Fazeli, V. Kanjirakkad, and C. A. Long

Abstract The flow and heat transfer inside HP compressor rotating disc cavities are buoyancy driven and are known to be extremely difficult to predict. The coupled nature of the disc thermal behaviour and the adjacent fluid flow within the disc cavity and shroud could form several interesting flow structures that are governed by the thermal boundary conditions.This makes the numerical modelling of such flows very challenging. Despite the use of URANS by several researchers to predict disc cavity flow-field, many LES studies reported in the literature argue that the overall structure and the small-scale features present in such flows cannot be captured by URANS due to turbulence modelling limitations. It has also been argued that URANS predicts advection of locally generated eddy viscosity to regions where it is unphysical and that it is unable to capture the buoyancy influences, stratification, rotation, and flow transition occurring within heated rotating cavities. All the while, URANS has been successfully used to capture the formation of single or multiple radial arms and the associated vortex pairs, and the formation of near-shroud flow structures such as the existence of circumferential vortex streaks. This paper presents comparisons between simulation results using URANS and LES and compare them with experimental data obtained from the Sussex multiple cavity rig. Both flow structure and thermal data are compared to further understand the differences between the two simulation methodologies from a thermal designer's perspective.

CUFS-2024-PAPER ID, Cambridge, United Kingdom, 4-5 March 2024.

S. M. Fazeli (✉) · V. Kanjirakkad · C. A. Long
Thermo-Fluid Mechanics Research Centre, School of Engineering and Informatics, University of Sussex, Falmer, England
e-mail: s.fazeli@sussex.ac.uk

V. Kanjirakkad
e-mail: v.kanjirakkadr@sussex.ac.uk

C. A. Long
e-mail: c.a.long@sussex.ac.uk

J. C. Tyacke and N. R. Vadlamani (eds.), *Proceedings of the Cambridge Unsteady Flow Symposium 2024*, https://doi.org/10.1007/978-3-031-69035-8_20

1 Introduction

The flow and heat transfer within the cavity between the adjacent discs of a gas turbine HP compressor are complex due to the large thermal gradients that exists between the disc shrouds and the disc bore air. The cavity flow thermal behaviour is critical in controlling the disc radial growth. Engine designers have the unenviable challenge of estimating and allowing for safe blade tip clearances especially during engine transients such as take-off, landing or a hot-restart.

The shroud and disc in a rotating cavity are hotter than the air of the axial throughflow at the lower radii of the disc. This causes the development of buoyancy-induced flow regime in the outer region of the cavity which in turn results in large scale pairs of cyclonic and anti-cyclonic circulations (low pressure and high pressure regions respectively) inside a heated rotating cavity. Figure 1 shows a simplified diagram of an HP compressor disc arrangement with axial throughflow of cooling air (a) and a schematic of the non- axisymmetric flow structure inside a rotating cavity (b).

The quest for improved propulsion efficiency has lead engine manufacturers to move towards ultra-high bypass ratio engines with small cores. Distributed propulsion concepts would also require small core engines in the future. The management of tight tip clearances will be a challenge as the core size reduces. Therefore, understanding and controlling the aerothermal behaviour of disc cavities will be of paramount interest.

Although it has been argued that URANS (Unsteady Reynolds averaged Navier-Stokes simulation) cannot accurately predict the flow structure and features in rotating cavities in detail due to the limitation of turbulence modelling required in this method, this method has been widely used in the literature since the advent of high-performance computers. It was reported, notably by practitioners of LES (Large Eddy Simulation), that former technique; is not suitable for compressor cavity flows [1], causes advection of locally generated eddy viscosity to regions where it is unphysical [2] and that it is unable to capture the buoyancy influences, stratification, rotation and flow transition [3]. On the other hand, Puttock-Brown et al. [4, 5] investigated near-shroud flow structures and observed the existence of circumferential vortex

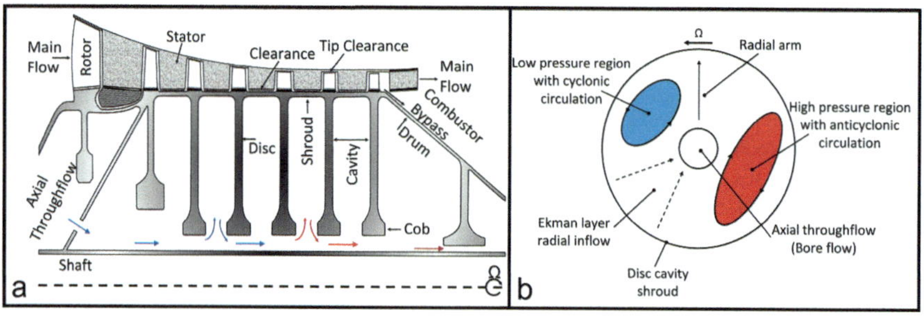

Fig. 1 **a** Simplified diagram of high-pressure compressor rotor with axial throughflow of cooling air ($r - z$), **b** Schematic of flow structure in heated rotating cavity with axial throughflow ($r - \phi$)

streaks. However their results showed an under prediction of the heat-flux by 31%. Fazeli et al. [6, 7] also investigated rotating cavity flow structure and compared the LDA (Laser Doppler Anemometry) experimental data with URANS calculations. It was stated that the CFD (Computational Fluid Dynamics) results had a promising agreement with the velocity components as well as with the frequency of vortex pairs inside the cavity. A similar study of heat transfer and flow structures using URANS is discussed in Dweik et al. [8, 9] respectively.

Tan et al. [10] compared URANS and LES simulations based on the experimental data of [11] and found that the level of errors in the disc heat transfer prediction is similar however the range of heat flux fluctuations is remarkably higher than those in URANS. Therefore, it is thought that a comparison between URANS and LES results based on experimental conditions from the Sussex multiple cavity rig (MCR) rig and their comparison with the experimental data will provide further elucidate on the usefulness of predictions (e.g., velocities, Nusselt number) from a turbine designer's perspective.

2 Previous Work

For a rotating cavity with an axial throughflow, the following dimensionless parameters are used; the axial and rotational Reynolds numbers, Re_z and Re_θ, the Rossby number Ro and since free convection can be significant the Grashof number, Gr. The buoyancy parameters and normalised temperature are $\beta \Delta T$, and Θ respectively. These are defined in Table 1.

Tucker [12] and Long and Tucker [13] used 3D unsteady laminar CFD on a simple heated rotating cavity similar to that used by Farthing et al. [14] with a cavity inlet to shroud radius ratio (a/b) of 0.1 and a cavity aspect ratio (width/height) of $G = 0.267$ for modest dimensionless parameters; $R_z = 2180$ and $R_\theta = 1.3 \times 10^4$. Four different radial temperature profiles obtained from two symmetrically heated discs were investigated. The numerical results revealed that the axial flow would enter the cavity in one or two radial arms and form with two regions of cyclonic and anticyclonic circulations, which was consistent with the observation by Farthing et al. [15].

Tucker [16] explored the temporal behaviour of the three-dimensional flow in a heated rotating cavity with axial throughflow and found good agreement with Farthing et al. [15] with regards to the cavity flow structure (vortex pair and radial arm).

Table 1 Governing dimensionless parameters for cavity flows

$Re_z = \dfrac{\rho W d_h}{\mu}$	$Ro = \dfrac{W}{\Omega a} = \dfrac{b^2}{2a(a - rs)} \dfrac{Re_z}{Re_\theta}$	$\beta \Delta T = \dfrac{(T_{sh} - T_{in})}{T_{in}}$
$Re_\theta = \dfrac{\rho \Omega b^2}{\mu}$	$Gr = \dfrac{\rho \Omega^2 b \beta \Delta T L^3}{\mu^2}$	$\Theta = \dfrac{T(r) - T_{in}}{T_{sh} - T_{in}}$

Sun et al. [17] used the experimental data of Long et al. [18] and Alexiou [19] to compare URANS calculations using RNG $k - \epsilon$ turbulence model and Large Eddy Simulation (LES) calculations. The LES prediction of the Nusselt numbers (Nu) on the shroud was found to be accurate while the RANS calculations showed under-predicted values. The comparison also showed that although there was good agreement with RANS data at the higher radii in terms of the ratio of relative tangential velocity to disc speed, with a core flow rotating slower than the discs, LES showed a better performance overall.

Tan et al. [10] conducted 3D steady RANS, URANS and LES calculations on three cases (2–4) presented in [11] and compared their results with experimental data. It was found that the URANS method, with a standard $k - \epsilon$ model and enhanced wall treatment, showed the best results. Although both URANS and LES predicted the circulation zones in the cavity and showed reasonable agreement with heat transfer measurements, the steady RANS failed to attain the flow structure in the cavity and only provided an acceptable prediction of the heat transfer.

Atkins and Kanjirakkad [20] used an in-house hybrid RANS/LES method to investigate the flow structure inside the rotating cavity for two cases at $Re_z = 1.1 \times 10^5$, $Re_\theta = 3.1 \times 10^6$, $\beta \Delta T = 0.15$ and 0.32. The numerical results showed that the flow turned from a relatively stable and stratified regime to an unstable Rayleigh-Bénard convection type cells with enhanced mixing in the cavity for increasing $\beta \Delta T$. It was shown that as the $\beta \Delta T$ value is increased, two radial arms were formed to provide the path for cold air from the bore regions to travel radially outwards and for hot air to leave the shroud radially inwards.

Pitz et al. [21] used Large Eddy Simulations (LES) to simulate the buoyancy-induced flow structure in a closed cavity with heated or adiabatic discs. The results showed the flow structure to have an important effect on the unsteady laminar Ekman boundary layer near the discs. The velocity contours showed that the flow structure inside the cavity consisted of pairs of counter-rotating cells. The temperature contours indicated that the temperature field was uniform throughout the cavity core, which is constrained by the shaft and shroud thermal boundary layers. The results were consistent with the DNS results reported by [22].

Kouwa et al. [23] employed Very Large Eddy Simulations (VLES) using Lattice-Boltzmann method to numerically model the experimental data reported by [24]. They studied two operating conditions with different rotation speeds. The range of simulated working parameters reported are; $Re_z = 1.6 \times 10^5$, $Re_\theta = 2.0 \times 10^5$ and 2.0×10^6, $Ro = 4.64$ and 46.4. The instantaneous near wall temperature contours on the upstream and downstream discs showed that the cold air enters the cavity through a radial arm near the disc upstream and travels radially outward to the shroud, with the counter-rotating cells formed on the sides of the radial arm. The hot air flows radially inwards to exit the cavity.

Hickling and He [3] used a new LES-CHT (conjugate heat transfer) method to simulate a rotating heated cavity with axial throughflow. The method uses a multi-scale frequency domain approach. Two studies were reported with different materials. A low/high conductivity disc's materials were respectively used for axially heated and radially heated by shroud in a rotating cavity. They found that the conditions of how the cavity is heated can significantly influence the formation of small-scale flow structure inside the cavity (near the shroud and near the disc). The comparisons of two cases revealed that the buoyancy-driven streaks were formed when the case with a heated shroud. For âŁ˜the axially-heatedâŁ™ case, at higher Nu, the mixing of hot near-disc air and the cold radially outward flow in the core caused the formation of a vortical structure in the radial-tangential plane.

3 Test Facility

The Sussex Multiple Cavity Rig (MCR) has been used to investigate the heat transfer and flow structure within the disc cavities that are typical of aero-engine HP compressors. The current test rig is a 70% scaled replica of a typical high pressure compressor internal air system. A comprehensive details of the test rig is reported in [6]. The simulations use measured steady state thermal conditions from the MCR as boundary conditions. For this purpose the âŁ˜steady stateâŁ™ is considered to be achieved when all air and metal thermocouples register a temperature variation of less than 0.2 K over a period of 5 minutes.

4 Numerical Methodology

A numerical study using URANS and LES was conducted in order to aid the interpretation of the flow field and heat transfer for a case with the following non-dimensional parameters: $Ro = 0.16$, $Re_z = 2.47 \times 10^4$, $Re_\theta = 2.85 \times 10^6$, $Gr = 8.7 \times 10^{11}$, and $\beta \Delta T = 0.34$. To achieve the best agreement with the experimental conditions, the flow domain for the current CFD modelling included the entire upstream flow regions and two cavities. ICEM-CFD 17.1 and Ansys Fluent 19.2 are employed to generate a high-resolution grid and to simulate the turbulent flow within the MCR. The relative frame of reference is used to solve the velocity components.

For URANS calculations, the $k - \omega$ SST turbulence model is used which covers both $k - \epsilon$ and $k - \omega$ models and benefits from automatic near wall treatment that switches to low-Reynolds number formulation in the viscous sublayer [9]. The Boussinesq approximation was also used to accommodate buoyancy effects. Second-order temporal and spatial discretisation is used by the pressure-based solver.

A second order upwind scheme along with the SIMPLEC algorithm were used for coupling the velocity and pressure in both URANS and LES calculations. The Smagorinsky-Lilly model from the CFD package Ansys Fluent was used (SGS model) in order to employ the Boussinesq hypothesis. The eddy-viscosity (μ_t) is modelled here which depends on the mixing length (L_s) for Subgrid scales, density and the Smagorinsky or Lilly constant values [25]. Although Lilly [26] proposed an average value for C_s for homogeneous isotropic flows, further research have shown that Lilly derived a value for Smagorinsky constant which was found unsuitable for flows near solid boundaries.

$$\mu_t \propto \rho L_s^2 |\bar{S}| \ \& \ L_s \propto C_s \tag{1}$$

The numerical time step of the transient simulations is equivalent to 1° of a revolution and the solution was run for at least 100 complete revolutions for URANS and 40 complete revolutions in LES. The convergence of each time step was completed when the equation residuals reached below 1×10^{-4}. The area-average heat fluxes on the cavity surfaces were monitored for convergence. The time taken for URANS solution (≈ 100 revs) and for LES (≈ 40 revs) was about 56 and 80 days respectively while 96 CPUs on the University of Sussex HPC were used.

4.1 Domain Geometry and Specifications

To achieve the best match with the experimental conditions most components of the rig inlet flow path were simulated as they contained circumferentially non-uniform features. The first two cavities in the flow direction (from right to left in the rig) are also modelled. The meridional and isometric view of the CFD geometry in Fig. 2.

A 2D multi block grid, including 2.5×10^4 nodes, was created and revolved over 360 degrees in the circumferential direction around the rig axis in order to generate a high-resolution 3D domain for each cavity. There are over 9 million grid points within each cavity. For LES calculation the number of mesh elements was more by a factor of 1.5 compared to the one used for URANS that was described in [6]. A number of studies [4, 8, 10] that employed URANS to predict flow and heat transfer in a similar sized cavity use mesh counts in the region of 0.8 to 6.5 million cells. For those studies that employ LES [23, 30, 31] the number of cells in the cavity range from 2.4 to 11 million. The current study uses over 9 million cells for the URANS

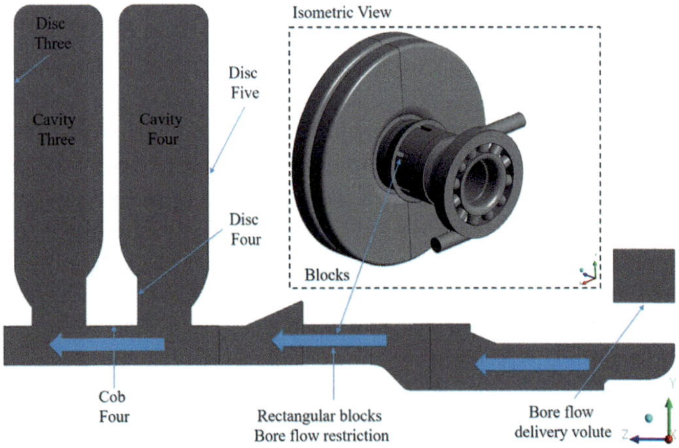

Fig. 2 Meridional along with isometric view of the CFD geometry

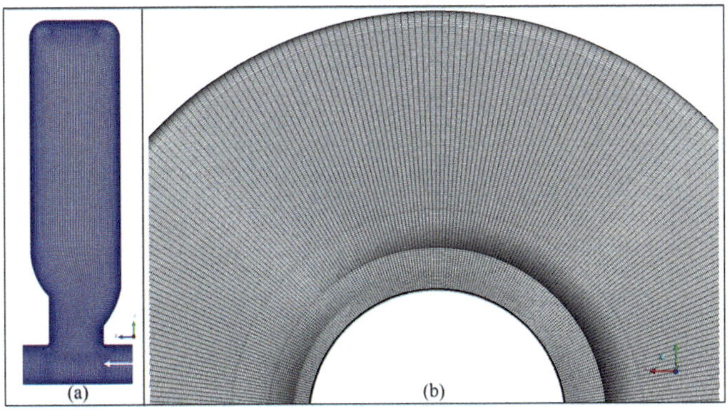

Fig. 3 Meridional view of the computational grid (**a**); circumferential view of the domain walls (**b**)

and 13.5 million cells for the LES. In the current study, the URANS mesh resolution is much higher than those reported, the LES mesh in the current study uses similar but slightly higher cell count. The method of Dweik et al. [8] was used to create 25 points in the first 2 mm normal to the surface with the first point located 0.005 mm from the walls (expansion ratio of 1.1). A logarithmic bunching algorithm is then used in order to reach a high-resolution boundary layer with $y^+ < 1$ on the cavity walls and shroud region. The time steps used for both URANS and LES are well within those recommended based on the smallest cell size. It should be noted that in order to reduce the computational time, the upstream region is of lower resolution. Figure 3 depicts the meridional view ($r - z$ plane) of the computational domain and the circumferential view ($r - \phi$ plane) of the domain walls.

The measured temperatures from the experiments are used to define the boundary conditions for all surfaces on the model. An isothermal wall boundary condition was used for the upstream section; these temperatures were extracted from measured conditions. To define the radial temperature profiles, the curve fitting method is used. Due to the absence of sufficient numbers of thermocouples, particularly downstream of disc 5 (cavity 4) and in the upstream inlet region, temperature values from a matched thermal model were used. A total pressure inlet boundary is used while the static pressure is specified at the outlet to match the mass flow rate. The disc of interest in this paper is disc-3 and the results shown are mostly from cavity-3 mid-axial plane and the upstream of disc-3.

5 Results

Figure 4 compares the normalised radial velocity ($V_r/V_r{}^*$) contours over the shroud surface and mid-axial plane respectively. Here V_r is the radial velocity and $V_r{}^*$ is the maximum theoretical radial velocity at the cavity mouth defined as $V_r{}^* = \dot{m}/2\pi\rho a s_c$.

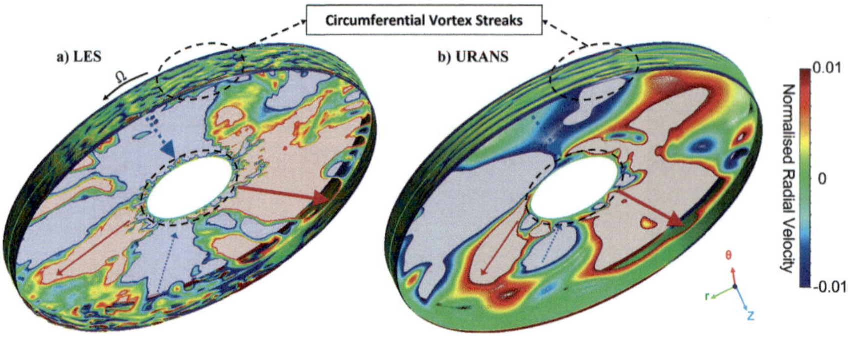

Fig. 4 Contour plots (iso-surface) of instantaneous normalised radial velocities at the shroud and the mid-plane; **a** LES, **b** URANS (for clarity of the contour scale used for the shroud is $1/25^{th}$ of that for the cavity mid-plane)

$\dot{m}$ is bore mass flow rate, a is the disc bore radius and s_c is the axial disc spacing between cobs at the cavity inlet. Consistent with [6], all contour plots presented in this paper show the $r-\phi$ plane as viewed from downstream looking towards the inlet. Although the URANS results show the regions of negative and positive radial velocities that are created by the cyclonic and anticyclonic circulations (seen annotated) inside the cavity, the weaker and stronger structures are evidently shown and distinguished at 3, 6, 9, and 12 o'clock in the $r-\phi$ plane by LES calculations. Note that the LES is able to capture the small-scale features in the high shear regions near the bore compared to URANS.

As the cavity flow evolves through more revolutions and the circumferential pressure gradient necessary for the air to travel inside the cavity (cold air in the bore-hot air around the shroud) is created by the cyclonic and anticyclonic circulations, the Rayleigh-Bénard like structures starts to emerge in the $r-\phi$ plane and the circumferential streaks appear near the shroud region. As seen in Fig. 4a, LES shows these streaks in detail and they are formed clearer than those in URANS, however URANS has a reasonable prediction of the location and formation of the structures. A comprehensive discussion of the differences between LES and URANS in terms of predicting the flow structures and heat transfer within the rotating cavity is presented in the following sections.

5.1 Aerodynamic Flowfield Comparisons

Figure 5 shows the comparisons of normalised radial and relative circumferential velocities and reduced static pressure from the URANS and LES calculations. The view shown is that of the radial-circumferential ($r-\phi$) plane of the mid-axial location ($z/s = 0.5$) of cavity-3. It is worth noting that all data presented in circumferential view are in the relative frame of reference. Figure 5a represents the instantaneous

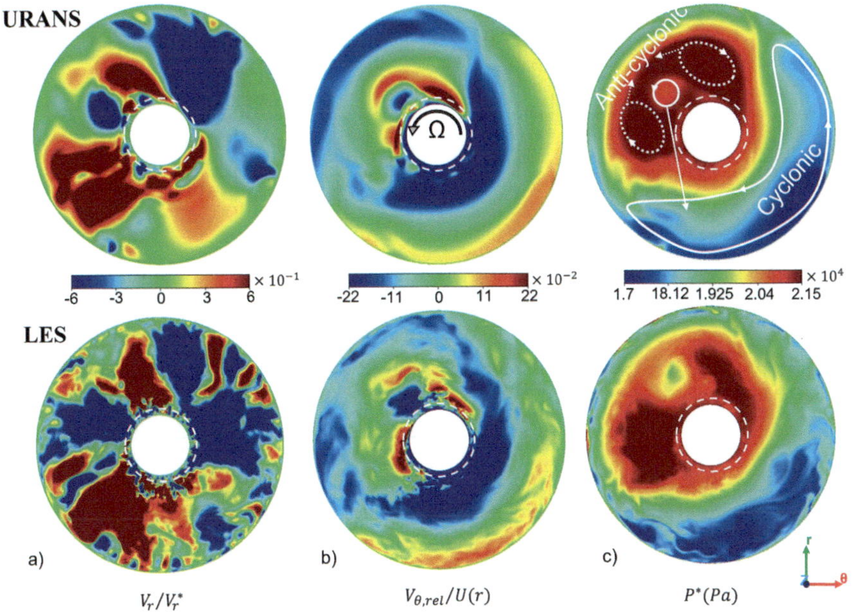

Fig. 5 Comparison of instantaneous **a** radial velocity, **b** relative circumferential velocity, and **c** reduced static pressure in the mid-plane of cavity-3 in the $r - \phi$ plane

normalised radial velocity captured in both URANS and LES calculations. Although it seems that some of the regions where a combination of high negative and positive radial velocities exist are shorter in their radial extent than the others, there are minimum two pairs of these regions that extend from the cob to the shroud (see annotations). Their existence was experimentally corroborated in [6, 7]. These produce two radial arms where the cold air from the bore region travels radially outward to the shroud and transport the hot air from the shroud back to the lower radii. These appear more refined in the LES result (e.g., note the small-scale âŁ˜Rayleigh-BénardâŁ™ like cells near the shroud located at 6 o'clock in the LES result in Fig. 5a. This will be discussed in the following section.

The cyclonic and anticyclonic circulations are the regions where the reduced static pressure ($P^* = P - \frac{1}{2}\rho\Omega^2 r^2$) is low and high respectively. The circulations produce alternative regions of lower ($V_\theta < \Omega r$) and higher ($V_\theta > \Omega r$) relative tangential velocity. This suggests that a slip is introduced in relation to the solid body rotation. This will help the air flow to travel radially outward and inward within the cavity. Figure 5b and c show the normalised relative tangential velocity and the reduced static pressure respectively. The reduced static pressure can be assumed as equivalent to static pressure corrected for centrifugal pressure rise at each radial location. Although the LES calculation shows small-scale flow features particularly near the bore and the shroud, the number of vortex pairs predicted remain unchanged between the two types of calculations.

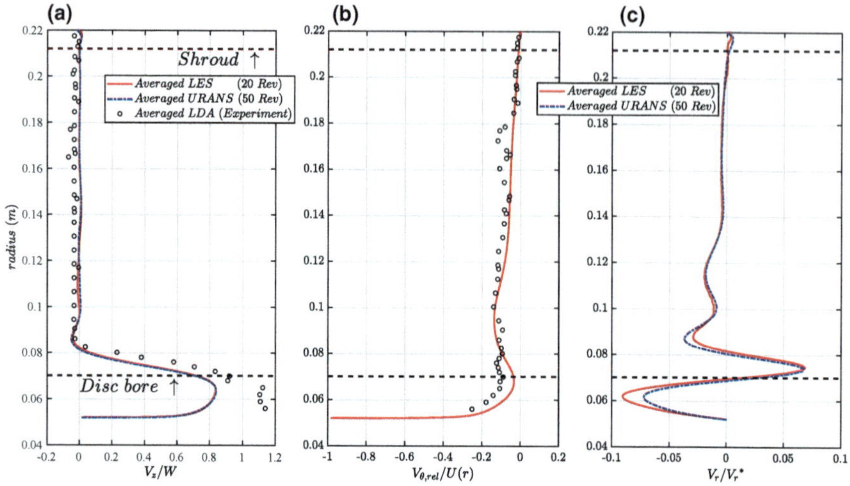

Fig. 6 Comparison of averaged data of normalised values of **a** axial velocity, **b** relative circumferential velocity, and **c** radial velocity with experimental results obtained from LDA in the mid-plane of cavity-3 ($z/s = 0.5$) in the $r - \phi$ plane

Figure 6 presents the comparison of averaged data acquired from numerical calculations with experimental data from LDA measurements. In this figure, W is bulk-average axial velocity at inlet to cavity, $U(r) = r\Omega$ is the disc speed at the corresponding radius.

The LDA measurement grid for the mid-plane inside the cavity includes 48 points from the bore to the shroud. Note that the LDA results discussed in this paper are the time-averaged results. The LDA only produced axial and circumferential components of velocity, therefore in the comparison does not include measured radial velocity. The data of numerical calculation is also averaged spatially and temporally on the cavity-3 axial mid plane over 90000 numerical 'measuring points' consisting of 250 radial and 360 (one degree apart) circumferential points. The URANS data is temporally averaged for 50 complete revolutions while the LES results is averaged for 20 revolutions. The variable values at '*inlet*' discussed in this paper are the area averages at the circumferential plane at inlet to cavity-3.

Although the level of agreement in normalised axial velocity graph is encouraging particularly for the data inside the cavity, there is some discrepancy between the experimental and numerical results in the bore region. The large averaging time of the test data compared to that of the CFD may be part of the reason for this discrepancy. The LDA is collected and averaged over 15 s at each measurement point. The time averages for the URANS calculation is collected over 50 disc revolutions worth only 0.375 s. For LES this is even shorter. Another reason might be the large local fluctuations in tangential velocity because of the block in the upstream domain which in turn causes high level of turbulence and mixing between the throughflow and the cavity flow. The agreement between the averaged data of URANS and LES is striking and both the numerical and the experimental results show that the axial velocity decays rapidly with increasing radius which is consistent with findings from previous work [6, 7, 15, 27]. Although the value of normalised axial velocity in the bore region is expected to be $V_z/W \approx 1$, the LDA results show higher value which is in agreement with five-hole probe experimental results reported in [28] and the LDA measurements reported in [6, 27]. This value is slightly under-predicted in both numerical calculations which is notable.

The normalised relative tangential velocity results show that although both experimental and numerical data follow the same trend, there are locations where the difference between the experimental data and numerical result is considerable. What is notable is that the flow rotates with a speed less than solid body rotation (negative $(V_{\theta,rel}/U(r))$ within the cavity. This value increases with increasing radius and near the shroud the averaged normalised tangential velocity is nearly zero which corresponds to solid body rotation. This results it consistent with literature.

Figure 6c compares the normalised radial velocity at the axial mid plane of cavity-3. The radial velocity is non-zero near the cobs where the cavity fluid interfaces with the bore flow. This shows the effect of the radial inflow and outflow at mid axial plane. The positive value of the normalised radial velocity at the mouth of the cavity (dashed line) shows that there is a net radial outflow from the bore region into the cavity at this axial plane. A negative value of normalised radial velocity also indicates that a radial inflow exists. This means that the flow from the cavity travels radially inward and exits the cavity to mix with the axial bore flow. The difference of the radial velocity between the numerical calculations in the bore region and at the lower radii of the cavity results from the LES being able to capture very small-scale flow prevalent in these regions.

5.2 *Circumferential Vortex Streaks*

As discussed earlier, in the near shroud region of cyclonic circulations, a series of counter rotating circumferential vortices is formed around the cavity. These streaks have been called âŁ˜Rayleigh-BÃ©nard cellsâŁ™ in some literature, however, there has been yet no experimental evidence of this type of cells. In this paper, this specific structure is referred to as '$circumferential\ vortex\ streaks\ or\ cells'$. As these streaks have been captured using both URANS and LES in [2, 6, 29–31] they could indeed be observed experimentally in the future. This section is concerned with the LES prediction of this flow feature. Figure 7 presents the contour plots of radial velocity, density ratio ($\Delta\rho/\rho = \rho_{local} - \rho_{cav,in}/\rho_{cav,in}$) and normalised temperature over the shroud surface alongside contours of the cavity-3 mid-plane radial velocity. As the shroud vortex streaks are aligned circumferentially, the shroud surface contour plots can be used to visualise them. Although this structure is seen all around the circumference, for this certain time-step it only appears to stretch between 9 and 12 o'clock positions. They are expected to enhance the mixing of the near shroud and the cavity air.

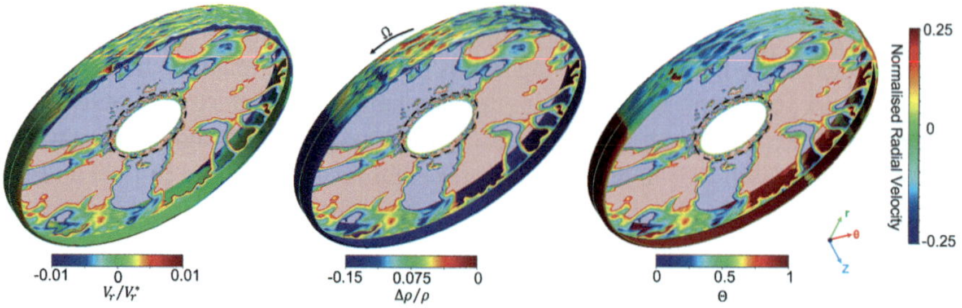

Fig. 7 Contour plots (iso-surface) of shroud normalised radial velocity, density ratio, and normalised temperature along with mid-plane ($z/s = 0.5$) of normalised radial velocity in circumferential view

5.3 *Thermal Flowfield Comparisons*

It was noted earlier that in a heated rotating cavity with axial throughflow, cold air enters the cavity in the radial arm and impinges on the shroud surface (radial outflow). Hot air from the shroud regions leaves the cavity in Ekman layers on the discs (radial inflow) that is created as a result of the circumferential pressure gradient created by the vortex pairs. Figure 8a shows the normalised temperature (Θ) on the mid axial plane of cavity three. A strong radial arm at 8 o'clock indicates that the flow migrates inside the cavity towards the shroud surface (radial outflow). Although the Ekman layers that transport the hot shroud fluid back towards the bore (radial inflow) is formed on the disc surfaces, the mid axial plane contours suggest that this may occur at the 1 o'clock position as a high temperature region is seen protruding radially inward from the shroud. A weaker pair of radial arms (inflow and outflow) occurs at 12 o'clock and 4 o'clock which is clearly captured by LES results but are much weaker and cannot produce large regions of cyclonic and anticyclonic regions. Instead these smaller plumes might be formed due to the unsteadiness generated by the rotating blocks that are located at the upstream of the rig.

The density changes ($\Delta\rho/\rho$) in the mid axial plane of cavity-3 is shown in Fig. 8b. The positive and near zero value of this variable in the bore region implies that the density of cold air is preserved in this region with comparison to the upstream inlet to the cavity. As seen, the cold air travels radially outward into the cavity 8 o'clock and a weak outward movement is also visible at 12 o'clock position. Once the colder air hits the shroud, the hot air from the shroud travels radially inwards on either side of the radial arm in order to satisfy mass continuity with in the cavity. The values of the static temperature in these regions are consistent with a cold outflow at 8 o'clock and a hot inflow at 1 o'clock.

Figure 8c demonstrates the instantaneous value of the Nusselt number over disc-3 (see Fig. 2). The Nusselt number contours predicted by both LES and URANS are consistent with the radial movement from the bore towards the shroud and back as described above. In the buoyancy dominated regions at the higher disc radius and near the shroud (particularly for a low Rossby number) small-scale vortex like

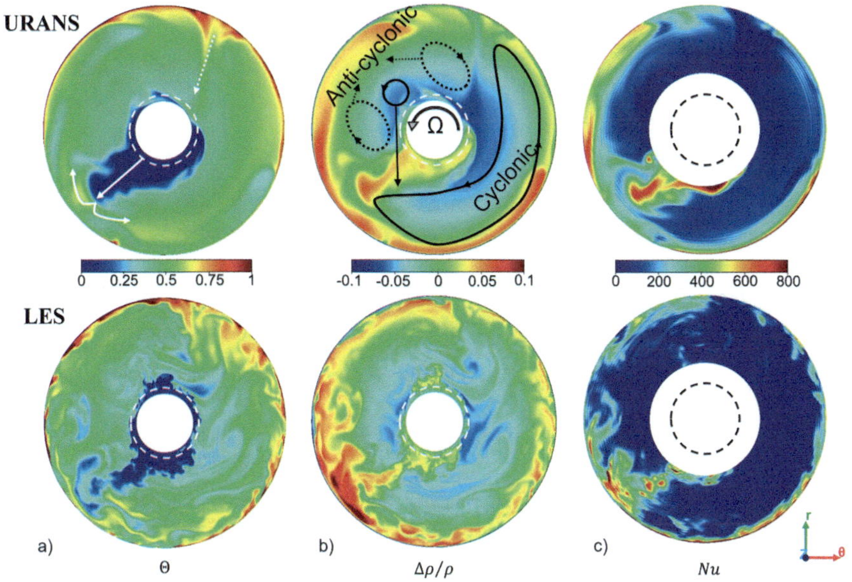

Fig. 8 Comparison of instantaneous **a** temperature ratio, **b** density ratio, and **c** Nusselt number in the mid-plane of cavity 3 ($z/s = 0.5$) for **a** and **b** and at the wall for **c** in circumferential view

structures are visible. These structures are thought to enhance the mixing of near the disc flow and the cavity air and therefore to increase the Nusselt number in this region. Figure 9 demonstrates the averaged results of the static temperature of air 4 mm away from the surface of disc-3 where the air thermocouples are installed in the experiment. Also shown are comparisons of the density ratio in the mid axial plane of cavity-3 and the Nusselt number of disc-3. As the air thermocouples in the experiment measure the total temperature in the rotating frame of reference, the corresponding static temperature for comparison as seen in Fig. 9 is derived using the equation below. The locations of the air thermocouples in relation to the disc are also shown for clarity.

$$T_{st} = T_{thermocouple} - \frac{\Omega^2 r^2}{2C_p} \quad (2)$$

As seen in Fig. 9a, there is exceptional agreement between the experimental data and the CFD calculations. The comparison between LES and URANS also shows a promising agreement albeit negligible variations at the cavity mouth and near the shroud regions compared to the mid cavity regions. One of the reasons for this could be the dominance of axial throughflow and buoyancy-induced effects at the lower and the higher radii, respectively. Another possible reason could be the mixing and the shear layer generation near the bore that the LES is better equipped to capture as it can resolve the small-scale features due to the flow breakdown.

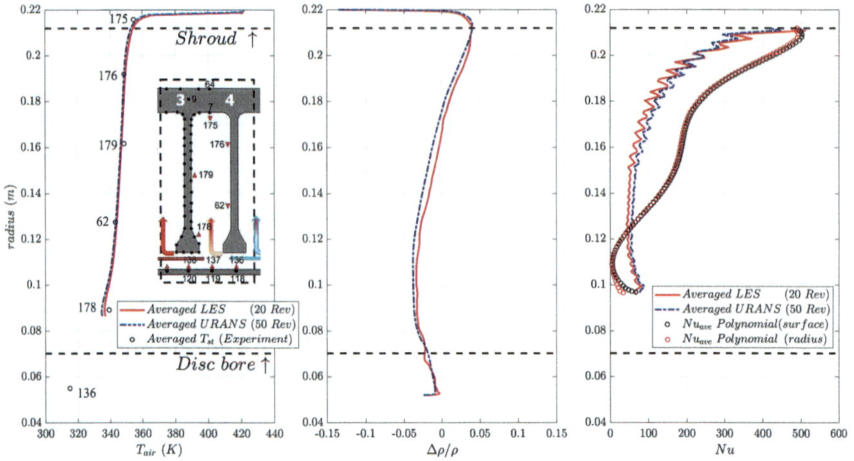

Fig. 9 Comparison of averaged data of normalised values of **a** temperature, **b** density ratio, and **c** Nusselt number with experimental results in the mid-plane of cavity-3 ($z/s = 0.5$) in circumferential view

Although Tucker et al. [32, 33] stated that the use of RANS modelling can prevent premature separation and show a weakness to recognise and capture predominantly laminar structures such as the cyclonic and anticyclonic structures, the present data along with the flow structure plots showed a good agreement with published literature. Hickling and He [3] concluded that URANS is unable to capture heat transfer enhancing streaks near the shroud. Ironically, such streaks were first reported in any published literature from the findings of a URANS study by Puttock-Brown et al. [4]. URANS's ability is further corroborated in the present study where near-shroud circumferential streaks are clearly visible in both URANS and LES predictions for the test case considered.

The averaged results of density change in the mid plane of cavity-3 is also shown in Fig. 9b. The are some variation of the this value particularly in the cavity however overall it shows good agreement between the LES and URANS data. There is a notable agreement on the shroud region which implies that the URANS is successful in capturing the circumferential vortex near the shroud region and the prediction of flow structure from the URANS calculation can be reliable.

Figure 9c shows the comparison of experimental and numerical Nusselt number on the surface of disc-3. Both methods capture similar values for the Nu especially in the regions above the cob. In relation the LES the URANS seems to marginally over-predict the Nu in the mid height regions whereas the opposite happens near the shroud. Both calculations under-predict the Nusselt number in comparison to the Nu values estimated using experimental disc surface temperature data. The methodology for estimating Nu from the experimental temperature data involves a PDE solution technique enhanced by a Monte-Carlo uncertainty analysis and this method is comprehensively discussed in [34].

6 Conclusions

Numerical simulations using both LES and URANS methodologies were conducted to simulate selected experiments in the Sussex MCR facility. Experimental data of cavity air and disc surface temperatures and LDA derived axial and tangential velocity components within the heated rotating cavity with axial throughflow were used to validate the current simulations.

The comparison of the radial distribution of the axial and radial velocities between the experiments and simulations reveal that these velocities are nearly zero in the core of the cavity. The predicted circumferential velocity component shows a good agreement with the experimental data. Both LES and URANS show good agreement with the corresponding LDA and thermal test data suggesting that the turbulence model can be used with an acceptable level of confidence. It can illustrate the flow structure in detail and show the unsteady flow features and their instabilities, however it might not successfully capture small-scale features within the cavity. Both numerical methodologies have been successful in capturing the vortex pair and the radial arm(s) inside the cavity. The LES and URANS were also able to capture the circumferential vortex streaks and their locations in the circumferential view near the shroud. These structures are able to enhance the mixing and heat transfer, resulting in a significant rise in the Nusselt number (Nu) at the higher radii and particularly on the shroud region. There is a need to devise an experimental methodology to prove the existence of the shroud circumferential streaks. With CFD routinely being used for design purposes by engineers, there is a question about the usefulness of high fidelity LES and DNS, if URANS is able to capture the major aerothermal events of relevance from a disc cavity and secondary flow design perspective.

References

1. Tucker, P.: Computation of unsteady turbomachinery flows: Part 2-LES and hybrids. Prog. Aerosp. Sci. **47**, 546–569 (2011)
2. Yao, M., He, L.: Implicit discontinuous Galerkin solution on unstructured mesh for turbine blade secondary flow. J. Turbomach. (2019). https://doi.org/10.1115/1.4045551
3. Hickling, T., He, L.: Some observations on the computational sensitivity of rotating cavity flows. J. Eng. Gas Turbines Power (2021). https://doi.org/10.1115/1.4049824
4. Puttock-Brown, M., Rose, M., Long, C.: Experimental and computational investigation of Rayleigh-Bénard flow in the rotating cavities of a core compressor (2017). https://doi.org/10.1115/GT2017-64884
5. Puttock-Brown, M.: Experimental and numerical investigation of flow structure and heat transfer in gas turbine hp compressor secondary air systems (2018)
6. Fazeli, S., Kanjirakkad, V., Long, C.: Experimental and computational investigation of flow structure in buoyancy dominated rotating cavities. ASME. J. Eng, Gas Turbines Power (2020)
7. Fazeli, S., Kanjirakkad, V., Long, C.: Experimental and computational investigation of flow structure of buoyancy induced flow in heated rotating cavities. J. Glob. Power Propul. Soc. No. GPPS-CH-2020-0037 (2020)

8. Dweik, Z., Briley, R., Swafford, T., Hunt, B.: Computational study of the unsteady flow structure of the buoyancy-driven rotating cavity with axial throughflow of cooling air. ASME Paper No. GT2009-59969
9. Dweik, Z., Briley, R., Swafford, T., Hunt, B.: Computational study of the heat transfer of the buoyancy-driven rotating cavity with axial throughflow of cooling air. ASME Paper No. GT2009-59978
10. Tan, Q., Ren, J., Jiang, H.: Prediction of flow features in rotating cavities with axial throughflow by RANS and LES (2009). https://doi.org/10.1115/GT2009-59428
11. Bohn, D., Deutsch, G., Simon, B., Burkhardt, C.: Flow Visualisation in a Rotating Cavity with Axial Throughflow. ASME Paper No. 2000-GT-0280 1–8, (2000)
12. Tucker, P.: Numerical and experimental investigation of flow structure and heat transfer in a rotating cavity with an axial throughflow of cooling air (1993)
13. Long, C., Tucker, P.: Numerical computation of laminar flow in a heated rotating cavity with an axial throughflow of air. Int. J. Numer. Method. Heat Fluid Flow **4**, 347–365 (1994). https://doi.org/10.1108/EUM0000000004043
14. Farthing, P., Long, C., Owen, J., Pincombe, J.: Rotating Cavity with axial throughflow of cooling air: heat transfer. ASME J. Turbomach. **1**, 229–236 (1992)
15. Farthing, P., Long, C., Owen, J., Pincombe, J.: Rotating cavity with axial throughflow of cooling air: flow structure. ASME J. Turbomach. **1**, 237–246 (1992)
16. Tucker, P.: Temporal behavior of flow in rotating cavities. Numer. Heat Transf. Part A Appl. **41**, 611–627 (2002)
17. Sun, Z., Lindblad, K., Chew, J., Young, C.: LES and RANS investigations into buoyancy-affected convection in a rotating cavity with a central axial throughflow. J. Eng. Gas Turbines Power **129**, 318–325 (2006). https://doi.org/10.1115/1.2364192
18. Long, C., Alexiou, A., Smout, P.: Heat transfer in H.P. compressor gas turbine internal air systems measurements from the peripheral shroud of a rotating cavity with axial throughflow. 2nd International Conference On Heat Transfer, Fluid Mechanics And Thermodynamics (HEFAT). (2003)
19. Alexiou, A.: Flow and heat transfer in gas turbine H.P Compressor internal air systems (2000)
20. Atkins, N., Kanjirakkad, V.: Flow in a rotating cavity with axial throughflow at engine representative conditions. ASME Paper No. GT2014-27174, 1-14 (2014)
21. Pitz, D., Chew, J., Marxen, O.: Large-eddy simulation of buoyancy-induced flow in a sealed rotating cavity. J. Eng. Gas Turbines Power **141**, (2018). https://doi.org/10.1115/1.4041113
22. Pitz, D., Chew, J., Marxen, O., Hills, N.: Direct numerical simulation of rotating cavity flows using a spectral element-fourier method. J. Eng. Gas Turbines Power **139**, (2017). https://doi.org/10.1115/1.4035593
23. Kouwa, J., Iso, Y., Polidoro, F., Gautier, S.: Very-large eddy simulations of disk heat transfer in a rotating cavity using lattice-boltzmann method (2018). https://doi.org/10.1115/GT2018-76832
24. Long, C.: Disk heat-transfer in a rotating cavity with an axial throughflow of cooling air. Int. J. Heat Fluid Flow **15**, 307–316 (1994)
25. Smagorinsky, J.: General circulation experiments with the primitive equations. Mon. Weather Rev. **91**, 99 (1963)
26. Lilly, D.: A proposed modification of the Germano subgrid-scale closure method. Phys. Fluids A **4**, 633–635 (1992). https://doi.org/10.1063/1.858280
27. Long, C., Miche, N., Childs, P.: Flow measurements inside a heated multiple rotating cavity with axial throughflow. Int. J. Heat Fluid Flow **28**, 1391–1404 (2007)
28. Fox, E., Puttock-Brown, M., Davies, S.: Design and development of a five-hole probe calibrator and traverse to investigate ingestion in rotating cavities of HP compressors. (2022). https://doi.org/10.1115/GT2022-80940
29. Puttock-Brown, M., Rose, M.: Formation and evolution of Rayleigh-Bénard streaks in rotating cavities (2018). https://doi.org/10.1115/GT2018-75497
30. Saini, D., Sandberg, R.: Large-eddy simulations of high rossby number flow in the high-pressure compressor inter-disk cavity. J. Turbomach. (2021). https://doi.org/10.1115/1.4050951

31. Hickling, T., He, L.: LES-CHT for a rotating cavity with axial throughflow (2022). https://doi.org/10.1115/GT2022-82982
32. Tucker, P., Eastwood, S., Klostermeier, C., Jefferson-Loveday, R., Tyacke, J., Liu, Y.: Hybrid LES approach for practical turbomachinery flows-Part I: hierarchy and example simulations. J. Turbomach. **134** (2011). https://doi.org/10.1115/1.4003061
33. Tucker, P., Eastwood, S., Klostermeier, C., Xia, H., Ray, P., Tyacke, J., Dawes, W.: Hybrid LES Approach for Practical Turbomachinery Flows-Part II: Further Applications. J. Turbomach. (2011). https://doi.org/10.1115/1.4003062
34. Fazeli, S., Kanjirakkad, V., Long, C.: Investigation of buoyancy-driven heat transfer in HP compressor disc cavities. AIAA Propulsion And Energy 2021 Forum (2021). https://doi.org/10.2514/6.2021-3484

Analysis of Separation and Transition on a Novel High-Speed Low Pressure Turbine Cascade

Maxime Borbouse, Gustavo Lopes, Sergio Lavagnoli, Michel Rasquin, and Koen Hillewaert

Abstract The transonic aerodynamics of a low pressure turbine (LPT) cascade, mimicking geared turbofan conditions during cruise, is currently under study using direct numerical simulation. In the first step of this endeavour, transition, separation and loss generation were examined at Mach 0.7, 0.9, and 0.95, with a constant Reynolds number of 70×10^3 based on the true chord, in clean inlet conditions. The aim is to identify the impact of sonic conditions and choking on self-induced transition, generated purely by flow features around the blade. The pressure distribution was slightly overestimated with respect to experimental conditions. Separation occurs over the suction side (SS) at 58% and 69% of blade length for the first two Mach numbers. Skin friction is low in the Mach 0.95 case, but unexpectedly separation does not take place for this case. The choking of the passage was considered as influencing factor. A supersonic region followed by a shock appears at Mach 0.9 or higher, and a fully choked passage is found at 0.95. Separation bubbles with turbulent reattachment are formed on the pressure side (PS) between 0% and 50% of the blade length. Separation on the SS was identified as laminar separation long bubble, while reversed transition develops on the PS after reattachment. Wake losses increase with higher Mach numbers, and turbulence is more prominent at lower Mach numbers. This investigation provides insights into the aerodynamic characteristics of the linear low pressure turbine cascade, addressing separation, transition, losses, and wake behavior in sonic conditions at low Reynolds numbers.

CUFS-2024-21, Cambridge, United Kingdom, 4-5 March 2024.

M. Rasquin · K. Hillewaert
Cenaero, Rue des frères Wright 29, 6041 Gosselies, Belgium

G. Lopes · S. Lavagnoli · K. Hillewaert
Turbomachinery and Propulsion Department, von Kàrmàn Institute for Fluid Dynamics, Chaussée de Waterloo 72, 1640 Rhode-St-Genèse, Belgium

M. Borbouse · K. Hillewaert (✉)
Aerospace and Mechanics Department, Université de Liège, Allée de la découverte 9, 4000 Liège, Belgium
e-mail: koen.hillewaert@uliege.be

J. C. Tyacke and N. R. Vadlamani (eds.), *Proceedings of the Cambridge Unsteady Flow Symposium 2024*, https://doi.org/10.1007/978-3-031-69035-8_21

1 Introduction

Over the past decades, there has been significant attention for improving fuel efficiency of turbofan engines. The thermodynamic efficiency of the cycles was consistently increased by better aerodynamic, thermal and material performance of the components, allowing higher pressure ratios and temperatures; whereas propulsive efficiency is continuously being improved by increasing engine bypass ratio [1].

In classical turbofan designs, the fan is directly installed on the low pressure spool. As the fan rotation speed is limited by the constraint of keeping the fan tip speed to transonic or slightly supersonic conditions to avoid excessive shock losses, the low pressure turbine (LPT) operates at very low rotation speeds, resulting in high stage counts. The geared turbofan (GTF) concept resolves this issue by introducing a fixed ratio between fan and LP spool rotation speed, providing better conditions for the LPT and the low pressure compressor (LPC) - if present. The GTF engine architecture offers several advantages. The primary focus is therefore, as highlighted by Kurzke [1], the substantial reduction in the number of stages, especially in the LPT, achievable with the GTF. This decoupling of the fan and LPT enables the LPT to turn faster leads to so-called high-speed LPT, which feature a much higher work per stage than the corresponding low-speed LPT. As a consequence much less stages are required, leading to a strong reduction of engine weight, overall fuel consumption, and operational costs, making it highly desirable for the next generation of aircraft.

In high-speed LPT, conditions arise that are very different from the low-speed LPT of classical typical turbofans, namely combining transonic conditions to very low Reynolds numbers. These conditions have a substantial impact on loss mechanisms and blade aerodynamics, as they are significantly influenced by compressibility effects, which can be critical. Research presented in several papers, such as [2, 3], highlights that the SS boundary layer (BL) of the blade contributes the most to the profile loss, accounting for up to 85% of the total losses. In such conditions, where the Reynolds number is low, there is a risk of laminar separation occurring on the suction side, leading to a significant increase in losses. To mitigate the potential separation issue, researchers have explored the wake-induced transition, which has been extensively detailed in various works like [4–8].

The occurrence of separation manifests itself in two primary ways: open separation without reattachment, leading to high losses, or closed bubbles of various nature. The control of this separation can be achieved through the passage of wakes or by modifying the blade characteristics, such as front or aft loading. Transition to different separation modes can occur naturally, through bypass, separated flow, reverse flow or as mentioned earlier, induced by wake passing. The reattachment of the flow is facilitated by this transition and results in the formation of a separation bubble. A comprehensive classification of different types of separation bubbles is provided in [9], while more detailed studies on transition and separation are presented in [10]. The objective of investigated blade aerodynamics is to comprehend the BL state at the point of separation to better understand the ensuing transition it undergoes.

The subject of this investigation involves the aerodynamics of a high-speed LPT, particularly in the context of the GTF engines, with a focus on cruise conditions. RANS computations face challenges in resolving turbulent transition in turbomachinery flows. To accurately model the transition in the shear layer and trigger reattachment in order to capture the laminar separation bubble, high-fidelity methods like Direct Numerical Simulation (DNS) are required.

This work aims at filling the gap in the available experimental and simulation data for on- and off-design conditions in high-speed LPTs and to validate the method used for studying these flows by comparing experimental and DNS data.

2 Numerical Setup

The chosen level of fidelity for the simulations is DNS, which computes all scales of the flow with precision as we aim to study the impact of compressible effects on auto-induced transition, i.e. the transition resulting from perturbations generated by the flow around the airfoil itself.

2.1 Numerical Method

DNS comes with a significant computational cost and cannot be performed on standard computers. The numerical resources required for DNS simulations can be estimated to be of the order of Re_c^3 to $Re_c^{3.5}$ [11], using the true chord c as reference length.[1] Thus, high-performance computers are essential for conducting DNS for the studied LPT conditions.

To optimize the numerical cost and increase accuracy, higher-order methods are employed. In this paper, the solver ArgoDG (see [12, 13]), which implements the discontinuous Galerkin method (DGM), was used. The DGM has favorable properties and advantages for solving elliptic and hyperbolic problems as it combines the strength of finite element method (FEM), i.e. high accuracy on unstructured methods, and finite volume method (FVM), i.e. conservation and stability for convection-dominated problems, making it suitable for the computational fluid dynamics (CFD) simulations for DNS and LES. The spatial discretization combines 4th order accurate discontinuous Galerkin method for convection with the symmetric interior penalty discontinuous Galerkin (SIPDG) for diffusion. No shock capturing strategy was employed.

[1] Starting from the grid point estimate $Re^{9/4}$ for Chapman's estimate to the revised version of $Re_c^{37/14}$ by Choi, assuming the number of time steps scales as the number of grid points per direction due to CFL constraints to maintain precision.

Finally, the implicit second-order accurate backwards difference time-integration scheme was used, which was solved using a Newton-GMRES iterative strategy, preconditioned with a single precision block Jacobi preconditioner.

2.2 Test Case and Boundary Conditions

The fluid properties are considered to be those of an ideal gas, with specific heat ratio $\gamma = 1.4$ and gas constant $R = 287.06\,\mathrm{J\,kg^{-1}K^{-1}}$. Viscosity and conductivity depend on temperature following Sutherland's law [16].

The computational domain is depicted in Fig. 1a. The geometric parameters of the passage are given in Table 1. To ensure that the solution remains unaffected by the domain size, appropriate inlet and outlet boundaries are set far from the blade, specifically at distance of $2 \times c_{ax}$ and $2.4 \times c_{ax}$ from the blade leading edge (LE) and trailing edge (TE) respectively. For the linear turbine cascade and its self-induced transition, this configuration is justified. Additionally, having the outlet located further from the inlet allows for a larger wake region to be analyzed.

Three computations were performed for a fixed outlet Reynolds number. The outlet Mach number was varied between 0.70, 0.90 and 0.95. This choice, along with the Reynolds number of 70×10^3, allows to reproduce subsonic, nearly choked and fully choked conditions during the typical cruise conditions encountered in GTFs. The incidence measured in the experiment [14, 15] was used to determine the inlet flow angles imposed at the inlet of the computational domain. The flow conditions are given in Table 2. Note that the computations were performed without turbulence injected at the inlet.

The computational domain includes seven boundaries: inlet, outlet, 2 pitchwise periodic surfaces, 2 spanwise periodic (or extrusion) planes, and finally the blade.

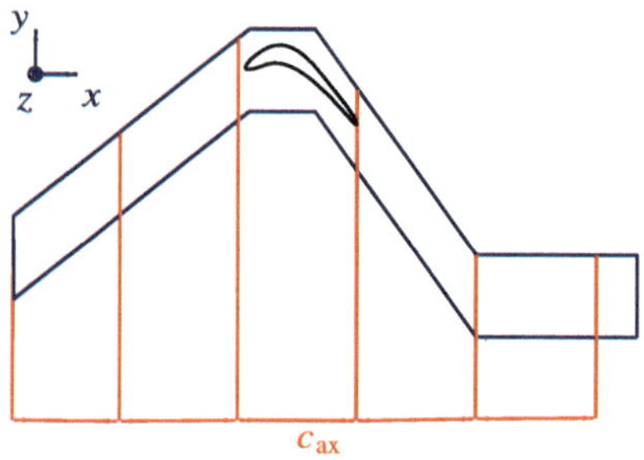

(a) Computational domain.

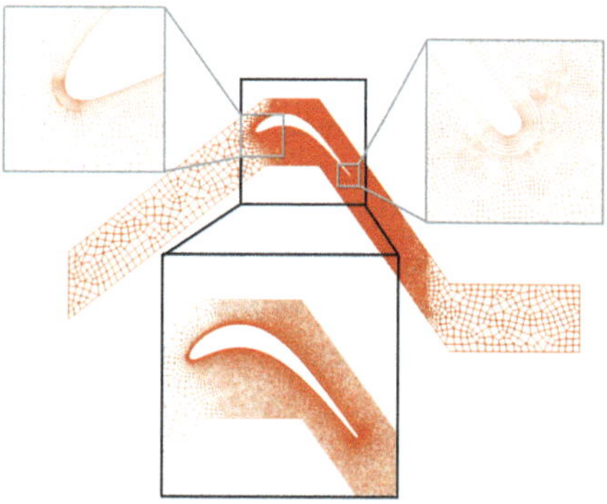

(b) Mesh definition.

Fig. 1 Domain and mesh definition in the periodic plane

Table 1 Geometric parameters

Parameter	Symbol	Value	Units
True chord	c	52.28	[mm]
Axial chord	c_{ax}	47.61	[mm]
Pitch	g	32.95	[mm]
Span		165	[mm]
TE radius	$d_{\mathrm{TE}}/2$	0.435	[mm]
Throat opening	o	19.4	[mm]
Inlet angle	$\beta_{\mathrm{m,in}}$	37.3	[deg]
Outlet angle	$\beta_{\mathrm{m,out}}$	53.8	[deg]
Stagger angle	ζ	24.4	[deg]

Table 2 Flow conditions

Quantity	Symbol		Value		Units
Is. out. Reynolds number[a]	$\mathrm{Re}_{\mathrm{s,out}}$		70×10^3		[-]
Is. out. Mach number[a]	$M_{\mathrm{s,out}}$	0.70	0.90	0.95	[-]
Inlet total pressure	p°_{in}	10779.39	9500.24	9318.31	[Pa]
Outlet static pressure	p_{out}	7771.16	5617.12	5213.22	[Pa]
Inlet total temperature	T°_{in}		300		[K]
Flow angle	α		36.3		[°]

[a] Is.: Isentropic; out.: outlet

Periodic BCs were employed in the pitch- and spanwise direction; the spanwise extent of the domain is $0.07 \times c$. At the blade surface, no-slip and adiabatic conditions are enforced, meaning that velocity and energy transfer are set to zero. Total conditions and flow angles are imposed at the inlet, while at the outlet, only static pressure is fixed. The values are summarized in Table 2. A sponge layer is added near the outlet, with spatially varying strength,

$$\sigma = \tan(x - 1.55) \cdot \mathrm{step}(1.55), \tag{1}$$

which penalises the instantaneous flow u with respect the time average $\overline{u}$ by adding the source term

$$S = \sigma(u - \overline{u}) \tag{2}$$

to the equations in order to prevent outgoing waves and perturbations to reflect back into the domain. In other word, taking the LE as origin, i.e. $x = 0$ at LE, the sponge layer is situated at a distance of $2.25 \times c_{\mathrm{ax}}$ downstream the TE.

Finally, it is important to note that no injected turbulence was used at the inlet. This is is a deliberate choice, as a first step in the analysis of the flow which should allow us to decouple auto-induced transition from that initiated by incoming turbulence. Later on the impact of inlet turbulence and incoming wakes (modeled by moving bars) will be studied.

2.3 *Resolution*

The final mesh is shown in Fig. 1b, being the one that achieved convergence. It was constructed using Gmsh 4.5.6 [17].

The mesh consists of an unstructured region far from the blade and a structured O-type mesh close to the blade with a thickness of 0.76 mm. The element size close to the blade is set at 0.38% of c, with an exponential growth rate of 1.2 in order to satisfy the resolution criterion on $n+$ and $s+$.[2] Three refinement zones were identified: the LE, TE, and the base region. In the LE and TE regions, the element sizes are 0.08% of c, while in the base flow, they are 0.76% of c. In the passage refinement zone extending from LE to $1.5 \times c_{ax}$, the cell sizes are 1.5% of c, with the maximum element size being 7.5% of c in regions where high precision is not necessary. The base flow refinement zone, extending from TE up to $0.5 \times c_{ax}$, has a reference cell size of 0.75% of c. The mesh elements adjacent to the surface conform to it using curved elements of order 2 and a resolution order of 3. The mesh was extruded in the z direction on 10 layers, resulting in spanwise elements of size 7% of c, as abovementioned. The final mesh contains 5337 hexahedra and 16981 two-dimensional quadrangles.

The grid was designed with the objective of achieving the highest spatial resolution around the blade.The typical mesh resolution in wall unit for a DNS is 10, 5 and 1 in respectively the streamwise, spanwise and wall normal direction. However, in this specific analysis, the wall units can be relaxed, ranging between 15 and 30 since the solution interpolation in the element uses cubic interpolants, which means that the first node is actually placed at a length scale that is divided by 3. The wall resolution is provided in Fig. 2.

The temporal resolution is $\Delta t = 0.1$ µs. Each time step is solved within a non-linear tolerance of 10^{-4} with respect to the initial residual, after an initial extrapolation. The typical CFL number is of the order of 10^{-1} at midchord up to 1 near the edges.

3 Results

The flow fields computed for distinct Reynolds and Mach numbers, encompassing a low Reynolds number (70×10^3) and diverse Mach numbers, including low (0.70),

[2] Since the solution is interpolated using cubic interpolants, the effective expansion rate with respect to the elementwise internal resolution corresponds to $\sqrt[3]{1.2} \approx 1.06$.

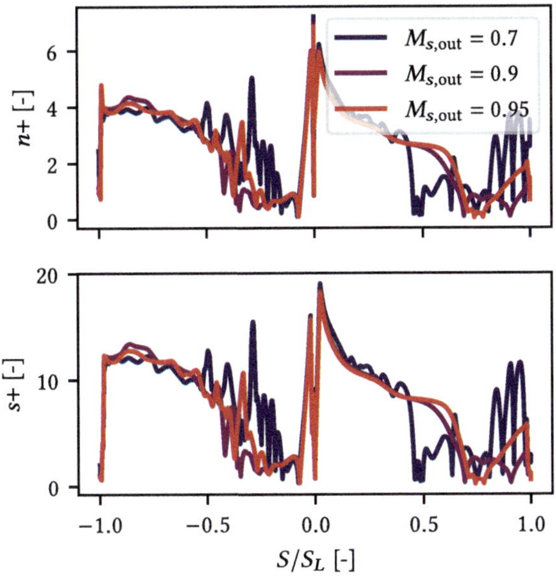

Fig. 2 Instantaneous distribution of wall resolution, (top) $n+$ and (bottom) $s+$, along blade PS (negative abscissa) and SS (positive abscissa) for $\text{Re}_{s,\text{out}} = 70 \times 10^3$ and $M_{s,\text{out}} = 0.7, 0.9$ and 0.95

moderate (0.90), and high (0.95) Mach regimes are presented. This wide spectrum of scenarios facilitates an in-depth exploration of flow dynamics in both on- and off-design conditions, offering insights into the influence of compressibility.

The flow field was analysed using paraview, extended with a specific plugin developed to analyse high-order mesh and solutions fields saved under the Gmsh format [49]. The snapshots of the density gradient normalized by ρ in Fig. 3 reveals distinct acoustic perturbation behaviors, delineating differences in flow nature.

Notably, the low Mach number case displays downstream and upstream propagation of acoustic perturbations due to vortex shedding. This reflective property manifests as "V" shapes on the blade suction side. Conversely, the presence of a supersonic region near the throat inhibits such behavior in higher Mach cases, particularly in the choked passage of the high Mach case.

3.1 Blade loading

Figure 4 presents the time and spanwise averaged isentropic Mach distribution. Experimental data is added whenever available.

The computed pressure distribution on the PS all exhibit a plateau which is not visible in the experimental data. This may be attributed to the experimental setup having an inlet turbulence level of $\text{Tu} = 2.5\%$, while there is no injected turbulence in the numerical simulations. This difference in turbulence levels affects the mixing and could lead to the avoidance of separation in the experimental conditions. Furthermore, turbulence appears to be significant at low Mach numbers, playing a crucial role in

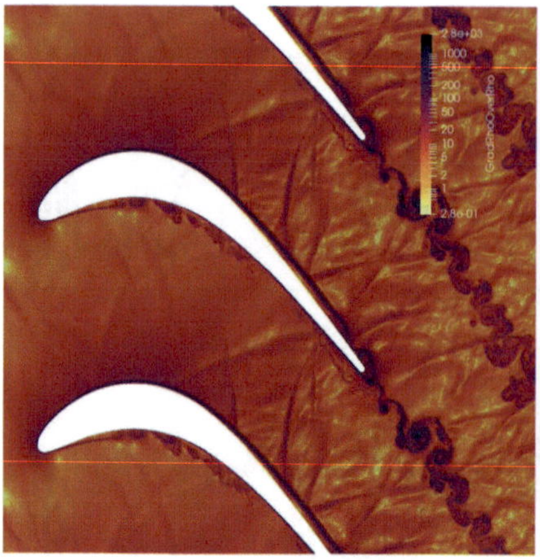

(a) $M_{s,\mathrm{out}} = 0.7$.

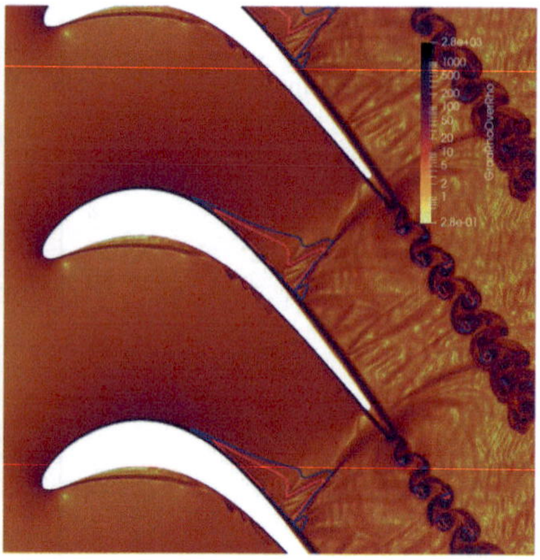

(b) $M_{s,\mathrm{out}} = 0.9$.

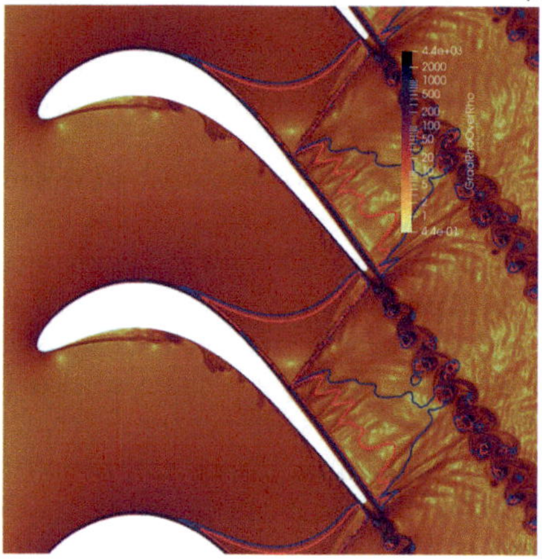

(c) $M_{s,\mathrm{out}} = 0.95$.

Fig. 3 Contours of the normalized density gradient for $\mathrm{Re}_{s,\mathrm{out}} = 70 \times 10^3$ and $M_{s,\mathrm{out}} = 0.7, 0.9, 0.95$

the extent of the separation bubble. The lower turbulence levels in the numerical simulations result in lower turbulent spot production, leading to a larger extent of transition. Overall, the numerical results match well with the remaining part of the PS pressure distribution, although it is slightly underestimated.

For the SS results, a good agreement is observed with experimental results for cases with Mach numbers of 0.90 and 0.95. However, a small discrepancy is observed for the low Mach number case as well as for the high Mach number case between $S/S_L = 0.5$ and 0.9 : the numerical is slightly lower than the experimental isentropic Mach number distribution. Additionally, the pressure recovery seen in the experimental results is not evident in the numerical results for this case. The lower loading

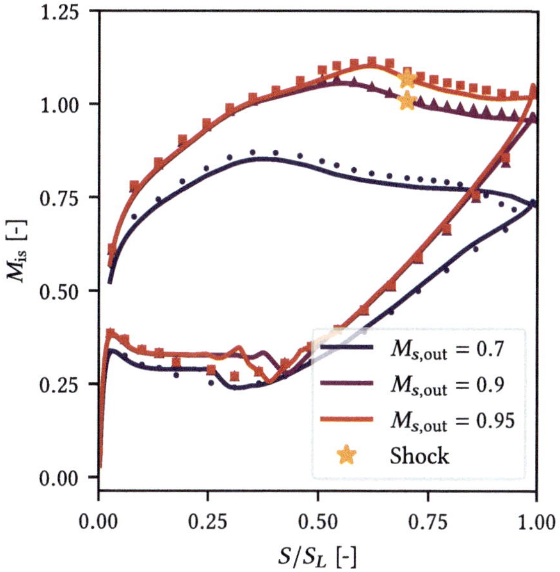

Fig. 4 Time and spanwise averaged isentropic Mach number distribution along the blade PS (bottom curve) and SS (top curve) for $Re_{s,out} = 70 \times 10^3$ and $M_{s,out} = 0.7, 0.9$ and 0.95. The symbols represent experimental data (see [14, 15]). Also, the time averaged shock location for high Mach numbers is shown. The symbols •, ▲ and ■ respectively represent the experimental measurements for $M = 0.7, 0.9$ and 0.95

observed from the LE to the laminar separation is attributed to an increase of the extent of the separation bubble due to the absence of inlet turbulence in the DNS. As detailed by Mayle [2], a large separation bubble can trigger the modification of the loading all the way back to the LE.

Increasing the Mach number leads to a shift in the velocity peak, resulting in the development of an adverse pressure gradient over a smaller portion of the blade. Consequently, the pressure rise between the suction peak and the TE is smaller than expected due to the very similar evolution at the front part of the blade SS up to the suction peak at $M_{s,out} = 0.90$.

As the boundary layer does not become turbulent, this results in a larger separation bubble, ultimately impacting the SS loading all the way to the TE, as seen in previous works such as [3].

3.2 *Boundary Layer Status*

Figure 5 presents the time and spanwise averages of the skin friction $c_{f,s}$ along the PS and SS for all three cases. Note that the skin friction is taken in the axis system attached to the blade in order to have a better comparison with the experimental data. For all three cases, similar trends are observed. Separation is located on the SS at $M_{s,out} = 0.70$ at $S/S_L \approx 0.584$ and at $M_{s,out} = 0.90$ at $S/S_L \approx 0.688$. The skin friction exhibits a conventional pattern, with a decrease in the front part, separation or very low values, stagnation in the rear part after separation (if any), and a slight increase near the TE. The shift in separation between the low and moderate Mach

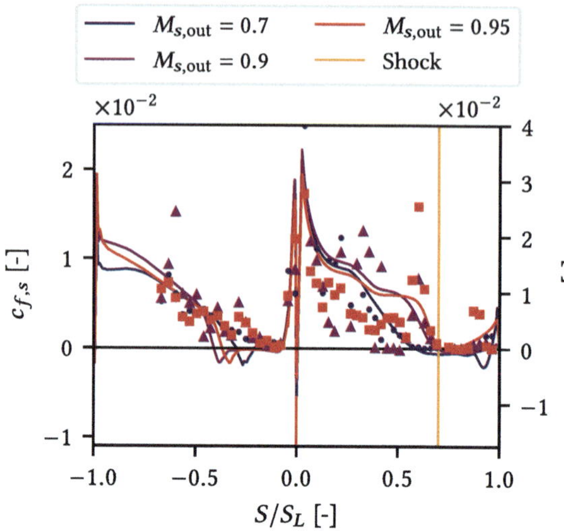

Fig. 5 Time and spanwise averaged skin friction coefficient distribution along the blade PS (negative abscissa) and SS (positive abscissa) for $\mathrm{Re}_{s,\mathrm{out}} = 70 \times 10^3$ and $M_{s,\mathrm{out}} = 0.7$, 0.9 and 0.95. The symbols represent experimental quasi-wall shear stress (see [14, 15]). The symbols •, ▲ and ■ respectively represent the experimental measurements for $M = 0.7$, 0.9 and 0.95. Also, the time averaged shock location for high Mach numbers is shown

number cases is due to the shift in velocity peak at higher Mach numbers, resulting in stronger gradients. For the low Mach case, the flow seems to reattach near the TE, while for $M_{s,\mathrm{out}} = 0.90$, no reattachment is observed. At $M_{s,\mathrm{out}} = 0.95$, the skin friction reaches very small values but never becomes negative, indicating that the BL does not separate. This unexpected behavior of the flow for high Mach numbers may be explained by the smaller pressure rise for these conditions. As discussed earlier, the shift in velocity peak and the shock restrict the pressure gradient on a smaller portion of the blade, such that the momentum deficit is smaller before the shock and the BL remains attached.

On the PS, the bubble is also observed, consistently with the previous findings. The bubble is smaller at low Mach numbers than at $M_{s,\mathrm{out}} = 0.90$ and 0.95. The skin friction increases over the PS until the TE for each case, driven by the favorable pressure gradient acting on the fluid in this region, potentially leading to relaminarization.

The main behavior of the skin friction is well-matched to the measured quasi-wall shear stress at low Mach, but at the intermediate Mach number, the experimental values are more oscillating over the SS. However, the PS is still well-represented, capturing separation accurately in most cases.

It is interesting to note that, with inlet turbulence, as in the experimental test case, the separation bubble would surely be shortened on the SS; but in contrast we should expect strong fluctuations of the skin friction, maybe due to high sensitivity of the shock position to passing turbulent flow features. Moreover, as it is observed in experiments [14, 15], the BL does not separate on the PS which let think that the separation bubble on the PS would also be shortened or even suppressed.

Figure 6 displays the integral parameters for the three conditions, i.e., the displacement, momentum loss and energy loss thicknesses, respectively,

$$\delta^* = \int_0^\infty \left(1 - \frac{\rho}{\rho_e}\frac{u}{U_e}\right) \mathrm{d}y,$$

$$\theta = \int_0^\infty \frac{\rho}{\rho_e}\frac{u}{U_e}\left(1 - \frac{u}{U_e}\right) \mathrm{d}y, \tag{3}$$

$$k = \int_0^\infty \frac{\rho}{\rho_e}\frac{u}{U_e}\left(1 - \left(\frac{u}{U_e}\right)^2\right) \mathrm{d}y$$

with ρ_e the local free stream density and U_e the local free stream velocity.

Across the pressure side, a consistent pattern emerges for each case. Post-separation, the displacement thickness exhibits notable increments. These reached values are comparatively lower at $M_{s,\mathrm{out}} = 0.70$, in line with predictions. However, the growth of momentum thickness follows a delayed trajectory compared to that of displacement thickness. Initially, the shape factor mirrors the trends of the displacement thickness. Subsequently, as the displacement thickness sharply decreases, the momentum thickness aligns closer to it, signifying the onset of transition. In the region of reattachment, the shape factor falls below 2, approaching approximately 1.4, indicative of a fully turbulent flow. This concurs with Mayle [2] and Hatman and Wang [10] propositions that reattachment results from transition while the initial separation is laminar.

Post reattachment, the shape factor gradually rises, approaching values around 2 at the trailing edge. This observation suggests the occurrence of reverse transition in this region. Such a reverse transition is contingent on a favorable pressure gradient, as emphasized in the works of Hatman and Mayle [2, 10]. This reverse transition pattern is consistently exhibited across all Mach numbers.

The boundary layer parameters mirror a similar evolution in response to separation along the suction side. At $M_{s,\mathrm{out}} = 0.70$, the shape factor is in close proximity to 2 pre-separation, undergoes an increment due to separation, reaches a peak value, and subsequently declines towards the trailing edge. Notably, the influence of the bubble displacement effect is detectable slightly upstream of separation, indicated by the shape factor rise around 50% of the suction side, even though separation occurs at 58%. The shape factor decline and the decrease in displacement thickness signify the initiation of transition. However, turbulent conditions are not prevalent before the trailing edge on average. At $M_{s,\mathrm{out}} = 0.90$, analogous observations can be made. Nonetheless, transition is even less advanced than at lower Mach numbers. For $M_{s,\mathrm{out}} = 0.95$, as previously discussed, no separation manifests. However, a distinct thickening of both displacement thickness and shape factor is noticeable. This effect can be attributed to the shock, occurring within the same region as the observed thickening. The shape factor gradient post-shock steepens, reaching a maximum before diminishing, without approaching 2 before the trailing edge. This pattern alludes to the initiation of transition for this case as well. Furthermore, the shock appears to inhibit separation. This is reflected in the increasing collapse of values

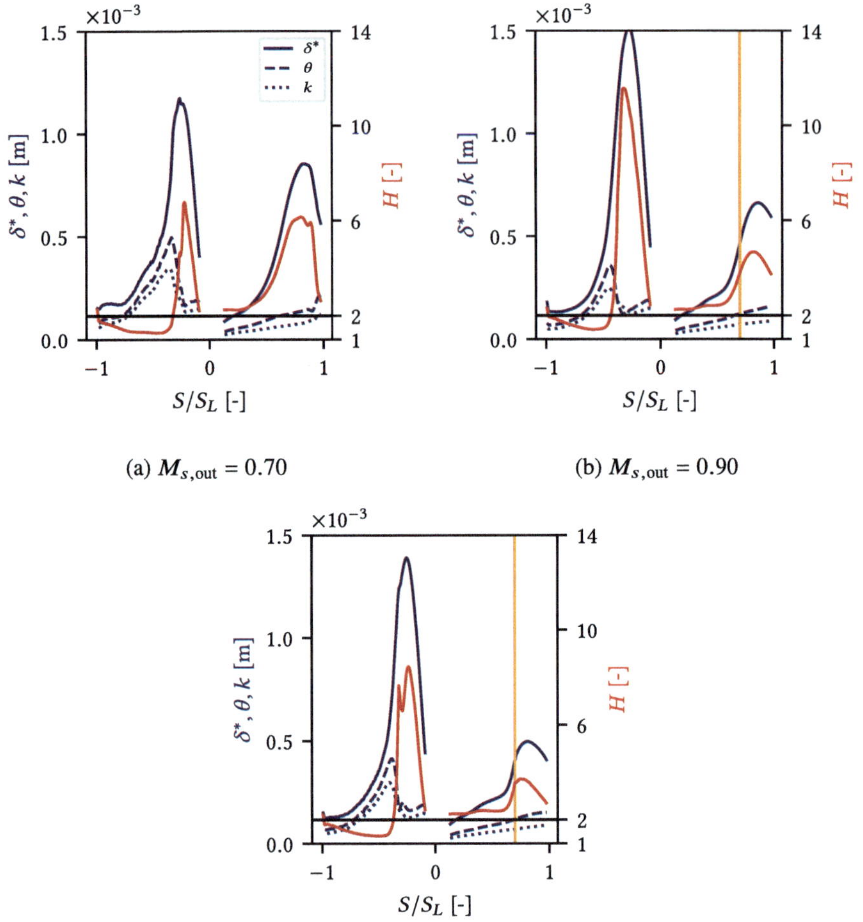

(a) $M_{s,\text{out}} = 0.70$

(b) $M_{s,\text{out}} = 0.90$

(c) $M_{s,\text{out}} = 0.95$

Fig. 6 Boundary layer integral parameters for $\text{Re}_{s,\text{out}} = 70 \times 10^3$, $M_{s,\text{out}} = 0.7$, 0.9 and 0.95. Also, the time averaged shock location for high Mach numbers is shown

upstream of the sonic region (similar, albeit milder, behavior can be observed for $M_{s,\text{out}} = 0.90$), which may contribute to a reduced pressure gradient.

3.3 *Wake Analysis*

Figure 7 presents the loss in terms of the total pressure defect. It reveals that the wake is wider for lower Mach numbers. This could be attributed to the vortex street being more disordered for the low Mach case, indicating a faster turbulent decay. The

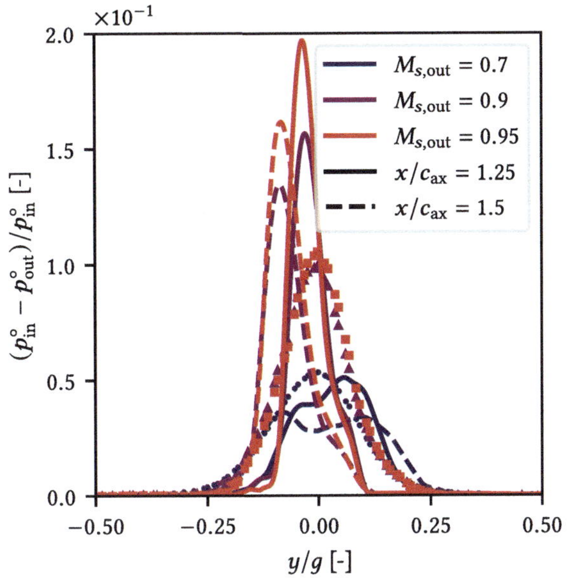

Fig. 7 Time average of loss coefficient, in terms of total pressure loss, at two different axial locations for $Re_{s,out} = 70 \times 10^3$ and $M_{s,out} = 0.7, 0.9$ and 0.95. The symbols represent experimental data at $x/c_{ax} = 1.5$

wake for $M_{s,out} = 0.70$ is associated with larger displacement thickness on the SS and smaller displacement thickness on the PS compared to higher Mach numbers, leading to decrease in wake thickness with increasing Mach number (see Fig. 6).

For the two highest Mach number cases, the decay of the wake is slower, characterized by a wake composed of two staggered rows of large eddies. These eddies contain significant amounts of macroscopic energy, which are convected downstream (see Figs. 3a–c, 9, 10 and 11). This idea of a much faster decay at low Mach is supported by the different shapes of the curves between the two downstream locations; the wake spreads, but its shape also changes. However it is not observed at $M_{s,out} = 0.90$ and $M_{s,out} = 0.95$.

While the computed results and experimental data are of the same order, there is a small discrepancy, which can be attributed to the issue with the inlet turbulence level.

The total temperature variation and its root mean square value in the wake are provided in Fig. 8. This gives a more complete insight into loss mechanisms as it can be interpreted as creation of entropy in the wake. The followed trends are similar to above mentioned results. This wake signature can be attributed to energy separation phenomena, as detailed in [18–21].

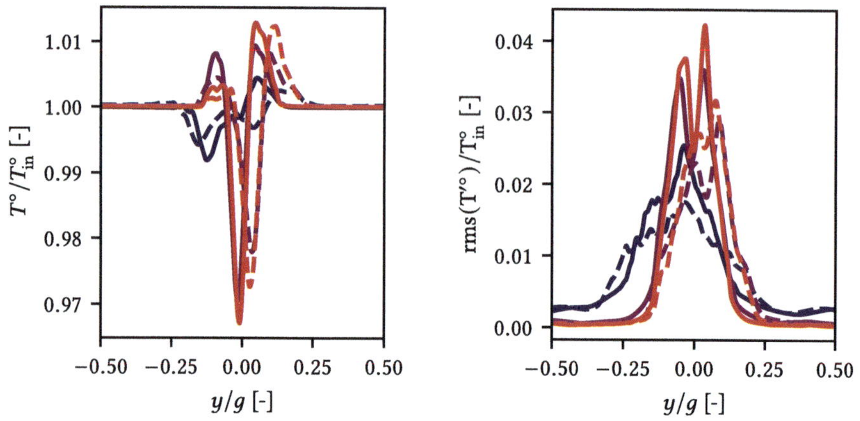

(a) Total temperatures. (b) RMS of total temperature.

Fig. 8 Time average of total temperature and RMS of total temperature at two different axial locations for $\text{Re}_{s,\text{out}} = 70 \times 10^3$ and $M_{s,\text{out}} = 0.7$, 0.9 and 0.95. For the legend, refer to Fig. 7

3.4 *Space-Time Diagrams*

Figures 9, 10, and 11 depict squared velocity perturbations normalized by outlet isentropic velocity squared. The first quantity offers insight into turbulent kinetic energy, i.e.,

$$k_{\text{d}} = \frac{1}{2} \times \frac{u'^2 + v'^2 + w'^2}{U_{\text{is}}^2} \tag{4}$$

while the second relates to non-diagonal terms of the Reynolds stress tensor, i.e.,

$$k_{\text{nd}} = \frac{1}{2} \times \frac{u'v' + v'w' + w'u'}{U_{\text{is}}^2}, \tag{5}$$

where u', v' and w' respectively represent the x, y and z components of the velocity perturbation and U_{is} is the outlet isentropic velocity. These figures complement each other and provide valuable context for a spectral analysis. The contours in Fig. 9, for instance, illustrate passing vortices in the wake. Their intensity is higher at the wake center, corresponding with findings from Fig. 3a–c.

Contours in Figs. 9, 10 and 11 reveal a 0.1 ms shift between planes, with vortices swiftly succumbing to turbulence-induced destruction. The presence of non-diagonal components is less pronounced than diagonal ones, although their magnitude is comparable. Overall, these contours align with previous results, indicating developed turbulence decay. Figure 10 mirrors previous observations, showing a more organized wake compared to low Mach. Notably, intensity of diagonal and non-diagonal components is two to three times lower, implying less decay at higher Mach. This

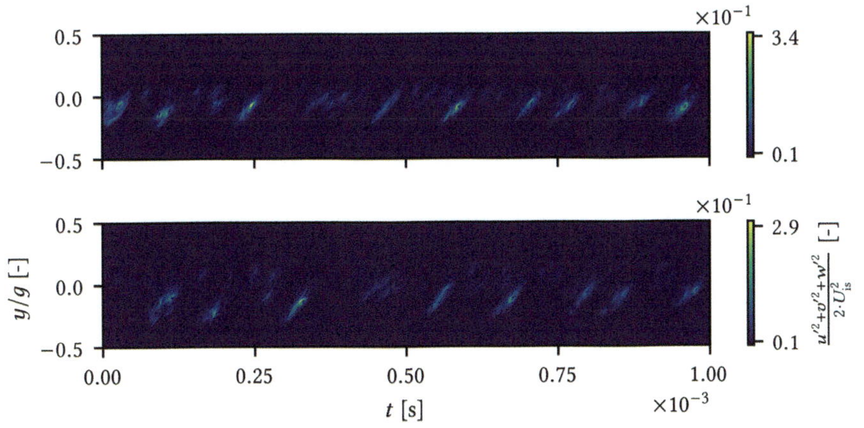

(a) Diagonal terms at x/c_{ax} = 1.25 (top) and x/c_{ax} = 1.5 (bottom).

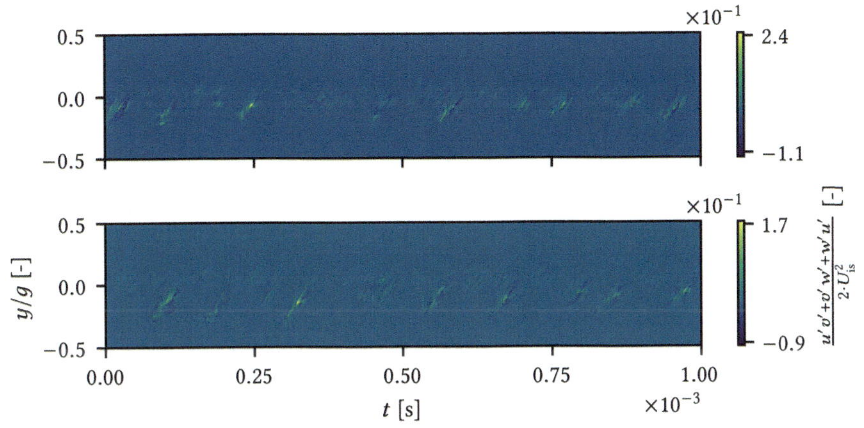

(b) Non diagonal terms at x/c_{ax} = 1.25 (top) and x/c_{ax} = 1.5 (bottom).

Fig. 9 Turbulent activity, represented by $1/2 \times (u'^2 + v'^2 + w'^2)/U_{\text{is}}^2$ (a), and $1/2 \times (u'v' + v'w' + w'u')/U_{\text{is}}^2$ (b), for $\text{Re}_{s,\text{out}} = 70 \times 10^3$ and $M_{s,\text{out}} = 0.7$

points to higher shedding frequencies at elevated Mach numbers. In Fig. 11, the same trends persist. The wake appears to narrow with increasing Mach number, particularly evident in the outer regions.

Space-time diagrams of wall shear stress at midspan is provided in Figs. 12, 13 and 14.

For the case with low Mach number, the separation position along the suction side displays temporal variability, as does the reattachment location. This is supported by the observations in Fig. 5, which indicates low values of $c_{f,s}$ at the trailing edge. These results, in conjunction with the space-time diagram, suggest periodic bursting

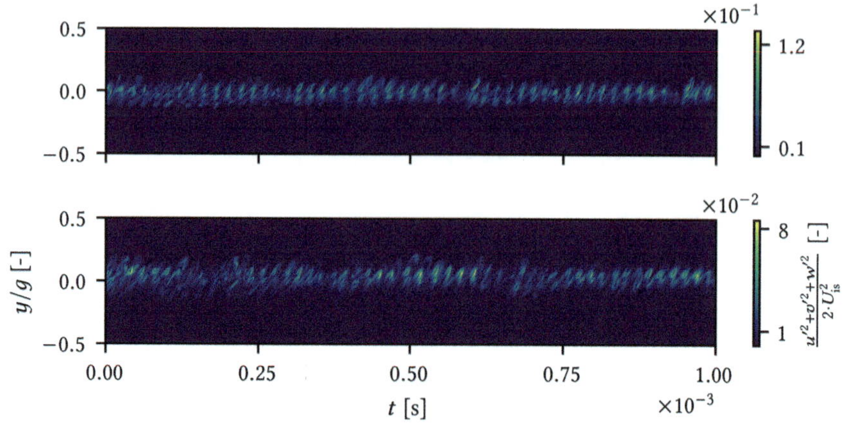

(a) Diagonal terms at $x/c_{\mathrm{ax}} = 1.25$ (top) and $x/c_{\mathrm{ax}} = 1.5$ (bottom).

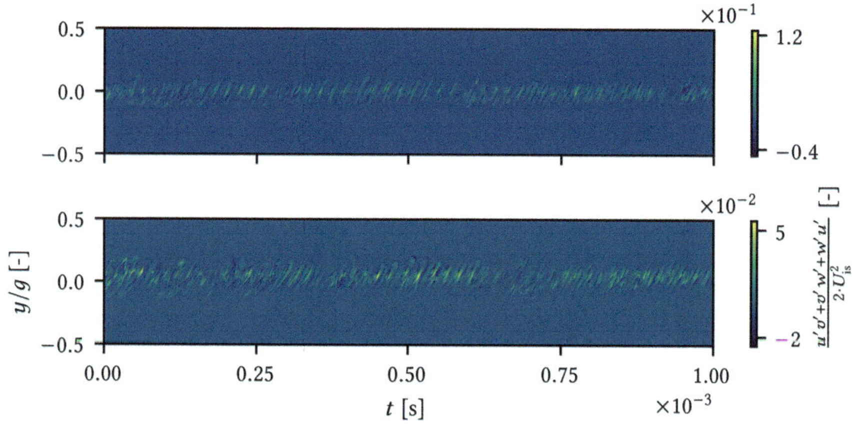

(b) Non diagonal terms at $x/c_{\mathrm{ax}} = 1.25$ (top) and $x/c_{\mathrm{ax}} = 1.5$ (bottom).

Fig. 10 Turbulent activity, represented by $1/2 \times (u'^2 + v'^2 + w'^2)/U_{\mathrm{is}}^2$ (a), and $1/2 \times (u'v' + v'w' + w'u')/U_{\mathrm{is}}^2$ (b), for $\mathrm{Re}_{s,\mathrm{out}} = 70 \times 10^3$ and $M_{s,\mathrm{out}} = 0.9$

of the separation bubble, wherein certain perturbations propagate upstream due to subsonic velocities. The presence of disorganized vortices detaching and flowing downstream in the front of the pressure side, as seen in Fig. 3a–c, further corroborates this observation, aligning with the conclusions from flow visualization that acoustic perturbations can propagate upstream.

For the moderate Mach number case, the variation in separation position is less pronounced compared to the low Mach case, with the reattachment location exhibiting a gentler variability. The separation location aligns well with the average shock

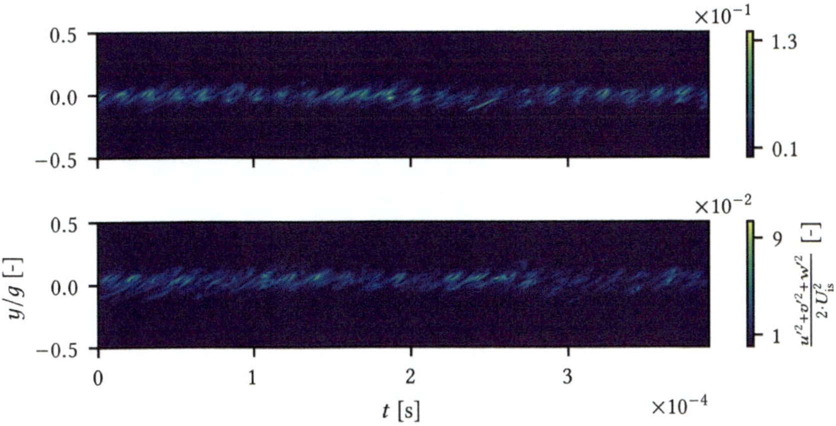

(a) Diagonal terms at x/c_{ax} = 1.25 (top) and x/c_{ax} = 1.5 (bottom).

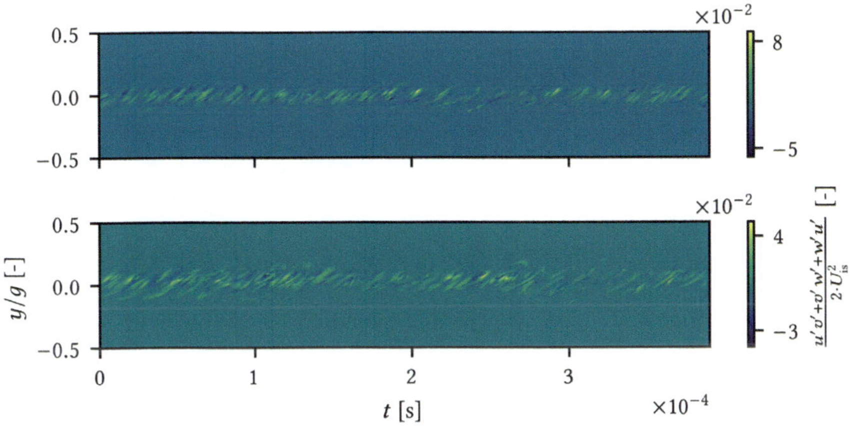

(b) Non diagonal terms at x/c_{ax} = 1.25 (top) and x/c_{ax} = 1.5 (bottom).

Fig. 11 Turbulent activity, represented by $1/2 \times (u'^2 + v'^2 + w'^2)/U_{is}^2$ (a), and $1/2 \times (u'v' + v'w' + w'u')/U_{is}^2$ (b), for $\mathrm{Re}_{s,out} = 70 \times 10^3$ and $M_{s,out} = 0.95$

location. On the pressure side, a cleaner separation bubble is observed, coupled with vortex shedding at reattachment.

In the high Mach number case, separation conditions on the suction side are rarely attained. Small bubbles occasionally form after the average shock location, yet their extent remains limited. Similar behavior is observed on the pressure side.

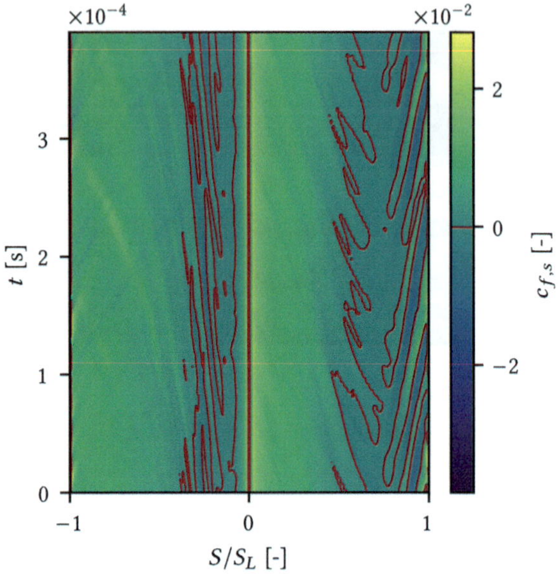

Fig. 12 Space-time diagrams of the skin friction coefficient $c_{f,s}$ along blade PS (negative abscissa) and SS (positive abscissa) at midspan for $\mathrm{Re}_{s,\mathrm{out}} = 70 \times 10^3$ and $M_{s,\mathrm{out}} = 0.7$, monitored during a time interval of 0.4 ms

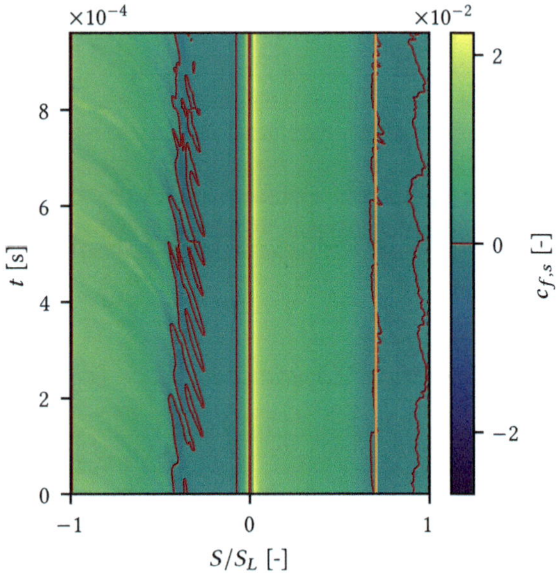

Fig. 13 Space-time diagrams of the skin friction coefficient $c_{f,s}$ along blade PS (negative abscissa) and SS (positive abscissa) at midspan for $\mathrm{Re}_{s,\mathrm{out}} = 70 \times 10^3$ and $M_{s,\mathrm{out}} = 0.9$, monitored for 1 ms. The time averaged shock position is displayed

4 Conclusions

In summary, this study offers valuable insights into the intricate interplay of separation, self-induced transition, choking and shock phenomena in high-speed LPT operating under GTF conditions during cruise.

Fig. 14 Space-time diagrams of the skin friction coefficient $c_{f,s}$ along blade PS (negative abscissa) and SS (positive abscissa) at midspan for $\mathrm{Re}_{s,\mathrm{out}} = 70 \times 10^3$ and $M_{s,\mathrm{out}} = 0.95$, monitored for 0.4 ms. The time averaged shock position is displayed

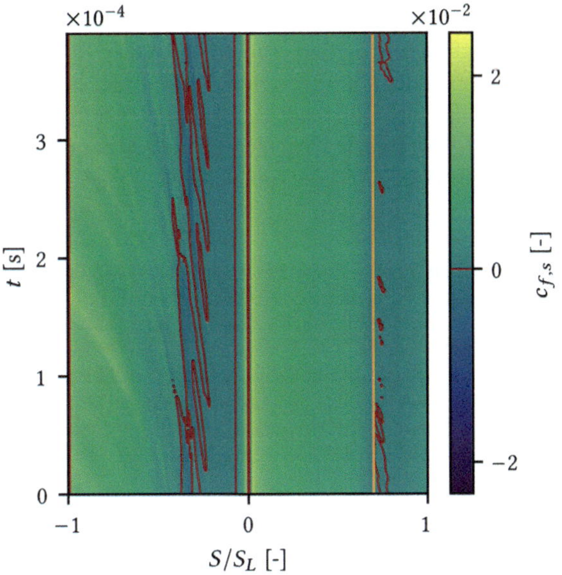

Flow visualization unveiled separation occurring on the suction side at Mach 0.70 and 0.90, and over the pressure side for all cases. The wake at lower Mach numbers exhibited broader and more chaotic behavior compared to other cases. Isentropic Mach number distributions are closely aligned with experimental data, except for the pressure side separation bubble that was not anticipated by experiments due to the absence of turbulence injected at the inlet of the numerical simulations. Increasing Mach number correlated with a shift in velocity peak towards the TE. Skin friction affirmed separation occurring roughly at 58% and 68% of suction side length for Mach 0.70 and 0.9, while no average separation manifested at higher Mach numbers. Separation position concurred with the average shock location, and as Mach number increased, the distribution converged towards the shock, resulting in diminished pressure gradient and absence of separation. Analysis of boundary layer profiles and integral parameters confirmed laminar separation on the suction side, with transition occurring around the point of maximum displacement. Turbulent conditions did not manifest prior to the TE. On the pressure side, laminar separation with turbulent reattachment and subsequent reversed transition was evident across all Mach numbers until the TE. The laminar separation long bubble mode was identified as dominant on the suction side. Investigation of the wake through downstream planes highlighted a correlation between Mach number and losses, and indicated a narrower wake at higher Mach numbers. Pronounced turbulent decay was observable at lower Mach numbers. The computations undertaken within this study harnessed DGM in tandem with SIPDG, offering higher order precision with coarser mesh resolutions compared to alternative methods.

In the future, the injection of turbulence at the inlet and the effect of passing wakes generated by rotating bars should be studied more closely.

To conclude, this research endeavor successfully highlighted the intricacies of separation, transition, and wake dynamics in high-speed LPT within the context of the SPLEEN project. The outcomes augment our comprehension of the mechanisms governing boundary layer phenomena, encompassing separation, transition on the suction and pressure sides, turbulent decay, and wake losses.

Acknowledgements Computational resources have been made available by the Consortium des Équipements de Calcul Intensif (CÉCI), funded by the Fonds de la Recherche Scientifique de Belgique (F.R.S.-FNRS) under Grant No. 2.5020.11 and by the Walloon Region, on the Tier-1 supercomputer zenobe of the Fédération Wallonie-Bruxelles, infrastructure funded by the Walloon Region under grant agreement no. 1117545. Additionally, we acknowledge the support of the SPLEEN project funded by the European Union's Horizon 2020 research and innovation program. Special thanks also go to Safran for their valuable collaboration.

References

1. Kurzke, J.: Fundamental differences between conventional and geared turbofans. Volume 1: Aircraft Engine; Ceramics; Coal, Biomass and Alternative Fuels; Controls, Diagnostics and Instrumentation; Education; Electric Power; Awards and Honors (2009)
2. Mayle, R.E.: The 1991 IGTI scholar lecture: the role of laminar-turbulent transition in gas turbine engines. J. Turbomach. **113**(4), 509–36 (1991)
3. Pacciani, R., et al.: A CFD study of low reynolds number flow in high lift cascades. ASME Turbo Expo 2010: Power for Land, Sea, and Air (2010)
4. Cobley, K., et al.: Design of new three stage low pressure turbine for the bmw rolls-royce BR715 turbofan engine. Volume 1: Aircraft Engine; Marine; Turbomachinery; Microturbines and Small Turbomachinery (1997)
5. Curtis, E.M., et al.: Development of blade profiles for low-pressure turbine applications. J. Turbomach. **119**(3), 531–38 (1997)
6. Haselbach, F., et al.: The application of ultra high lift blading in the BR715 LP turbine. J. Turbomach. **124**(1), 45–51 (2001)
7. Howell, R., et al.: Boundary layer development in the BR710 and BR715 LP turbines-the implementation of high-lift and ultra-high-lift concepts. J. Turbomach. **124**(3), 385–92 (2002)
8. Zhang, X.F., et al.: Unsteady boundary layer studies on ultra-high-lift low-pressure turbine blades. Proc. Ins. Mech. Eng. Part A: J. Power Energy **219**(6), 451–60 (2005)
9. Lou, W., Hourmouziadis, J.: Separation bubbles under steady and periodic-unsteady main flow conditions. J. Turbomach. **122**(4), 634–43 (2000)
10. Hatman, A., Wang, T.: A prediction model for separated-flow transition. J. Turbomach. **121**(3), 594–602 (1999)
11. Choi, H., Moin, P.: Grid-point requirements for large eddy simulation: Chapman's estimates revisited. Phys. Fluids **24** (2012)
12. Hillewaert, K.: Development of the discontinuous Galerkin method for high resolution, large scale CFD and acoustics in industrial geometries. Université catholique de Louvain (2013)
13. Carton De Wiart, C.: Towards a discontinuous galerkin solver for scale-resolving simulations of moderate reynolds number flows, and application to industrial cases. Université catholique de Louvain (2014)

14. Simonassi, L., et al.: An experimental test case for transonic low-pressure turbines - part I : Rig Design, instrumentation and experimental methodology. In: Volume 10B : Turbomachinery - Axial Flow Turbine Aerodynamics ; Deposition, Erosion, Fouling, and Icing ; Radial Turbomachinery Aerodynamics (2022)
15. Simonassi, Loris et al.: An experimental test case for transonic low-pressure turbines - part 2 : Cascade aerodynamics at on- and off-design Reynolds and Mach numbers. In: Volume 10B : Turbomachinery - Axial Flow Turbine Aerodynamics ; Deposition, Erosion, Fouling, and Icing ; Radial Turbomachinery Aerodynamics (2022)
16. Sutherland, W.J.: LII. the viscosity of gases and molecular force. The London, Edinburgh and Dublin Philos. Mag. J. Sci. **36**(223), 507–31 (1893)
17. Geuzaine, C., Remacle, J.-F.: Gmsh: a three-dimensional finite element mesh generator with built-in pre- and post-processing facilities. Int. J. Numer. Meth. Eng. **79**(11), 1309–1331 (2009)
18. Eckert, E., Weise, W.: Messungen der Temperaturverteilung auf der Oberfläche schnell angeströmter unbeheizter Körper. Forschung auf dem Gebiete des Ingenieurwesens. **13**(6), 246–254 (1942)
19. Carscallen, W.E., Currie, T.C., Hogg, S.I., Gostelow, J.P.: Measurement and computation of energy separation in the vortical wake flow of a turbine nozzle cascade (1999)
20. Sieverding, C.H., Ottolia, D., Bagnera, C., Comadoro, A., Brouckaert, J.F., Desse, J.M.: Unsteady turbine blade wake characteristics. J. Turbomach. **126**(4), 551–559 (2004)
21. Hodson, H.P., Dawes, W.N.: On the interpretation of measured profile losses in unsteady wake-turbine blade interaction studies (1998)
22. Sieverding, C., Manna, M.: A review on turbine trailing edge flow. Int. J. Turbomach. Prop. Power **5**(2), 10 (2020)
23. Vázquez, R., et al.: The influence of Reynolds number, Mach number and incidence effects on loss production in low pressure turbine airfoils. Proceedings of GT2006, ASME Turbo Expo 2006: Power for Land, Sea and Air (2006)
24. Vázquez, R., Torre, D.: The effect of mach number on the loss generation of LP turbines. ASME Turbo Expo 2012: Turbine Technical Conference and Exposition (2012)
25. Vera, M.T., Howard P.H.: Low speed vs high speed testing of LP turbine blade-wake interaction. The 16th Symposium on Measuring Techniques in Transonic and Supersonic Flow in Cascades and Turbomachines (2002)
26. Benyahia, A., et al.: Prediction of separation-induced transition on high lift low pressure turbine blade. ASME 2011 Turbo Expo: Turbine Technical Conference and Exposition (2011)
27. Coull, J.D., et al.: Velocity distributions for low pressure turbines. J. Turbomach. **132**(4), ASM International (2008)
28. Houtermans, R., et al.: Aerodynamic performance of a very high lift low pressure turbine blade with emphasis on separation prediction. Trans. ASME **126**(3), 406 (2004)
29. Michálek, J., et al.: Aerodynamic performance of a very high lift low pressure turbine airfoil (T106C) at low reynolds and high mach number with effect of free stream turbulence intensity. J. Turbomach. **134**(6), ASM International (2010)
30. Sommerfeld, A.: Ein Beitrag zur hydrodynamische Erklärung der turbulenten Flüssigkeitsbewegungen, Proceedings of the 4th International Congress of Mathematicians, Rome **III**, pp. 116-124 (1908)
31. Bellows, W.J.: An experimental study in leading edge separating-reattaching boundary layer flows (1985)
32. Bolyn, A.: Detailed flow analysis of a transonic low pressure turbine at low turbulence levels. University of Liège (2020)
33. Börner, M., Niehuis, R.: Dynamics of shock waves interacting with laminar separated transonic turbine flow investigated by high-speed schlieren and surface hot-film sensors. J. Turbomach. **143**(5), ASM International (2021)
34. *CECI*. www.ceci-hpc.be/clusters.html#zenobe
35. Dähnert, J., et al.: Transition mechanisms in laminar separated flow under simulated low pressure turbine aerofoil conditions. J. Turbomach.**135**(1), ASM International (2011)

36. Davidson, P.: Turbulence: An Introduction for Scientists and Engineers. Oxford UP, USA (2004)
37. Dolejší, V., Miloslav F.: Analysis and applications to compressible flow. Springer, Discontinuous Galerkin Method (2015)
38. Gaster, M.: The structure and behaviour of laminar separation bubbles (1969)
39. Gault, D.E.: An experimental investigation of regions of separated laminar flow (1955)
40. Hodson, H.P., Dominy, R.G.: Three-dimensional flow in a low-pressure turbine cascade at its design condition. J. Turbomach. **109**, 177–85 (1987)
41. Hodson, H.P., Dominy, R.G.: The Off-Design Performance of a Low-Pressure Turbine Cascade. J. Turbomach. **109**(2), 201–209 (1987)
42. Jean, H:: Aerodynamic design of low pressure turbines. AGARD Lecture Series, **187** (1989)
43. Khateeb, M.: Impact of Operating conditions on the transitional flow in a low-pressure fast turbine cascade. University of Liège (2021)
44. Michelassi, V., et al.: Compressible direct numerical simulation of low-pressure turbines-Part II: effect of inflow disturbances. J. Turbomach. **137**(7), (2014)
45. Pacciani, R., Marconcini, M., Arnone, A., et al.: An assessment of the laminar kinetic energy concept for the prediction of high-lift, low-reynolds number cascade flows. Proc. Inst. Mech. Eng. Part A: J. Power Energy **225**(7), 995–1003 (2011)
46. Roberts, W.B.: Calculation of laminar separation bubbles and their effect on airfoil performance. AIAA J. **18**(1), 25–31 (1980)
47. Sandberg, R.D., et al.: Compressible direct numerical simulation of low-pressure turbines-Part I: methodology. J. Turbomach. **137**(5), ASM International (2014)
48. Vera, et al.: The effect of Mach number on LP turbine wake-blade interaction. Unsteady Aerodynamics, Aeroacoustics and Aeroelasticity of Turbomachines, 203-16 (2006)
49. Rasquin, M., Bauer, A., Hillewaert, K.: Scientific post hoc and in situ visualisation of high-order polynomial solutions from massively parallel simulations. Int. J. Comput. Fluid Dyn. 171–180 (2019)

Flight Stream Effects on Installed Jet Noise with Wing-Flap and Fuselage

James C. Tyacke, Z.-N. Wang, and P. G. Tucker

Abstract Reduction of jet noise through use of higher bypass ratios has significantly reduced isolated jet noise and installation effects generally, have become of higher relative importance. Two flight stream velocities for an installed jet engine including pylon and a wing with deployed flap, connected to a fuselage are contrasted. The successful framework developed previously is extended to this challenging configuration. Key flow and noise data are validated against measurements in a blind test. A significant decrease in turbulence kinetic energy levels below the wing and within the shear layer downstream of the flap is found with associated decrease in noise. The authors believe the framework is general enough for inclusion of landing gear and tail-plane.

CUFS-2024-35, Cambridge, United Kingdom, 4-5 March 2024.

J. C. Tyacke (✉)
Department of Mechanical and Aerospace Engineering, Brunel University London, London, Uxbridge UB8 3PH, UK
e-mail: james.tyacke@brunel.ac.uk

Z.-N. Wang
School of Engineering, University of Birmingham, Birmingham B15 2TT, UK
e-mail: z.n.wang@bham.ac.uk

P. G. Tucker
Cambridge University Engineering Department, University of Cambridge, Cambridge CB2 1PZ, UK
e-mail: pgt23@cam.ac.uk

J. C. Tyacke and N. R. Vadlamani (eds.), *Proceedings of the Cambridge Unsteady Flow Symposium 2024*, https://doi.org/10.1007/978-3-031-69035-8_22

Nomenculture

Acronyms/Labels

CFL	Courant-Friedrichs-Lewy
F16	16° flap angle
FO	Flyover
FS	Flight stream
FW-H	Ffowcs Williams-Hawkings
L-R	LES-RANS
LES	Large-eddy simulation
PS	Pressure side
RANS	Reynolds-Averaged Navier-Stokes
SGS	Sub-grid scale
SL	Sideline
SPL	Sound pressure level (dB)
SS	Suction side

Greek

β	LES-RANS transition zone fraction
Δ	Spatial increment (m), LES filter width (m)
δ	Boundary layer thickness (m)
δ_{99}	99% freestream velocity boundary layer thickness (m)
δ_{ij}	Kronecker delta
$\langle\,\rangle$	Time mean
μ	Viscosity m^2/s
∇	Laplacian
ϕ	Polar angle (degrees), flow variable
ρ	Density ($\mathrm{Kg/m}^3$)
τ_{ij}	Modelled stress tensor
θ	Azimuthal angle (degrees)
ε	Smoothing parameter

Roman

$'$	Fluctuation
$\mathbf{u}$	Cartesian velocity vector (m/s)
$\mathbf{x}$	Cartesian coordinate vector (m)
-	Spatial filter

$\overline{L}$	Linear modelled stress
$\overline{NL}$	Nonlinear modelled stress
A	Jacobian matrix
C_s	Smagorinsky model constant
C_α	Alpha model constant
D	Diameter (m)
d_{RANS}	RANS layer extent (m)
d_{wall}	Wall normal distance (m)
f	Frequency (Hz), LES-RANS blending function
F_{ij}	Edge flux
k	Turbulence kinetic energy (m^2/s^2)
P	Pressure (Pa)
R, r	Radius (m)
S_{ij}	Rate of strain tensor
St	Strouhal number
T	Temperature (K)
t	Time (s)
t^*	Non-dimensional time
Ti	Turbulence intensity
U	Velocity scale (m/s)

Sub/super-script

0	Total
b	Bypass
c	Core
i,j,k,l	coordinate indices
mean	Time mean
rms	Root mean squared

1 Introduction

The EU Flightpath 2050 report states a 65% perceived noise level reduction relative to year 2000 levels as a goal [1]. Reduction of jet noise through use of higher bypass ratios has significantly reduced isolated jet noise and installation effects generally, have become of higher relative importance. Engine airframe coupling is now if crucial importance and will remain so for future architectures. The effects of jet stream proximity to lifting surfaces has hence received growing attention.

An experimental campaign with flight stream was performed by Mengle et al. [2–4] systematically measuring and studying noise sources from isolated and installed round and serrated nozzles. Brown [5] includes finite span effects of jet-flat-plate interaction into empirical models. Thomas et al. [6] study the overall effect of coaxial nozzle serrations, pylon, evelon deflection and application of acoustic liners. Mengle [7] also study the effect of reducing the gulley height and increasing flap angle between coaxial round and serrated jet nozzles under flight conditions, noting the appearance of two peaks in the installed spectra. Although much has been learned from experiments, compromises and simplifications in problem definition and corrections to data are commonplace.

Numerical simulation can offer detailed data anywhere in the flowfield within ever shortening design cycles. It is of great interest to industry to begin replacing measurements with simulation for certification.

Georgiadis and DeBonis [8] provide a good overview of both RANS and LES modelling of jets. With the lack of a single reliable RANS model, there is a requirement to resolve the large scale eddies responsible for turbulence noise production.

For a 3D wing and ultra-high bypass ratio (bypass ratio=15) nozzle with flight stream, Tyacke et al. [9] show up to 20dB increase (dependent on flap deflection) in mid-low frequencies over an isolated nozzle. For the same conditions, Wang et al. [10] show serrated nozzles weaken jet-wing interaction and can reduce noise by 5dB at the peak frequency. Unsteady surface pressures, calculated turbulence lengthscales and 2nd and 4th order space-time correlations indicate generally that reduced turbulence lengthscale and intensity are responsible for lowered noise [10, 11]. Vogel et al. [12] also study a coaxial jet (bypass ratio=5) with pylon and a single stream jet near ($1 < h/D < 3$) a flat plate. Source localisation is performed using beamforming identifying sources in the wake of the pylon, at the nozzle lip and plug. An increase in low frequency noise was observed for the installed cases.

In order to extend and further test the authors' previous methodology, we contrast two flight stream velocities for an installed jet nozzle (bypass ratio=10.5) with a wing with deployed flap, connected to a fuselage with pylon. The latter two geometric components are in addition to previous studies. To the authors' knowledge, the presence of an entire fuselage has not been modelled before.

The case setup and boundary conditions are set out in Sect. 2. In Sect. 3 the previous successful framework employing zonalisation of Large-Eddy Simulation and Reynolds-Averaged Navier-Stokes turbulence modelling (LES-RANS), hybridised structured-unstructured meshing and Ffowcs Williams-Hawkings (FW-H) surface definition is extended to this challenging configuration. Section 4 shows validation of LES-RANS against flow (for F1 only due to lack of measured data) and acoustics measurements (F1 and F2), with exploration of flow features responsible for variations in noise sources.

2 Case Setup

2.1 Geometry and Boundary Conditions

The geometry is shown in Fig. 1. The nacelle extends upstream as per the experiments. The half-fuselage geometry received was reflected in the x-z plane to create

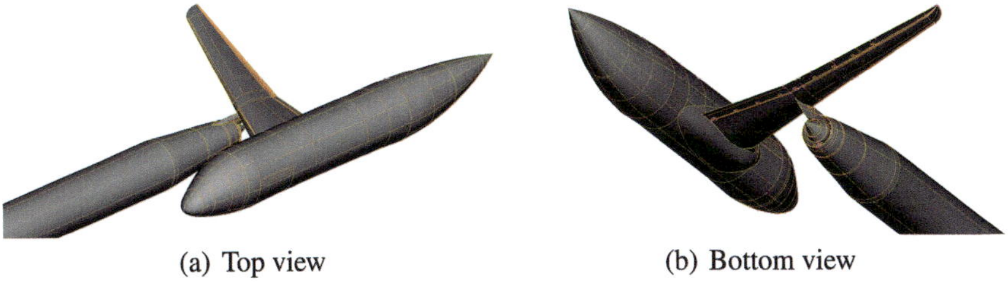

Fig. 1 Geometry consisting of a coaxial nozzle, pylon, wing, flap and fuselage

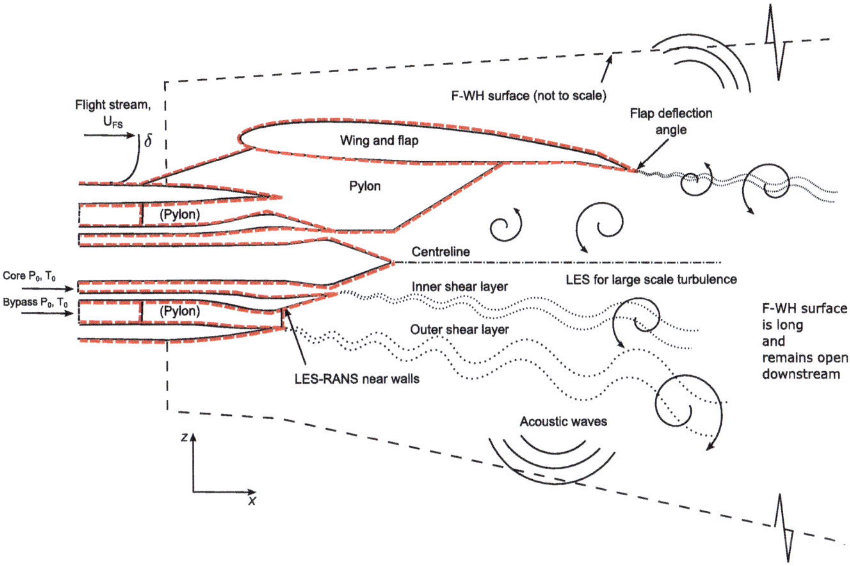

Fig. 2 Case setup schematic indicating boundary conditions and geometry features in the $y = 0$ plane

a watertight geometry with no discontinuities. There is no wing-stub or mounting plate as in the measurements.

Figure 2 indicates the main features of the geometry, modelling, flow and boundary conditions used. The bypass and core fluid stream inlets and the outer flight stream are indicated. Freestream conditions are applied at the flight stream inlet and radial far-field. Downstream, a subsonic outlet is used at atmospheric pressure. Near walls, as indicated by the red dashed lines, a RANS layer is used to model near wall streaks. Downstream, LES is used to resolve large scale turbulent structures. The near field acoustic waves generated are recorded on the permeable FWH surface.

Table 1 provides a summary of the nozzle conditions as indicated in Fig. 2 and are identical for both flight stream cases. At the bypass and core stream inlets, total pressure and total temperature conditions are imposed. $P_a = 101325$ Pa and

Table 1 Surface URANS layer thickness

	Bypass	Core
P_0/P_a	1.37	1.20
T_0/T_a	1.20	2.42

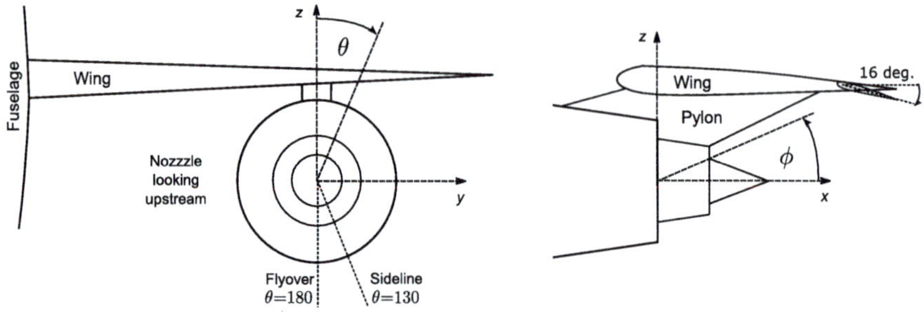

Fig. 3 Coordinate system showing azimuthal (left) and polar (right) coordinates

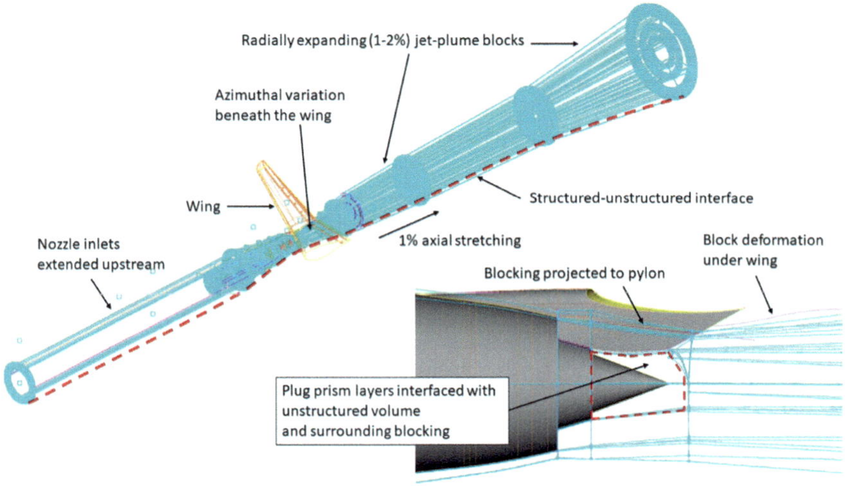

Fig. 4 Jet structured mesh blocking

$T_a = 288.15$ K. At the far field, ambient conditions and a flight stream of M $= 0.15$, and M $= 0.3$, denoted F1 and F2 respectively. The bypass nozzle diameter $D = 0.1857$ m. The fully expanded velocities are $U_c = 267.6$ ms^{-1} and $U_b = 244.5$ ms^{-1}. The wing is tapered and swept representing a typical civil aircraft wing. The deployed flap angle is 16°, denoted F16. The coordinate system used is shown in Fig. 3 along with the azimuthal locations of the sideline (SL) and flyover (FO) profiles.

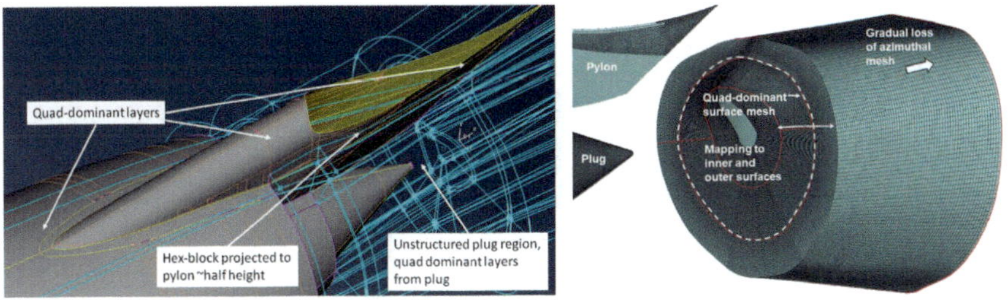

(a) (b)

Fig. 5 Jet unstructured mesh content, **a** Prism layer generation near the pylon, **b** Azimuthal variation beneath the wing

2.2 *Mesh Generation*

The mesh was generated using ANSA (Beta-CAE). Wall-normal mesh spacings are at approximately $y^+ = 1$ or below. Figure 4 shows the structured mesh blocks that cover most of the jet plume and nozzle geometry. Challenging zones are dealt with by inserting regions of unstructured and semi-structured mesh. These are near the pylon-nozzle intersection, the plug and azimuthal variation beneath the wing. These are highlighted in Fig. 4.

At the nozzle lips 310Δ are used circumferentially. Downstream of the flap, and upstream of the pylon, this reduces to 230Δ and 160Δ respectively.

As indicated in 4, near the pylon, a section of the structured mesh blocking is removed and replaced with a quad-dominant surface mesh, extruded normally, interfacing with the structured mesh cells. The remaining structured mesh is projected onto the side of the pylon (5a). For the plug, quad-dominant surface mesh is mapped normally to create prism layers. The remaining void is filled with tetrahedral cells, joining with the innermost extent of the mesh.

Figure 5b indicates the procedure for the reduction of azimuthal nodes as the turbulent scales grow downstream. Unstructured quad-dominant mesh is generated at a radial location mid-way between an inner and outer radial boundary and projected in both directions causing a ratio of 2 stretching or squeezing in tangential mesh spacing. The final mesh presented in Fig. 6 consists of 201 million cells (approx. 112 and 89 million hexahedral and tetrahedral cells respectively).

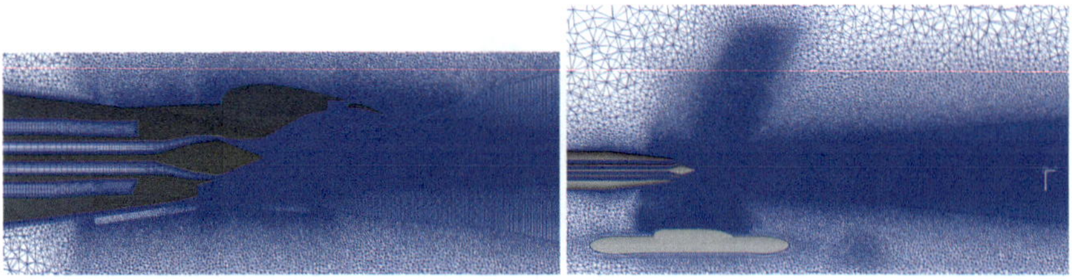

Fig. 6 Hybrid mesh **a** $y = 0$ and **b** $z = 0$ planes

3 Numerical Methods

3.1 Spatial Discretisation

An unstructured finite volume, edge-based compressible solver is used. A second order kinetic energy preserving (KEP) discretisation [13] is used in space with numerical smoothing adapted for jet simulations. The solver is has been extensively validated [14–18]. The KEP cell face flux with additional numerical smoothing is given by Eq. (1). Previous tests using homogeneous decaying turbulence with hexahedral, tetrahedral and prism elements showed the KEP scheme was less sensitive to cell type than the previous Roe scheme enabling the use of the hybrid meshing strategy presented.

$$F_{ij} = \frac{1}{2}\left(\rho_i + \rho_j\right) \times \frac{1}{2}\left(u_i + u_j\right) \times \frac{1}{2}\left(\phi_i + \phi_j\right) - \frac{1}{2}\varepsilon|A|[\nabla_i^2\phi - \nabla_j^2\phi] \quad (1)$$

The parameter ε is used to minimise numerical dissipation in the jet flow and FW-H region and is increased along with mesh expansion to form sponge zones near the far-field boundaries preventing wave reflections. Additional smoothing is applied to remove smaller scales prior to mesh coarsening, such as the azimuthal variation, larger turbulence structures still being resolved on the coarser mesh. This zone is indicated in Fig. 7a.

3.2 Temporal Discretisation

Second order backward difference implicit time discretisation was used with dual time-stepping (explicit 4 stage Runge-Kutta). The time step was set according to the smallest axial spacing at the nozzle lip giving a global acoustic CFL number below 1. Flow was developed for $150T^*$ ($T^* = D_b/U_b$) and data recorded for $180T^*$.

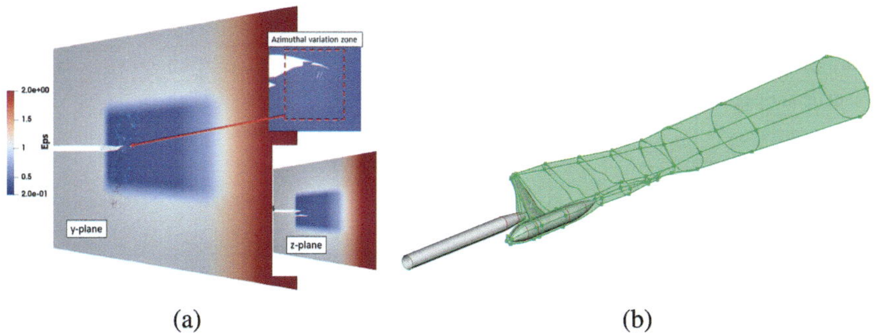

Fig. 7 **a** Numerical smoothing applied in the computational domain, **b** FW-H surface definition (innermost)

3.3 Turbulence Modelling

The approach taken for turbulence modelling is nominally the same as previous studies [9, 14–16, 19, 20]. This is based on zonalisation of URANS to model near wall streaks, reducing computational cost at high Reynolds numbers and reverting to LES at greater wall distances to resolve larger scale noise generating turbulence. The Spalart-Allmaras model [21] is employed in RANS regions.

In LES regions, a mixed nonlinear model is used [22] combining the linear WALE and nonlinear alpha SGS models to provide dissipation and model anisotropy respectively. The WALE model [23] is chosen to remedy known limitations of the Smagorinsky model [24]. The SGS filter width is defined as $\Delta_{SGS} = (\Delta_x \Delta_y \Delta_z)^{(1/3)}$ to appropriately capture shear layer development. After approximately $(x/D = 10)$, $\Delta_{SGS} = max(\Delta_{max}, (\Delta_x \Delta_y \Delta_z)^{(1/3)})$ in order to suitably model the more isotropic flow structures on the anisotropic mesh downstream. A Hamilton Jacobi equation [25] is used to compute the wall distance which is used to define a blending function f, of the turbulent viscosity μ_t, resulting in $f = 1$ and $f = 0$ for LES and URANS respectively. The overall model is described by Eqs. (2)–(4), where d_{RANS} is the URANS layer thickness and β defines the size of the URANS to LES transition zone as a fraction of d_{RANS}.

$$\tau_{ij}^{model} = \overline{L} + \overline{NL} \tag{2}$$

$$\overline{L} = \frac{\tau_{kk}\delta_{ij}}{3} + 2[f \cdot \mu_{SGS} + (1 - f) \cdot \mu_{RANS}]S_{ij} \tag{3}$$

$$\overline{NL} = \rho C_\alpha \Delta_{SGS}^2 f\left(\frac{\partial u_i}{\partial x_l}\frac{\partial u_l}{\partial u_j} + \frac{\partial u_i}{\partial x_l}\frac{\partial u_j}{\partial u_l} + \frac{\partial u_l}{\partial x_i}\frac{\partial u_l}{\partial u_j}\right) \tag{4}$$

$S_{ij} = 0.5(\partial u_i/\partial x_j + \partial u_j/\partial x_i)$ is the strain rate tensor and the blending function is given as

Table 2 Surface URANS layer thickness

Surface	d_{RANS} (m)	d_{RANS} (wall units)
Nacelle + Fuselage	3.0×10^{-2}	3000
Bypass annulus	2.7×10^{-4}	100
Core annulus + Plug	8.3×10^{-4}	100
Pylon	1.0×10^{-3}	100
Wing + Flap	2.7×10^{-4}	100

$$f = min[max(\frac{d_{wall} - (1 - \beta)d_{RANS}}{\beta d_{RANS}}, 0), 1] \tag{5}$$

The URANS layer thickness on geometric surfaces is defined locally as per Table 2. In order to reproduce the thick boundary layer on the nacelle and fuselage, d_{RANS} is extended relative to other surfaces, which have proven to work well with thinner URANS layers due to the dominance of intense shear at the nozzle lips. This is also consistent with cases where internal geometry may be included such as fans and vanes, producing duct turbulence as [15]. A (thin) URANS layer also prevents separation on the plug as noted by Eastwood et al. [17].

3.4 FW-H Surface Generation

An FW-H surface is defined using turbulence intensity as a proxy for Reynolds stresses in the Lighthill stress tensor, vorticity to avoid vortical motion crossing the surface and wall distance to avoid geometric intersections as per [9]. As there is no incoming turbulence the surface encloses the nozzle geometry and fuselage upstream of the wing. Further surfaces are offset from this and data recorded for later interrogation. This is not possible with co-processing. The innermost surface enclosing these bounds is shown in Fig. 7b.

The authors suggest the approach is to a great extent, generic. It appears the addition of an empennage would not introduce new issues. Bluff-bodies generating wake flow such as landing gear or perhaps the empennage may be introduced using Immersed Boundary Methods without significant meshing effort so long as pressure waves are preserved and reflected. Further details of unsteady data collection and processing are presented by Wang et al. [26].

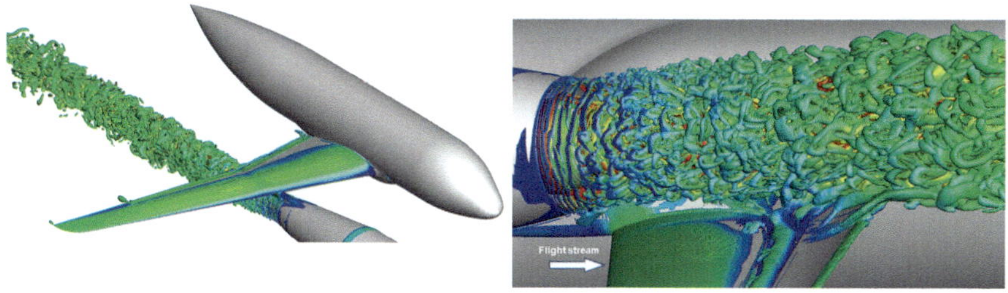

(a) (b)

Fig. 8 Instantaneous Q-criterion iso-surfaces coloured by axial velocity: **a** large scale jet plume, **b** below the wing

4 Results

4.1 Instantaneous Flow

Instantaneous flow of the jet plume is shown in Fig. 8a indicating a range of resolved scales and interaction with the wing. Wing-pylon interaction is shown in more detail from below the wing in Fig. 8b. Rapid breakdown of initial flow structures from the nozzle lip into 3D turbulence is visible before the flow proceeds along the pylon and is deflected by the deployed flap, flap recess flow and wing downwash.

4.2 Mean Flowfields

Figure 9 shows mean u-velocity contours. Typical flow features are evident including potential cores from the bypass and core streams, blockage from the pylon within the bypass duct, jet shear layer interaction with the wing and flap and downstream decay in velocity with radial spreading. Reduced shear is visible for F2 relative to F1, due to the visible doubling of the external flight steam velocity. This lengthens the jet plume and increases the Reynolds number of external geometry.

Figure 10 compares velocity profiles between cases and measured PIV data (F1 only due to data availability) in the y and z-planes. Excellent agreement is achieved both near and downstream of the nozzle. As noted, F2 increases velocity downstream and broadens the profiles but trends are essentially identical.

4.3 Turbulence Statistics

Figure 11 shows mean contours of TKE for both cases. Figure 11b (F2) shows a general decrease of TKE within the jet shear layers relative to the F1 condition

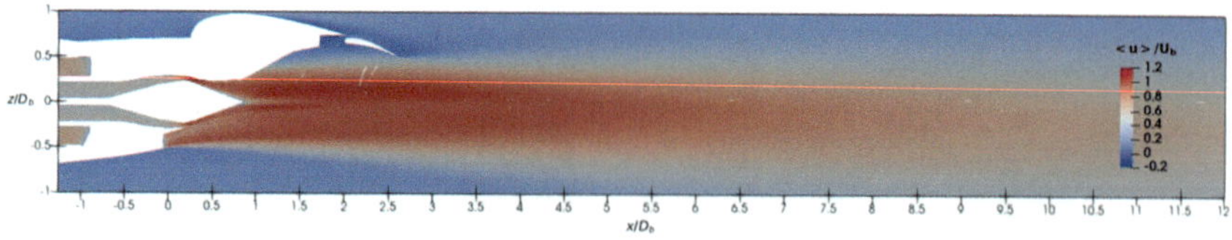

(a)

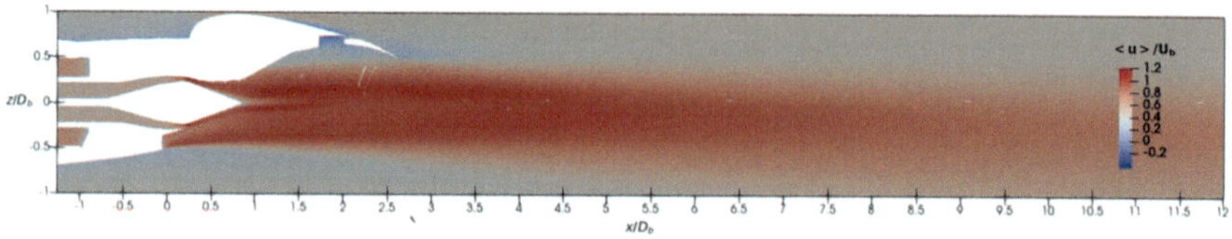

(b)

Fig. 9 Mean axial velocity contours y=0 plane: **a** F1, **b** F2

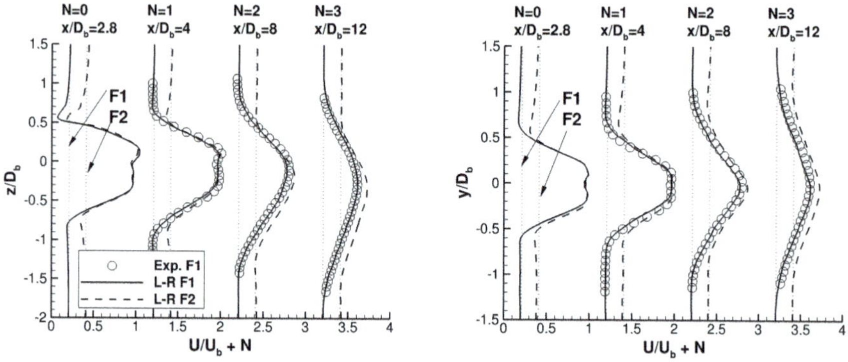

(a) (b)

Fig. 10 Mean axial velocity profiles at: **a** y=0 plane, **b** z=0 plane

(Fig. 11a). This is particularly noticeable downstream of the pylon ($x/D_b = 1$) and the flap region $2.5 < x/D_b < 5$.

Figure 12 again compares the F1 case with PIV measurements and provides detail on LES-RANS F1 and F2 cases. Peaks and profile shape are captured well downstream in both the y and z-planes. In Fig. 12a F2 TKE peaks have reduced magnitude and reduced z-position relative to F1, particularly downstream of the flap (circa. $0.25 < z/D_b < 0.5$ indicating a downward deflection of jet plume turbulence due to downwash through higher flight stream.

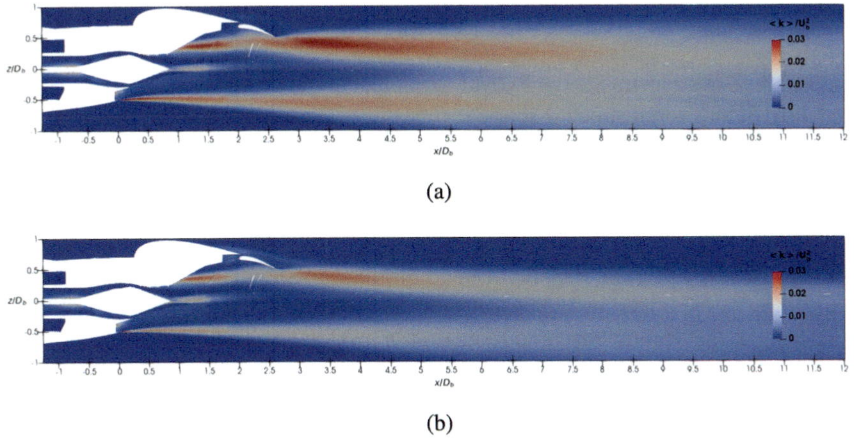

Fig. 11 Mean TKE contours y=0 plane: **a** F1, **b** F2

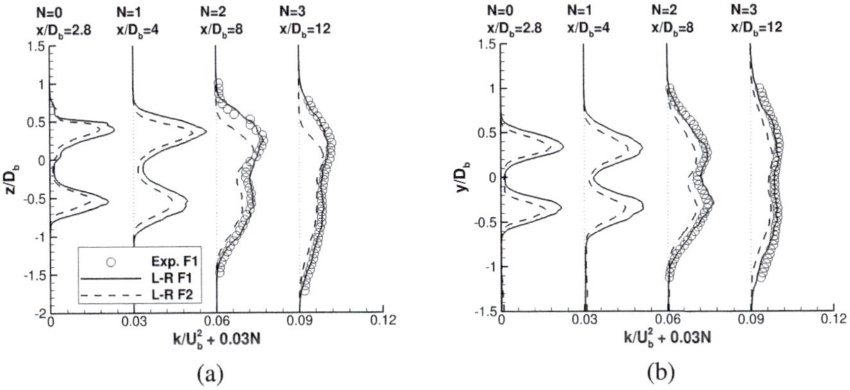

Fig. 12 Mean TKE profiles at: **a** y=0 plane, **b** z=0 plane

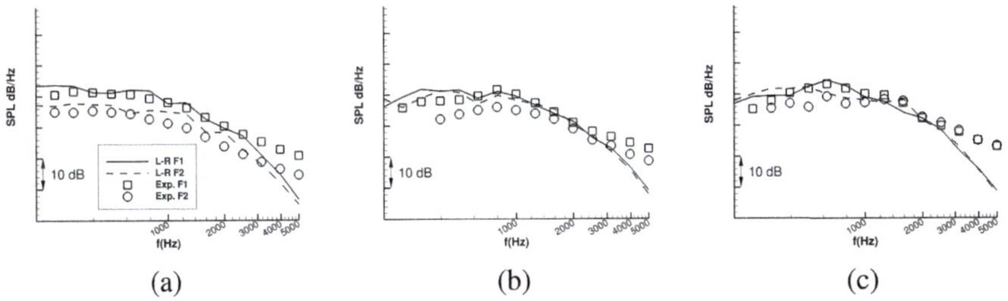

Fig. 13 FO SPL spectra for F1 and F2 cases: **a** FO $\phi = 30$, **b** FO $\phi = 90$, **c** FO $\phi = 120$

4.4 Acoustics

Figure 13 shows spectra for the flyover azimuthal angle for three polar angles 30, 90 and 120°. The overall spectral shape is represented as per the measurements for F1 and F2 cases. At $\phi = 30^\circ$, the noise reduction due to increased flight stream velocity is predicted well across a wide frequency range. Noise reduction at $\phi = 90^\circ$ appears underestimated in the low frequencies. The reason for this is under further investigation given the significant decreases in TKE. At $\phi = 120^\circ$, low frequency noise is reduced near the peak at 1000 hz in line with measurements. Overall reductions in noise may be explained by the reduction in turbulence near the flap responsible for scattering at the trailing edge and reduction in mixing noise in the shear layers. The LES-RANS becomes under-resolved at frequencies higher than $\approx 2500\,hz$ where numerical and measured data diverge. At mid-frequencies, numerical and measured data are similar.

5 Conclusions

LES-RANS has been used to predict flow and noise from a complex geometry including coaxial jet, pylon, wing, flap and fuselage. Two flight streams were contrasted. Mean flow and turbulence statistics were validated for flight stream 1 (F1). The second (increased) flight stream (F2) showed typical traits expected due to lowered shear and hence turbulence production leading to lower noise levels. At lower frequencies the difference between the F1 and F2 cases was not as great as measured at a polar angle of 90°, however overall spectral shape and trends were reflected. Given the great deal of geometric and flow complexity, the predictive accuracy over a range of frequencies is notable.

Acknowledgements The authors would like to thank the EPSRC for the computational time made available on the UK supercomputing facility ARCHER/ARCHER2 via the UK Turbulence Consortium (EP/R029326/1). The authors would like to acknowledge the financial support of the research by the UK ATI program ACAPELLA. The authors are also grateful to the UK Government for supporting the programs SYMPHONY and HARMONY during which the model-scale data was acquired in the QinetiQ NTF, and Rolls-Royce plc for facilitating access to these data.

References

1. European Commission.: Flightpath 2050 Europe ' s Vision for Aviation. Tech. Rep. (2011). https://doi.org/10.2777/50266
2. Mengle, V., Elkoby, R., Brusniak, L., Thomas, R.H.: Reducing propulsion airframe aeroacoustic interactions with uniquely tailored chevrons: 1. Isolated nozzles. 12th AIAA/CEAS aeroacoustics conference (27th AIAA Aeroacoustics Conference) **2467**, (2006). https://doi.org/10.2514/6.2006-2467

3. Mengle, V., Elkoby, R., Brusniak, L., Thomas, R.H.: Reducing propulsion airframe aeroacoustic interactions with uniquely tailored chevrons: 2. Installed nozzles. 12th AIAA/CEAS aeroacoustics conference (27th AIAA aeroacoustics conference) **1**, (2006). https://doi.org/10.2514/6.2006-2434
4. Mengle, V., Brusniak, L., Elkoby, R., Thomas, R.H.: Reducing propulsion airframe aeroacoustic interactions with uniquely tailored chevrons: 3. Jet-flap interaction. 12th AIAA/CEAS aeroacoustics conference (27th AIAA aeroacoustics conference) **2435**, (2006). https://doi.org/10.2514/6.2006-2435
5. Brown, C.: Including finite surface span effects in empirical jet-surface interaction noise models. In: AIAA Scitech (AIAA, 2016), p. 14 https://doi.org/10.2514/6.2016-0006
6. Thomas, R.H., Czech, M.J., Doty, M.J.: High bypass ratio jet noise reduction and installation effects including shielding effectiveness. 51st AIAA aerospace sciences meeting including the new horizons forum and aerospace exposition, AIAA 2013 (2013). https://doi.org/10.2514/6.2013-541
7. Mengle, V.: The effect of nozzle-to-wing gulley height on jet flow attachment to the wing and jet-flap interaction noise. Aiaa **2011–2705**, 5 (2011). https://doi.org/10.2514/6.2011-2705
8. Georgiadis, N.J., DeBonis, J.R.: Navier–stokes analysis methods for turbulent jet flows with application to aircraft exhaust nozzles. Prog. Aerosp. Sci. **42**(1), 377 (2006). https://doi.org/10.1016/j.paerosci.2006.12.001
9. Tyacke, J.C., Wang, Z.N., Tucker, P.G.: LES–RANS of installed ultra-high-bypass-ratio coaxial jet aeroacoustics with flight stream. AIAA J. **57**(3), 1215 (2019). https://doi.org/10.2514/1.J057057.
10. Wang, Z.N., Tyacke, J.C., Tucker, P.G.: LES–RANS study of serrated nozzle jet aeroacoustics for an installed ultra-high-bypass-ratio aeroengine. AIAA J. **59**(10), 4155 (2021). https://doi.org/10.2514/1.J060447
11. Tyacke, J.C., Wang, Z.N., Tucker, P.G.: Noise source, length and time scale distributions in installed jets with a flight stream. In: AIAA/CEAS aeroacoustics conference (AIAA, Atlanta, Georgia, 2018), p. 11. https://doi.org/10.2514/6.2018-3619
12. Vogel, P., Bin, J., Sinha, N.: LES predictions of jet noise for a pylon-mounted dual stream nozzle and jet surface interactions. In: ASME Turbo Expo 2016 (ASME, Seoul, South Korea, 2016), pp. V02AT41A012–V02AT41A012. https://doi.org/10.1115/GT2016-58175
13. Jameson, A.: J. Sci. Comput. **34**(2), 188 (2007). https://doi.org/10.1007/s10915-007-9172-6
14. Tyacke, J.C., Naqavi, I.Z., Wang, Z.N., Tucker, P.G., Boehning, P.: J. Turbomach. **139**(8), 081003 (2017). https://doi.org/10.1115/1.4035662
15. Tyacke, J.C., Mahak, M., Tucker, P.G.: Large-scale multifidelity, multiphysics, hybrid reynolds-averaged navier–stokes/large-eddy simulation of an installed aeroengine. J. Propul. Power **32**(4), 997 (2016). https://doi.org/10.2514/1.B35947
16. Naqavi, I.Z., Wang, Z.N., Tucker, P.G.: Far-field noise prediction for jets using large-eddy simulation and Ffowcs Williams–Hawkings method. Int. J. Aeroacoustics **15**(8), 1 (2016). https://doi.org/10.1177/1475472X16672547
17. Eastwood, S., Tucker, P., Xia, H., Dunkley, P., Carpenter, P.: Large-eddy simulations and measurements of a small-scale high-speed coflowing jet. AIAA J. **48**(5), 963 (2010). https://doi.org/10.2514/1.44534
18. Xia, H., Tucker, P.G., Eastwood, S.: Large-eddy simulations of chevron jet flows with noise predictions. Int. J. Heat Fluid Flow **30**(6), 1067 (2009). https://doi.org/10.1016/j.ijheatfluidflow.2009.05.002
19. Eastwood, S., Tucker, P., Xia, H., Klostermeier, C.: Developing large eddy simulation for turbomachinery applications. Phil. Trans. R. Soc. A **367**(1899), 2999 (2009)
20. Eastwood, S.: Hybrid LES - RANS of Complex Geometry Jets. Phd thesis, University of Cambridge (2009)
21. Spalart, P.R., Allmaras, S.R.: A one-equation turbulence model for aerodynamic flows. La Recherche Aérospatiale **1**(1), 5 (1994). http://www.mendeley.com/research/a-oneequation-turbulence-model-for-aerodynamic-flows/

22. Liu, Y., Tucker, P.G., Kerr, R.M.: Linear and nonlinear model large-eddy simulations of a plane jet. Comput. Fluids **37**, 439 (2008). https://doi.org/10.1016/j.compfluid.2007.02.005
23. Nicoud, F., Ducros, F.: Subgrid-scale stress modelling based on the square of the velocity. Flow Turbul. Combust. **62**, 183 (1999). https://doi.org/10.1016/j.jcp.2004.10.018
24. Smagorinsky, J.: General circulation experiments with the primitive equations. I. The basic experiment. Mon. Weather Rev. **91**(3), 99 (1963)
25. Tucker, P.G.: Differential equation-based wall distance computation for DES and RANS. J. Comput. Phys. **190**(1), 229 (2003)
26. Wang, Z.N., Tyacke, J.C., Tucker, P.G., Boehning, P.: Parallel computation of aeroacoustics of industrially relevant complex-geometry aeroengine jets. Comput. Fluids **178**, 166 (2019). https://doi.org/10.1016/j.compfluid.2018.04.039

Mitigating Distortion in Subsonic Intakes Using Vortex Generator Jets

Adrin Issai Arasu, Sumit Sarvankar, and Nagabhushana Rao Vadlamani

Abstract Flow separation on the intake lip of a gas-turbine engine is usually encountered when the engine operates under the off-design condition of strong crosswinds. Turbulent breakdown of the separated shear layer results in an increased flow distortion and the total pressure loss at the fan face adversely affecting the performance of the fan. This paper investigates the effectiveness of vortex generator jets (VGJs), an active flow control strategy, to mitigate flow distortion over an intake under strong crosswinds. High-fidelity eddy-resolving simulations are carried out on a quasi-3D (Q3D) duct, which mimics the dynamics of the flow over a full 3D axisymmetric intake at a significantly lower computational cost. Two jet configurations are examined: vertical jets and pitched jets (angled) with the VGJs placed at three locations: the windward side of the intake lip, the throat, and the diffuser section. The results show a reduction in the fan-face distortion coefficient by 12.7% using vertical jets and 6% using pitched jets, thereby demonstrating the superiority of vertical jets over pitched jets.

1 Introduction

The intake lips of modern aircraft engines with high bypass ratios are generally slimmer than those of conventional intakes [1, 2]. However, when the aircraft is operating at high angles of incidence or in strong crosswinds, the flow over these

CUFS-2024-24, Cambridge, United Kingdom, 4-5 March 2024.

A. I. Arasu (✉) · S. Sarvankar · N. R. Vadlamani
Department of Aerospace Engineering, Indian Institute of Technology Madras, Chennai, India
e-mail: ae19d006@smail.iitm.ac.in

S. Sarvankar
e-mail: sarvankarsumit@gmail.com

N. R. Vadlamani
e-mail: nrv@smail.iitm.ac.in

J. C. Tyacke and N. R. Vadlamani (eds.), *Proceedings of the Cambridge Unsteady Flow Symposium 2024*, https://doi.org/10.1007/978-3-031-69035-8_23

intake lips is prone to separation [3]. The following distortion that is transferred to the fan face decreases the fan's performance and efficiency. In extreme crosswinds, unsteadiness of the flow can even cause stall and fatigue stresses on the fan blades.

In the literature, very few studies explored flow control strategies to mitigate distortion and enhance intake performance under off-design conditions. These include the usage of both passive flow control devices like boundary layer trips, wedges, microvane generators, etc., and active flow control devices like plasma actuators and vortex generator jets (VGJs). Although the active flow control devices require an additional power source for operation, they provide more flexibility than the passive flow control approaches as they can be activated and deactivated as needed. Wakelam et al. [4, 5] conducted an extensive set of experiments with boundary layer trips and vortex generator jets on the windward side of the intake lip to reduce distortion. At low velocities, the pre-transitional fluctuations caused by these devices modify the boundary layer state upstream of the separation point, encouraging mixing within the boundary layer and sustaining an attached flow at the leading edge of the intake lip. Additionally, their findings show that vortex generator jets (VGJs) eliminate hysteresis behavior at low speeds and reduce flow distortion at higher speeds. Vadlamani and Tucker [6] demonstrated the potential of a dielectric barrier discharge (DBD) plasma actuator to reduce the separation of a two-dimensional intake lip when exposed to crosswinds, albeit at low speeds. Recently Jia et al. [7] carried out experiments at moderate crosswind speeds and angles with and without microsecond pulsed surface DBD actuators. They report a $\approx 12\%$ decrease in the distortion coefficient with flow control. Further experiments by Zhang et al. [8] with nano-second DBD actuators show a $30 - 40\%$ reduction in the distortion coefficient at higher actuation frequencies. On the other hand, steady Reynolds Averaged Navier Stokes (RANS) simulations by Sumit et al. [9] indicate that the microsecond pulsed surface DBD actuators are effective in suppressing distortion at low speeds while these are ineffective at higher speeds.

The flow over the intake lip is complex as it experiences severe acceleration around the intake lip, relaminarization, separation over the intake lip and in the diffuser, separation-induced transition, etc. which are challenging to capture using the RANS-based methods [10]. Very few studies in the literature captured the complex dynamics of unsteady flow over the intake lip under strong crosswinds using eddy-resolving simulations [11, 12]. In addition, to the authors' knowledge, there aren't any scale-resolving studies that explored the potential of VGJs for flow control over intakes. This paper hence investigates the efficacy of VGJs to reduce the flow distortion over the intake lip when subjected to strong crosswinds. We present the outcomes of high-fidelity eddy-resolving simulations on quasi-3D intake ducts with multiple VGJs. Additionally, we analyze the influence of VGJ configurations with vertical and pitched jets on flow control.

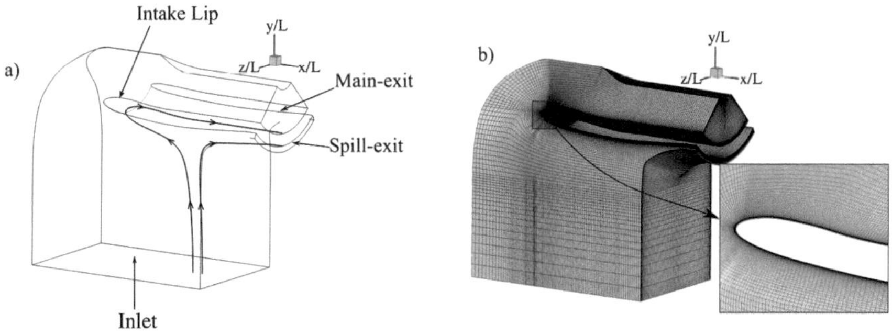

Fig. 1 **a** Computational domain and boundary conditions **b** Multi-block structured mesh around the Q3D duct

2 Numerical Framework

2.1 Computational Domain and Boundary Conditions

A Quasi 3D (Q3D) duct accurately replicates the flow physics observed on a full 3D intake at a significantly lower computational cost [13]. Scale-resolving simulations are hence feasible on the Q3D configuration when compared to the full annulus 3D intake. The Q3D intake configuration considered in this study is in line with the Q3D duct used in the experiments by Wakelam [5] and the computations of Yeung et al. [13]. This duct is created by extracting the stream tube from a full 3D axisymmetric intake RANS simulation (see [13] for more details). Figure 1a shows the computational domain. The streamlines in Fig. 1a show that the flow approaching the intake stagnates on the windward side of the intake lip at approximately $x/L = 0$. Subsequently, the majority of the flow enters the engine and leaves through the main exit, while the remainder of the flow leaves through the spill exit. This allows for the growth of a new boundary layer on the windward side of the intake lip downstream of the stagnation point.

Table 1 presents the flow and grid parameters. Subsonic inflow boundary condition [14] is specified at the inlet, with a total pressure of 101325 Pa and a total temperature of 294.4 K. The direction of the incoming crosswind flow is perpendicular to the inflow plane, along the y/L-axis. A viscous no-slip condition is imposed on the surface of the intake lip, while a convective outflow is used at the main and spill exits to prevent any reverberations caused by the vortices leaving the exit [15]. The static pressures at the main and spill exits are set to 90 kPa and 92.5 kPa, respectively. The side walls and spinner walls are designed to replicate the effect of three-dimensional stream tube contraction experienced by the flow in a 3D engine. It ensures that the pressure distribution over the intake lip is accurately represented under attached flow conditions. Hence, the boundary layers are resolved only on the intake surface, while the side walls and spinner are treated as free-slip boundaries. This approach not only provides the necessary stream tube contraction and pressure distribution over the

Table 1 Flow and grid parameters

Parameters	Values
Reynolds number, Re_L	220000
Fan-face Mach Number, M	0.4
Lip length, L	185.4 mm
Fanface diameter, H	315.6 mm
Computational domain range	
x/L	−2 to 1.615
y/L	−2.958 to −0.003
z/L	−0.901 to 0.901
Location of fan	$x/L = 0$
Grid size	
Leeward side	$\approx$ 50.5 million
Windward side	$\approx$ 32 million
Grid resolution in wall units	
$\Delta x^+ \leq 56$ $\Delta y^+ \leq 1.5$ $\Delta z^+ \leq 15$ $\Delta t^+ \leq 2 \times 10^{-5}$	

intake lip but also reduces the computational cost. Figure 1b shows the structured multiblock mesh used to discretize the domain. All the simulations reported in this paper are carried out on a mesh comprising of $\approx 82 \times 10^6$ nodes with a grid resolution of $\Delta x^+ \leq 56$, $\Delta y^+ \leq 1.5$ and $\Delta z^+ \leq 15$ in viscous wall units.

2.2 Vortex Generator Jet Parameters

The details of the VGJs are presented in Table 2. As demonstrated in Fig. 2a, b, the VGJs are placed at three different locations on the intake surface. The outer jets are located on the windward side of the lip at $x/L = -0.9$. The throat jets and diffuser jets are located on the leeward side at $x/L = -0.9$ and $x/L = -0.58$ respectively. All the jets have a skew angle of 90° and a jet diameter of $D/L = 0.054$. The jets are spaced at a distance of $1.7D$ along the circumference of the intake duct. The eddy-resolving simulations are performed on a structured mesh comprising 82 million grid points. Each of the outer, diffuser, and throat jets are resolved with $\approx$ 620, 300, and 1500 points, respectively. The mesh resolution of the VGJs is in accordance with the simulations carried out by Postl et al. [16] and is fine enough to capture the shear layer development and instabilities in flow. Following Postl et al. [16], a Poiseuille-type velocity profile with a cos^3 distribution shape function is imposed at the inlet of each VGJ as follows:

$$v(r) = V_{jet} \times cos^3 \left[\pi \frac{r}{D} \right] \tag{1}$$

Table 2 Controlled case parameters

Parameters	Outside jets	Throat jets	Diffuser jets
Jet diameter, D	$0.054L$	$0.054L$	$0.054L$
Jet spacing, S_j	$1.7D$	$1.7D$	$1.7D$
Jet center locations	$x = -0.9L$	$x = -0.9L$	$x = -0.58L$
Grid Resolution (Single VGJ)	≈ 620	≈ 1500	≈ 300
Number of jets	13	11	11
Skew angle	90°	90°	90°
Pitch angle (Vertical Jets)	90°	90°	90°
Pitch angle (Angled Jets)	45°	45°	30°

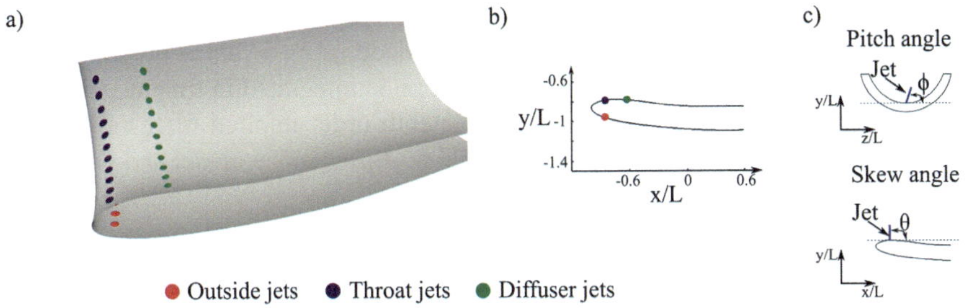

Fig. 2 **a** Schematic of the intake showing the jet distribution along the circumference; **b** Location of the VGJs at three streamwise locations; **c** Representation of the jet orientation

Here, V_{jet} is the amplitude of the jet velocity, and r is the radial distance from the center of the VGJ. Preliminary simulations (not shown here) with different jet-blowing ratios ($B = V_{jet}/U$ ranging from 0.5–1.5) showed an effective flow control with increasing values of B. While $B = 1.5$ provided the best control, test cases with $B < 1$ were ineffective due to the weaker jet shear layer and inadequate wall-normal mixing. Hence, the velocity boundary condition at all the jet locations is specified by setting a jet-blowing ratio of $B = 1.5$. Three simulations are reported in this paper. These include the baseline case without flow control, and two VGJ configurations, one with vertical jets and the other with pitched jets. The schematic in Fig. 2c illustrates the pitch and skew angles of the jet. For the test case with pitched jets, the pitch angle of the outer and throat jets is set to 45°, while the diffuser jets are angled at 30° to the surface.

2.3 *Solver Details*

COMPSQUARE, an in-house high-order structured code, is utilized to perform eddy-resolution simulations. This multi-block structured compressible flow Navier-Stokes solver employs high-order explicit/compact schemes for spatial discretization and a fourth-order four-stage Runge-Kutta scheme for time integration [17–19]. After every time step, high-frequency numerical errors are eliminated by means of high-order Pade-type implicit filters. The solver is optimized to scale on multi-GPU and CPU systems using OpenACC and MPI. Communication between multiple blocks at the block interfaces is established through a Characteristic Interface Boundary Condition (CIBC). This boundary condition is non-reflective and matches the primitive variables on the left and right faces of the block interface. To impose CIBC, the Navier-Stokes equations in generalized coordinates are recast in a characteristic form in the direction normal to the interface. The left/right convection term in the resulting quasi-linear wave equation is adjusted on the basis of the sign of the convection speed. The incoming convection term is corrected while the outgoing term is retained (see [20–22] for more details). For the simulations reported in this paper, both the inviscid and viscous fluxes are discretized using a fourth-order, centered explicit Finite Difference approach. The solution is integrated in time using a four-stage, fourth-order Runge-Kutta scheme. A tenth-order implicit filter with a filter coefficient of $\alpha_f = 0.45$ is used to eliminate high-frequency errors and ensure the stability of the solver.

3 Results and Discussion

In this section, we first examine the sensitivity of results to the grid resolution followed by a discussion on the dynamics of the uncontrolled flow and its validation. Subsequently, the efficacy of two VGJ configurations is demonstrated using the instantaneous and time-averaged flow characteristics.

3.1 Grid Sensitivity

To demonstrate grid independence, two grids were chosen: a coarse grid with 45 million grid points and a finer grid with 82 million grid points. Two parameters are used to quantify the flow distortion due to crosswinds. These include the radial variation of the normalized pressure, P_{norm}, at the fan face and the variation of the isentropic Mach number, ISM, over the intake lip. The normalized pressure, which measures the pressure loss due to crosswind, is expressed as:

$$P_{norm} = \frac{P_0 - P}{P_{atm} - P} \tag{2}$$

while the isentropic Mach number ISM is given as:

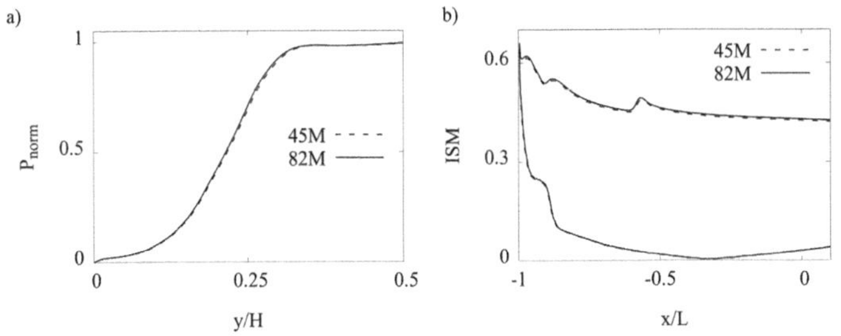

Fig. 3 Grid Independence study: **a** Radial variation of normalized pressure at the fan-face; **b** Streamwise variation of Isentropic Mach number at $z/L = 0$; for the vertical jet configuration

$$ISM = \sqrt{\left\{\frac{2}{\gamma - 1} \times \left[\left(\frac{P_{atm}}{P}\right)^{\frac{\gamma-1}{\gamma}} - 1\right]\right\}} \tag{3}$$

The test case with vertical jets is chosen to demonstrate grid independence. For both the two grids, Fig. 3a, b respectively compare the radial variation of the P_{norm} at the fan face and the streamwise variation of the ISM on the intake lip at $z/L = 0$. The flow separating over the intake lip and at the diffuser due to strong adverse pressure gradients contribute to the total pressure loss at the fan face. These losses are restricted to a wall-normal extent of $y/H < 0.3$. The ISM variation in Fig. 3b shows that the flow on the windward side of the lip stagnates at $x/L = 0$. Subsequently, a significant amount of the flow accelerates over the intake lip. The ISM over the leading edge of the intake increases to ≈ 0.65 and saturates to ≈ 0.4 downstream of the diffuser section ($x/L \geq -0.5$). It is evident from Fig. 3 that the results are grid-independent. However, all the simulations reported in this paper are carried out on a finer mesh with 82 million points.

3.2 *Uncontrolled Flow*

The flow is initiated from the flow field of coarse grid simulations with a Reynolds number of 220000 and a fan-face Mach number of 0.4. Figure 4a shows the instantaneous isosurface of Q with a value of 500, coloured by the streamwise velocity. The windward side of the intake experiences a stagnation of the crosswind flow, with the majority of it directed towards the intake and the remainder passing through the spill exit. As the flow approaches the intake lip, it accelerates along the curved surface of the intake. At the leading edge of the intake lip, the laminar shear layer separates due to an adverse pressure gradient. The Kelvin-Helmholtz rollers, which span uniformly throughout the circumference of the Q3D duct until $x/L = -0.98$, are visible in the separated shear layer. Between $x/L = -0.98$ and -0.96, the 2D KH rollers break into part-span rollers due to flow perturbations and the flow transitions to turbulence.

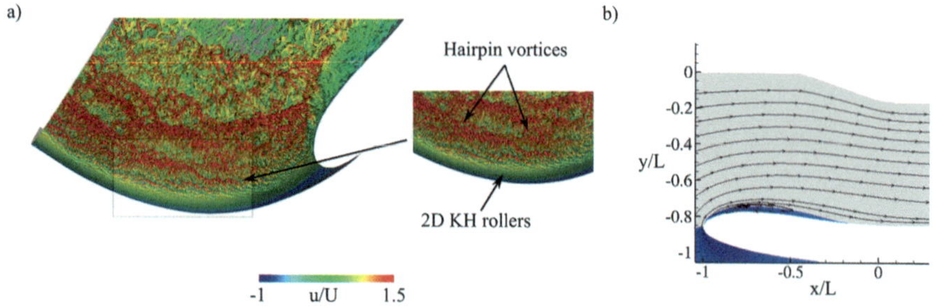

Fig. 4 **a** Flow visualisation of coherent structures using instantaneous Q-isosurface (Q=500) colored by streamwise velocity **b** Contour of streamwise velocity showing regions of reverse flow with velocity streamlines overlaid

The separated shear layer undergoes minimal wavering and the transition is abrupt with no vortex pairing in the flow field. Although this phenomenon is comparable to the typical laminar boundary layer separating over a flat plate (see Spalart and Strelets [23]), we observe a more abrupt boundary layer transition due to the interaction of the reversed flow with detached shear layer [12]. Hairpin vortices form in the region between the separated shear layer and the recirculation zone. These vortices stretch as they convect downstream and eventually breakdown into smaller structures, resulting in an energy cascade to much finer scales. This enhances wall normal mixing and momentum transfer, promoting attached flow downstream of the fan face.

After flushing out the initial transients for 4 throughflows, the flow variables are averaged for over 16 throughflows to ensure the convergence of the second-order statistics. Here, time for one throughflow is defined as $T = L/U$ and comprises ≈ 50000 time steps with $\Delta t = 2 \times 10^{-5}$. The contours of the time-averaged velocity in Fig. 4b illustrates the streamwise extent of the separation bubble on the leeward side of the intake lip, which is indicated by the regions of negative streamwise velocity (reverse flow). The separation bubble starts at the leading edge and extends to the fan face. The maximum height of the separation bubble $S_b/L = 0.09$ is observed at $x/L = -0.5$.

Figure 5a shows the variation of pressure coefficient along the streamwise direction over the intake lip. Three distinct regimes are notable: (1) an acceleration over the intake lip, (2) a constant pressure region between $-1 \leq x/L \leq -0.4$, which indicates flow separation, and (3) a pressure recovery during flow reattachment. Figure 5b plots the radial variation of normalized pressure at the fan-face for the uncontrolled case. Experimental data of Wakelam [24] is also overlaid for comparison. It is worth noting that the experiments were conducted at almost $4.75\times$ higher Reynolds number than the computations. Nevertheless, it is apparent from the plot that the radial extent of distortion at the fan face is identical with almost 40% of the passage along the wall-normal direction being affected by flow distortion.

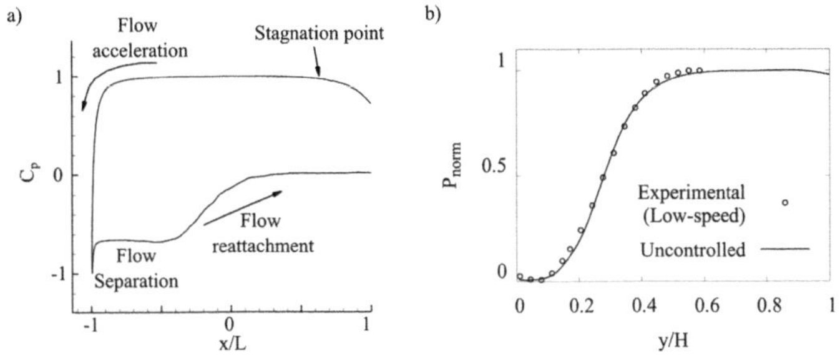

Fig. 5 **a** Streamwise variation of pressure coefficient over the intake surface at $z/L = 0$ plane **b** Radial variation of normalized pressure at the fan face

3.3 Controlled Flow

Simulations with flow control are conducted with the outer jets, throat jets, and diffuser jets as illustrated in Fig. 2. Several jets were placed along the circumference of the intake with a spacing of $1.7D$. As mentioned in Sect. 2.2, two different jet configurations: vertical jets and pitched jets, are tested in an effort to reduce the flow distortion at the fan face. Figure 6 shows the instantaneous Q (=500) isosurfaces colored with streamwise velocity on the windward and leeward sides of the intake for the test cases with flow control. For both jet configurations, a cylindrical jet shear layer forms at the exit of the outer jets. This shear layer is then distorted on the windward side due to the incoming flow leading to the formation of a counter-rotating vortex pair (CVP). Initially, the CVP is symmetrical about the jet axis, but it distorts rapidly away from the surface. As illustrated in the figure, the flow on the windward side of the jet shear layer experiences an adverse pressure gradient as it moves around the jets, resulting in the formation of spanwise horseshoe vortices. Multiple wake vortices form between the wall and the jet shear layer which extend to the leeward side of the intake. These flow characteristics are in agreement with those of a Jet in a Crossflow (JICF) reported in literature [25–27]. As a result of the curved surface of the intake, the jet shear layer moves faster on the windward side than on the leeward side. The Q-iso surfaces show that the jet shear layer destabilizes close to the leading edge of the intake lip resulting in the formation of a recurrent train of hairpin vortices. The hairpin vortices interact with the detached shear layer at the leading edge, resulting in an earlier onset of transition. It is worth noting that the secondary structures resulting from the vertical jets are relatively stronger than those of pitched jets and hence sustain until the diffuser section thereby providing better flow control.

For all the test cases, Fig. 7a depicts the time-averaged contours of normalized pressure at the fanface, P_{norm}. For the uncontrolled case, that the radial extent of distortion (see the P_{norm} distribution) is similar all along the circumference. However, for the test cases with flow control, the distortion is relatively larger on the midplane ($z/L = 0$ plane) as the jets tangentially convect the flow towards the center. With

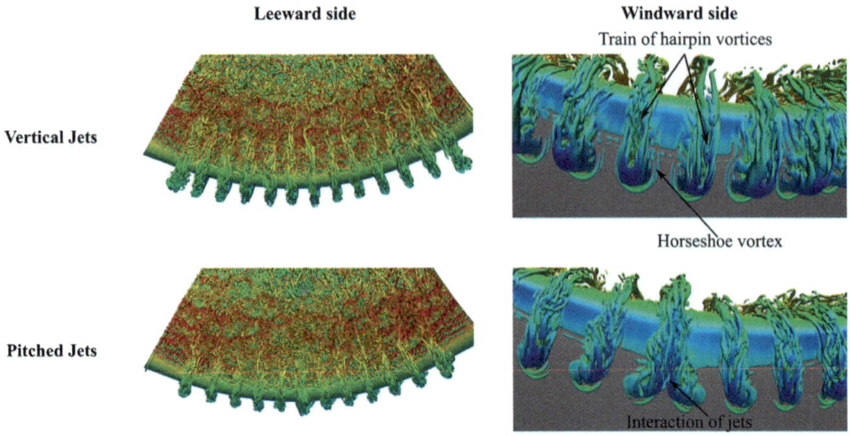

Fig. 6 Flow visualisation of coherent structures using instantaneous Q-isosurfaces (Q = 500) colored by streamwise velocity

pitched jets, the interaction of the counter vortex pairs (see inset plot in Fig. 6) of adjacent jets resulted in increased losses at the midplane. Figure 7b further compares the sector-averaged (over $\pm 5°$ about the midplane) mean off-diagonal stresses. A threshold value of $-u'v'/U^2 = 0.001$ is generally used to identify the transition onset in various experiments and simulations [28–30]. The transition onset is close to the leading edge of the intake lip for all the cases. However, the controlled cases transition relatively earlier (at $x/L = -1$) than the uncontrolled case (at $x/L = -0.98$). The mean turbulent production is also relatively lower with flow control than in the uncontrolled case. In particular, the mean turbulent production is lower for the test case with vertical jets, resulting in better flow control than the pitched jets. The highest $-u'v'/U^2$ for the uncontrolled case is observable between $0 \le x/L \le -0.7$. This is attributed to the momentum transfer and turbulent mixing promoted by the breakdown of the separated shear layer. As noted earlier, with pitched jets, the interaction between the center jets results in a stronger CVP thereby increasing the stresses downstream of the jets away from the wall. In contrast, with vertical jets, the peak stresses are observed behind the jets but are significantly lower in magnitude thus demonstrating the efficacy of vertical jets in mitigating distortion at the fan face.

The streamwise variation of the skin friction coefficient, C_f, plotted in Fig. 8a, reveals that the flow separates near the leading edge (depicted using negative C_f values) for all the cases. The extent of the separation bubble is between $-0.98 \le x/L \le 0$ for the uncontrolled case. For the controlled cases, two positive peaks in the C_f are notable at the locations of throat and diffuser jets. Notice that the flow remains attached until the fan face for the test case with vertical jets. With pitched jets, despite a local flow reattachment behind the throat and diffuser jets, the flow remained separated at the fan face. Figure 8b compares the radial variation of P_{norm} on the midplane ($z/L = 0$). The vertical jets marginally mitigated the distortion when compared to the uncontrolled case. In contrast, the pitched jets increased the

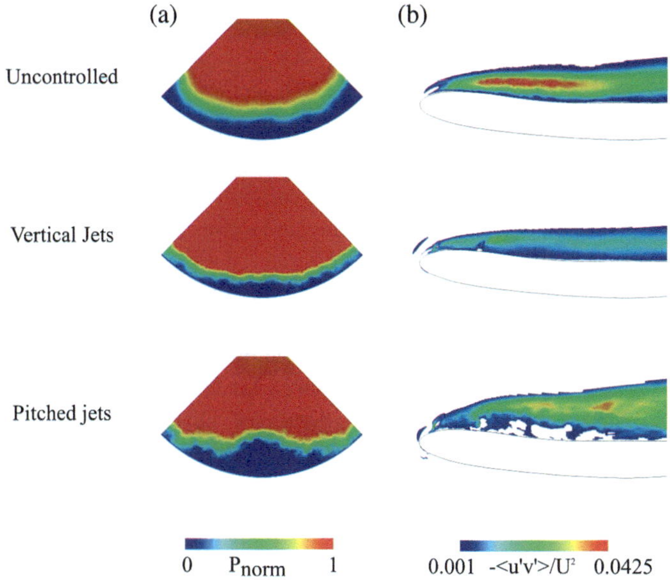

Fig. 7 **a** Distribution of normalized pressure at the fanface; **b** Sector-averaged mean off-diagonal stresses

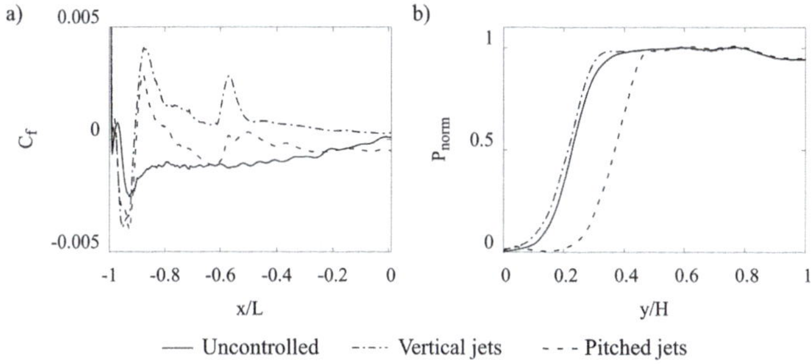

Fig. 8 Comparison of **a** streamwise variation of the skin friction coefficient, C_f, and **b** radial variation of the normalized pressure at fan-face

radial extent distortion by 40% at the fan face due to the interaction of the CVP between adjacent jets. On the other hand, the distortion coefficient is used to quantify the flow distortion at the fan face over the entire sector. It is expressed as $DC = (P_{0,inlet} - \overline{P_0})/(\overline{P_0} - \overline{P})$ where $\overline{P_0}$ and $\overline{P}$ are the area-weighted stagnation and static pressures at the fan face respectively and $P_{0,inlet}$ is the stagnation pressure at the inlet. The DC value for the uncontrolled case is 0.63 while for the cases with pitched and vertical jets is 0.59 and 0.55 respectively. Hence, the vertical jets effectively mitigated the distortion at the fan face thereby reducing the distortion coefficient by $\approx 12.7\%$. Despite an increased distortion on the midplane due to pitched jets, it is interesting to note that the overall distortion across the sector has decreased by 6%.

4 Conclusion

The characteristics of flow over an intake lip under severe crosswinds are studied using high-fidelity scale-resolving simulations. Simulations are carried out on quasi-3D intake ducts which mimic the prominent dynamics of the flow over a full 3D axisymmetric intake at a significantly lower computational cost. The possibility of mitigating flow separation over the intake lip and diffuser sections using vortex generator jets (VGJ) is explored. Two jet configurations are examined: vertical jets and pitched jets (angled) with the VGJs placed at three locations: the windward side of the intake lip, the throat, and the diffuser section. For the uncontrolled case, the crosswind flow stagnates on the windward side of the lip, accelerates around the leading edge, and separates due to an adverse pressure gradient. The 2D Kelvin-Helmholtz rollers rapidly break into part-span rollers due to secondary instabilities and the flow transitions to turbulence. For the controlled cases with VGJs, the distortion transfer to the fan face has been relatively reduced. In particular, the vertical jets are found to effectively mitigate the distortion reducing the distortion coefficient by 12.7% when compared to the $\approx 6\%$ reduction using pitched jets. With pitched jets, the interaction between the counter-rotating vortex pair (CVP) of the adjacent jets resulted in increased losses at the midplane. Hence, the pitched jets that are situated away from the center performed better than those at the midplane resulting in an overall reduction in the sector distortion coefficient.

Acknowledgements The authors acknowledge NVIDIA for generously donating the A100 GPU card through the Academic Hardware Grant Program. We also thank P.G. Senapathy Centre, IIT Madras, and the National PARAM Supercomputing Facility (NPSF) for providing the computing resources on the PARAM Siddhi cluster as part of the National Supercomputing Mission. Dr. Nagabhushana Rao Vadlamani also acknowledges Science and Engineering Research Board (SERB) for funding the project under MATRICS Scheme (MTR/2022/000807).

References

1. Peters, A., Spakovszky, Z.S., Lord, W.K., Rose, B.: Ultrashort nacelles for low fan pressure ratio propulsors. J. Turbomach. **137**(2), 021001 (2015)
2. Vadlamani, N.R., Cao, T., Watson, R., Tucker, P.G.: Toward future installations: mutual interactions of short intakes with modern high bypass fans. J. Turbomach. **141**(8), 081013 (2019)
3. Cao, T., Vadlamani, N.R., Tucker, P.G., Smith, A.R., Slaby, M., Sheaf, C.T.: Fan-intake interaction under high incidence. J. Eng. Gas Turbines Power **139**(4), 041204 (2017)
4. Wakelam, C., Hynes, H., Hodson, T., Evans, S., Chanez, P.: Separation control for aeroengine intakes, part 1: low-speed investigation of control strategies. J. Propul. Power **28**(4), 758–765 (2012)
5. Wakelam, C.T., Hynes, T.P., Hodson, H.P., Evans, S.W., Chanez, P.: Separation control for aero engine intakes, part 2: high-speed investigations. J. Propul. Power **28**(4), 766–772 (2012)
6. Rao, V.N., Tucker, P.: Intake lip separation control using plasma actuators, In: 16th international symposium on transport phenomena and dynamics of rotating machinery **1**, 1–4 (2016)

7. Jia, Y., Liang, H., He, Q., Su, Z., Song, G.: Flow separation control of nacelle in crosswind by microsecond pulsed surface dielectric barrier discharge plasma actuator, Flow, Turbulence and Combustion, pp. 1–21 (2021)
8. Zhang, D., Liang, H., Yang, H., Su, Z., Zhang, C., Liu, S.: Flow separation control of nacelle inlets in crosswinds by dielectric barrier discharge plasma actuation, In: Actuators **12**(6). MDPI, p. 229 (2023)
9. Sarvankar, S.S., Issai Arasu, A., Vadlamani, N.R.: Crosswind aerodynamic analysis using quasi 3d ducts, In: Turbo Expo: Power for Land, Sea, and Air, **86113**. American Society of Mechanical Engineers, V10CT33A018 (2022)
10. Oriji, U.R., Tucker, P.G.: Modular turbulence modeling applied to an engine intake. J. Turbomach. **136**(5), 051004 (2014)
11. Oriji, U.R., Yang, X., Tucker, P.G.: Hybrid rans/iles for aero engine intake, In: Turbo Expo: Power for Land, Sea, and Air, **45578**. American Society of Mechanical Engineers, V01AT01A032 (2014)
12. Vadlamani, N.R., Tucker, P.: Eddy resolving simulations of intake under crosswinds, In: Direct and Large-Eddy Simulation XI. Springer, pp. 523–529 (2019)
13. Yeung, A., Vadlamani, N.R., Hynes, T., Sarvankar, S.: Quasi 3d nacelle design to simulate crosswind flows: merits and challenges. Int. J. Turbomach. Propul. Power **4**(3), 25 (2019)
14. Rodriguez, D.L., Aftosmis, M.J., Nemec, M.: Formulation and implementation of inflow/outflow boundary conditions to simulate propulsive effects, In: 2018 AIAA Aerospace Sciences Meeting, p. 0334 (2018)
15. Miyauchi, T., Tanahashi, M., Suzuki, M.: Inflow and outflow boundary conditions for direct numerical simulations. JSME Int. J. Ser. B **39**(2), 305–314 (1996)
16. Postl, D., Balzer, W., Fasel, H.: Control of laminar separation using pulsed vortex generator jets: direct numerical simulations. J. Fluid Mech. **676**, 81–109 (2011)
17. Vaid, A., Vadlamani, N.R., Sivaramakrishnan Malathi, A., Gupta, V.: Dynamics of bypass transition behind roughness element subjected to pulses of free-stream turbulence, Physics of Fluids (2022)
18. Sivaramakrishnan Malathi, A., Nardini, M., Vaid, A., Rao Vadlamani, N., Sandberg, R.D.: Profile loss reduction of high-lift turbine blades with rough and ribbed surfaces. J. Turbomach. **145**(2), 021001 (2022)
19. Nampelly, G., Malathi, A.S., Vaid, A., Vadlamani, N.R., Rengarajan, S., Kontis, K.: Surface roughness effects on cavity flows, Flow, Turbulence and Combustion 1–25 (2022)
20. Kim, J.W., Lee, D.J.: Characteristic interface conditions for multiblock high-order computation on singular structured grid. AIAA J. **41**(12), 2341–2348 (2003)
21. Kim, J., Lee, D.: Generalized characteristic boundary conditions for computational aeroacoustics, part 2. AIAA J. **42**(1), 47–55 (2004)
22. Peers, E., Zhang, X., Kim, J.W.: Patched characteristic interface condition for high-order multiblock aeroacoustic computation. AIAA J. **48**(11), 2512–2522 (2010)
23. Spalart, P.R., Strelets, M.K.: Mechanisms of transition and heat transfer in a separation bubble. J. Fluid Mech. **403**, 329–349 (2000)
24. Wakelam, C.T.: Aero engine intake separation control, Ph.D. dissertation, University of Cambridge (2009)
25. Acarlar, M., Smith, C.: A study of hairpin vortices in a laminar boundary layer. part 2. hairpin vortices generated by fluid injection. J. Fluid Mech. **175**, 43–83 (1987)
26. Mahesh, K.: The interaction of jets with crossflow. Annu. Rev. Fluid Mech. **45**, 379–407 (2013)
27. Majander, P., Siikonen, T.: Large-eddy simulation of a round jet in a cross-flow. Int. J. Heat Fluid Flow **27**(3), 402–415 (2006)
28. Volino. R.J., Hultgren, L.S.: Measurements in separated and transitional boundary layers under low-pressure turbine airfoil conditions, In: Turbo Expo: Power for Land, Sea, and Air, **78569**. American Society of Mechanical Engineers, V003T01A065 (2000)
29. Hu, H., Yang, Z.: An experimental study of the laminar flow separation on a low-reynolds-number airfoil. J. Fluids Eng. **130**(5), 051101 (2008). https://doi.org/10.1115/1.2907416
30. Buchmann, N., Atkinson, C., Soria, J.: Influence of znmf jet flow control on the spatio-temporal flow structure over a naca-0015 airfoil. Exp. Fluids **54**, 1–14 (2013)

Multi-fidelity Simulation of Trailing Edge Cutback Flows

Ruiqin Wang, Xin Yan, and Zhong-Nan Wang

Abstract A novel multi-fidelity approach of both turbulence and geometry is developed to study the tailing edge cutback flows. Large-eddy simulation is zonalised around the blade trailing edge to predict coolant mixing, while the Immersed Boundary Method (IBM) is used for modelling pin fins in the upstream internal passage. This reduces the overall computational cost while retaining the high-fidelity solution in the region of interest and allowing the parametric study of pin-fin geometry and arrangement without generating new meshes. The multi-fidelity approach has been fully validated with the experimental data and then demonstrated to study the effects of pin-fin shapes on the unsteady cutback flows and cooling performance.

Keywords Turbulence modelling · IBM · Heat transfer · Gas turbine

1 Introduction

The demand for high efficient gas turbines keeps increasing the turbine inlet temperature, which has far exceeded the blade material limits [1]. Aerodynamic performance requires thin blade trailing edge (TE) and it therefore suffers an extremely high thermal load. Applications of internal cooling structures are constrained due to the thin blade trailing edge, so the trailing edge cutback design is widely adopted to provide effective cooling. In this configuration, the blade pressure surface is partially cut off, allowing the cold air to be ejected from a slot to cool the exposed cutback surface. To enhance the structural strength, pin fins or ribs are usually arranged in the slot with lands configured in the downstream cutback region [2]. They significantly affect the cooling effectiveness of cut-back TE. It is challenging to predict

R. Wang · Z.-N. Wang (✉)
College of Engineering and Physical Sciences, University of Birmingham, Birmingham B15 2TT, UK
e-mail: z.n.wang@bham.ac.uk

R. Wang · X. Yan
College of Energy and Power, Xi'an Jiaotong University, Xi'an 710049, Shang'xi, China

J. C. Tyacke and N. R. Vadlamani (eds.), *Proceedings of the Cambridge Unsteady Flow Symposium 2024*, https://doi.org/10.1007/978-3-031-69035-8_24

this complex turbulence and associated heat transfer in these geometrically complex configurations. In this paper, a cost-effective but accurate method, which integrates multi-fidelity turbulence and geometry modelling in one solution, will be developed to investigate the effects of the pin-fin shape on the turbine TE cooling performance.

1.1 Turbulence Modelling

In the trailing edge cutback region, strong mixing occurs between mainstream hot gas and the ejected coolant, leading to large-scale unsteady coherent structures. They are found to play an important role in trailing edge heat transfer [3]. It is therefore desirable to capture these large-scale flow structures from the first principle to improve cooling performance prediction. In the hierarchy of turbulence modelling [4], the simulation can be classified from RANS to LES, and finally to direct numerical simulation (DNS), depending on which turbulent scales are resolved. The simulation cost increases with the resolved content. Large-eddy simulation (LES) has been demonstrated with great success in resolving the large-scale mixing structures for predicting the cutback TE cooling effectiveness [5–7]. However, the cost of LES is not fully affordable for large-scale industrial applications. Hybridising LES and RANS becomes a cost-efficient way for practical applications. With such methods, LES is preferred in the region downstream of the lip to capture the unsteady mixing while RANS is used in the region where the flow is relatively steady and attached to the wall. The DES type of simulation, the first and the most popular method for hybird LES-RANS, has been used for predicting the flow and cooling performance in the TE cutback region [8, 9]. Compared with RANS/URANS, better agreement is achieved with the experimental data [2] on TE cooling effectiveness. Despite this, the definition of LES region is highly dependent on the mesh and it is not fully clear until the simulation is performed. The suitable mesh to achieve the desired division of LES and RANS regions is not usually known priori, which is part of the so-called 'grey area' problem. In this context, hybridizing LES and RANS in a more controlled way is needed to reduce this uncertainty, which motivates the development of the zonal LES (ZLES) approach in this paper.

1.2 Geometry Modelling

Another challenge of numerically predicting TE cutback flows is the mesh generation of complex geometries, e.g. multiple pin fins or ribs inside the cooling slots. To ensure high mesh quality, it usually takes a significant amount of time to generate the body-fitted grids for these complex internal pin fins. This has been one of the main reasons that numerical research has been rarely performed on realistic TE cooling configurations. Similar to turbulence modelling, the hierarchy is also present in geometry modelling [4]. Geometry can be either fully resolved by body-fitted meshes or approximately modelled by the body force, e.g. immersed boundary method (IBM) [10]. According to TE flow characteristics, the wakes behind these pin fins have a significant impact on the downstream cutback flow and heat transfer. Therefore, IBM could be beneficially used for modelling the pin-fin wakes to alleviate the meshing efforts. Such a strategy will facilitate a rapid parametric study of turbulator configurations without regenerating meshes. This forms another motivation for the research presented in this paper.

1.3 Scope of This Paper

In this paper, a multi-fidelity approach, a synergy of ZLES and IBM, is developed to provide a cost-effective solution to the TE cutback flow simulation. It enables us to focus the major computational cost in the regions that need high-fidelity simulations, i.e. mesh-resolved geometry and turbulence, while treating the rest of the region with relatively low fidelity simulations, i.e. RANS and IBM. The rest of the paper is organised as follows. Firstly, the multi-fidelity simulation approach is introduced in detail. Then, it is validated with experimental results on canonical TE cutback cases. Finally, the mutli-fidelity simulation is employed to investigate the effects of pin-fin shapes on the TE cooling performance with realistic lands structure included in the cutback area.

2 Methodology

2.1 Numerical Method

The multi-fidelity approach is developed in the in-house code KEPLES [11]. It is an unstructured, vertex-based finite volume solver designed for eddy-resolving simulations of compressible flows. For unsteady simulation, the dual time stepping method [12] is used, allowing a large global time step to efficiently advance the solution in time. Within each physical time step, pseudo time stepping is performed to drive the pseudo time derivatives towards zero using explicit four-stage Runge-Kutta method [13]. When a steady solution is reached at pseudo-time level, the solution is marched to the next physical time step. The convective flux is discretised with the KEP (kinetic energy preserving) scheme [14], which reduces numerical dissipation for LES without losing numerical stability. A second-order central scheme is used for the discretization of viscous fluxes with flow gradients calculated using the Green-Gauss method.

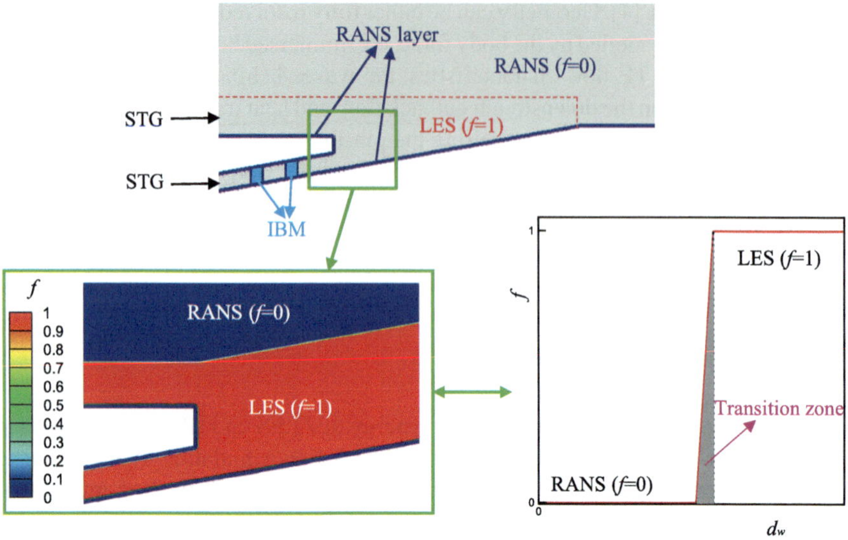

Fig. 1 Schematics of multi-fidelity simulation strategies for cutback TE

2.2 *Multi-fidelity Approach*

The multi-fidelity approach for TE cutback flow is sketched in Fig. 1. It is achieved by exploring the modelling hierarchy of both turbulence and geometry [4] to reduce the computational cost while achieving a high-fidelity solution in the region of interest. In the multi-fidelity turbulence modelling stragegy, LES is zonalised in the TE cutback region to capture the coolant mixing with hot gases for accurate prediction of cooling effectiveness, while RANS is applied in the rest of the region to reduce the computational cost with reasonable accuracy. A thin RANS layer with $y^+ \sim 100$ is used as a wall model to avoid the cost of resolving the near-wall turbulent streaks. At the inlet of LES zones, velocity fluctuations are generated with the synthetic turbulence method based on a modified von Karman spectra with Reynolds stress as input. The multi-fidelity geometry modelling is a synergy of the mesh-resolved geometry and the immersed boundary method (IBM). In the mesh-resolved cooling slot, IBM is used for the pin fins to generate wake flows for the downstream lands that is resolved by body-conformal mesh. This enables the study of pin-fin geometry and spatial arrangement for the cooling of the downstream cutback lands without changing the computational mesh.

2.2.1 Multi-fidelity in Turbulence

The viscous tensor can be written as $\tau_{ij} = \tau_{ij}^{lam} + \tau_{ij}^{turb}$, where τ_{ij}^{lam} is the stress of molecular viscosity and τ_{ij}^{turb} is the turbulent stress introduced by the modelling. The ZLES is achieved by blending at the turbulent stress level.

$$\tau_{ij}^{turb} = (1 - f)\,\tau_{ij}^{RANS} + f\tau_{ij}^{LES} \tag{1}$$

where the blending function f is used to switch between RANS and LES., τ_{ij}^{RANS} is the Reynolds stress provided by RANS model and τ_{ij}^{LES} is the sub-grid stress (SGS) tensor provided by LES SGS model. Therefore, it allows any combination of RANS and LES SGS models used for ZLES. In our simulations, the SST turbulence model [15] and the wall-adapting local eddy-viscosity (WALE) model [16] are adopted. The blending function f is set to 0 in the RANS region, and 1 in the LES region, shown in Fig. 1. To mitigate the grey area, a sharp transition between LES and RANS is achieved via f. In addition, the synethic turbulence [17] is used at the inlet of LES region to increase the turbulent mixing at the RANS and LES interface near the wall, leading to a rapid transition from RANS to LES. Near the wall, a thin RANS layer ($d_w < d_{RANS}$) is used as a wall model to further reduce the cost of ZLES. The blending function f can be defined as a function of wall distance d_w.

2.2.2 Multi-fidelity in Geometry

As discussed before, the body-conformed mesh is combined with the IBM for multi-fidelity geometry modelling. The IBM is used to model the pin fins in the internal passage. The wall boundaries are enforced by a body force term S_{IBM} in the source S of the governing equation. The IBM source is taken in proportion to the difference between the desired and the computed velocities in the solid region.

$$\boldsymbol{S_{IBM}(U)} = \frac{\chi}{\xi}\begin{Bmatrix} 0 \\ \rho(u_1^D - u_1) \\ \rho(u_2^D - u_2) \\ \rho(u_3^D - u_3) \\ \rho(E^D - E) \end{Bmatrix} \tag{2}$$

where, χ is a mask function used to distinguish between the fluid region and solid region, and it is set as 1 inside the solid region and 0 elsewhere. ξ is the penalization parameter and it is specified as the local pseudo time step Δt^* [18]. The desired velocities u_i^D in the solid region are set to zero. For the walls where the temperature is constant, the desired temperature T^D is imposed to calculate the target internal energy $E = \rho c_v T^D + 0.5\rho u_i^D u_i^D$ where c_v is the specific heat coefficient at constant volume.

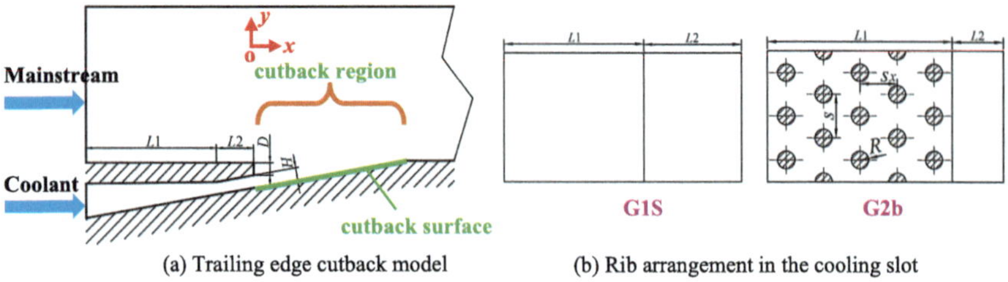

Fig. 2 Illustration of trailing edge cutback models and rib arrangements

3 Validation

A trailing edge cutback model based on Martini et al.'s experiment [19], shown in Fig. 2a, is used to validate the multi-fidelity approach. The validation using G2b configuration with a five-row staggered pin-fin array is presented here.

Figure 1 illustrated the blending function f_d distribution and the boundary condition settings. A thin RANS layer is set near the wall below $y^+ \approx 100$. The boundary conditions follow the LDA (Laser Doppler Anemometry) experiment in [19] which measures the velocity and the Reynolds shear stress distributions in the cutback region. Within each physical time step, 20 inner iterations (pseudo time steps) are adopted to achieve a $2 \sim 3$ orders of residual magnitude drop of residuals. The average flow field was then obtained over a period of $\frac{60D}{u_{hg}}$.

Two sets of meshes are generated for numerical investigations. One is a body-fitted mesh set with mesh-resolved pin-fin array. The other is using IBM to model the pin fins. Except for the pin-fin region, grids are kept the same elsewhere. The case with blowing ratio of 1.1 is investigated following the LDA experiment where the velocity is measured at various x location downstream of the lip.

Figures 3 and 4 show the time-averaged streamwise velocity and Reynolds shear stress profiles predicted by the body-fitted meshes and the IBM. The predictions using the body-fitted mesh agree well with the experimental data. Unlike the G1S case, the Reynolds stress is much better predicted by ZLES with the peak at $y^*/H = 1$ and $x/H = 1$ captured. The wake from the pin fins dominates the coolant and hot gas mixing and mitigates the effects of slot inlet turbulence that was not measured. This further verifies the accuracy of the ZLES method. Compared with the body-fitted mesh, the IBM produce a very close prediction. This suggests the combination of ZLES and IBM can be used to accurately predict the mixing process of hot gas and coolant.

Figure 5 presents the corresponding laterally-averaged film cooling effectiveness on cutback predicted by both the body-fitted meshes and IBM. The cooling effectiveness predictions of both the IBM and the body-fitted mesh agree quantitatively well with the experimental data. Moreover, the cooling effectiveness value predicted by IBM also stays close to that predicted by the body-fitted mesh with the maximum relative error less than 2.5%. This fully demonstrates the accuracy and validity of the developed multi-fidelity approach in both turbulence and geometry, using ZLES and IBM, to predict TE cutback flows.

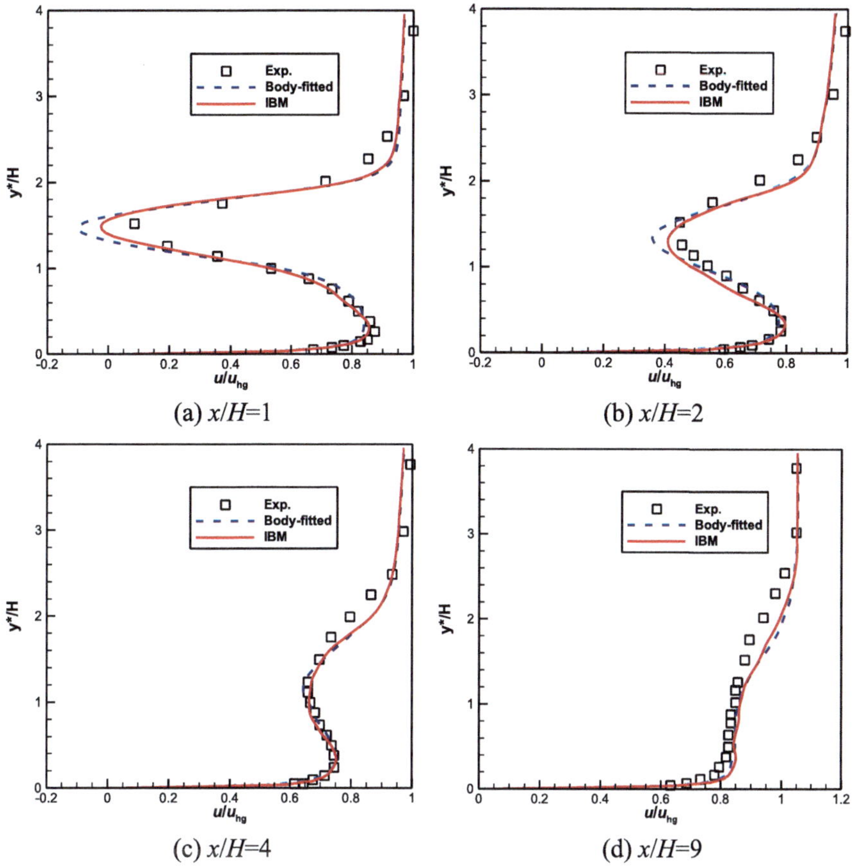

(a) x/H=1 (b) x/H=2

(c) x/H=4 (d) x/H=9

Fig. 3 Streamwise velocity profiles in the cutback region of G2b

4 Effect of Pin-Fin Shape on TE Cutback Cooling

Based on the G2b model, the effects of pin-fin shapes on the TE cutback flow and cooling performance are investigated with the validated multi-fidelity approach. Land extensions are added considering the practical cutback design. Figure 6 shows the computational meshes for the TE cutback model with lands. The total mesh node number is 8.74 million with only local grid density satisfying the LES mesh requirements in the cutback trailing edge region. Figure 7 shows the three investigated pin-fin shapes. They are cylindrical (G2b pin fins), elliptical pin fins, and c-e (i.e. cylindrical-elliptical) pin fins. All these pin fins are arranged as staggered arrays. The multi-fidelity strategies, boundary settings, the physical time step, and the data collection are all kept the same as the G2b model in Sect. 3.

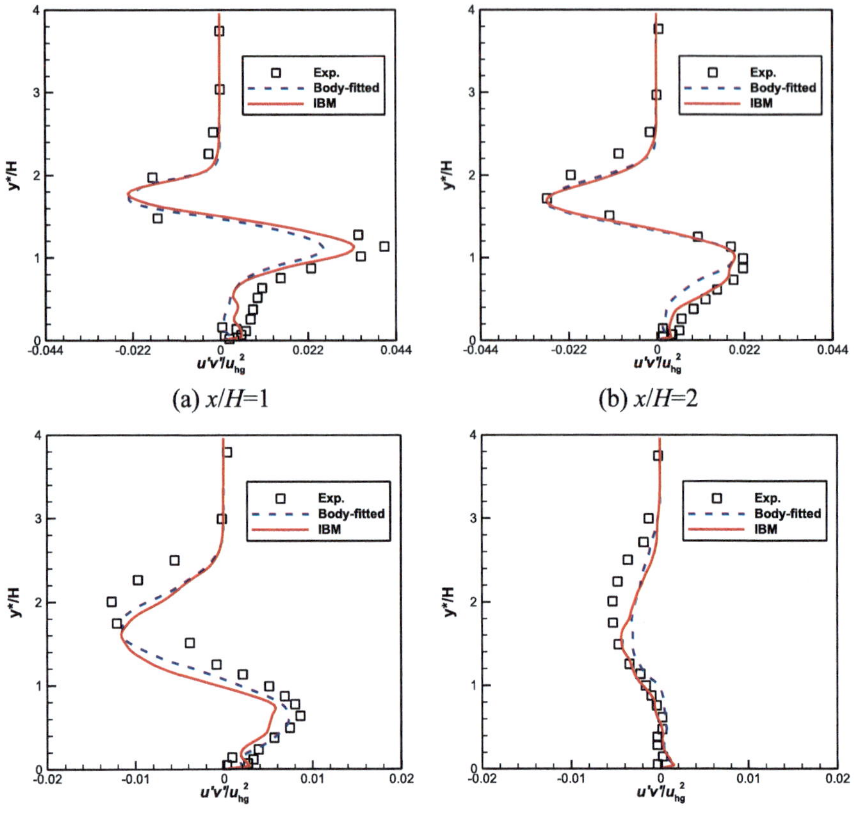

(a) x/H=1 (b) x/H=2

(c) x/H=4 (d) x/H=9

Fig. 4 Reynolds shear stress profiles in the cutback region of G2b

Figure 8 shows the laterally-averaged film effectiveness on cutback for the three kinds of pin-fin shapes. For all three pin-fin shapes, the film cooling effectiveness shows a shape decrease immediately downstream of the lands. This is attributed to the lack of coolant right behind the lands. Among the three pin-fin shapes, the elliptical pin fins lead to the highest level of film cooling effectiveness on the cutback, followed by the c-e pin fins, and the cylindrical pi-fins lead to the lowest film cooling effectiveness on the cutback. At the end of lands (x/H=11.5), the laterally-averaged film cooling effectiveness is 0.892, 0.852, 0.812 for the elliptical, c-e, and cylindrical pin fins, respectively. Compared with the cylindrical pin fins, the elliptical pin fins increase the cooling effectiveness value by about 9.85% at the end of the lands (x/H=11.5).

The time-averaged film cooling effectiveness distributions are shown on the cutback surface for the three pin-fin shapes in Fig. 9. For all pin-fin shapes, the film cooling effectiveness is uniform along the lateral direction except for the region

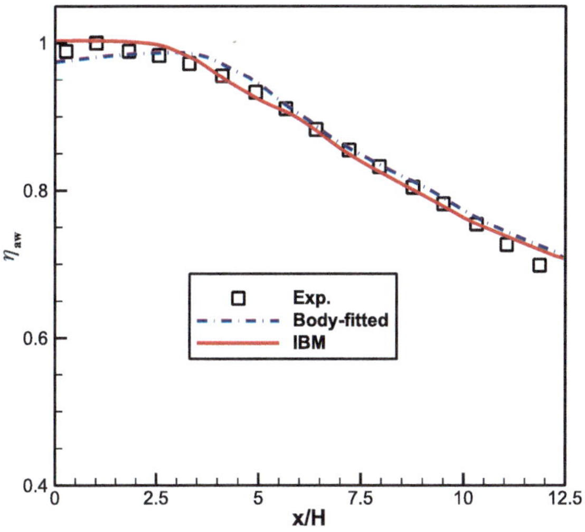

Fig. 5 Laterally-averaged film cooling effectiveness on cutback for G2b

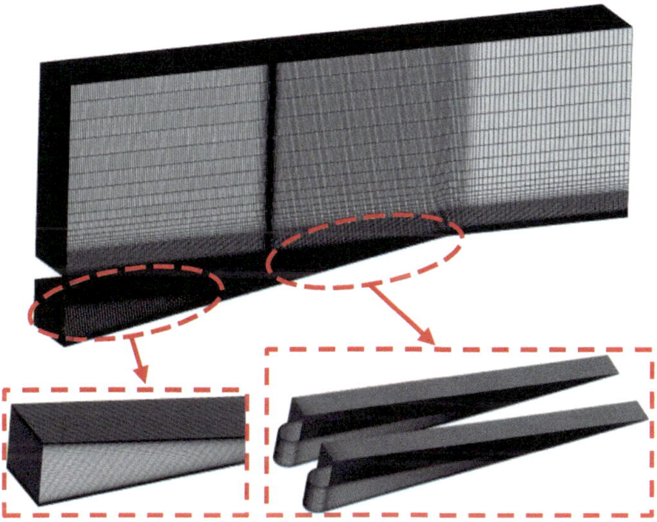

Fig. 6 Computational meshes for G2b model with land extensions

downstream of lands ($x/H > 11.5$). Compared with the cylindrical and c-e pin fins, the elliptical pin fins have a better film cooling performance. The cooling effectiveness is 0.9 on the cutback surface between lands and this trend continues downstream of the lands ($x/H > 11.5$).

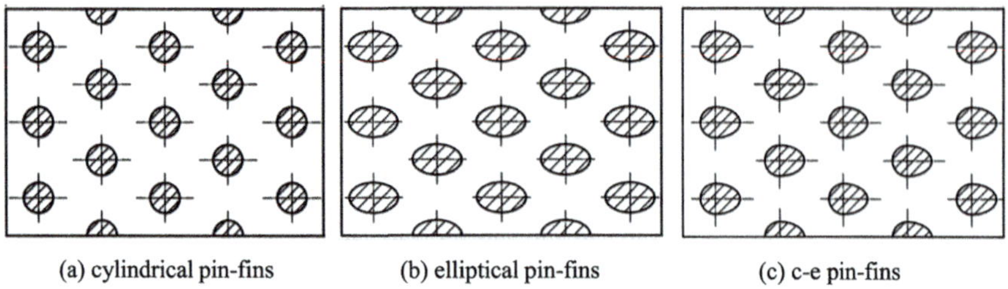

Fig. 7 Illustrations of pin-fin shapes

Fig. 8 Laterally-averaged film cooling effectiveness on cutback with three pin-fin shapes

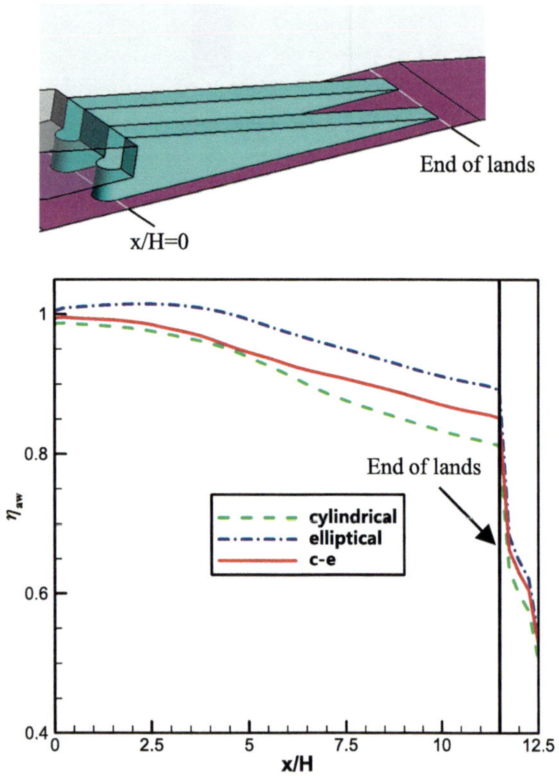

This can be explained by the time-averaged turbulent intensity $T_i = \sqrt{\frac{1}{3}(\overline{u'^2} + \overline{v'^2} + \overline{w'^2})}/u_{hg}$ in the cooling slot, shown in Fig. 10. Among three pin-fin shapes, the cylindrical pin-fin shape leads to the highest turbulence intensity T_i level in the slot, followed by the c-e and the elliptical shapes. In other words, the coolant flow unsteadiness is much less induced by the elliptical design as compared to the c-e and cylindrical designs. The coolant turbulence intensity is closely related to the cutback cooling effect. When the coolant turbulence intensity is lower, the

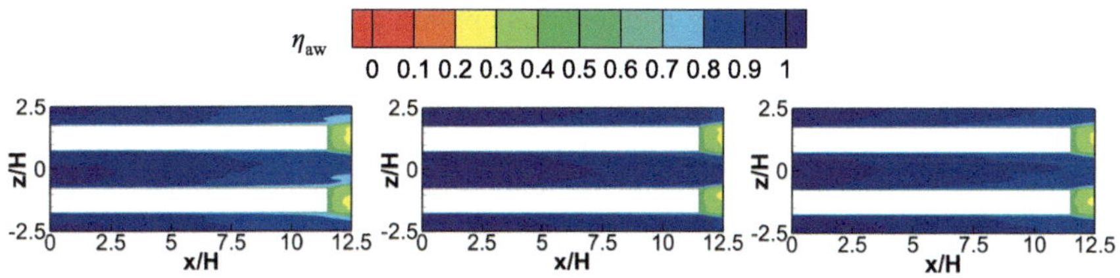

Fig. 9 Time-averaged film cooling effectiveness distributions on cutback under three pin-fin shapes

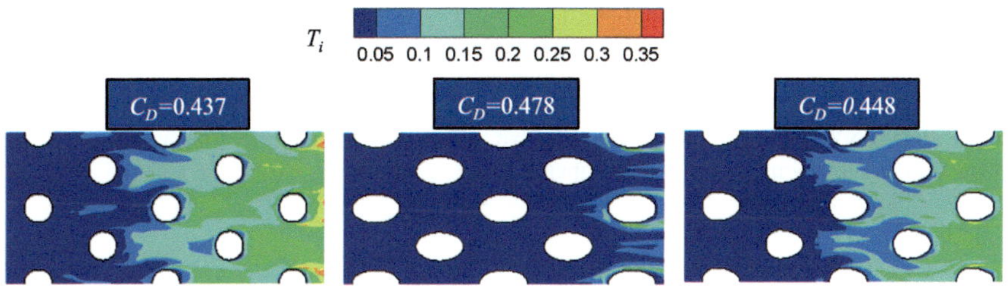

Fig. 10 Turbulence intensity distributions in the slot for three pin-fin shapes

coolant flow is more stably attached to the cutback, preventing the penetration of the hot mainstream. Hence, the cutback surface is better protected by the slot coolant.

5 Conclusions

In this paper, a novel multi-fidelity approach of both turbulence and geometry is developed to facilitate the simulation of TE cutback flows. Based on the experimental data, the reliability and high accuracy of the multi-fidelity approach are fully validated. With the developed multi-fidelity approach, the effects of pin-fin shapes and arrangements with land extensions on the unsteady cutback flow and cooling performance are then investigated with physical insights obtained. The main conclusions are drawn as follows:

1. The multi-fidelity approach is a synergy of the ZLES for turbulence and the IBM for geometry. the ZLES focus the major solution cost in the region where high fidelity is required, while the IBM is used to alleviate the meshing effort for the pin fins in the cooling slot. It has been fully validated with experimental data on both flow field and wall temperature (film cooling effectiveness on the cutback surface.

2. The developed multi-fidelity method was applied to study the effect of pin-fin shape on the thermal loading of cutback surface. Among the three studied pin-fin shapes, the elliptical pin fins are found with the highest film cooling effectiveness on the cutback. Compared with the two pin fins, the elliptical pin fins significantly reduce the coolant turbulence intensity, making the coolant flow from the slot more resistant to the entrainment of hot mainstream.

Acknowledgements RW is very grateful for the Innovative Leading Talents Scholarship (No. 0321) sponsored by the Graduate School of Xi'an Jiaotong University to support her spend one year study by the University of Birmingham as a visiting scholar. The authors are grateful for the computational time provided on the BlueBEAR HPC of the University of Birmingham. Last but not least, ZNW would like to express his most heartfelt gratitude to Professor Paul Tucker for his inspiration and guidance provided over the years. This paper is prepared to celebrate his 60th Birthday.
Declarations RW: developing and implementing the method, performing simulation, drafting&revising the paper; XY: supervising the result discussion and reviewing&commenting the paper draft; ZNW: conceptualising research plan, developing the method, supervising the implementation and reviewing&revising the paper draft.

References

1. Sciences, E., Medicine, et al.: Advanced technologies for gas turbines. National Academies Press, ??? (2020)
2. Martini, P., Schulz, A., Bauer, H.J.: Film cooling effectiveness and heat transfer on the trailing edge cutback of gas turbine airfoils with various internal cooling designs. J. Turbomach. **128**(1), 196–205 (2005)
3. Medic, G., Durbin, P.: Unsteady effects on trailing edge cooling. J. Heat Transfer **127**(4), 388–392 (2005)
4. Tucker, P.G., Wang, Z.-N.: Eddy resolving strategies in turbomachinery and peripheral components. J. Turbomach. **143**(1), 010801 (2021)
5. Schneider, H., Terzi, D., Bauer, H.-J.: Large-eddy simulations of trailing-edge cutback film cooling at low blowing ratio. Int. J. Heat Fluid Flow **31**(5), 767–775 (2010)
6. Schneider, H., Von Terzi, D., Bauer, H.J.: Turbulent heat transfer and large coherent structures in trailing-edge cutback film cooling. Flow Turbul. Combust. **88**(1), 101–120 (2012)
7. Schneider, H., Von Terzi, D.A., Bauer, H.J., Rodi, W.: Coherent structures in trailing-edge cooling and the challenge for turbulent heat transfer modelling. Int. J. Heat Fluid Flow **51**, 110–119 (2015)
8. Martini, P., Schulz, A., Bauer, H.-J., Whitney, C.F.: Detached eddy simulation of film cooling performance on the trailing edge cutback of gas turbine airfoils. J. Turbomach. **128**(2), 292–299 (2005)
9. Ivanova, E., Ledezma, G., Wang, G., Laskowski, G.M.: Experimental and numerical investigations of the heat transfer and flow field in a trailing edge cooling geometry: Part 2-les and hybrid rans/les study. In: Turbo Expo: Power for Land, Sea, and Air, **56727**, 05–12041 (2015). American Society of Mechanical Engineers
10. Sotiropoulos, F., Yang, X.: Immersed boundary methods for simulating fluid-structure interaction. Prog. Aerosp. Sci. **65**, 1–21 (2014)
11. Wang, Z.-N.: Research on low-dissipative numerical scheme and flow physics for complex turbulent flows in turbomachinery. Phd thesis, Tsinghua University, Beijing, China (2014)
12. Jameson, A.: Time dependent calculations using multigrid, with applications to unsteady flows past airfoils and wings. In: 10th Computational Fluid Dynamics Conference (1991)

13. Jameson, A., Schmidt, W., Turkel, E.: Numerical solution of the euler equations by finite volume methods using runge kutta time stepping schemes. AIAA (1981)
14. Jameson, A.: Formulation of kinetic energy preserving conservative schemes for gas dynamics and direct numerical simulation of one-dimensional viscous compressible flow in a shock tube using entropy and kinetic energy preserving schemes. J. Sci. Comput. **34**, 188–208 (2008)
15. Menter, F., Ferreira, J., Esch, T., Konno, B.: The sst turbulence model with improved wall treatment for heat transfer predictions in gas turbines. gas turnine congress (international) proceedings. tokyo. paper no. Tokyo. Paper (IGTC2003), 059 (2003)
16. Ducros, F., Nicoud, F., Poinsot, T.: Wall-adapting local eddy-viscosity models for simulations in complex geometries. Numer. Method. Fluid Dyn. VI **6**, 293–299 (1998)
17. Shur, M.L., Spalart, P.R., Strelets, M.K., Travin, A.K.: Synthetic turbulence generators for rans-les interfaces in zonal simulations of aerodynamic and aeroacoustic problems. Flow Turbul. Combust. **93**, 63–92 (2014)
18. Kou, J., Joshi, S., Hurtado-de-Mendoza, A., Puri, K., Hirsch, C., Ferrer, E.: Immersed boundary method for high-order flux reconstruction based on volume penalization. J. Comput. Phys. **448**, 110721 (2022)
19. Martini, P.: Filmkühlung hochbeanspruchter turbinenschaufelhinterkanten: wärmeübergang und strömung im nahfeld praxisbezogener ausblasespalte. PhD thesis, Karlsruher Institut für Technologie, Berlin, Germany (2008)

Beyond Navier-Stokes

Towards Full Molecular Gas Dynamics Simulations of Complex Flows via the Boltzmann Equation

Tarik Dzanic and Luigi Martinelli

Abstract This work explores the capability of simulating complex fluid flows by directly solving the Boltzmann equation. Due to the high-dimensionality of the governing equation, the substantial computational cost of solving the Boltzmann equation has generally limited its application to simpler, two-dimensional flow problems. Utilizing a combination of high-order spatial discretizations and discretely-conservative velocity models along with their highly-efficient implementation on massively-parallel GPU computing architectures, we demonstrate the current ability of directly solving the Boltzmann equation augmented with the BGK collision model for complex, three-dimensional flows. Numerical results are presented for a variety of these problems including rarefied microchannels, transitional and turbulent flows, and high-speed atmospheric re-entry vehicles, showcasing the ability of the approach in accurately predicting complex nonlinear flow phenomena and non-equilibrium effects.

1 Introduction

For the vast majority of problems in fluid dynamics, numerical techniques have broadly relied on the solution of the Navier–Stokes equations to give insight into the dynamics of fluid flows. Embedded in these governing equations is the assumption that the fluid can be treated as continuum, allowing for the use of the macroscopic conservation laws (i.e., conservation of density, momentum, and energy) to characterize the flow. However, in certain applications ranging from microflows to hypersonic aeronautics, this assumption can start to break down as the flow begins to experience strong thermodynamic non-equilibrium effects, such that it is necessary to revert to

CUFS-2024-10, Cambridge, United Kingdom, 4-5 March 2024.

T. Dzanic (✉) · L. Martinelli
Princeton University, Princeton, NJ 08544, USA
e-mail: tdzanic@princeton.edu

L. Martinelli
e-mail: martinel@princeton.edu

J. C. Tyacke and N. R. Vadlamani (eds.), *Proceedings of the Cambridge Unsteady Flow Symposium 2024*, https://doi.org/10.1007/978-3-031-69035-8_25

more general governing equations derived from the kinetic theory of gases. These kinetic schemes can offer a more robust approach which can accurately describe these complex flow physics while seamlessly recovering the hydrodynamic equations in the continuum limit.

In this work, we explore the capability of simulating complex fluid flows by directly solving the Boltzmann equation for molecular gas dynamics which underpins the macroscopic behavior of the flow. In this approach, the flow is simply represented through the evolution of a scalar particle distribution function from which a macroscopic flow state can be recovered that is equally valid across both the rarefied and continuum flow regimes. While this approach can be highly advantageous, owing both to its generalizability and the numerical simplicity of the primarily linear governing equation, its application to complex fluid flows has been very limited primarily as a result of the substantial computational cost of directly solving the Boltzmann equation. This cost can be attributed to two main sources, dimensionality and collision modeling. For the former, the phase space in the governing equation, which represents a probability density of a particle existing at some location with some velocity, can require discretizations of up to six dimensions which results in a rapidly increasing computational cost with respect to resolution. For the latter, the approximation of the particle collision process can be extremely computationally intensive, requiring the numerical evaluation of integrals of even higher dimensionality. As a result, directly solving the Boltzmann equation has typically been limited to simpler, often two-dimensional, flow problems, and the extension to more complex three-dimensional flows has generally been considered to be computationally intractable.

While the computational cost of solving the Boltzmann equation can be very restrictive, the approach may present some unique benefits which motivate its development and application. Aside from its typical use in rarefied gas dynamics, for which only kinetic descriptions of the flow are truly valid, the use of the Boltzmann equation in other flow regimes can potentially lead to enhancing our understanding of fundamental flow flows and offer more robust and accurate approaches for complex flow problems. For flows in the low-speed continuum regime, where the hydrodynamic equations can accurately describe the flow, the solution of the Boltzmann equation encodes the flow dynamics through the linear evolution of a high-dimensional particle distribution function which represents the flow in a manner that is inaccessible from the solution of the hydrodynamic equations. As such, the Boltzmann equation for these flows can offer a radically different perspective for analyzing fundamental flow problems such as transition to turbulence and present opportunities for alternate approaches to turbulence modeling. Furthermore, the Boltzmann equation can offer a robust approach for simulating the multi-scale nature of high-speed flows in re-entry and hypersonic conditions as it can accurately approximate the flow physics across the entire range of flow regimes encountered and may provide a better framework for simulating the complex high-temperature aerothermodynamic effects which are encountered in these applications.

To this end, this work presents an overview of some developments introduced by the authors for drastically reducing the computational cost of solving the Boltzmann equation, enabling its application to complex three-dimensional flow problems that were previously intractable. These improvements can broadly be categorized by three separate advancements: 1) the use of high-order spatial discretizations which allow for higher-fidelity approximations that reduce the required resolution; 2) their combination with discretely-conservative collision models which avoid the large computational cost of directly computing the collision operator and memory requirements associated with ensuring macroscopic conservation; and 3) the efficient implementation of these numerical methods on modern massively-parallel GPU computing architectures. A brief summary of this numerical approach is presented in Sect. 2, after which an overview of numerical results obtained by this approach is shown Sect. 3. Conclusions are then drawn in Sect. 4 along with a discussion on potential future development and applications.

2 Methodology

The Boltzmann equation can be represented as a scalar conservation law with a linear transport term and nonlinear source term, given as

$$\partial_t f(\mathbf{x}, \mathbf{u}, t) + \mathbf{u} \cdot \nabla f = \mathcal{C}(f, f'), \tag{1}$$

where $\mathbf{x} \in \Omega^{\mathbf{x}}$ is the physical space in a physical domain $\Omega^{\mathbf{x}} \subseteq \mathbb{R}^d$ for some spatial dimension d, $\mathbf{u} \in \Omega^{\mathbf{u}}$ is the associated velocity space in a velocity domain $\Omega^{\mathbf{u}} \in \mathbb{R}^m$ for some velocity dimension $m \geq d$, $f(\mathbf{x}, \mathbf{u}, t) \in \mathbb{R}$ is a scalar particle distribution function, and $\mathcal{C}(f, f')$ is the collision operator that models the effects of particle interactions [1]. The distribution function represents a phase space probability density for a particle existing at a given location $\mathbf{x}$ with a given velocity $\mathbf{u}$. From this distribution function, the conserved flow variables $\mathbf{Q}(\mathbf{x}, t)$ can be recovered through its moments as

$$\mathbf{Q}(\mathbf{x}, t) = [\rho, \rho\mathbf{U}, E]^T = \int_{\mathbb{R}^m} f(\mathbf{x}, \mathbf{u}, t)\, \boldsymbol{\psi}(\mathbf{u})\, \mathrm{d}\mathbf{u}, \tag{2}$$

where ρ is the density, $\rho\mathbf{U}$ is the momentum vector, E is the total energy, and $\boldsymbol{\psi}(\mathbf{u}) := [1, \mathbf{u}, (\mathbf{u} \cdot \mathbf{u})/2]^T$ is the vector of collision invariants.

To reduce the cost of directly computing the collision operator, an approximate collision model introduced by [2], known as the Bhatnagar–Gross–Krook (BGK) model, was instead used. For this model, collision is approximated as a relaxation process to thermodynamic equilibrium, represented as

$$\mathcal{C}(f, f') \approx \frac{g(\mathbf{x}, \mathbf{u}, t) - f(\mathbf{x}, \mathbf{u}, t)}{\tau}, \tag{3}$$

where $g(\mathbf{x}, \mathbf{u}, t)$ is the local thermodynamic equilibrium distribution and τ is the relaxation time scale. For a monatomic particle, the equilibrium distribution function is given by the Maxwellian

$$g(\mathbf{x}, \mathbf{u}, t) = \frac{\rho(\mathbf{x}, t)}{[2\pi\theta(\mathbf{x}, t)]^{d/2}} \exp\left[-\frac{\|\mathbf{u} - \mathbf{U}(\mathbf{x}, t)\|_2^2}{2\theta(\mathbf{x}, t)}\right], \tag{4}$$

where $P = (\gamma - 1)(E - \frac{1}{2}\rho\mathbf{U}\cdot\mathbf{U})$ is the pressure and $\theta = P/\rho$ is a scaled temperature, and the collision time is given by

$$\tau = \frac{\mu}{P}, \tag{5}$$

where μ is the dynamic viscosity which can be set adaptively based on some temperature-based viscosity law (e.g., power law, Sutherland's law).

2.1 Discretization

We present here a very brief overview of the numerical approach used to solve the Boltzmann–BGK equation. For a more in-depth description, the reader is referred to [3, 4]. We utilize a nodal spatial and velocity discretization, where each degree of freedom corresponds to a given location in the spatial and velocity domains, i.e.

$$f_{ij} = f(\mathbf{x}_i, \mathbf{u}_j). \tag{6}$$

A schematic of this discretization is shown in Fig. 1. In this approach, the transport (spatial) and collision (velocity) terms are decoupled, resulting in a separate linear advection equation for each velocity node $\mathbf{u}_j$ with a spatially-independent nonlinear source term for each spatial node $\mathbf{x}_i$.

Spatial discretization. To approximate the particle transport portion of the Boltzmann equation, we utilize high-order discontinuous spectral element methods [5], specifically the flux reconstruction scheme of [6], to discretize the spatial domain instead of standard low-order finite volume methods. These numerical schemes, which possess the geometric flexibility of finite volume methods while retaining the arbitrarily high-order accuracy and efficiency of finite difference methods, can achieve the equivalent accuracy of highly-resolved low-order schemes with substantially fewer degrees of freedom, drastically reducing the computational cost and memory requirements of accurately resolving complex fluid flows.

In this approach, the solution within each element $\Omega_k^{\mathbf{x}}$ is represented by a high-order interpolating polynomial across a set of N_s solution nodes $\mathbf{x}_i^s$ as

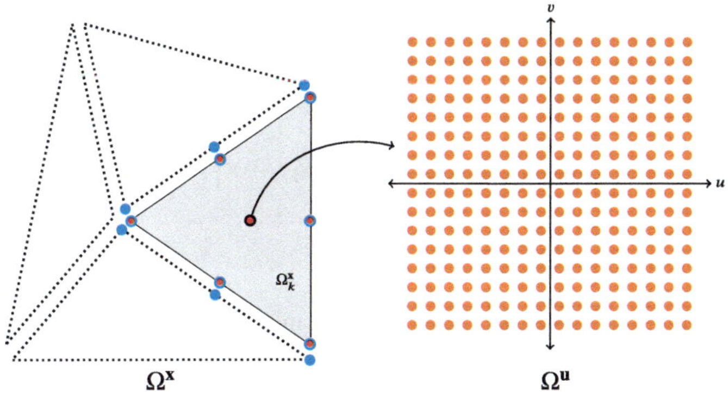

Fig. 1 Schematic of a two-dimensional phase space discretization using an unstructured spatial domain $\Omega^{\mathbf{x}}$ with $\mathbb{P}_2$ elements and a velocity domain $\Omega^{\mathbf{u}}$ with $N_v = 16^2$. Circles denote the spatial solution nodes (red), interface flux nodes (blue), and velocity space nodes (orange), respectively

$$f(\mathbf{x}) = \sum_{i=1}^{N_s} f(\mathbf{x}_i^s)\phi_i(\mathbf{x}), \tag{7}$$

where $\phi_i(\mathbf{x})$ is the nodal interpolating polynomial associated with the given solution node. Without loss of generality, we present this spatial discretization in terms of an arbitrary velocity node $\mathbf{u}_0$.

The flux is calculated via flux reconstruction methodology as a collocation projection with an interface correction term as

$$\mathbf{F}(\mathbf{x}) = \mathbf{u}_0 f(\mathbf{x}) + \sum_{i=1}^{N_f} \left[F_i^I - \mathbf{u}_0 \cdot \mathbf{n}_i f(\mathbf{x}_i^f)\right] \mathbf{g}_i(\mathbf{x}), \tag{8}$$

where $\mathbf{x}_i^f$ is a set of N_f interface flux nodes, $\mathbf{n}_i$ is their associated outward-facing normal vector, and $\mathbf{g}_i$ is their associated correction function which is chosen to recover the nodal discontinuous Galerkin approach [5–7]. Furthermore, F_i^I is the upwind-biased common interface flux, i.e.,

$$F_i^I = \begin{cases} u_n f_i^-, & \text{if } u_n > 0, \\ u_n f_i^+, & \text{else,} \end{cases} \tag{9}$$

where $u_n = \mathbf{u}_0 \cdot \mathbf{n}_i$ and the superscripts $-$ and $+$ denote the interior value and the exterior value of the solution at the interface, respectively.

As these high-order schemes do not typically preserve a maximum principle, they do not guarantee that the distribution function remains strictly positive which is physically inconsistent for a probability measure and can result in the divergence of the numerical scheme as it may not ensure a strictly-positive macroscopic density and temperature. Therefore, the spatial discretization is augmented with the high-order, positivity-preserving limiter of [8], which contracts the solution to the element-wise mean $\bar{f}$ if the distribution function attains a negative value at any spatial node, i.e.,

$$\hat{f}(\mathbf{x}_i) = \bar{f} + \beta \left[f(\mathbf{x}_i) - \bar{f} \right], \tag{10}$$

where

$$\beta = \min \left[\left| \frac{\bar{f}}{\bar{f} - f^{\min}} \right|, 1 \right] \quad \text{and} \quad f^{\min} = \min f(\mathbf{x}_i) \ \forall \ i \in \{1, ..., N_s\}. \tag{11}$$

As a result of this limiting procedure, the positivity of the distribution function and, by extension, the macroscopic density and temperature, are ensured while retaining the high-order accuracy of the underlying numerical scheme.

Velocity discretization. To approximate the particle collision portion of the Boltzmann equation, the velocity space is represented as a finite subset of the infinite velocity domain $\Omega^{\mathbf{u}} \subset \mathbb{R}^m$, which is discretized by a uniform Cartesian grid of N_v velocity nodes, shown on the right-hand side of Fig. 1. The extent of the velocity domain is taken as some factor of the maximum thermal velocity in the problem, where a factor of 4 results in 99% of the distribution function being contained within $\Omega^{\mathbf{u}}$ for near-Maxwellian distributions.

To evaluate the equilibrium distribution function g corresponding to f, it is necessary to compute the moments of the f. This is approximated via a discrete integration operator $\mathbf{M}$ with strictly-positive entries, i.e.,

$$\mathbf{M} \cdot \mathbf{f} \approx \int_{\mathbb{R}^m} f(\mathbf{u}) \, \mathbf{du}, \qquad \mathbf{Q} = \mathbf{M} \cdot \left[\mathbf{f} \otimes \boldsymbol{\psi} \right]. \tag{12}$$

As the nodes in the velocity domain are uniformly distributed and the distribution functions are smooth and compactly supported, the spectral convergence of the trapezoidal rule makes it an ideal choice for quadrature [9]. However, since this integration operator cannot exactly represent the integration of the moments, these integration errors result in discrete conservation errors in the solution as the distribution function the solution is relaxing to via the BGK operator does not possess the same macroscopic state, i.e.,

$$\mathbf{M} \cdot \left[\mathbf{f} \otimes \boldsymbol{\psi} \right] \neq \mathbf{M} \cdot \left[\mathbf{g} \otimes \boldsymbol{\psi} \right]. \tag{13}$$

While these conservation errors can be contained to a reasonable tolerance with increased resolution in the velocity domain, it was shown in [3, 4] that the primary source of approximation error in the scheme stemmed from the conservation error and that if discrete conservation could be ensured, accurate approximation of particle collision could be obtained with much fewer degrees of freedom in the velocity domain.

This discretely-conservative velocity model was achieved through the discrete velocity model (DVM) approach of [10]. In this approach, a discrete equilibrium distribution function is sought which satisfies the discrete compatibility condition (i.e., possesses the same moments as the distribution function) and satisfies the discrete form of Boltzmann's H-theorem. It was shown in [10] that this discrete distribution function is represented by a Maxwellian formed around a perturbed macroscopic state $\mathbf{Q}'$ which converges to the true macroscopic state $\mathbf{Q}$ in the limit of infinite velocity resolution. However, as there does not exist a closed-form expression for this perturbed state, it must be computed numerically via a nonlinear optimization process. Due to the low dimensionality of the optimization problem and the presence of a closed-form expression of the Jacobian, a discretely-conservative approximation of the BGK operator could be achieved easily and efficiently with as few as two iterations of Newton's method. For an in-depth overview of this numerical approach, the reader is referred to [4], Sect. 3.4. As a result of this discrete conservation property, it was shown that the required resolution for accurately approximating complex fluid flows could be decreased by up to two orders of magnitude [3, 4], requiring as few as eight velocity nodes per dimension.

2.2 Implementation

The presented numerical approach was implemented within PyFR [11], a high-order flux reconstruction solver which can efficiently target massively-parallel GPU computing architectures. Due to the space-velocity decoupling of the transport and collision operators and the compute intensive nature of the discrete velocity model, the numerical scheme could be very efficiently parallelized for GPU computing. As the purpose of this numerical approach is to target spatially well-resolved flows tending towards direct numerical simulation where the collision operator is not excessively stiff compared to the transport term, the use of explicit time stepping schemes was deemed preferable due to the many computational advantages they offer for GPU computing.

For the numerical results to be presented, computations were performed on up to 80 32 GiB NVIDIA V100 GPUs with up to approximately 10^{11} total degrees of freedom. The numerical experiments indicate that for three-dimensional flows, the total computational cost for simulating flows via the Boltzmann equation is roughly one to two orders of magnitude higher than an equivalently-resolved Navier–Stokes simulation [3]. While this additional cost is somewhat substantial, these algorithmic improvements still make it entirely feasible to simulate complex fluid flows on modern compute hardware. In comparison, standard low-order finite volume approaches with nodal velocity models would require roughly an additional two orders of magnitude more computational effort for an equivalent level of resolution.

3 Results

In this section, an overview of selected numerical experiments previously presented by the authors [3, 4] is shown along with some novel results. A common application of the Boltzmann equation (and kinetic schemes in general) is for microchannels, where the small characteristic length scales result in very low Reynolds number flows, yielding strong rarefaction and non-equilibrium effects which cannot be accurately modeled using the standard Navier–Stokes equations. These geometries and flow conditions are commonly encountered in biomedical engineering applications and micro-electro mechanical system (MEMS) devices. One of the validation studies of the numerical approach was performed for the rarefied bent microchannel case of [12]. A subset of the results of the numerical experiment is shown in Fig. 2 as contours of pressure with velocity streamlines at varying Knudsen numbers. Non-equilibrium effects can be observed as the increasing Knudsen number results in higher degrees of slip velocity at the wall, affecting the structure of the flow around the concave and convex corners. These results showed excellent agreement with the reference data and showcased how a discretely-conservative velocity model can accurately resolve non-equilibrium effects with very few degrees of freedom. For example, in this experiment, a resolution of $N_v = 8^2$ was found to be sufficient to be converged in the velocity domain. At this resolution, the extension to three-dimensional microchannels was performed with only a factor of three increase in the computational cost in comparison to a standard Navier–Stokes approach.

While the application of the Boltzmann to rarefied gas dynamics is quite typical, its use for more complex continuum flows has been severely limited. Due to the improved efficiency and accuracy of the numerical approach, the simulation of transition to turbulence for a three-dimensional compressible Taylor–Green Vortex at $Re = 1600$ was carried out. To the authors' knowledge, this was the first simulation of a three-dimensional turbulent flow performed by directly solving the Boltzmann equation. It was shown that the molecular gas dynamics equations could accurately

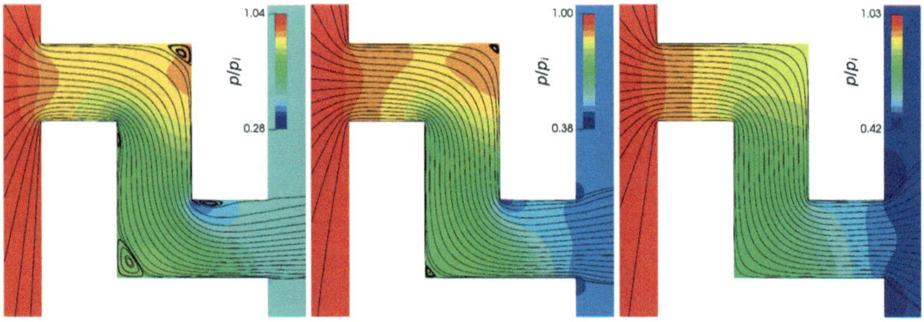

(a) $Kn = 0.01$ (b) $Kn = 0.02$ (c) $Kn = 0.05$

Fig. 2 Contours of normalized pressure overlaid with velocity streamlines for the bent microchannel problem at $Kn = 0.01$ (left), $Kn = 0.02$ (middle), and $Kn = 0.05$ (right) computed using a $\mathbb{P}_3$ approximation

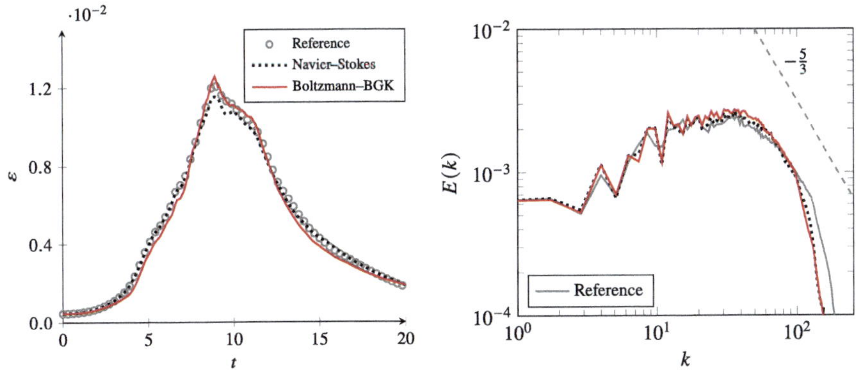

(a) Enstrophy (b) Kinetic energy spectra

Fig. 3 Dissipation measured by enstrophy (left) and turbulent kinetic energy spectra at $t = 10$ (right) for the compressible Taylor-Green vortex at $Re = 1600$ computed using a $\mathbb{P}_3$ approximation

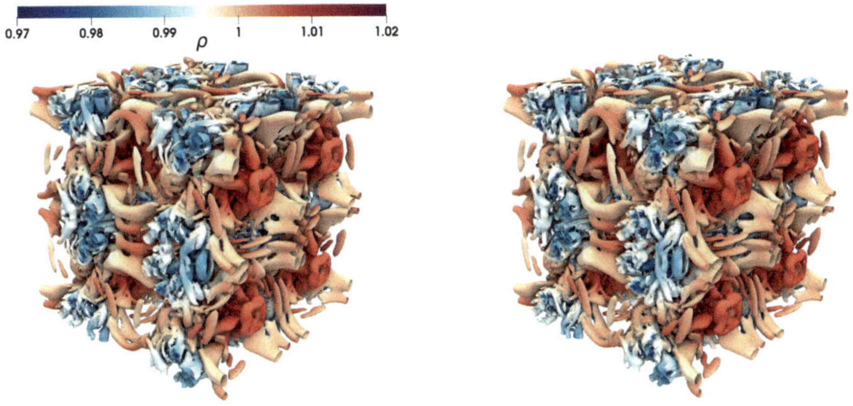

(a) Navier–Stokes (b) Boltzmann–BGK

Fig. 4 Isosurfaces of Q-criterion colored by density at $t = 10$ for the compressible Taylor–Green vortex at $Re = 1600$ computed using a $\mathbb{P}_3$ approximation

predict nonlinear flow phenomena such as transition to turbulence consistently with the hydrodynamic equations, as shown through the enstrophy-based dissipation and turbulent kinetic energy spectra in Fig. 3. Furthermore, nearly identical approximation of the turbulent flow structures was obtained with the Boltzmann–BGK equation as with the Navier–Stokes equations, shown through the Q-criterion isosurfaces in Fig. 4.

Although the Taylor–Green vortex demonstrates that the Boltzmann–BGK approach can accurately predict nonlinear flow phenomena such as transition to turbulence, the lack of wall-fluid interactions in the problem greatly reduces the complexity of the flow. For the Boltzmann equation, the correct choice and validity

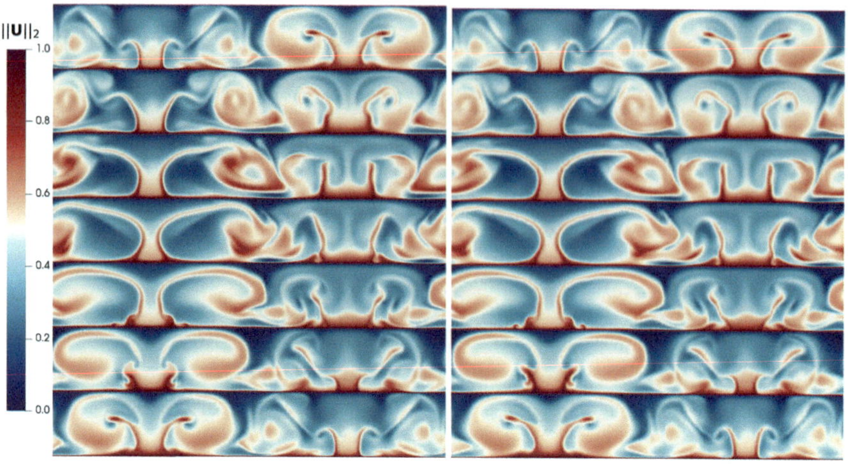

Fig. 5 Contours of velocity magnitude for the three-dimensional transitional Taylor–Couette flow problem at $t = 22$ on angularly equispaced cross-sections from $\theta = 0$ (top) to $\theta = \pi$ (bottom) computed using a $\mathbb{P}_5$ approximation

of wall boundary conditions is still somewhat of an open problem. A comprehensive validation of the effects of wall boundary conditions on momentum transfer for complex wall-bounded fluid flows was performed by the authors [3], with one of the numerical experiments evaluating these effects on the three-dimensional transitional Taylor–Couette flow [13]. For this problem, the flow is driven by concentric rotating cylinders, with the initially laminar flow transitioning to a turbulent state [14]. Much like with the Taylor–Green vortex, this was, to the authors' knowledge, the first simulation of a three-dimensional wall-bounded turbulent flow performed by directly solving the Boltzmann equation, which was obtained using 4.7 billion degrees of freedom with approximately a 36 times larger computational cost than the Navier–Stokes approach. It was seen that the shear-induced transition to turbulence could be accurately resolved by the Boltzmann–BGK approach, with a comparison of the velocity and vorticity contours shown in Figs. 5 and 6, respectively. Notably, it was found that to resolve these effects consistently with the Navier–Stokes equations, accurate prediction of particle transport was significantly more important than than accurate prediction of particle collision, i.e., a highly-resolved spatial domain was much more important than a highly-resolved velocity domain. These findings suggest that high-order schemes may be especially well-suited for simulating complex flows via the Boltzmann equation and that directly solving the Boltzmann equation can be performed at a reasonable cost compared to the Navier–Stokes equations as a result of the relatively few degrees of freedom necessary in the velocity domain.

A more rigorous validation of the effects of wall-boundary conditions was performed through the simulation of the flow around an SD7003 airfoil at a Reynolds number of $Re = 60,000$ and angle of attack of $\alpha = 8°$, which was obtained using

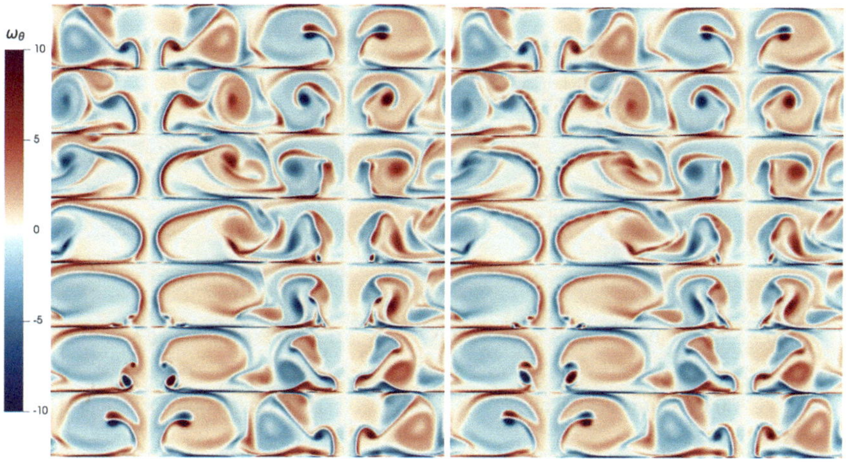

(a) Reference (b) Boltzmann–BGK

Fig. 6 Contours of vorticity along the cross-section normal direction for the three-dimensional transitional Taylor–Couette flow problem at $t = 22$ on angularly equispaced cross-sections from $\theta = 0$ (top) to $\theta = \pi$ (bottom) computed using a $\mathbb{P}_5$ approximation

88.3 billion degrees of freedom with approximately a 62 times larger computational cost than the Navier–Stokes approach. This case exhibits a laminar separation bubble that subsequently transitions to turbulence and reattaches into a turbulent boundary layer and wake [15], which makes it highly sensitive to the flow conditions and difficult to accurately resolve. A comparison of the average surface pressure coefficient and skin friction coefficient as predicted by the Boltzmann–BGK approach and the Navier–Stokes approach is shown in Fig. 7. Good agreement between the two approaches was seen, both for the surface pressure coefficient and skin friction coefficient. Furthermore, a visualization of the instantaneous Q-criterion and average streamwise velocity contours as predicted by the Boltzmann–BGK approach is shown in Fig. 8. The laminar separation bubble and the transition of the shear layer can be clearly seen through the Q-criterion isosurfaces and velocity contours, which showcases complex flow phenomena which can be accurately resolved by the Boltzmann–BGK approach.

While the previous experiments focused on flows in the low Mach regimes, the use of the Boltzmann equation for high-speed flows also offers some notable advantages, such as the ability to directly resolve shock structures and strong aerothermodynamic effects. To demonstrate the ability of the Boltzmann–BGK approach for high-speed flows, we consider the flow around spacecraft in atmospheric re-entry conditions. In particular, the flow around a three-dimensional AS-202 Apollo capsule at a Mach number of 22.7 and a Reynolds number of 43,000 was simulated, which is, to the authors' knowledge, the first high-order simulation of a three-dimensional hypersonic flow obtained by the direct solution of the Boltzmann equation. The surface slip velocity contours/streamlines and a cross-section of the contours of density are shown

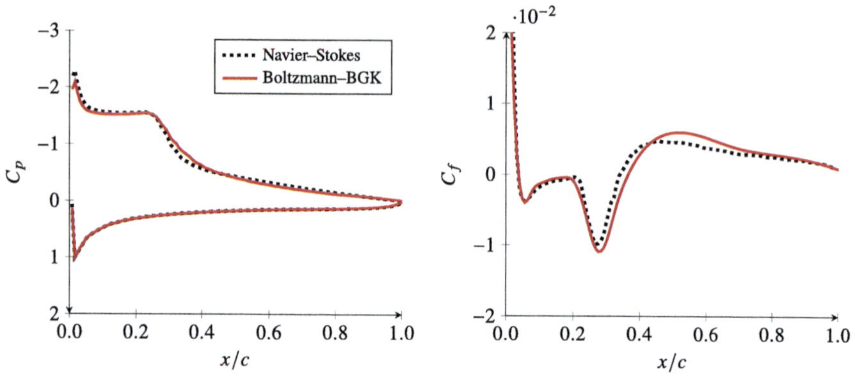

(a) Pressure coefficient

(b) Skin friction coefficient

Fig. 7 Surface pressure coefficient (left) and suction-side skin friction coefficient (right) for the $Re = 60{,}000$ SD7003 airfoil case computed with the Navier–Stokes equations (black) and the Boltzmann–BGK equation (red) using a $\mathbb{P}_4$ approximation

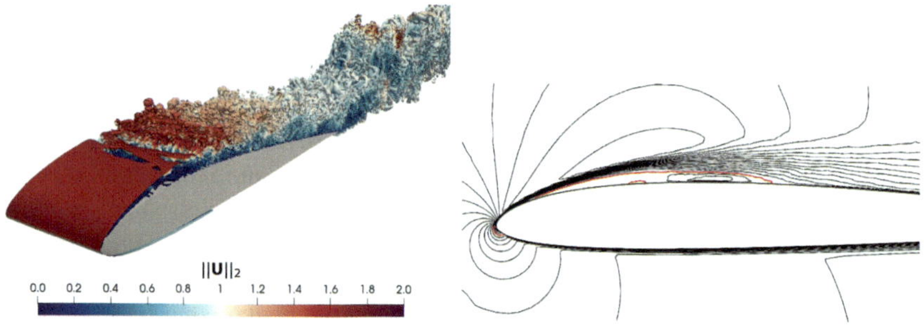

(a) Q-criterion

(b) Streamwise velocity

Fig. 8 Instantaneous Q-criterion isosurfaces colored by velocity magnitude (left) and average streamwise velocity contours, equispaced on the range [-0.2, 1.5], (right) for the $Re = 60{,}000$ SD7003 airfoil case computed with the Boltzmann–BGK approach using a $\mathbb{P}_4$ approximation

in Fig. 9. The strong detached bow shock can be clearly seen in the density contours which is resolved by the high-order numerical scheme *without any ad hoc numerical shock capturing approach*. Furthermore, the wall slip velocity shows the local non-equilibrium nature of the flow, with complex unsteady behavior at the wall that would not be accurately predicted with the governing equations of continuum fluid dynamics.

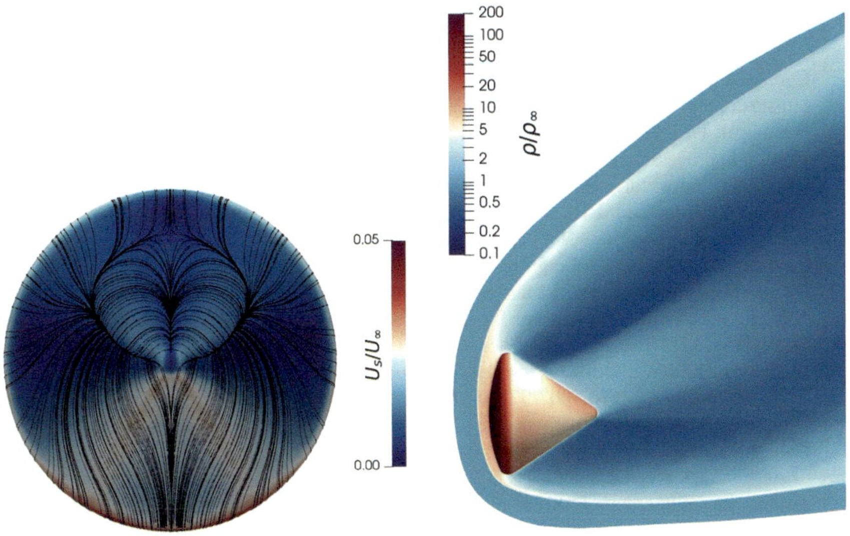

(a) Surface slip velocity (b) Density cross-section

Fig. 9 Surface slip velocity contours overlaid with slip velocity streamlines (left) and density contours on the cross-section $z = 0$ (right) for the Mach 22.7 Apollo re-entry capsule case at $Re = 43,000$ computed with the Boltzmann–BGK approach using a $\mathbb{P}_3$ approximation

4 Conclusion

We present a brief overview of a numerical approach for simulating complex flows through directly solving the Boltzmann equation augmented with the BGK collision model. Through the combination of highly-efficient high-order spatial discretizations, discretely-conservative velocity models, and massively-parallel GPU computing, the approach allows for the simulation of complex three-dimensional flows ranging from rarefied microchannels to high-speed re-entry vehicles which were previously intractable. It was shown that complex nonlinear flow phenomena such as transition to turbulence as well as thermodynamic non-equilibrium effects could be accurately computed through this molecular gas dynamics approach. Furthermore, these simulations could be performed at a reasonable computational cost in comparison to the Navier–Stokes equations. The results of this work presents opportunities for entirely novel perspectives and approaches for complex flow problems, such as the analysis and study of fundamental flow phenomena through the evolution of a phase space distribution function and the use of a unified numerical framework for simulating flow problems that span a variety of flow regimes.

References

1. Cercignani, C.: The Boltzmann equation and its applications. Springer New York (1988). https://doi.org/10.1007/978-1-4612-1039-9
2. Bhatnagar, P.L., Gross, E.P., Krook, M.: A model for collision processes in gases. I. Small amplitude processes in charged and neutral one-component systems. Phys. Rev. **94**(3), 511–525 (1954). https://doi.org/10.1103/physrev.94.511
3. Dzanic, T., Witherden, F.D., Martinelli, L.: Validation of wall boundary conditions for simulating complex fluid flows via the Boltzmann–BGK equation: momentum transport and skin friction (2023a). https://doi.org/10.48550/arXiv.2308.00700
4. Dzanic, T., Witherden, F.D., Martinelli, L.: A positivity-preserving and conservative high-order flux reconstruction method for the polyatomic Boltzmann-BGK equation. J. Comput. Phys. **486**, 112146 (2023). https://doi.org/10.1016/j.jcp.2023.112146
5. Hesthaven, J.S., Warburton, T.: Nodal discontinuous Galerkin methods. Springer New York (2008). https://doi.org/10.1007/978-0-387-72067-8
6. Huynh, H.T.: A flux reconstruction approach to high-order schemes including discontinuous Galerkin methods. In: 18th AIAA computational fluid dynamics conference. American Institute of Aeronautics and Astronautics (2007). https://doi.org/10.2514/6.2007-4079
7. Trojak, W., Witherden, F.D.: A new family of weighted one-parameter flux reconstruction schemes. Comput. Fluid **222**, 104918 (2021). https://doi.org/10.1016/j.compfluid.2021.104918
8. Zhang, X., Shu, C.-W.: On maximum-principle-satisfying high order schemes for scalar conservation laws. J. Comput. Phys. **229**(9), 3091–3120 (2010). https://doi.org/10.1016/j.jcp.2009.12.030
9. Trefethen, L.N., Weideman, J.A.C.: The exponentially convergent trapezoidal rule. SIAM Rev. **56**(3), 385–458 (2014). https://doi.org/10.1137/130932132
10. Mieussens, L.: Discrete velocity model and implicit scheme for the BGK equation of rarefied gas dynamic. Math. Models Methods Appl. Sci. **10**(08), 1121–1149 (2000). https://doi.org/10.1142/s0218202500000562
11. Witherden, F.D., Farrington, A.M., Vincent, P.E.: PyFR: An open source framework for solving advection–diffusion type problems on streaming architectures using the flux reconstruction approach. Comput. Phys. Commun. **185**(11), 3028–3040 (2014). https://doi.org/10.1016/j.cpc.2014.07.011
12. Ho, M.T., Li, J., Su, W., Wu, L., Borg, M.K., Li, Z., Zhang, Y.: Rarefied flow separation in microchannel with bends. J. Fluid Mech. (2020). https://doi.org/10.1017/jfm.2020.585
13. Taylor, G.I.: Stability of a viscous liquid contained between two rotating cylinders. Proceedings of the Royal Society of London. Series A, Containing Papers of a Mathematical and Physical Character, **102**(718): 541–542 (1923). https://doi.org/10.1098/rspa.1923.0013
14. Wang, Z.J., Jourdan, E.: Benchmark for scale-resolving simulation with curved walls: the Taylor Couette flow. Adv. Aerodyn. **3**(1), (2021). https://doi.org/10.1186/s42774-021-00071-0
15. Garmann, D.J., Visbal, M.R., Orkwis, P.D.: Comparative study of implicit and subgrid-scale model large-eddy simulation techniques for low-Reynolds number airfoil applications. Int. J. Numer. Meth. Fluids **71**(12), 1546–1565 (2012). https://doi.org/10.1002/fld.3725

Efficient System Decomposition of High Order Partial Differential Equations

Rob Watson

Abstract A linear second order partial differential equation, which has historically been used for computing the approximate distance to the nearest solid domain boundary, is decomposed into a first order system of linear equations. The importance of the details of the decomposition on the stability, accuracy, and convergence speed of the discretised system are investigated. Inferences are drawn about the requirements of the decomposed system which ensure that it results in the discretised system being stable and convergent under explicit time stepping. A multiobjective optimisation is undertaken to find the Pareto front which balances the accuracy with which the discretised system represents the underlying equations against the condition number of the system, which serves as a surrogate for convergence speed. From this, optimised first order decompositions are generated, the best of which is found to reduce the numerical error to 10% of that found in the baseline case, and the number of time steps required to converge to 30% of the baseline.

1 Introduction

In recent years, significant effort has been expended on developing software which accurately and efficiently solves partial differential equations of the type:

$$\frac{\partial \mathbf{u}}{\partial t} + \nabla \cdot \mathbf{f}(\mathbf{u}) = \mathbf{s} \tag{1}$$

Much of this effort has been driven by the needs of the aerospace industry and the quest for highly accurate computational fluid dynamics simulations. The computational expense of these calculations can be enormous, so researchers have striven for ways to both reduce this load and to spread it more effectively across multiple

CUFS-2024-PAPER ID, Cambridge, United Kingdom, 4-5 March 2024.

R. Watson (✉)
Department of Aeronautical and Automotive Engineering, Loughborough University, Leicestershire, UK
e-mail: r.a.watson@lboro.ac.uk

J. C. Tyacke and N. R. Vadlamani (eds.), *Proceedings of the Cambridge Unsteady Flow Symposium 2024*, https://doi.org/10.1007/978-3-031-69035-8_26

processes which can operate simultaneously. This kind of optimisation has involved input from the disciplines of physics, engineering, mathematics and computer science. The outputs of this research effort are powerful tools which can be rapidly and efficiently deployed on supercomputing hardware at a massive scale [1, 3–5, 19].

However, there is also considerable demand for solving PDEs outside of the (relatively) narrow confines of fluid dynamics, with varieties appearing in medical and biological settings, in optical applications, and fundamental physics at the quantum level—amongst many others. It would be convenient and beneficial to both researchers and industry working across this array of applications to have a general-purpose numerical PDE solver—and ideally this solver would take advantage of the vast body of research and development which has gone into the aforementioned software.

To be a little more precise, the CFD software mentioned above does not solve Eq. (1), at least, not exactly. Instead, they tend to solve:

$$\frac{\partial \mathbf{u}}{\partial t} + \nabla \cdot \mathbf{f}(\mathbf{u}, \nabla \mathbf{u}) = \mathbf{s} \tag{2}$$

due to the dependence of the Navier-Stokes equations on the transport of momentum and energy by diffusion. This turns out to be mildly problematic—it requires a preliminary loop to compute the approximate gradients of the variables (typically either using some form of Gauss-Green or Least Squares), and can introduce numerical issues into the solution.

Some other sets of partial differential equations beyond the familiar Navier-Stokes, when written in conservative form, have flux term dependencies on even higher order gradients, such as the Korteweg-de-Vries-Burgers equation (second order) [2], which has found applications in modelling ion acoustic waves in plasma physics, or the Cahn-Hilliard equation (third order) [6], which is used to model the separation of phases in binary fluids. Computing the gradient terms for these kinds of equations in the manner of those traditionally used for the Navier-Stokes equations would require rewrites to the code base, and would compound the numerical issues introduced—particularly in situations when the fluxes dependent on these higher order gradients are the dominant transport mechanism.

In this paper, we explore the potential for rewriting these equations into first order form. We attempt to do this by restricting the problem to very simple linear situations, in which the flux function is independent of the solution. We do this for simplicity, but this approach, although restrictive, does turn out to be capable of throwing some light upon some interesting issues.

2 From Higher Order to First Order

A lot of effort has been expended in recent years to move towards higher order methods in computational fluid dynamics, which seem to be able to produce numerical solutions more efficiently than more traditional lower order methods. It feels a little unusual, therefore, to be instead attempting to move instead from a higher order towards a lower. Nevertheless, let us take as an example a very simple Poisson equation, with the right hand side function being simply set to negative one. For simplicity, we only concern ourselves with the one dimensional case:

$$\frac{\partial}{\partial x}\frac{\partial \phi}{\partial x} = -1 \tag{3}$$

which makes it obvious that we have a second order partial differential equation. Even this extremely simple PDE has found important applications—serving as the basis for a range of approaches to solving for the boundary-normal distance in computational domains [15–17]. It is possible to rewrite this second order equation as a system of first order equations. One possible way of achieving this could be [18]:

$$\begin{aligned} \frac{\partial p}{\partial x} &= -1 \\ \frac{\partial \phi}{\partial x} &= p \end{aligned} \tag{4}$$

Solutions of which obviously also satisfy Eq. (3). The way in which this process could be extended to two or more dimensions is also immediately apparent. Given that we have had to do a little bit of work to get here, it is reasonable to ask: why bother? The first part of this answer is that it is now possible to attempt to use a putative "general-purpose" solver designed for attacking systems of equations of the form of Eq. (1) to solve a significantly wider variety of problems. Nishikawa has noted that decompositions of this form can improve the stability of solution, such that larger explicit time steps can be used to solve it [10], and reducing the number of time steps to convergence—something which has been confirmed for this specific wall distance problem [18]. This has proven to be true using different discretisation schemes [7, 11], for more complex nonlinear problems [9], and even for third order systems [8]. There appear to be, however, further benefits, which are explored in more detail over the next sections.

3 Richardson Iteration

The forward Euler method is the most straightforward approach for advancing an ordinary differential equation in time. It requires no new exchange of information, no further iteration, no knowledge about prior states, no tuned constants, no extra storage, and has no dependency on the ability to solve a subsidiary system of equations. As such, it is trivial to program effectively, and trifling to parallelise extremely efficiently. The general form of the method is given by:

$$\mathbf{u}^{n+1} = \mathbf{u}^n + \Delta t \times \mathbf{r}(\mathbf{u}^n) \tag{5}$$

where $\mathbf{r}(\mathbf{u}^n) = \mathbf{s} - \mathbf{f}(\mathbf{u}^n)$ is the residual of the right hand side of the system to be advanced. The connections between the forward Euler method and Richardson iteration are well known, but it is worth briefly revisiting them here for completeness. Richardson iteration [12] is a stationary iterative approach to solving linear systems of equations, such that $\mathbf{Au} = \mathbf{b}$. A single Richardson iteration consists of:

$$\mathbf{u}^{n+1} = \mathbf{u}^n + \omega \times \left(\mathbf{b} - \mathbf{Au}^n\right) \tag{6}$$

With the forward Euler time step size, Δt, and the Richardson iterator step size, ω, set to the same value, the equivalence between the two methods is obvious. The Richardson iterator will converge to an approximate solution as long as the spectral radius of the iteration matrix $\mathbf{C} = \mathbf{I} - \omega\mathbf{A}$ is less than 1. From the point of view of PDEs, this is equivalent to saying that the system will converge using forward Euler to a steady state. As an aside, this gives a mildly interesting alternative interpretation of a discretised system. Consider the one dimensional linear advection equation with periodic boundaries and an advection velocity of one:

$$\frac{\partial \phi}{\partial t} + \frac{\partial \phi}{\partial x} = 0 \tag{7}$$

Discretising this equation in space with first order upwinding, and discretising in time with the forward Euler method leads to a very diffusive but stable solution, as shown in Fig. 1, with the solution taken after 20 time units. There is a maximum time step restriction on the method, which scales with the Courant number, $\Delta t \propto \Delta x$. As this is a linear problem, however, we can cast it into the form of Eq. (6) and solve it via Richardson iteration. The first order upwinding discretisation in space results in a very sparse (0.25%) matrix $\mathbf{A}$. Running the iteration until the number of iterations multiplied by ω is equal to 20 leads to an identical solution in Fig. 1, as we expect, given the equivalence of the methods. Since this is a linear problem, and we have direct access to the matrix $\mathbf{A}$, we can use it to dig a little deeper. For this problem, $\mathbf{A}$ is a Toeplitz matrix, with values of $^1/_{\Delta x}$ on the main diagonal and $^{-1}/_{\Delta x}$ on the first lower diagonal, with zeros elsewhere. The Richardson iteration matrix, $\mathbf{C} = \mathbf{I} - \omega\mathbf{A}$ is therefore also a Toeplitz matrix of a similar form, with a set of n repeated eigenvalues given by $\lambda_n = 1 - {}^{\omega}/_{\Delta x}$. This means that the spectral radius of the the iteration matrix will be less than one as long as the iterative step size is less than $2\Delta x$. On an infinite mesh, then, all modes in the solution decay at an equal rate.

The periodic boundary condition, however, introduces a value of $^{-1}/_{\Delta x}$ into the upper right corner of matrix $\mathbf{A}$, however, which introduces a characteristic length into the problem. This has a dramatic effect on the evolution of the solution. The first thing of note is that the eigenvalues of the iteration matrix are no longer equal and a simple function of the grid spacing. The spectrum of eigenvalues is shown in Fig. 2a for three different values of the step size. The first thing to note from this figure is that the magnitude of all the eigenvalues are ≤ 1. This remains true until ω is increased

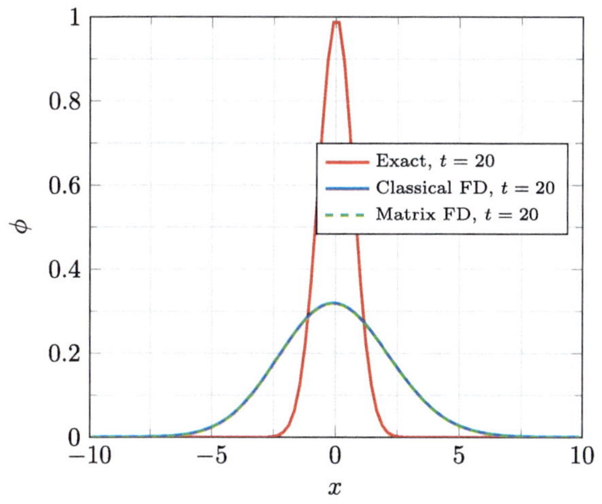

Fig. 1 The linear advection equation, Eq. (7), at $t = 20$

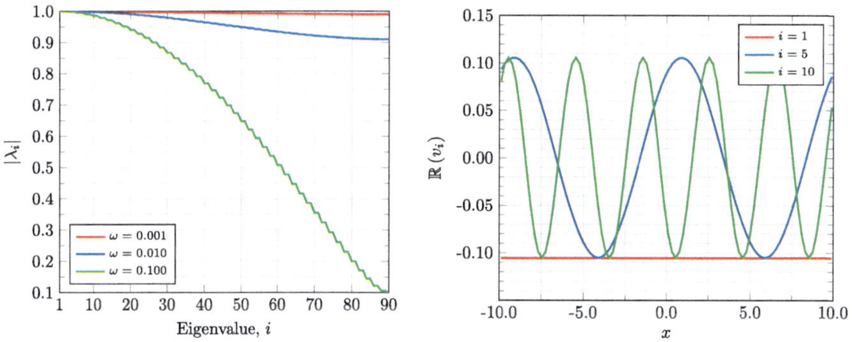

(a) The absolute eigenvalues

(b) The real part of selected eigenvectors

Fig. 2 The eigenvalues and eigenvectors of the Richardson iteration matrix

past Δx. Beyond this value, the sign of the diagonal of the iteration matrix changes, and all eigenvalues become greater than one, indicating that the Richardson iterator becomes unstable across all modes—as we see with the forward Euler method with a time step above the critical CFL condition.

Below this critical value, however, the magnitude of the largest eigenvalue is always equal to 1, regardless of the step size. All the other eigenmodes decay except for this one, which stubbornly persists throughout the calculation. Examining the eigenvector in question reveals that it is the uniform mode. Further examination of the modes suggests, not unexpectedly, that higher frequencies are damped more aggressively than low frequencies, as shown in Fig. 2b.

The lack of any damping on one particular mode leads us to the conclusion that the discretised system does not have a unique solution—that $\mathbf{A}$ is uninvertible. We might expect this, as a conservative discretisation automatically leads to each row in the matrix summing to zero. In fact, **A** is rank deficient with a nullity of one, and in its null space lies the vector of ones. Practically, this means that the sequence does converge, but to a non-unique constant value which will be equal to the mean of the integral of the initial condition over the periodic region, which is indeed what we see if we allow the simulation to run on until it converges.

We conclude, then, that the solution of the forward Euler, first-order upwind discretisation of the linear advection equation can be interpreted as an incompletely converged solution to a linear system via Richardson iteration. More importantly, we may also conclude that it is possible to learn some interesting behaviours of the system by explicitly forming the matrix of coefficients—and this motivates us to apply this procedure to systems of PDEs resulting from decomposition of equations with higher order gradients into first order systems.

As with all iterative schemes, the condition number of the coefficient matrix is an important guide to how quickly the error will decay with each iteration. Lower condition numbers are associated with faster convergence. For Richardson iteration, then, we have two key parameters of interest—the spectral radius of the iteration matrix, **C**, must be less than 1, and the condition number of the coefficient matrix, **A**, must be as small as possible.

4 Second Order Systems

Let us return to Eq. (3). There are a plethora of ways to decompose this equation into a first order system. Perhaps the most obvious is that given in Eq. (4), but there are infinite alternatives available, all of which are equivalent to the original wall distance Poisson equation, but with different discretisations and therefore different paths of their solution. As all of these sets of equations represent linear systems, we are interested here in whether we can look at the structure of their corresponding **A** and **C** matrices to determine if any of the possible first order decompositions are more or less effective than others.

Forming the matrix of coefficients, **A**, requires that a spatial discretisation scheme be specified. To keep this straightforward, and to ensure that it is easy to recover the separate coefficient matrix, a node-centred finite volume approach is used, with piece-wise constant (first order) data within each cell, and an exact first order upwinded flux constructed at the interfaces. This naturally restricts the generality of the conclusions, but we shall try not to let this hold us back from making sweeping generalisations. The interface flux is given by diagonalising the flux Jacobian, and then using this in the interface calculation to achieve the exact first order upwinding as:

$$\mathbf{f}_{L,R} = \frac{1}{2}\left(\mathbf{f}_R + \mathbf{f}_L\right) - \frac{1}{2}\mathbf{W}\left|\Lambda\right|\mathbf{W}^{-1}\left(\mathbf{u}_R - \mathbf{u}_L\right) \tag{8}$$

There is no need for an approximate Riemann solver such as those of Roe [13] or Rusanov [14] since the equations are linear and there is no unknown interface state.

4.1 Decomposing Second Order Systems

When we decompose a second order system in this manner, we introduce a new auxiliary variable, p. The automatic choice is to set this auxiliary variable to be equal to the gradient of the primary variable of interest, ϕ. However, this is not a unique choice. More generally, we could decompose Eq. (3) as:

$$\begin{aligned} \frac{\partial g(\phi, p)}{\partial x} &= -1 \\ \frac{\partial \phi}{\partial x} &= g(\phi, p) \end{aligned} \tag{9}$$

To maintain the linear nature of the equations, g must be a linear function of ϕ and p, so, $g = C_1\phi + C_2 p$. For now, however, we choose to restrict ourselves to $g = p$. This keeps things manageable, maintains p as a fairly intuitive variable, and avoids the need to change the boundary conditions with each calculation. Future investigations will explore the performance of alternative choices. This simplicity also means that the resulting equations and Richardson iterations can be rewritten in the form of Eq. (1) as:

$$\mathbf{P}\frac{\partial \mathbf{u}}{\partial t} + \mathbf{P}\nabla \cdot \mathbf{f}(\mathbf{u}) = \mathbf{Ps} \tag{10}$$

where:

$$\mathbf{u} = \begin{pmatrix} \phi \\ p \end{pmatrix} \quad , \quad \mathbf{f} = \begin{pmatrix} p \\ \phi \end{pmatrix} \quad , \quad \mathbf{s} = \begin{pmatrix} -1 \\ p \end{pmatrix} \tag{11}$$

and where $\mathbf{P}$ is a 2×2 preconditioning matrix. The resulting linear source terms which appear on the right hand side of Eqs. (4), (9), and (10) can be absorbed into the matrix of coefficients after the PDEs have been discretised. These source terms change the nature of the system considerably, moving away from the more familiar conservative behaviour. It is the interaction between these sources and the first order transport terms which dictate the behaviour of the system, and which affect the two key parameters for the performance of our Richardson iteration. In the following section, we use numerical experiments to explore how different choices as to the decomposition influence these values.

4.2 Decomposed Performance

To conduct some preliminary numerical experiments, random precondition matrices, $\mathbf{P}$, were constructed, and the performance of the resultings system explored. Boundary conditions were set for the problem as $\phi = 0$ at $x = \pm 10$, although these are essentially unimportant—once the type of boundary condition is specified, changing the boundary values only impacts the constant vector, **b**. With the boundaries specified to the values above, the differential equation can be trivially solved analytically to give at exact solution of $\phi = 50 - \frac{x^2}{2}$.

This test was repeated 1×10^5 times, with the random entries in **P** chosen to lie between ±1000. This results in a wide array of systems with different characteristics. The values of **P** which are of interest to use are those which converge under Richardson iteration, and those which give meaningful approximations to the analytical solution. The first test merely requires that the spectral radius of the iteration matrix is less than one. The second is more arbitrary. As we have deliberately focused on (small) linear systems, we do not need to rely on iterative methods to provide a solution. We can thus compare the direct solution of the system to the true analytical solution to get a measure for how accurately each discretised system is replicating the underlying problem. Figure 3 shows the population sorted by the its direct L_2 error. There is a clear knee at around an error of 50, after which the error grows quickly, and we choose the beginning of this as our cutoff for 'reasonable accuracy'. Just over 23, 000 of the systems tested achieved both these criteria.

Figure 4 illustrates what turned out to be the key criteria in determining whether or not a system would fall into the category of being both reasonably accurate and convergent under Richardson iteration. The first thing of note is that all of the relevant systems occur within the lower left quadrant—that is, the product of the eigenvalues is negative, and the bottom right entry in the preconditioning tensor must also be negative. These seem to be necessary but not quite sufficient conditions, with 6% of the systems which had both $\lambda_1\lambda_2 < 0$ and $\mathbf{P}_{2,2} < 0$ being not both accurate and convergent. Almost all of these troublesome systems had either relatively small absolute values of $\mathbf{P}_{2,2}$ or small and similar values of eigenvalue of the flux Jacobian.

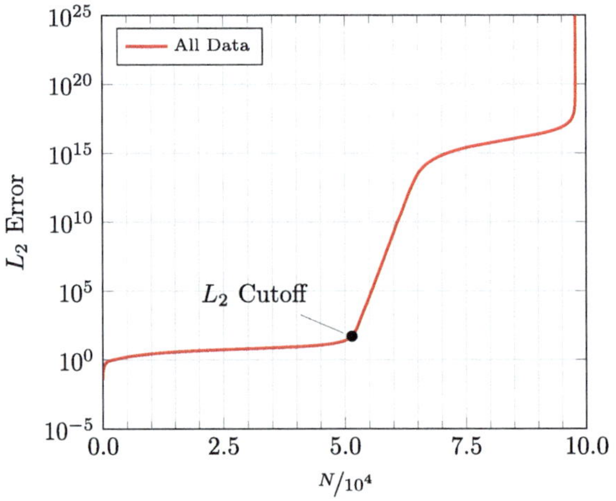

Fig. 3 The L_2 error of the population as a whole

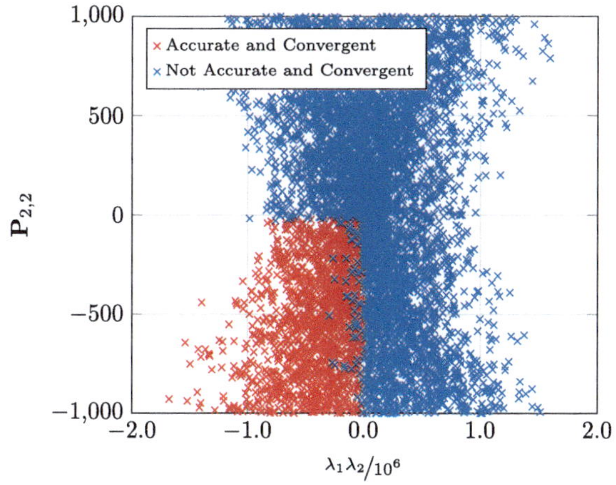

Fig. 4 Finding accurate and convergent systems

Of course, these two measures are not independent of each other, since the eigenvalues of the flux Jacobian are in part determined by $\mathbf{P}_{2,2}$. This is suggestive of the idea that the scaling of the source terms and the convective terms must be in some way compatible. It also means that the case of naive decomposition, with the precondition matrix set to the identity matrix, $\mathbf{P} = \mathbf{I}$, does not in fact converge.

Since the two eigenvalues of the flux Jacobian must have opposite signs for the solution to be convergent and reasonably accurate, it follows that such systems must necessarily be hyperbolic. There are occasionally some differing definitions for hyperbolic systems in the literature, so for the avoidance of doubt we are here defining a hyperbolic system as one in which the Jacobian of the flux terms is diagonisable (*i.e.* it has real and distinct eigenvalues). This criteria is obviously satisfied for any system in which $\lambda_1\lambda_2 < 0$, as any complex eigenvalues will come out as conjugate pairs since both $\mathbf{P}$ and ${}^{\partial \mathbf{f}}/_{\partial \mathbf{u}}$ have exclusively real entries. The fact that we require our distinct eigenvalues to possess different signs is likely attributable to the boundary conditions, which are specified as Dirichlet values on ϕ at each end, and so we want the information to be able to travel in both directions across the domain.

As long as the system converges, we are really interested in minimising both the error relative to the analytical solution and the condition number of the system. In Fig. 5, these variables are plotted against each other for all the relevant systems we have found. While the two do have a positive correlation, it is weak, and in terms of minimising both, we are interested more in the lower left area of the plot. The smallest error found with the tested schemes was found to have a condition number of around 1×10^4, whereas the scheme with the lowest condition number found to have 225, albeit with an error almost a factor of 60 higher.

Figure 6a shows the converged solutions, $\mathbf{u}$, of a high condition number, low error system and a low condition number, relatively high error system. The differences are fairly dramatic, with the more accurate solution being essentially indistinguishable

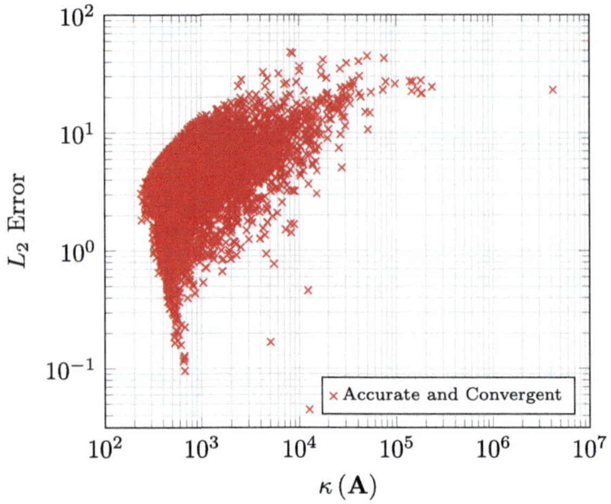

Fig. 5 The spectrum of possible analytical errors and condition numbers

from the exact analytical solution. Given that this is a first order solver with only 50 grid points, this is quite impressive. In contrast, the low condition number system has a peak value around 10% lower than the analytical solution. However, this is not quite the end of the story. These simulations were carried out with the largest stable value of the iterator step size in both cases, and their convergence histories are shown in Fig. 6b. The convergence error is calculated relative to the direct solution of the linear system, rather than to the analytical solution of the PDEs. The low condition number (225) solution does converge initially extremely rapidly—within the first two thousand iterations it has dropped six orders of magnitude towards the direct solution of the system. After this point, the convergence rate drops dramatically, but it continues to head downwards. The high condition number case, which was vastly more accurate compared to the analytical solution, took more like 3×10^7 iterations before it converged three orders of magnitude. At this point, the convergence rate again drops to zero and stalls completely. As expected, high condition numbers lead to very slow convergence speeds and low converged accuracy.

This apparent trade-off in behaviour between our two key parameters of interest can be characterised as a Pareto front. A multi-objective optimiser, aimed at determining this front, was used to find its approximate location. This front is shown in Fig. 7, superimposed on the originally randomly generated population. The trade-off between the requirements is clear—it is possible to have extremely accurate simulations with poor condition numbers which will converge slowly, or simulations with excellent convergence properties which will be considerably less accurate.

To get an idea of the behaviours which underlie the systems which fall on the Pareto front, the eigenvalues of the flux Jacobian, and the value of the $\mathbf{P}_{2,2}$, which controls the linear source term, are plotted in Fig. 8a. Two clusters of the Pareto-optimal solutions are immediately obvious—this is confirmed by the use of k-means

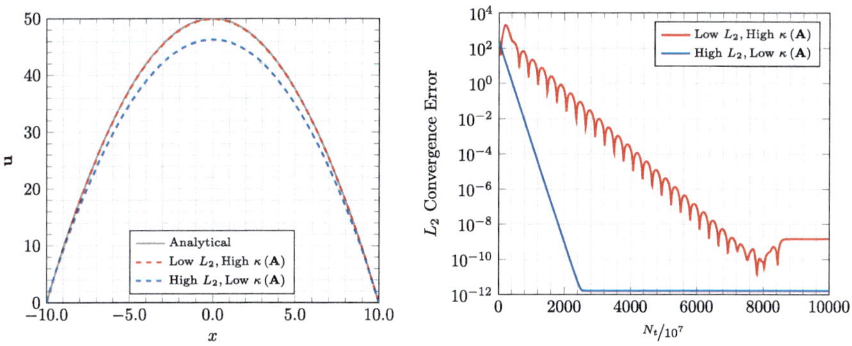

(a) Solution (b) Convergence history

Fig. 6 Using extreme values of accuracy and condition number

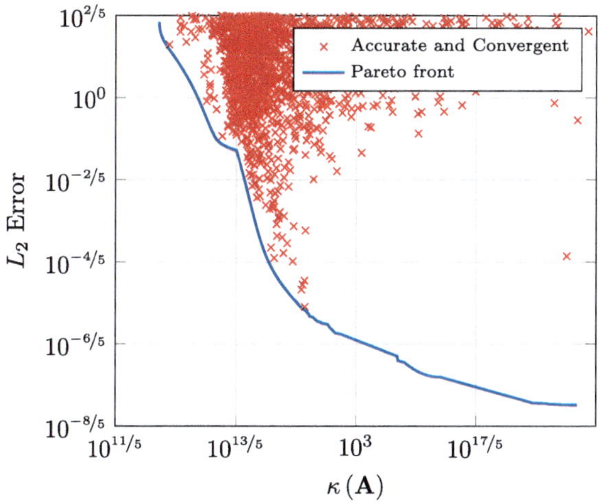

Fig. 7 Pareto front showing the trade-off between conditioning and accuracy

clustering to divide the solutions between the two sets with $k = 2$. It is a little difficult interpret the three dimensional plot on the page, but cluster two has values of λ_1 which are relatively small and positive, with a significant majority of values lying between $2.0 \leq \lambda_1 \leq 3.5$, and values of λ_2 which are relatively small and negative, with the vast majority of values lying between $-4.5 \leq \lambda_2 \leq -2.5$. The values of $\mathbf{P}_{2,2}$ are small and negative, with almost all lying in the range $-0.35 \leq \mathbf{P}_{2,2} \leq -0.12$. The other cluster flips the sign of the eigenvalues around, with $-3.0 \leq \lambda_1 \leq -1.5$ and $2.0 \leq \lambda_2 \leq 3.0$. The values for $\mathbf{P}_{2,2}$ in this cluster are also more negative, running between $-1.05 \leq \mathbf{P}_{2,2} \leq -0.33$.

It is also instructive to re-plot the Pareto front, whilst dividing it between the two clusters. This front too divides fairly exactly between the clusters, with the low accuracy but well conditioned systems corresponding to cluster 1, and the high

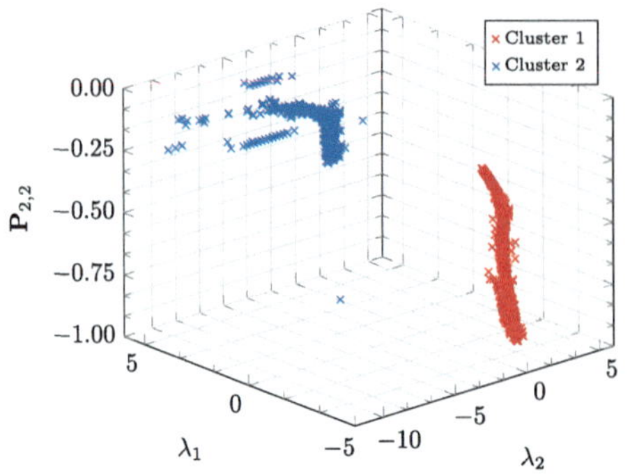

(a) Eigenvalues and $\mathbf{P}_{2,2}$ values

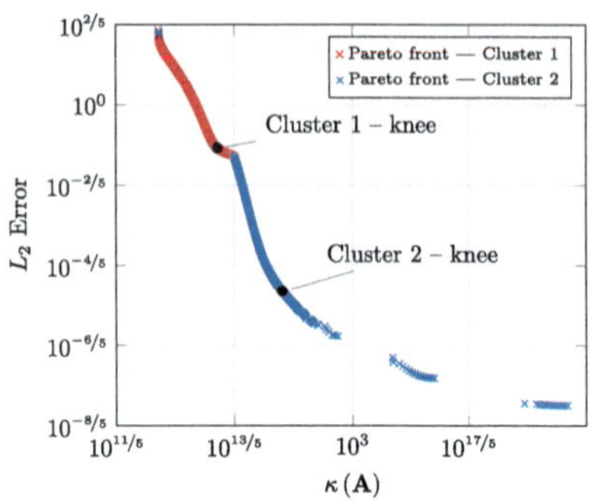

(b) Pareto front

Fig. 8 Identifying Pareto optimal clusters

accuracy but more poorly conditioned systems belonging to cluster 2. Each of these clusters on the front also possesses a bit of a knee, where reasonable accuracy might be obtained with reasonable conditioning.

The system at the cluster one knee point converges both in slightly fewer iterative steps ($N_c \approx 2050$) and to a considerably lower L_2 error ($L_2 = 0.6062$) than the originally selected low condition number system from the original population ($N_c \approx 2550$, $L_2 = 2.6371$). The system at the cluster two knee point is actually found to perform slightly worse in terms of both accuracy and convergence than the original randomly generated high accuracy system ($N_c \approx 10500$ vs. $N_c \approx 8650$, and $L_2 = 0.1467$ vs. $L_2 = 0.0452$), although it does converge more accurately to its direct solution. The difference in the accuracy of the solutions is obviously explained by their relative position on the Pareto front. The difference in the convergence speeds is more subtle—condition number is statistically linked to faster convergence speeds for different initial conditions, but this is not necessarily true for individual cases. It is expected that the system from the cluster 2 knee would outperform the convergence speed of the randomly generated high accuracy system were a range of initial conditions tested.

As noted, the original naive system decomposition, in which the preconditioning matrix equals the identity matrix is not convergent, since the sign of the linear source term is wrong. This means that we can't use this as a baseline case. Perhaps the nearest we can come to a meaningful baseline is by setting $\mathbf{P} = -\mathbf{I}$, which corresponds to form of the equations that the author has previously used for similar cases.

Figure 9a, b show the cluster one and cluster two optimal results and residuals, respectively, compared to the $\mathbf{P} = -\mathbf{I}$ baseline case. Clearly, the optimised systems perform significantly better than the naively decomposed baseline system. In particular, the system from cluster 1 both converges in a third of the time, and to a result which much more closely approximates the analytical solution ($L_2 = 5.9747$ vs. $L_2 = 0.6062$). The cluster 2 system takes almost twice as long to converge as the baseline, but is far more accurate ($L_2 = 5.9747$ vs. $L_2 = 0.1467$).

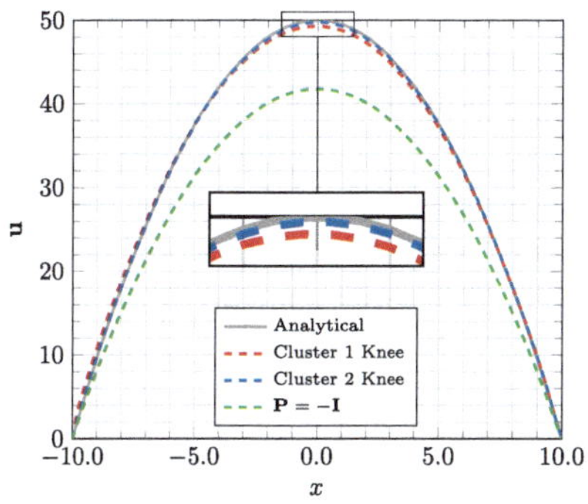

(a) Solution

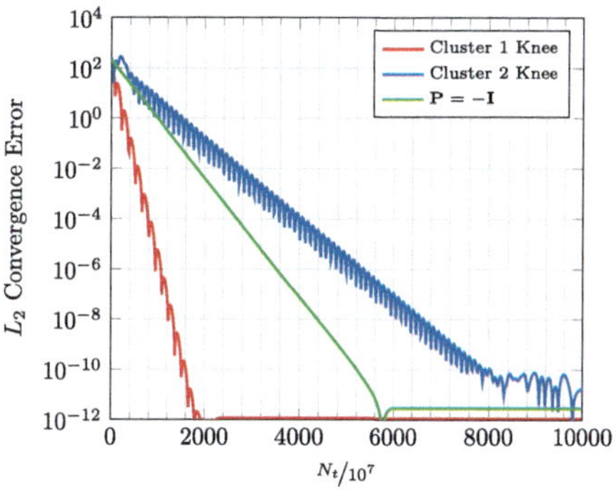

(b) Convergence history

Fig. 9 Using Pareto optimal systems

4.3 Grid Effects

Since this is a linear system, we have only considered the accuracy and convergence of our systems to a single solution. As noted before, this means that changes to the boundary conditions are only represented at the ends of the condition vector—and thus the solution accuracy is independent of the boundary conditions specified.

However, there is another common way in which the problem may be changed, which is to change the grid spacing. To that end, we want to be sure that our improved systems continue to perform reasonably on meshes other than the one for which they were optimised. The optimisation was carried out on a $N_x = 50$ mesh. Hence, we also try the methods on meshes of $N_x = 100$, $N_x = 250$, and $N_x = 500$, and compare their performances to the baseline. These results on other meshes are summarised in Table 1.

On the increasingly fine meshes, the improved performance of the optimised systems is maintained, even after the refinement reaches a factor of 10. We can also use this data to confirm two things. First, that we have maintained the expected order of accuracy of the solution —- and we do indeed see first order accurate reduction in the L_2 error norm with reducing grid spacing. Secondly, we can confirm that we

Table 1 Performance on other meshes

	Naive		Cluster 1		Cluster 2	
Mesh N_x	N_c	L_2	N_c	L_2	N_c	L_2
50	6000	5.9747	2350	0.6062	9700	0.1467
100	12100	3.2484	4600	0.3081	16000	0.0751
250	30650	1.3714	10850	0.1251	37150	0.0313
500	64600	0.6987	21150	0.0629	68500	0.0159

have a linear scaling of Δx with N_c—one of the prime motivations for rewriting the Poisson equation as a set in the first place. For first order accurate solvers, it is likely the balance will favour systems which improve the accuracy of the simulation, since the increase in computational cost incurred to reduce error through increasing the mesh density is high. For higher order accurate solvers, it is possible that the balance will favour systems which reduce convergence time, since the cost of reducing error through mesh increases is relatively low. This underlines the general thrust of the paper—that it is beneficial to match the formulation of your problem to match the solver you are using.

5 Conclusions

In this paper, we have determined first order systems which are equivalent to the second order Poisson equation for wall distance. Rewriting as a first order system can achieve considerable improvements in performance. Conditions on these systems have been found which it is suggested are necessary for them to converge plausibly under fixed point iteration to the analytical solution: the eigenvalues of the flux Jacobian must be real and have alternating signs, which also necessitates that the systems are hyperbolic; and the linear component of the source term in the auxiliary equation must also be negative. This means that the simple explicit time marching techniques build into almost all PDE solvers will be effective at converging to a steady state.

It has also been shown that there is considerable scope for optimising the original first order decomposition to make it better suited to the discretisation scheme being used—preconditioning them not to better represent the physics, but rather to better represent the underlying discretisation. This led to not only faster convergence, reducing the number of iterations to convergence by almost two thirds from the baseline naive case, but also to improved accuracy, reducing the L_2 error norm to 10% of its original baseline value. Two families of optimal solution were found—one which tended to have higher condition numbers and excellent accuracy, and one which tended to have lower condition numbers at the expense of reduced accuracy.

These findings open up some potentially fruitful lines of investigation for solvers making use of the first order decomposition approach to solving higher order PDEs. It is noted that the details of the decomposition into the equivalent system can have a dramatic effect on the performance of the solver.

References

1. Bartholomew, P., Deskos, G., Frantz, R.A., Schuch, F.N., Lamballais, E., Laizet, S.: Xcompact3d: an open-source framework for solving turbulence problems on a cartesian mesh. SoftwareX **12**, 100550 (2020)
2. Canosa, J., Gazdag, J.: The korteweg-de vries-burgers equation. J. Comput. Phys. **23**(4), 393–403 (1977)

3. Cantwell, C., Moxey, D., Comerford, A., Bolis, A., Rocco, G., Mengaldo, G., Grazia, D.D., Yakovlev, S., Lombard, J.E., Ekelschot, D., Jordi, B., Xu, H., Mohamied, Y., Eskilsson, C., Nelson, B., Vos, P., Biotto, C., Kirby, R., Sherwin, S.: Nektar++: an open-source spectral/*hp* element framework. Comput. Phys. Commun. **192**, 205–219 (2015)
4. Economon, T.D., Palacios, F., Copeland, S.R., Lukaczyk, T.W., Alonso, J.J.: SU2: an open-source suite for multiphysics simulation and design. AIAA J. **54**(3), 828–846 (2016)
5. Fischer, P., Lottes, J., Tufo, H.: Nek5000 (2007)
6. Kim, J., Lee, S., Choi, Y., Lee, S.M., Jeong, D.: Basic principles and practical applications of the cahn-hilliard equation. Math. Probl. Eng. **2016**, 1–11 (2016)
7. Lou, J., Li, L., Luo, H., Nishikawa, H.: Reconstructed discontinuous galerkin methods for linear advection-diffusion equations based on first-order hyperbolic system. J. Comput. Phys. **369**, 103–124 (2018)
8. Mazaheri, A., Ricchiuto, M., Nishikawa, H.: A first-order hyperbolic system approach for dispersion. J. Comput. Phys. **321**, 593–605 (2016)
9. Nakashima, Y., Watanabe, N., Nishikawa, H.: Hyperbolic navier-stokes solver for three-dimensional flows. In: 54th AIAA Aerospace Sciences Meeting. American Institute of Aeronautics and Astronautics (2016)
10. Nishikawa, H.: A first-order system approach for diffusion equation. II: Unification of advection and diffusion. J. Comput. Phys. **229**(11), 3989–4016 (2010)
11. Nishikawa, H.: Alternative formulations for first-, second-, and third-order hyperbolic navier-stokes schemes. In: 22nd AIAA Computational Fluid Dynamics Conference. American Institute of Aeronautics and Astronautics (2015)
12. Richardson, L.F.: The approximate arithmetical solution by finite differences of physical problems involving differential equations, with an application to the stresses in a masonry dam. Philosophical Transactions of the Royal Society of London. Series A, Containing Papers of a Mathematical or Physical Character **210**(459-470), 307–357 (1911)
13. Roe, P.: Approximate riemann solvers, parameter vectors, and difference schemes. J. Comput. Phys. **43**(2), 357–372 (1981)
14. Rusanov, V.: On difference schemes of third order accuracy for nonlinear hyperbolic systems. J. Comput. Phys. **5**(3), 507–516 (1970)
15. Tucker, P.: Assessment of geometric multilevel convergence robustness and a wall distance method for flows with multiple internal boundaries. Appl. Math. Model. **22**(4–5), 293–311 (1998)
16. Tucker, P.: Differential equation-based wall distance computation for DES and RANS. J. Comput. Phys. **190**(1), 229–248 (2003)
17. Tucker, P.G.: Prediction of turbulent oscillatory flows in complex systems. Int. J. Numer. Meth. Fluids **33**(6), 869–895 (2000)
18. Watson, R.A., Trojak, W., Tucker, P.G.: A simple flux reconstruction approach to solving a poisson equation to find wall distances for turbulence modelling. In: 2018 Fluid Dynamics Conference. American Institute of Aeronautics and Astronautics (2018)
19. Witherden, F., Farrington, A., Vincent, P.: PyFR: an open source framework for solving advection-diffusion type problems on streaming architectures using the flux reconstruction approach. Comput. Phys. Commun. **185**(11), 3028–3040 (2014)

Magnetohydrodynamic Turbulence in Bounded Liquid Metal Flows

Francesco Fico, Hao Xia, and Ivan Langella

Abstract Turbulent alteration in bounded liquid metals flow within a uniform transverse magnetic field are examined using Large-eddy simulation (LES) methodology. The effect of magnetic field on turbulence suppression are analysed for a concentric pipe and a square duct. The calculations are run at Reynolds numbers (based on bulk velocity and hydraulic diameter) of 5600 and 8900, and for Hartmann numbers, $Ha \leqslant 125$. Anisotropic effects, causing turbulence to deviate from its conventional state, are observed for the concentric pipe for $Ha = 60$ and $Ha = 120$. This effect appears to be related to the appearance of coexisting quasi-2D and 3D turbulence states. Turbulence suppression and regeneration is studied for a duct flow for various wall conductivity parameter c_w. It is shown that highly conductive Hartmann walls can trigger turbulence in an otherwise laminar flow, apparently due to the local re-organization of electric currents.

1 Introduction

Liquid metals are of paramount importance in many engineering applications, including geothermal energy extraction, metallurgy, electronics cooling, and fusion reactors [12]. In particular, in fusion reactors, liquid metals are seen as promising cooling agents for various designs of blankets and divertor targets. Liquid metals are characterised by very small Prandtl numbers due to their very large molecular conductivity. This makes them able to exchange energy more efficiently and across smaller surfaces

CUFS-2024, Cambridge, United Kingdom, 4-5 March 2024.

F. Fico (✉) · H. Xia
Loughborough University, Loughborough LE11 3TU, UK
e-mail: f.fico@lboro.ac.uk

H. Xia
e-mail: h.xia@lboro.ac.uk

I. Langella
Delft University of Technology, Delft 292F+VC, Netherlands
e-mail: i.langella@tudelft.nl

J. C. Tyacke and N. R. Vadlamani (eds.), *Proceedings of the Cambridge Unsteady Flow Symposium 2024*, https://doi.org/10.1007/978-3-031-69035-8_27

than conventional fluids. Therefore, they are very appealing when the dimension and weight of the heat exchange devices is limited or when high thermal loads are present. An important property is that liquid metals respond to magnetic effect. When a magnetic field ($\mathbf{B}_0$) is applied to a conductive moving fluid, it induces a current density ($\mathbf{J}$) and the interaction of this current with the magnetic field generates a Lorentz force $\boldsymbol{\ell}_F$, defined (per unit volume) as:

$$\boldsymbol{\ell}_F = \frac{1}{\rho}(\mathbf{J} \times \mathbf{B}_0) \tag{1}$$

The Lorentz force interacts with the flow and alters the velocity field. In turbulent flows, the effect is even more relevant, as the magnetic field can alter the structure of the turbulence and in some cases can laminarise the flow [17, 19]. Consequently, the mixing phenomena and heat transfer characteristics in turbulent flows subjected to a magnetic field can significantly differ from those without a magnetic field. The magnetic interaction is ruled by the Hartmann (Ha) number and the Re/Ha ratio. The Hartmann number represents the ratio between electromagnetic and viscous forces, while the Re/Ha ratio provides a measure of the relation between inertia and electromagnetic force. Many studies have investigated magnetohydrodynamic (MHD) turbulence. The main findings of these studies can be summarized as follows:

1. Turbulence suppression [4, 10, 11, 14, 16, 19, 23], characterised as the anisotropic suppression of the Reynolds stress tensor components by the magnetic field;
2. Enlargement of the turbulent coherent structures [1, 5, 7, 8, 16], that has been observed in both near wall streaks and streamwise vortices;
3. Existence of different flow regimes [3, 16, 26, 28, 30, 31] depending on relative strength of inertia and electromagnetic force, i.e. the Re/Ha ratio.

In general, the above studies indicated preferential suppression of velocity fluctuations whose direction was parallel to the direction of the magnetic field. While, the existence of different flow regimes and the presence of turbulence special states were also observed. In particular, depending on the Re/Ha ratio, the flow field can be found in a special turbulent state, referred to as Quasi two-dimensional (Q2D) [13, 28] turbulence. The laminar-turbulent transition in MHD ducts and pipes was investigated by Zikanov et al. [32], finding that in transversal magnetic field, the critical Re/Ha values for transition from laminar to turbulent flow in the Hartmann boundary are bounded to $200 < Re/Ha < 400$. Smolentsev et. al. [27] presented a $Ha - Re$ diagram to identify the flow regime for non-conducting straight rectangular duct. These authors suggested that Q2D turbulence states are found in the range $65\sqrt{Ha} < Re < 200Ha$.

Despite the work mentioned, the majority of studies addressed only pipe or duct flows, whereas less conventional geometries have been under researched. On the other hand, the majority of study addressed only electrically insulated walls, as in although the effect of the wall conductivity completely alter the flow structure [33]. Recently, Blishchik and Kenjereš [2] showed that when the Hartmann walls are very conductive, turbulence can be regenerated.

In this work two topics are addressed, one concerning the anisotropic effects of the magnetic field in a concentric annulus, while the other extends the results obtained by [2] for low Re/Ha ratios.

2 Numerical Methods

An incompressible, electrically conductive, Newtonian fluid under uniform and static magnetic field for a concentric annulus, and a square duct with conductive walls is considered. Figure 1a shows a schematic diagram of the annulus geometrical characteristics, while 1b shows the square duct. Continuity and momentum equations of the incompressible MHD flow at low magnetic Reynolds number (Re_m) are, in Einstein's notation:

$$\frac{\partial \bar{u}_i}{\partial x_i} = 0, \tag{2}$$

$$\frac{\partial \bar{u}_i}{\partial t} + \frac{\partial}{\partial x_j}(\bar{u}_i \bar{u}_j) = -\frac{1}{\rho}\frac{\partial \bar{p}}{\partial x_i} - \frac{\partial \tau_{ij}^{sgs}}{\partial x_j} + \nu \frac{\partial^2 \bar{u}_i}{\partial x_j \partial x_j} + \ell_{F_i} \tag{3}$$

where $(\bar{\ })$ indicates the spatial filtering, and u_i, p, ρ, represent respectively velocity vector component, pressure, and density. The velocity vector components, $\mathbf{u}(u_x, u_r, u_\theta)$ are expressed respectively in axial, radial and azimuthal directions (annulus), while $\mathbf{u}(u, v, w)$ are used for the square duct. The term $\boldsymbol{\ell}_F = (1/\rho)\bar{\mathbf{J}} \times \mathbf{B}_0$ is the Lorentz force, where $\mathbf{J}$ is the electric current density and $\mathbf{B}_0$ is the imposed magnetic field, oriented at $\theta = 0$ for the annulus (and positive y for the duct), as shown in Fig. 1.

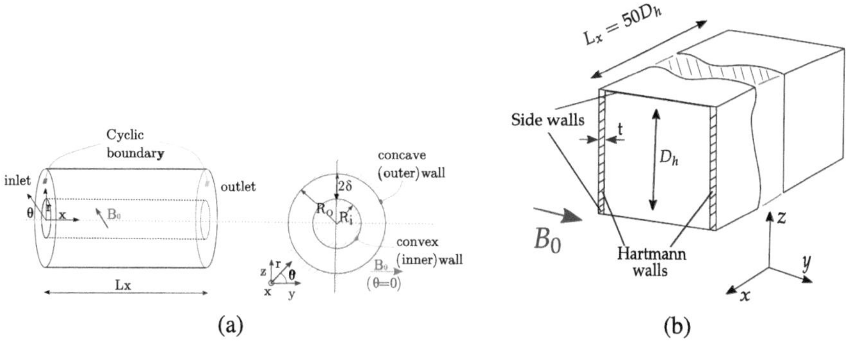

Fig. 1 Concentric annulus with electrically insulated walls (**a**) and square duct with electrically conductive Hartmann walls (**b**)

Under the condition of low-Reynolds magnetic number (Re_m), the induced magnetic field $\mathbf{B}_i$ is small in comparison to the imposed one [6]; hence $\mathbf{B}_0$ and $\mathbf{u}$ are not coupled. Thus, the magnetic field affects the velocity field, and not vice-versa. In this condition, the inductionless formulation can be used, where the electric density current is computed using the Ohm's law (Eq. 4) and charge conservation (Eq. 5) equations:

$$\bar{J}_i = \sigma\left(-\frac{\partial \bar{\phi}}{\partial x_i} + \epsilon_{ijk}\bar{u}_j B_{0_k}\right), \tag{4}$$

$$\frac{\partial \bar{J}_i}{\partial x_i} = 0 \tag{5}$$

where ϕ is the electric potential, σ is the electric conductivity of the fluid, and ϵ_{ijk} is the Levi-Civita tensor (alternating unit tensor). Combining Eqs. (4) and (5), one obtains the Poisson equation for the electric potential:

$$\frac{\partial^2 \bar{\phi}}{\partial x_i^2} = \frac{\partial}{\partial x_i}(\epsilon_{ijk}\bar{u}_j B_{0_k}) \tag{6}$$

The subgrid-scale (SGS) stress tensor, τ_{ij}^{sgs} appears in Eq. (3) due to the filtering operation, and are defined as:

$$\tau_{ij}^{sgs} = \tau_{ij} - \frac{1}{3}\tau_{kk}\delta_{ij}, \qquad \tau_{ij} = \overline{u_i u_j} - \bar{u}_i \bar{u}_j \tag{7}$$

where δ_{ij} is the Kronecker delta. The SGS stresses are evaluated with wall-adapting local eddy-viscosity (WALE) model [22] as:

$$\tau_{ij}^{sgs} = -2\,\nu_{sgs}\,\overline{S}_{ij}, \tag{8}$$

where $\overline{S}_{ij}$ is the resolved strain rate tensor and ν_{sgs} is the SGS viscosity. Periodic boundary conditions are applied in the axial (streamwise) direction. No-slip boundary conditions are enforced at the walls for the velocity field. The walls are considered electrically insulated for the annulus, so that a zero-wall normal current density is imposed:

$$J_n|_{wall} = -\left.\frac{\partial \phi}{\partial n}\right|_{wall} = 0,$$

with n indicating the normal direction to the wall. For the square duct electrically conductive walls are considered. To consider the electric currents flowing in the walls, the geometrical solid walls are considered in the simulation as shown in Fig. 1b. The presence of solid walls is accounted for through a segregated approach. The computational domain is split into liquid and solid subdomains, where the two are solved separately and then coupled through a boundary condition. Consequently, also for the solid, the Poisson equation is solved:

$$\nabla \cdot (\sigma_s \nabla \phi_s) = 0 \quad (9)$$

where the subscript s indicates the solid domain. Along the fluid/wall interface, the conservation and continuity of the electric current density ($\boldsymbol{J}$) needs to be kept. This is achieved by imposing the following set of boundary conditions at the interface:

$$\sigma \frac{\partial \phi}{\partial n} = \sigma_s \frac{\partial \phi_s}{\partial n} \quad (10)$$

$$\phi = \phi_s \quad (11)$$

The quantity of electric current density passing though the interface is controlled by the wall conductivity parameter, expressed as:

$$c_w = \frac{\sigma_s t}{\sigma D_h} \quad (12)$$

where t is the solid wall thickness. At $c_w = 0$, the solid walls can be considered perfectly insulated.

The simulations are carried out with an in-house version of the open source code OpenFOAM [29], which is modified to include the inductionless approach. The pressure-velocity coupling is solved using PIMPLE methodology. The electric potential is computed using Eq. (6) and the electric current is estimated using a conservative interpolation following Ni et al. [21] approach. The discretized equations are integrated in time with backward second order implicit scheme. The spatial dis-

Table 1 Geometrical, physical and mesh values for concentric annulus

Data	
Re_{D_h}	8900
Pr	0.026
Re_m	0.001
Ha	0, 40, 60, 120
δ	$(R_o - R_i)/2$
L_x	25δ
R_i/R_o	0.5
Mesh	
$N_x\ N_r\ N_\theta$	(260, 80, 520)
$\Delta r^+_{\min}$	0.2
$\Delta r^+_{\max}$	5.3
$R_i \Delta\theta^+$	3.6
$R_o \Delta\theta^+$	7.2
Δx^+	14

Table 2 Geometrical, physical and mesh values for the square duct

Data	
Re_{D_h}	5600
Pr	0.0238
Re_m	0.00097
L_x	$50D_h$
t	$0.1D_h$
Mesh	
$N_x\ N_y\ N_z$	(1000, 100, 100)
Δx^+	19.5
$\Delta y^+\|_{(max)}$	6.25
$\Delta y^+\|_{(min)}$	0.25
$\Delta z^+\|_{(min)}$	0.25
$\Delta z^+\|_{(max)}$	6.25

cretization is made through pure central differences. The overall accuracy in both time and space is then 2nd order. The mesh characteristics are in Tables 1 and 2., where N_i, are the number of cells in the specific direction; while Δr^+, $R_{i/o}\,\Delta\theta^+$ and Δx^+ are the grid spacing in the radial, azimuthal and longitudinal directions. The flow field is initialized throughout a mixing layer, and moved by means of a body force. The flow is fully developed after a computational time of $t^* \approx 100$, where $t^* = D_h/U_b$ is the characteristic time. The temporal averages, denoted by $\langle\cdot\rangle$, are collected over a computational time of $600\,t^*$. Wall coordinates, represented using the subscript ($^+$), are normalized using the kinematic viscosity ν and the friction velocity $u_\tau = \sqrt{\tau_w/\rho}$, where τ_w is the wall shear stress. When global coordinates are used, they are normalized with D_h. For the square duct y and z are used, and the subscript ($*$) is omitted in the next sections.

3 Results and Discussion

In this section, the interaction between the magnetic field and the fluid flow is investigated, in order to shed light on how the flow statistics are affected by their local orientation to the magnetic field. When a magnetic field is present, a relevant effect on the decay of turbulence in the conducting fluid is observed. When $Re_m < 1$, the velocity and magnetic field are uncoupled [20] and the latter acts suppressing preferentially the Fourier components of the velocity field whose wave vector is parallel to $\mathbf{B}_0$ [19]. The reason for turbulence suppression is as follows. The magnetic field is responsible for the removal of the turbulent kinetic energy, necessary to make electric currents circulate. The electric currents lead to Joule dissipation, resulting in a rise of thermal energy that corresponds to a fall in kinetic energy [20]. The energy loss follows the timescale $\tau_J = (\sigma B^2/\rho)^{-1}$, that is, the characteristic timescale of the Joule effect. This timescale is very short compared to the dissipation by viscosity, that is usually related to the turn-over time of turbulent structures [6]. Another effect is that $\mathbf{B}_0$ acts anisotropically and opposes to the flow only in certain directions. The variation of turbulence structures is then dependent on the relative direction between $\mathbf{B}_0$ and the flow velocity. Validation of the numerical method is discussed in [9]. In Sect. 3.1, a brief overview of the anisotropic effect of the magnetic field is given for a concentric annulus. While, in Sect. 3.2 the effect of the wall conductivity is discussed for a square duct.

3.1 Concentric Annulus

In this section the effect of the magnetic field effect is studied at two Reynolds Hartmann numbers ratios, $\mathrm{Re}/\mathrm{Ha} = 148,\ 74$ for a concentric annulus. According to literature [27, 32], at these ratios, turbulence suppression and partial laminarization are expected to be observed. When Re/Ha decreases, the Lorentz force becomes more important, and the flow tends to laminarise, however as mentioned before, the turbulence suppression is not uniform, and quasi-2Dimensional turbulence states are observed for $\mathrm{Re}/\mathrm{Ha} = 148,\ 74$. The turbulence statistics are discussed in [9], following the analysis of the anisotropic effect are discussed. Following the idea of Lumley [18], the stress tensor can be divided in the traceless deviatoric part a_{ij} and an isotropic part. The normalized anisotropy tensor is defined as:

$$b_{ij} = \frac{a_{ij}}{2k} = \frac{\langle u_i u_j \rangle}{2k} - \frac{1}{3}\delta_{ij} \tag{13}$$

with k being the kinetic energy. The principal components of the anisotropic tensor are then obtained as:

$$I_1 = tr(\mathbf{H}), \tag{14}$$

$$I_2 = \frac{1}{2}(tr(\mathbf{H})^2 - tr(\mathbf{H}^2)), \tag{15}$$

$$I_3 = det(\mathbf{H}) \tag{16}$$

where tr and det refer respectively to the trace operator and the determinant. I_k refers to the first, second and third invariants of the normalized anisotropy tensor $\mathbf{H}$ (with b_{ij} as components). I_1 is always zero for incompressible flows, and the $I_2 = I_3 = 0$ condition represents an isotropic state.

Intermediate states are summarized in Table 3.

As shown in Fig. 2, where the anisotropic map is plotted along a line going from the inner wall to the centerline, the hydrodynamic case ($Ha = 0$) follows the expected pipe-like behaviour (two-component close to the wall and an isotropic state along the centreline [15]). On the other hand, the magnetic cases deviates from this behaviour,

Table 3 Description of turbulent anisotropy states over Lumley triangle [25]

Shape	Turbulent state	Invariants
	Isotropic	$I_3 = I_2 = 0$
	Two component axisymmetric	$I_3 = -\frac{1}{108}$, $I_2 = -\frac{1}{12}$
	Two component	$I_3 = -(I_2/3 + 1/27)$
	One component	$I_3 = \frac{2}{27}$, $I_2 = -\frac{1}{3}$
	Axisymmetric ($I_3 > 0$)	$I_3 = 2(-I_2/3)^{3/2}$
	Axisymmetric ($I_3 < 0$)	$I_3 = -2(-I_2/3)^{3/2}$

Each state is linked to the respective shape and value of the invariants

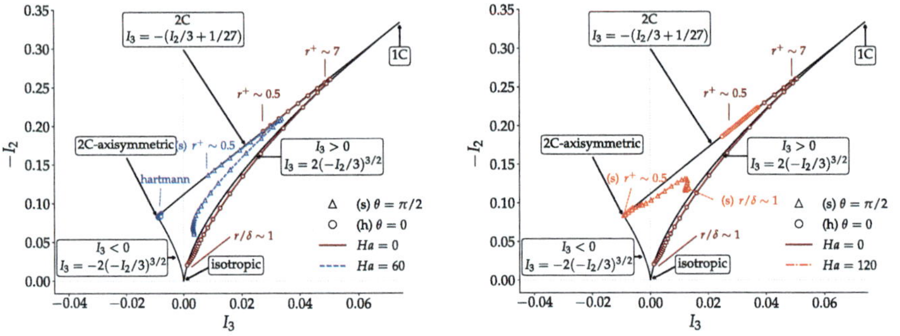

(a) (b)

Fig. 2 Anisotropic invariant map on a line going from inner wall to centerline at $\theta = 0$ and $\theta = \pi/2$ (orientation to the magnetic field.) [9]

and in particular two-dimension turbulent states are observed along the direction normal to the Hartmann wall. This is true for both cases $Re/Ha = 148$ ($Ha = 60$) and $Re/Ha = 74$ ($Ha = 120$). This brief analysis provides some insight in the interaction between magnetic field and turbulence. In Sect. 3.2, the effect of the wall conductivity parameter is discussed.

3.2 *Square Duct*

In this section, the MHD turbulence for a square duct with conductive Hartmann walls is studied. As opposite to the previous case in Sect. 3.1, due to the electrical conductivity of the solid walls, the electric currents can preferentially enter the solid, modifying their distribution within the fluid, i.e. changing the Lorentz force acting on the flow. Due to the wall electrical conductivity and the local reorganization of the electric current the flow can initially be laminarised and then transition again to turbulent when c_w is increased. In Fig. 3 the velocity profile for $Re/Ha = 265$ and $Re/Ha = 160$ are plotted along the y and z directions. The typical "M" shape appears

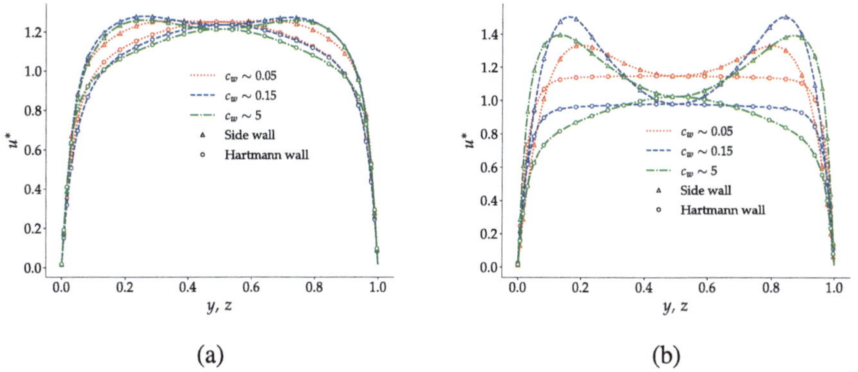

Fig. 3 Mean axial velocity. Re/Ha = 265 (**a**); Re/Ha = 160 (**b**)

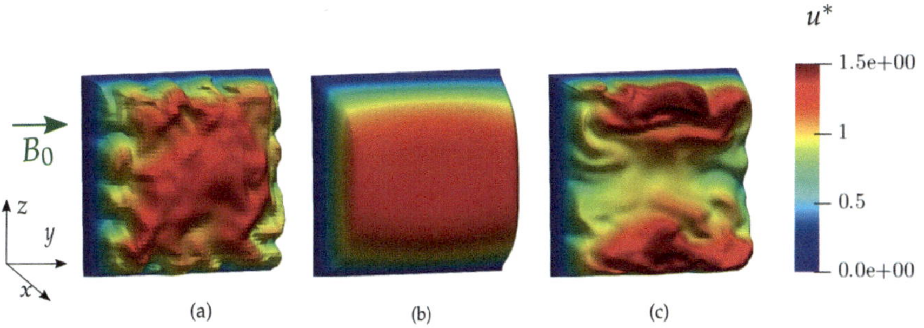

Fig. 4 Instantaneous axial velocity. Hydrodynamic case (**a**), electrically insulated walls ($c_w = 0$) at Re/Ha = 125 (**b**) and Hartmann electrically conductive walls ($c_w = 0.5$) at Re/Ha = 125 (**c**)

along z, and at $Re/Ha = 160$, is more pronounced. It can also be noticed that the peak move closer to the walls increasing c_w. Along y direction, the typical Hartmann velocity profile can be observed only for $Re/Ha = 160$ case, for $c_w = 0.05$ and $c_w = 0.15$. Following the previous analysis, a qualitative view of the flow is provided by Fig. 4, where the instantaneous axial velocity are plotted. It is clear that when the walls are electrically insulated, the flow is completely laminarised at Re/Ha = 125 (Fig. 4b), while increasing c_w 4c), turbulence is regenerated. The velocity distribution seen in Figs. 3 and 4 is the direct result of the re-organization of the electric currents within the duct. In Fig. 5, the electric currents density is plotted in an axial section for $c_w \approx 0.05$, $c_w \approx 0.15$, $c_w \approx 5$. It can be noticed that at low c_w, the electric currents tend to follow similar path to the insulated case [24], and mostly closing within the thin Hartmann boundary layers. Increasing c_w, the electric currents partially enter the solid walls. At $c_w = 5$, the electric currents enter perpendicularly the Hartmann walls, under this condition, the currents "loop" close to the Hartmann walls is not observed in the fluid (due to the electric currents closing within the solid).

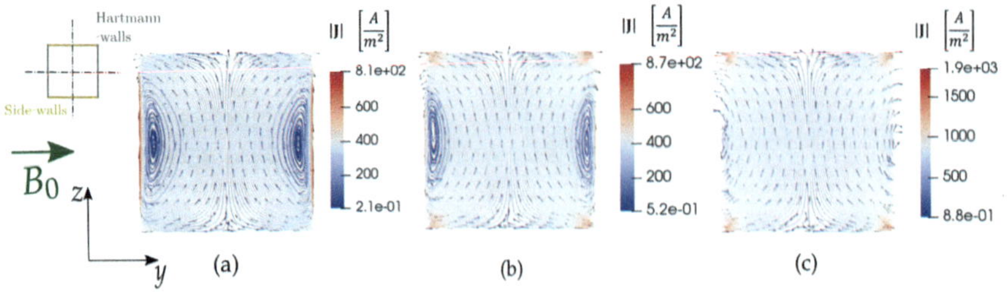

Fig. 5 Electric currents density magnitude in axial section. $c_w \approx 0.05$ (**a**), $c_w \approx 0.15$ (**b**) $c_w \approx 5$ (**c**). Note that the side walls are electrically insulated

So far, the effect of the wall conductivity parameter has been analysed through the first order statistics. A better representation of the effect on turbulence is given by looking at the main components of the Reynolds stress tensor. In Fig. 6, the r.m.s. of velocity fluctuations are plotted along y, z for the cases $Re/Ha = 265$ and $Re/Ha = 160$. In general, it can be seen that suppression is obtained for $c_w < 0.5$, with maximum turbulence suppression observed at $0.025 < c_w < 0.25$. On the other hand, at $c_w = 5$ turbulence is unsuppressed for all main Reynolds stress components. This indicates a consistent turbulence regeneration, apparently due to the local reorganization of the electric currents. Moreover, it can be noticed how turbulence suppression is effective a lower Re/Ha, with r.m.s. amplitude being 10% of the hydrodynamic case at $c_w \approx 0.05$ and $c_w \approx 0.15$. At $c_w = 5$, turbulence is regenerated also for the $Re/Ha = 160$ case. Similar results on the Reynolds stress are obtained also at lower Re/Ha (not shown here).

From the qualitative point of view, the flow structures due to high c_w are slightly different from the hydrodynamic one. In particular, looking at near walls streaks at $y^+ \approx 15$, in Fig. 7, it can be noticed some similarities between the two cases. However, the turbulent structures in Fig. 7b, seems to be slightly more elongated compared to the one in Fig. 7a. Finally, in Table 4 a summary of the turbulence "state" of the cases studied is provided. Turbulence regeneration is observed even at low Re/Ha ratio (up to 58 in this study), when $c_w > 0.5$. From a design perspective instead, it is interesting to look at Fig. 8, where the pressure gradient is plotted against the wall conductivity parameter, and at various Re/Ha. It can be noticed that the pressure losses increase almost logarithmically with c_w, and that its increment is larger decreasing Re/Ha. Overall the results provide a glimpse of the interaction between magnetic field and turbulence. It has been shown that the wall conductivity can completely change the flow structure, due to the local reorganization of the electric currents. Moreover, quantitative data about pressure losses has been provided for a square duct.

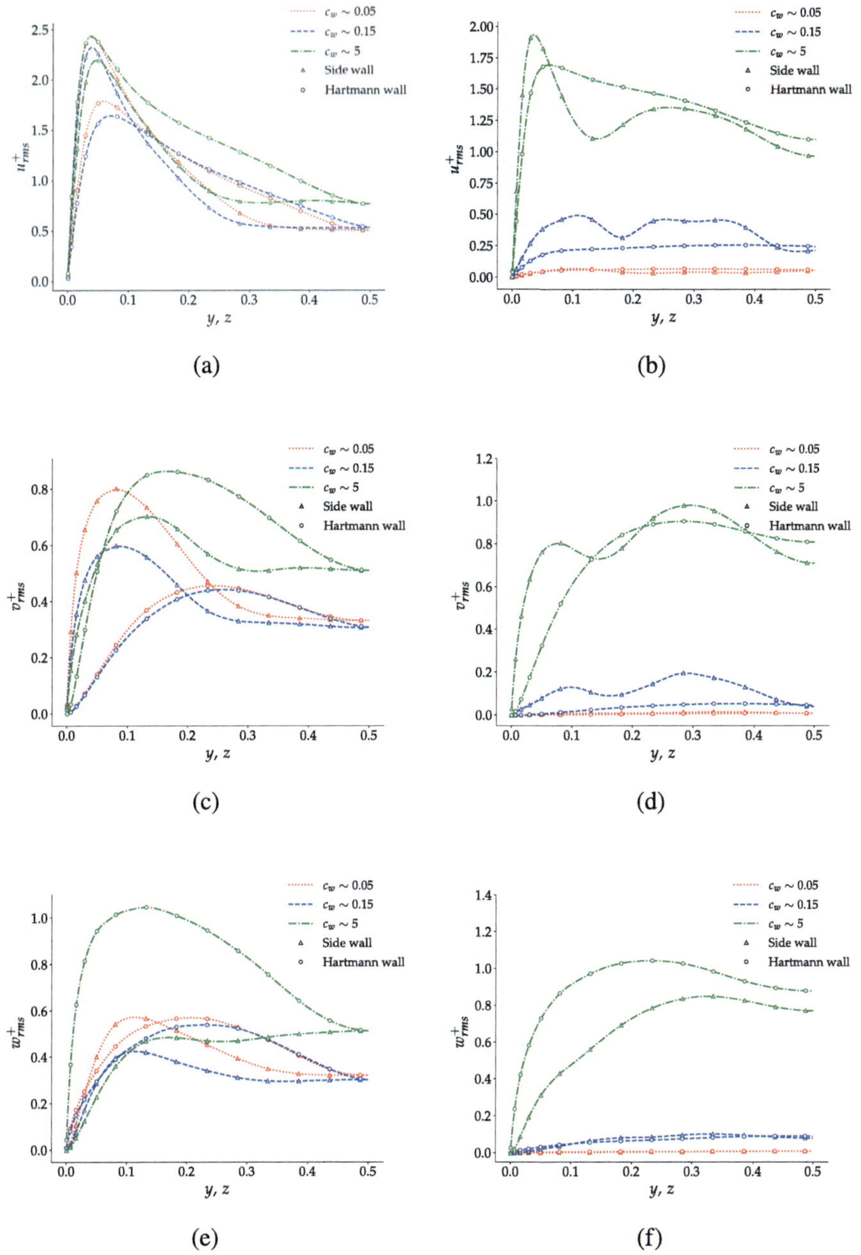

Fig. 6 r.m.s. of velocity fluctuations. Re/Ha = 265 (**a**), (**c**), (**e**); Re/Ha = 160 (**b**), (**d**), (**f**)

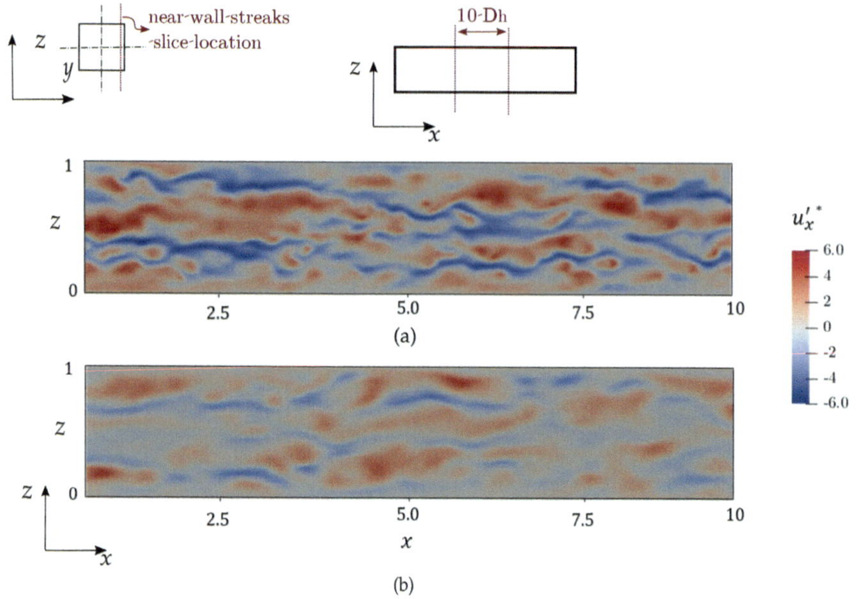

Fig. 7 Near wall (Hartmann wall) streaks at $y^+ = 15$. Hydrodynamic case (**a**), $Re/Ha = 125$ and $c_w = 0.5$ (**b**)

Table 4 Flow state dependence on Re/Ha and c_w for Hartmann electrically conductive walls

Re/Ha	c_w							
	0	0.05	0.1	0.15	0.2	0.25	0.5	5
264	turb	turb	turb	turb	turb	turb	turb	turb
160	turb	tran	tran	tran	turb	turb	turb	turb
124	lam	–	–	tran	–	tran	turb	turb
93	–	lam	–	lam	–	tran	turb	turb
59	–	lam	–	lam	–	–	lam	turb

4 Conclusions

In this article the effect of the magnetic field on turbulence has been explored for two geometries, a concentric pipe and a square duct. The effect of the wall conductivity on turbulence suppression and regeneration has been discussed. With the annulus, the anisotropic effect and the genesis of non-conventional turbulence has been observed. Whereas the square duct results highlight the complexity of the interaction of magnetic field and turbulence. In summary, it has been shown that:

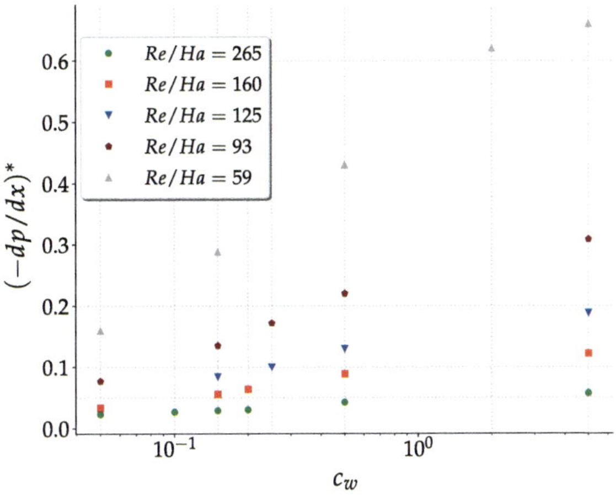

Fig. 8 Normalised pressure gradient in a square duct with electrically conductive Hartmann walls at various Re/Ha

- for a concentric annulus with insulated walls, presence of coexisting quasi-2D and 3D turbulence states in the same flow, depending on the circumferential position;
- for a square duct with electrically conductive Hartmann walls, turbulence regeneration at high c_w (even at low Re/Ha), due to the local reorganization of the electric currents.

Overall, this work showed some features of MHD turbulence. Moreover, quantitative data regarding the pressure drop in square ducts has been provided. It must be noted that turbulence regeneration is not observed when the four walls are equally conductive. Finally, this work does not take into account buoyancy forces, in light of nuclear fusion blanket design, these aspects should be considered in the future studies.

Acknowledgements The authors acknowledge the funding support from Engineering and Physical Science Research Council (EPSRC) under Grant No. EP/R513088/1 and UK Atomic Energy Authority (UKAEA). The authors are grateful for the use of Lovelace HPC at Loughborough University, Sulis Tier-2 at HPC Midlands+.

References

1. Alemany, A., et al.: Influence of an external magnetic field on homogeneous MHD turbulence. In: Journal de Mecanique (1979)
2. Blishchik, A., Kenjereš, S.: Turbulence suppression and regeneration in a magnetohydrodynamic duct flow due to influence of arbitrary electrically conductive walls. Phys. Fluids **34**(4), 045101 (2022). https://doi.org/10.1063/5.0084442
3. Boeck, T., Krasnov, D., Zienicke, E.: Numerical study of turbulent magnetohydrodynamic channel flow. J. Fluid Mech. **572**, 179–188 (2007). https://doi.org/10.1017/s0022112006003673

4. Burattini, P., Zikanov, O., Knaepen, B.: Decay of magnetohydrodynamic turbulence at low magnetic Reynolds number. J. Fluid Mech. **657**, 502–538 (2010). https://doi.org/10.1017/s0022112010001795
5. Chaudhary, R., Vanka, S.P., Thomas, B.G.: Direct numerical simulations of magnetic field effects on turbulent flow in a square duct. Phys. Fluids **22**(7), 075102 (2010). https://doi.org/10.1063/1.3456724
6. Davidson, P.A.: An introduction to magnetohydrodynamics. Cambridge New York: Cambridge University Press (2001). isbn: 9780521794879. https://doi.org/10.1017/CBO9780511626333
7. Davidson, P.A.: The role of angular momentum in the magnetic damping of turbulence. J. Fluid Mech. **336**, 123–150 (1997). https://doi.org/10.1017/s002211209600465x
8. Dong, S., Krasnov, D., Boeck, T.: Secondary energy growth and turbulence suppression in conducting channel flow with streamwise magnetic field. Phys. Fluids **24**(7), 074101 (2012). https://doi.org/10.1063/1.4731293
9. Fico, F., Langella, I., Xia, H.: Large-eddy simulation of magnetohydro-dynamics and heat transfer in annular pipe liquid metal flow. Phys. Fluids **35**(5), (2023). https://doi.org/10.1063/5.0143687
10. Gardner, R.A., Lykoudis, P.S.: Magneto-fluid-mechanic pipe flow in a transverse magnetic field. Part 1. Isothermal flow. J. Fluid Mech. **47**(4), 737–764 (1971). https://doi.org/10.1017/s0022112071001344
11. Hartmann, J., Lazarus, F.: Hg-dynamics II : experimental investigations on the flow of mercury in a homogeneous magnetic field. In: K. Dan. Vidensk. Selsk. Mat. Fys. Medd. (1937)
12. Kirillov, I.R., et al.: Present understanding of MHD and heat transfer phenomena for liquid metal blankets. Fusion Eng. Des. **27**, 553–569 (1995). https://doi.org/10.1016/0920-3796(95)90171-x
13. Knaepen, B., Moreau, R.: Magnetohydrodynamic turbulence at low magnetic reynolds number. Annu. Rev. Fluid Mech. **40**(1), 25–45 (2008). https://doi.org/10.1146/annurev.fluid.39.050905.110231
14. Krasnov, D., et al.: Magnetohydrodynamic turbulence in a channel with span-wise magnetic field. Phys. Fluids **20**(9), 095105 (2008). https://doi.org/10.1063/1.2975988
15. Krogstad, P., Torbergsen, L.E.: Invariant analysis of turbulent pipe flow. Flow Turbul. Combust. **64**(3), 161–181 (2000). https://doi.org/10.1023/a:1009996021533
16. Lee, D., Choi, H.: Magnetohydrodynamic turbulent flow in a channel at low magnetic Reynolds number. J. Fluid Mech. **439**, 367–394 (2001). https://doi.org/10.1017/s0022112001004621
17. Lehnert, B.: The decay of magneto-turbulence in the presence of a magnetic field and Coriolis force. Q. Appl. Math. **12**(4), 321–341 (1955). https://doi.org/10.1090/qam/67648
18. Lumley, J.L.: Computational modeling of turbulent flows. Adv. Appl. Mech. Elsevier, 123–176 (1979). https://doi.org/10.1016/s0065-2156(08)70266-7
19. Moffatt, H.K.: On the suppression of turbulence by a uniform magnetic field. J. Fluid Mech. **28**(03), 571 (1967). https://doi.org/10.1017/s0022112067002307
20. Moreau, R.J.: Magnetohydrodynamics. Springer Netherlands, Nov. 30, 1990. 336 pp. isbn: 0792309375. https://www.ebook.de/de/product/2242611/r_j_moreau_magnetohydrodynamics.html
21. Ni, M., et al.: A current density conservative scheme for incompressible MHD flows at a low magnetic Reynolds number. Part I: On a rectangular collocated grid system. J. Comput. Phys. **227**(1), 205–228 (2007). https://doi.org/10.1016/j.jcp.2007.07.023
22. Nicoud, F., Ducros, F.: Subgrid-scale stress modelling based on the square of the velocity gradient tensor. Flow Turbulence Combust. **62**(3), 183–200 (1999). https://doi.org/10.1023/a:1009995426001
23. Satake, S., Kunugi, T., Smolentsev, S.: Direct numerical simulations of turbulent pipe flow in a transverse magnetic field. J. Turbul. **3**, N20 (2002). https://doi.org/10.1088/1468-5248/3/1/020
24. Shercliff, J.A.: Steady motion of conducting fluids in pipes under transverse magnetic fields. Math. Proc. Cambridge Philos. Soc. **49**(1), 136–144 (1953). https://doi.org/10.1017/s0305004100028139

25. Simonsen, A.J., Krogstad, P.Å.: Turbulent stress invariant analysis: clarification of existing terminology. Phys. Fluids **17**(8), 088103 (2005). https://doi.org/10.1063/1.2009008
26. Smolentsev, S.: Physical background, computations and practical issues of the magnetohydrodynamic pressure drop in a fusion liquid metal blanket. Fluids **6**(3), 110 (2021). https://doi.org/10.3390/fluids6030110
27. Smolentsev, S., et al.: MHD thermofluid issues of liquid-metal blankets: phenomena and advances. Fusion Eng. Des. **85**(7), 1196–1205 (2010). https://doi.org/10.1016/j.fusengdes.2010.02.038
28. Sommeria, J., Moreau, R.: Why, how, and when, MHD turbulence becomes two-dimensional. J. Fluid Mech. **118**(1), 507 (1982). https://doi.org/10.1017/s0022112082001177
29. Weller, H.G., et al.: A tensorial approach to computational continuum mechanics using object-oriented techniques. Comput. Phys. **12**(6), 620 (1998). https://doi.org/10.1063/1.168744
30. Zikanov, O., Thess, A.: Direct numerical simulation of forced MHD turbulence at low magnetic Reynolds number. J. Fluid Mech. **358**, 299–333 (1998). https://doi.org/10.1017/s0022112097008239
31. Zikanov, O., et al.: Decay of turbulence in a liquid metal duct flow with trans- verse magnetic field. J. Fluid Mech. **867**, 661–690 (2019). https://doi.org/10.1017/jfm.2019.171
32. Zikanov, O., et al.: Laminar-turbulent transition in magnetohydrodynamic duct, pipe, and channel Flows. ASME. Appl. Mech. Rev. **66**(3), (2014). https://doi.org/10.1115/1.4027198
33. Zikanov, O., et al.: Mixed convection in pipe and duct flows with strong magnetic fields. Appl. Mech. Rev. **73**(1), (2021). https://doi.org/10.1115/1.4049833

<u>GPSR Compliance</u>

The European Union's (EU) General Product Safety Regulation (GPSR) is a set of rules that requires consumer products to be safe and our obligations to ensure this.

If you have any concerns about our products, you can contact us on ProductSafety@springernature.com

In case Publisher is established outside the EU, the EU authorized representative is:

Springer Nature Customer Service Center GmbH
Europaplatz 3
69115 Heidelberg, Germany

Batch number: 10755079

Printed by Printforce, the Netherlands